Our Precarious Habitat

Our Precarious Habitat

Fifteen Years Later

Melvin A. Benarde

Professor and Associate Director
Asbestos Abatement Center
College of Engineering
Temple University

WILEY

JOHN WILEY & SONS

New York · Chichester · Brisbane · Toronto · Singapore

Library of Congress Cataloging in Publication Data:

Benarde, Melvin A.
    Our precarious habitat / Melvin A. Benarde.
        p.   cm.
    Includes bibliographies and index.
    ISBN 0-471-61750-4
    1. Environmental health.   2. Pollution.   3. Man—Influence on
nature.   4. Man—Influence of environment.   I. Title.
    RA565.B45 1989
    363.1—dc19   88-29158
                                                              CIP

Printed in the United States of America

10 9 8 7 6 5 4 3 2 1

To my students—

For demonstrating the timelessness
of Chaucer's assertion "and gladly wolde he learne and gladly teche."

Foreword

In his *Politics of Pollution,** J. Clarence Davies III asked the question: What is pollution? It was not a rhetorical question. "The very definition of pollution," he wrote, "hinges on politics." That immediately subverts objectivity and places scientists on precarious ground. He went on to note that "pollution can not be defined with any scientific or mathematical finality. The definition hinges on the concept of human use. . . . It is dependent," he continued, "on the public's decision as to what it wants to make of the environment." If there was a homogeneous public, the problem would take on manageable proportions. Again, falling back on the political content, Davies remarks that "it becomes a political decision, a voicing by the community of its concept of the public interest." And finally, "underlying much popular discussion is the idea that pollution is the artificial befouling of the pure state of the environment. This is reflected in the many appeals to return our air—or water or soil—to its 'natural state'."

Unfortunately, I must concur that this concept of pollution appears to describe contemporary longing. Few ask what pure water or air is, or when if ever such a condition existed. Or, if it is even necessary or desirable. One must wonder what period in human history is being alluded to or longed for. Whatever, it bespeaks a time when things must have been better. To ask, better than what?, only prolongs the unanswerable.

For my purpose, it is not necessary to pursue this line of inquiry. I believe, with the benefit of hindsight, that both ideas can exist cheek by jowl as it were; a yearning for a primeval Garden of Eden, as well as an understanding that the real world is not just liveable but is in fact a remarkably healthy place. One is neither required to be an incurable optimist no oblivious of the problems that require attention, to make that statement.

* Pegasus-Bobbs-Merrill Co. Inc., 1975.

It is this second belief that is being staunchly resisted. It does not fit well with elitist notions of a "military/industrial complex," whose polluting excesses must surely be taking its dismal toll in human lives and health.

Fortunately, I am not so presumptuous as to believe that this single volume can change all that. However, I do believe in the Buddhist injunction that "a journey of a thousand miles begins with a single step." Therefore, my purpose in writing this book is to provide readers with a broad spectrum of accurate and reliable information which can assist in placing attitudes and beliefs about environmental impacts in clearer perspective.

Questions are raised throughout the book. They require discussion. Most are questions with which communities, state, and the federal government will have to come to grips sooner or later. Sooner is far better than later.

Unfortunately, science majors, both undergraduates and graduates, often find too many of the questions baffling. They are unprepared to deal with the economic, social, political, and psychological ramifications inherent in all environmental issues. They are often surprised to learn that these concerns, more often than not, take primacy over "hard" data. That is a fundamental lesson. And speaking of "hard" data, I have dealt at length with this in Chapter 15 and additionally in other chapters. It is a subject that cries out for extensive exploration.

Although this book can be entered at any point, concepts such as risk assessment, safety, and hazard, along with the translation of data obtained from animal studies projected to the human condition, are discussed in several chapters. Thus, the fullest understanding will come from a close reading of the entire book—or judicious use of the index.

In the current argot, my "bottom line" is mobilization—of public opinion—away from the fixed position it has assumed since *Silent Spring*.* My interest in taking on the subject after a hiatus of 14 years proceeds from the felt need to clarify issues that continue to bedevil us. I wanted to place before readers a tableau of evidence which would motivate them to initiate questioning. At that point, inquisitive minds may proceed to discover the real world: come to appreciate rather than fear it. Such a body of knowledge cuts against the "it is alleged" type of commentary so rampant today. It is too easy to repeat and accept environmental banalities. Some favorites are: "We are being slowly poisoned to death. Perhaps not so slowly." "Our lives hang in a precarious balance" and "We are ticking bombs" are other favorites. Questioning current environmental myths is an idea whose time has come.

I would also be delighted to hear or learn that as a consequence of reflecting on the ideas and statements in this book—incorporating the

* Rachel Carson, *Silent Spring*, Houghton Mifflin, Boston, 1962.

best available data—that people begin to question, to say "wait a minute, what's the evidence for that?" "Run that by me again" would be another appropriate response. That is what we need—questions, lots of questions—and accountability on the part of speakers and writers. A healthy skepticism is far more preferable than passive acceptance of much of the nonsense served up as our daily fare.

I believe too that the fever and emotion running so high during the 1970s has cooled. By the 1980s it was apparent that the predictions of the "Apocalytics" was so much rhetoric. With their prognostications of fatal consequences for planet Earth, they held a viselike grip on center stage. In his polemic, "There Isn't Much Time,"* George Wald, an otherwise eminent physiologist and Nobel Laureate, wrote the following: "All humanity now suffers from a series of exceedingly threatening new developments, all coming to a head at about the same time—close to the year 2000. I am one of those scientists who try as we will, hope as we do, that something in our view of things is wrong still find it difficult to see how the human race will get itself much past the year 2000. So there isn't much time."

Wald was deadly serious, and he saw no way out. "Though we already know enough to cope with all our major problems," He said, "I don't know one of those major problems that we can begin to cope with while *maximizing profits* [italics his]. And a society like ours which puts maximization of profits above all other considersations, is therefore heading for destruction."

At this point he warmed to his subject. "I still have young children at home," he indicated, "and I teach about 200 magnificent young people at Harvard. Though I desperately want to, I can not find any asurance that they will be in physical existence ten (1985), twenty (1995), twenty-five years (2000) from now." He truly believed life in the United States was hopeless, about to become extinct.

For Professor Wald, corporations were the villians. "I believe our so-called free world is now wholly controlled by such multinational super-enterprises as General Motors, Exxon, The Chase Manhattan Bank, ITT, Dutch Shell and British Petroleum." According to Wald, this power elite was "leading our society rapidly towards disaster."

Unfortunately, the remarks of the President of the National Academy of Sciences were lost on Wald. In the midst of the environmental "panic" of the 1970s, Dr. Philip Handler stated that "the nations of the world may yet pay a dreadful price for the behavior of scientists who depart from fact to indulge in hyperboles."

By the 1980s it was clear that evidence supporting predictions of doom and destruction was baseless.

---

* Progressive 39: 12 (1975).

Perhaps the approach of the millenium offers a period of pause, during which the scientific data obtained over the past 20 years can be assessed calmly. The substantially increased life span and life expectancy at birth are facts of life, at variance with much popular and preferred belief. Perhaps too, it is time "to march to a different drummer."

I believe this book provides that much needed assessment of the most up-to-date scientific findings concerning the impact of potential environmental risk factors on human health.

MELVIN A. BENARDE, PH.D.

Princeton, New Jersey
January 1989

Preface

Our *Precarious Habitat* departs from traditional treatment of environmental issues by apportioning in-depth scrutiny—including the known human health effects—to a group of specific problems. It eschews rhetoric and polemics, which unfortunately can insinuate themselves to the detriment of these vital issues. Instead, this book marshals the best available epidemiologic evidence to support relationships and conclusions about each of the issues considered.

This book emphasizes the need to go beyond simplistic associations. This is vital in order to establish the untoward effects of the many environmental concerns faced by our communities. Indicated too is the need for an interdisciplinary approach, if our many complex problems are to be solved. Such an approach mandates the participation of economists, political scientists, and communications experts along with an array of biological and chemical specialists. It may come as a surprise to some people to learn that economists, sociologists, and political scientists may have more to contribute by way of appropriate *management* of our many and diverse problems than the so-called hard scientists.

Given that faculty members seek to tailor their instructional goals to the particular needs of their students, the issues in this book can be pursued from either biochemical processes or their relationships to the economic and political realities existing in all communities.

Each chapter contains material permitting the instructor to deal with the topic on a variety of levels. Instructors should find its framework helpful in developing student projects, oral and written reports, as well as stimulating active classroom discussion—especially as it involves the panoply of disciplines that must impinge on rational decisionmaking. Furthermore, this book can be entered at any chapter—although links to other chapters are frequently indicated.

Depending on the approach taken, *Our Precarious Habitat* can be used at either the graduate or undergraduate level.

M.A.B.

Acknowledgment

It used to be said that "behind every great man, there is a great woman." Books are like that, but even more so. While I'm hardly so gauche as to claim greatness for what I've written, I do claim the behind-the-scenes involvement of many wonderful people—many, but not all, women.

For his insistence that a revised edition of *Our Precarious Habitat* was long overdue, Dr. Irwin Suffet, a collegue at Drexel University's Environmental Studies Institute, must be given pride of place; up front, at the head of the line. I was stubborn. But he continued to urge, and his urgings bore fruit. I'm glad he didn't give up on me.

Vivian Price came into my life one Saturday afternoon. A friend, Jens Glysing-Jensen, responding to my plea for top-notch secretarial assistance, suggested his former personal secretary. Vivian was extraordinary. Not only was she an accomplished typist, but she could spell. From often illegible scrawls, she produced a gem of a manuscript. I suspect that she and her husband Fred, a dandy human being, are glad this one is over. I brought chapters and sections of chapters at all hours, weekends and holidays. The neighbors must surely have wondered.

Diedre Harper and her co-workers at the Hagerty Library's interlibrary Loan Division, Drexel University, along with the efficient staff of the Science and Technology Division, helped me every step of the way, as did Betsy Tabas and Vicki Newton, of the Temple University library. I know they are all relieved this episode is over.

For some authors, the only good editor is a dead editor. David Huttco, my editor at W.W. Norton, and, thank goodness, very much alive, was not just a good editor, he was a great one. If Dr. Suffet is "at the head of the line," David is all over it. He had that uncanny ability to see in terms of organizational frameworks, to ask the right questions, and offer appropriate alternatives. David has my gratitude.

That lightning can strike twice became apparent when Jim Smith, Senior Editor at Wiley, took on and guided *OPH* to fruition. His support, knowledge, and understanding paved the way from manuscript to opus. And Rosalyn Farkas, Senior Production Editor, introduced me to a higher order of the copy editing and production craft. There cannot be too much praise for her efforts. Kudos for the technical quality of this book belong to her.

And for my wife, Anita, there will be a special place in heaven: for her uncomplaining lost weekends and holidays, for understanding that "the den" was off-limits—an untouchable mess that could go uncleaned for months on end—and for never questioning those visits to Vivian's.

Contents

Our Precarious Habitat

1
Webs of Causation:
The Environment as a System

Sir Percival Dinsmoor had to go on a long business trip. He left his wife and home in the care of his faithful servant, Putnam, and left for a remote area. After several months, he returned to a city, whereupon he placed a call to Putnam.

"How is everything?" he asked.

"Fine," replied the trusted employee. "Everything is just fine, except the cat died."

"The cat died!" exclaimed Sir Percival. "What happened to him?"

"Well," came the answer, "he was kicked by the horse when it ran out of the barn."

"How did the horse get out of the barn?"

"The wall fell down," answered the servant.

"What made the wall fall down?" asked Sir Percival, puzzled, for his barn had been well built.

"The fire," replied his man.

"What fire?"

"The barn burned down," said the servant.

1

"How did the barn catch fire?" asked the man, shocked.

"From the house."

"My God," said Sir Percival, "how did the house catch fire?"

"From the curtains."

"And how did the curtains catch fire?"

"From the oil lamps in the bedroom."

"What were oil lamps doing in the bedroom?" the man asked, since their bedroom was fully electric.

"We put them there for the funeral," said the servant.

"Whose funeral?" gasped the man.

"Your wife's," said the servant.

An ecologist reading this anecdote would surely smile. That the deaths of Sir Percival's cat and his wife are related fits snugly into the ecologist's concept that all things are interrelated.

Biologists, but especially ecologists, refer to aggregations of links as an ecosystem, which usually consists of all living species in an area and their surrounding physical environment. However, for ease of management or purposes of investigation, scientists set arbitrary boundaries.

The multiple relationships among the different populations within a community (the biological portion of an ecosystem) are often referred to as a web. A well-known web is the food web, or food chain, which indicates which animals feed upon others within a community. It is another way of indicating the flow of energy within a community. Figure 1 is an example of the elaborate set of interlocking links in a food chain. In this instance that of the chaparrel.* The chaparral provides direct sustenance to a wide variety of rodents and birds, who in turn supply energy, via the chaparral, to the bobcat at the very top of the chain whose only predator is human beings.

Food webs need not be as extensive as the chaparral's. The simplest of chains exists on the arctic tundra near Point Barrow, Alaska (Figure 2). Here, the grass supplies energy to the lemming, who in turn supplies it to the snowy owl, its predator. It is simple and direct.

But food webs can also be far more complex than that of the chaparral. They generally are when we consider countrywide and global involvement. On this scale, with a complexity of overwhelming proportions, a snag in one link of the chain of events can be widely destructive. In the following section, five specific instances of this destructive power and a more general one are discussed.

---

* Chaparral, from the Spanish *chaparro*, meaning evergreen oak, is a community of shrubby plants widely distributed in the southwestern United States. It is especially adapted to dry, hot summers and moist winters. Arthur S. Boughey of the University of California at Irvine set out the details of the chaparral ecosystem in 1971.

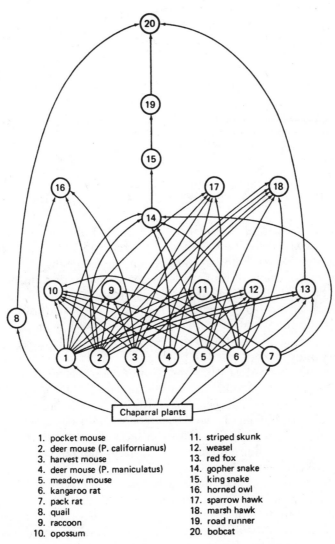

1. pocket mouse
2. deer mouse (P. californianus)
3. harvest mouse
4. deer mouse (P. maniculatus)
5. meadow mouse
6. kangaroo rat
7. pack rat
8. quail
9. raccoon
10. opossum
11. striped skunk
12. weasel
13. red fox
14. gopher snake
15. king snake
16. horned owl
17. sparrow hawk
18. marsh hawk
19. road runner
20. bobcat

Figure 1. The chaparral ecosystem. Reproduced by permission of Springer-Verlag Publishing Company, New York.

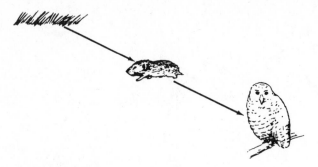

Figure 2. A simple food chain on the arctic tundra, Point Barrow, Alaska. Reproduced by permission of Holt, Rinehart and Winston, New York.

I.   THE AFRICAN SAHEL: AN ECOLOGIC DISASTER

The famine currently spreading through Africa's Sahel*—an area consisting of, from west to east, Mali, Niger, Chad, the Sudan, Ethiopia, Somalia, and Djiboute—poses both an immediate and ongoing crisis of vast proportions. Hundreds of thousands of people have died and thousands more are dying of starvation, disease, and dehydration.

From our current perspective, the sub-Saharan drought of the past 15 years has been the worst in 150 years (Figures 3 and 4). In this century, 1984 and 1985 have been the driest. Although nature is often harsh and hostile, people can aggravate the misery.

The famine in Africa illuminates the deep-seated problems that threaten the future of arid lands around the world. But the outline of the impending disaster was perceived only dimly 15 years ago. Climatologists had recorded the unusually wet weather of the 1950s, as well as the decades of sparse rainfall that followed. Deserts were literally moving. But the interplay of environmental forces either had not yet been grasped or was of little concern—or both.

Desertification, a process by which the productivity of the land is ruinously degraded by human abuse and natural forces acting in concert, is an important underlying cause of the famine that has already claimed hundreds of thousands of lives and decimated cattle populations.

It may be, however, that the *primary* cause of desertification is not drought but human exploitation of lands through overcultivation, overgrazing, poor irrigation practices, and deforestation. These occurrences

---

* Sahel is from the Arabic *Sahel* and refers to the transition zone between the Sahara and the humid Savanna to the south. This is a region of West Africa extending from Senegal to the Sudan. It forms a belt separating the arid Sahara from tropical West Africa and it should not be confused with the northern coastal band of hills in Algeria or the northern coastal plain of Tunisia, often referred to as the North African Sahel.

Figure 3. In drought and famine stricken Mauritania, a bedouin stops at an oasis to fill an inner tube with fresh water to take to his family. Courtesy of Associated Press.

Figure 4. Goats devour remaining bits of vegetation north of Nairobi in Kenya's parched northern region. Courtesy of Associated Press.

usually follow population growth that exceeds the carrying capacity of the land, whether caused by an increase in the birth rate or by the influx of people onto marginal lands that are ill suited to support them. As the population grows, farmers are forced to till poorer and poorer lands to provide enough food. In doing so, they reduce or eliminate the fallow periods needed to regenerate the soil. Every tree is cut down for firewood or building material, thereby increasing erosion of the land by wind and water, and cattle consume the ground vegetation so that the exposed land bakes hard under the sun and loses its ability to absorb the store water (1).

With the increase in soil temperature, the vital microorganisms needed for adequate plant growth are destroyed. Particularly affected are the nitrogen-fixing bacteria, without which the plant cannot complete its manufacture of food. Dust particles blown up from the denuded land scour the remaining plants or bury them. When rain does come, it runs off quickly or evaporates.

Drought and human activity are elaborately intertwined, and whether it is primarily humans or nature may not matter when the concerted efforts of both result in an ecological debacle such as this, which has been years in the developing. And this ecological debacle has direct human effects: chronic malnutrition, dehydration, and debilitating illness. These create additional stresses for a country in loss of productive workers to do the work that must be done to alleviate the situation as well as additional expense of costly medical care.

The current drought is exacerbating the negligence begun hundreds of years before. Since time immemorial camel caravans have cut down trees and bushes to make charcoal for fires and to feed camels. Settlements and military outposts have long been built atop sand dunes, causing destabilization and drift of the dunes. Commercial pressures eliminated the acacia (gum) tree forests of the western Sahel by destructive tapping.

Pacification of the Sahel by colonial powers unknowingly fostered desertification. With the cessation of incursions by raiding tribes, agricultural populations and herds of cattle, sheep, and goats expanded northward into marginal lands that had been previously closed to them. Their cattle were not adapted to the dry area sand, they required large amounts of scarce and precious water, and they often devoured perennial grasses, trampled seedlings, and compacted the soil. More recently, urbanization and transportation have allowed further encroachments on the resources of the surrounding countryside.

The current famine appears to have gained impetus by the excessive rains of the 1950s. The availability of more water than is otherwise present encouraged the expansion of both human and domestic animal populations into heretofore marginal areas. When the drought began, these lands could no longer support their population.

Even well-intentioned efforts have added to the problem. The drilling of wells to increase water supplies prompted some farmers to increase the size of their herds—well beyond the capacity of the land. Garret Hardin's *Tragedy of the Commons**** is nowhere more exemplified than here: the herds eat and trample the vegetation to the point of destruction. With the drought, the animals die—not of thirst but of lack of forage. "Biology," as Hardin has been trying unsuccessfully to tell us for almost 20 years, "requires a respect for carrying capacity." A little selfishness, he believes, will go a long way toward helping everyone (2).

In 1978 a United Nations conference adopted a plan to reverse desertification and even halt the insidious process by the year 2000. Politics being the slothful creature it is, especially on the international level, significant action was not taken. According to Earthscan, a London-based environmental organization, the technical knowledge with which to combat desertification exists, but the political, social, and economic means do not.

Our environment and planet remain an indefinable collection of interacting and interrelated forces, whose bewildering complexity continues to elude us. To understand and deal with so "simple" a problem as air pollution, for instance, it is necessary to consider and comprehend its relationship to meteorologic conditions, the effects of sunlight on organic chemicals, electric power generation and waste disposal, public transportation, human and animal health, agricultural practices, the chemistry of the air, and economic trends. These are a sampling of the relationships involved.

Although it would have been well nigh impossible to have predicted the demise of the family cat with Lady Dinsmoor's passing, the two can be seen as related. On the much larger national, regional, and global levels, however, interrelationships can be difficult to perceive and consequently are often overlooked or ignored. Moreover, political boundaries conspire to impede understanding and solutions, and cooperation between governments stagnates. Currently, these desperately poor countries face an additional peril as the most serious plague of locusts in over 30 years struck North Africa.

---

* Hardin's *Tragedy of the Commons* proceeds from a consideration of the village common loaded with cattle: "Once the carrying capacity is reached, uniquely destructive processes are set in train when any of the herdsmen contemplates increasing the size of his herd." For Hardin it proceeds this way: "As a rational being each herdsman seeks to maximize his gain. He asks, 'what is the utility to me of adding one more animal to my herd?' Given the palpable fact that the herdsman receives the entire profit from all transactions, a rational herdsman must conclude that the only sensible course for him to pursue is to add another animal to his herd. And another . . . and another. . . . Unfortunately for everyone, a similar conclusion is reached by all the herdsman. Each man is locked into a system that compels him to increase his herd without limit . . . pursuing his own best interest in a society that believes in the freedom of the commons." The "commons" obviously are the world's resources, and the tragedy is that it is clearly to an *individual's* advantage and best interest to make the most use of them. If all of us followed such a dictum, the resources would be rapidly exhausted—to the detriment of all. The tragedy of the famine areas of the Sahel is in major part just that.

In 1986 locusts were a major threat to East Africa. Then, civil strife in the Sudan and Ethiopia hampered control efforts. By April 1988, North Africa, especially the area of the Atlas Mountains, was under siege. But the refusal of Western countries to supply pesticides may further imperil the Sahel (3).

The locust swarms could well be blown along by a seasonal windshift that usually occurs in June. Once the hordes move south, they will destroy the remaining food crops in Senegal, Upper Volta, Mali, Niger, and Chad.

The size of the locust population is awesome. A square kilometer contains up to 50 million of them, capable of devouring 100,000 tons of vegetation during a night. Dieldrin, the most effective spray for desert use and capable of stopping the invasion, is not being used because of concern for its long-term effects on the environment (cf. Chapter 5). Without it, the Sahel is bracing for a new onslaught. Figures 5 and 6 show the locust swarms settled and in flight.

Figure 5. Settled swarms of the 3-in.-long banded desert locus *Shistocerca gregaria*, shown denuding an acacia tree in Morocco. Courtesy of the FAO, Rome.

Figure 6. A swarm of locusts being driven out of a cultivated area in Morocco. Courtesy of the FAO, Rome.

Just how this locust infestation reached its uncontrollable size without any of the surrounding countries realizing it is yet another cause of concern. The answer appears to lie in a combination of unusually wet weather that encouraged breeding, bureaucratic inefficiency, and civil strife that put large areas of Africa out of bounds to locust hunters. It may be 1990 before this plague comes under control.

The grim panorama unfolding in the Sahel is only the latest in a series of ecologic upheavals that suggests the interconnectedness of people and nature. El Niño is another.

II. EL NIÑO: "THE CHRIST CHILD"

An ecologic convulsion from which the place of people in the scheme of things is seen as only one thread in the total tapestry of life was witnessed and described by the ornithologist Robert Cushman Murphy. According to him, each year around Christmas, the dry desert regions along the northern coast of Peru are watered by rains brought by currents of warm air. These rains are so regular and so welcome in this otherwise arid land

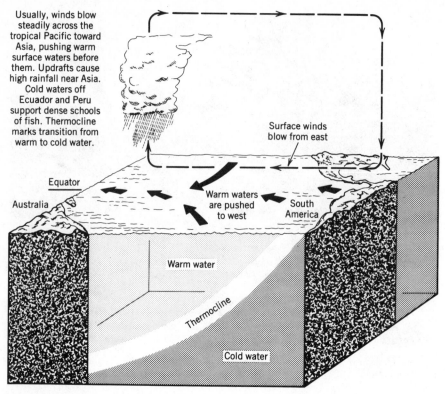

Usually, winds blow steadily across the tropical Pacific toward Asia, pushing warm surface waters before them. Updrafts cause high rainfall near Asia. Cold waters off Ecuador and Peru support dense schools of fish. Thermocline marks transition from warm to cold water.

Surface winds blow from east

Equator

Australia

Warm waters are pushed to west

South America

Warm water

Thermocline

Cold water

Figure 7. The most recent characterizations of the normal climatic conditions prevailing off the coasts of Ecuador and Peru. Reproduced by permission of *The New York Times*.

that they have come to be called *El Niño* ("the Christ child"). These rains sustain life through the many months of drought.

Although El Niño's arrival usually means abundance (some of the finest cotton in the world is grown in the northern coastal areas of Peru), it can also mean disaster. Two or three times a century, El Niño makes its way too far south and does not quickly return to its usual area (Figure 7). This change in position disrupts the ecological balance on land and sea so drastically that even humans cannot escape the consequences.

In 1925, the warm current ranged far south, killing the marine life adapted to the normally cold-water coast. Unable to outrace the warm current, fish in fantastic numbers perished and were cast up on the beaches. To the vast seabird population that frequented the coast, the death of the fish meant starvation. The bodies of countless dead and dying birds literally covered the beaches and blocked the harbors. The putrefaction was so great that hydrogen sulfide, a product of the microbial degradation of protein, was produced in quantities sufficient to blacken the paint on ships in the harbors.* The pollution of the sea with the dead birds caused an overabundance of nitrate nitrogen and gave rise to blooms of a microscopic marine protozoan that actually turned the water blood red.

On land, where rain fell for 5 months, the barren soil eroded, destroying the huts of the inhabitants. Without their supply of fish, the people began to starve. In the standing pools of water, mosquitos began to breed in large numbers and malaria flourished. Contamination of the water supply with human fecal matter, as a result of seepage into wells, precipitated an outbreak of typhoid fever. The rat population, cut off from its normal food supply began to die of plague and, shortly thereafter, so did the natives.

To the shock of the Peruvians, El Niño changed its pattern again in 1972. Instead of growing cold and fading in April as it normally does, it lingered on the coast. By August, the schools of anchovy that made Peru the world's largest producer of fishmeal had not returned. Record flooding destroyed crops and food prices soared.

In 1982–1983, El Niño struck again—with the strongest warming of the equatorial Pacific in this century. The warm water brought close to 600 mm of rain to Quito, Ecuador, during June, a "dry-season" month in which precipitation barely manages 10 mm. With this last appearance El Niño ranged farther south than ever before (Figure 8). The unseasonably warm weather front stretched 13,000 km (7800 mi) from the South American coast to the archipelagoes of the western Pacific, creating a drought that enveloped Australia. Over 1000 people died and staggering economic damage occurred in Australia, Asia, South America, and Africa.

This havoc resulted from an elevation in the temperature of the seawater of less than 5°C. It shows the chain that can link people, animals, and plants to the physical environment.

For the most part El Niño has been portrayed as an unmitigated natural disaster. Nevertheless, there have been several positive aspects to the 1982–1983 cycle. While the traditionally hunted fish populations were driven away by the warming of the offshore water, several tasty species, including skipjack, Spanish mackerel, yellowfin tuna, and bonito thrived, allowing fishermen to make up some of their losses. Although one shrimp species fell prey to an invasion of crabs, another boomed and was both sold in local markets and exported to the United States. Scallops soared in number and became a popular, inexpensive food in Lima (4).

Unfortunately for the British, their imports of pilchards (sardines) fell by 68%, while Chile, far to the south, was able to increase its sardine catch as pilchard stocks moved south to escape the warm water.

That environmental shifts can be double-edged swords is seen as well in Florida's current Kissimmee restoration project.

---

* Hydrogen sulfide reacts with lead in paint to form black lead sulfide.

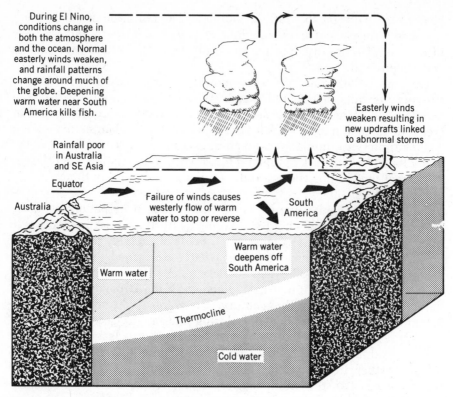

During El Nino, conditions change in both the atmosphere and the ocean. Normal easterly winds weaken, and rainfall patterns change around much of the globe. Deepening warm water near South America kills fish.

Easterly winds weaken resulting in new updrafts linked to abnormal storms

Rainfall poor in Australia and SE Asia

Equator

Australia

Failure of winds causes westerly flow of warm water to stop or reverse

South America

Warm water deepens off South America

Warm water

Thermocline

Cold water

Figure 8. The severely altered conditions of an El Niño occurrence. Reproduced by permission of *The New York Times.*

III.   FLORIDA'S KISSIMMEE RIVER

As he dropped a baby cypress tree into a water-filled hole on the banks of the Kissimmee River, Florida's governor Bob Graham said, "Our goal is, by the year 2000, the water system will look and function more as it did in 1900 than it does today" (July 1984). He was referring to one of the most disastrous projects ever undertaken by the Army Corps of Engineers. In 1964, the Corps spent $30 million on the construction of a 52-mi channel along the 98-mi Kissimmee, studded with locks every 10 mi. The pupose was to control the seasonal flooding that spilled over the river's banks, destroying property and jeopardizing tourism. Now $65 million would be spent by the South Florida Water Management District to undo the Corps' work and restore the river to its original course.

If all goes according to plan, over the ensuing 15 years the river will gradually rise and engulf the now artificially dry plains that surround it, transforming them back to the lush, mosquito-ridden swamps that they were for untold thousands of years.

The reason for the turnaround is the dire effects some believe the channel has had on southern Florida's complex ecological system. When the 200-ft wide, 30-ft deep channel was completed and the Kissimmee's annual overflow eliminated, 60% of its adjacent marshland, approximately 20,000 acres, shriveled up, taking with it plants and animals dependent on the wet–dry cycles. Bald eagles, which fed off the marsh fish, declined significantly. Mottled ducks and coots were reduced to a fraction of their former numbers and alligators disappeared from large areas. Environmentalists want the river and thus the wildlife.

Ranchers and tourists, however, prefer the change. Wild hogs, quail, and turkeys have replaced eagles and ducks; cattails and bottonbush, rather than arrowhead and maiden cane, dot the pools. With grass growing, cattle have a wide grazing range, Bass fairly jump into fishermen's nets. Mosquitoes have become almost unknown. For many it has become a recreational paradise.

The controversy raises several questions. Is there a best, or optimal, ecological condition for Kissimmee? If so, who is to say which it is? Is it one of the two described or one as yet unknown? How was the decision to revert to the original course made? Perhaps most important, from the view of the community, is why so few people appear to have participated in so crucial a decision. Were they aware of it? Were they concerned? What would it take to increase awareness and participation? Would any benefits ensue? These questions beg for answers, and they could be asked of many of the examples cited.

The problems created by El Niño are probably beyond our control. The devastation of Central Africa and the changes along the Kissimmee may not be. The Great Lakes, as we shall see, appear to be a problem in transition. The goal is to restore the Lakes to their former wholesome state, but lack of knowledge of the many complex interconnecting webs has so far thwarted efforts.

IV. THE GREAT LAKES

The Great Lakes, as shown in Figure 9, are no small gentle ponds, but the world's largest freshwater lake system.

Although direct discharge of industrial wastes is controlled under the Clean Water Act,* chemicals and heavy metals still enter the lakes. Lakes Superior, Huron, and Michigan appear to be receiving chemical pollutants from the atmosphere. How these pollutants got into the air is purely speculative. Evaporation from the spraying of agricultural chemicals, evaporation from landfills, and incomplete combustion and vaporization at industrial treatment systems are currently among best guesses.

* See Chapter 11, Water Pollution.

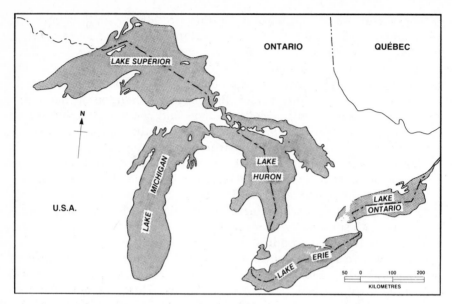

Figure 9. The relationship of the five Great Lakes and their points of connection.

Since the city of Duluth, Minnesota, built a new sewage-treatment plant, the St. Louis River, which empties into Lake Superior, is almost devoid of municipal pollution. As a consequence, the walleye pike has returned in force and sports fishing has increased markedly.

But with the cleaned-up river, the sea lamprey, *Petromyzon marinus* (the rock sucker), also returned. This prehistoric eel-like predator, a vestige of a bygone age, is not only spreading through the lakes but is again preying on the trout.

The sea lamprey is a primitive vertebrate that is not usually classified as a true fish. Neither is it an eel; although bearing a superficial resemblance, it has no jaws, no scales, and no swim bladder and does not have the paired fins and bones common to most fish. The destructive characteristic of the adult lamprey is its tooth-studded oral cavity. The lamprey's fearsome mouth with its lining of horny teeth and the toothed-tongue are shown in Figure 10. The type of severe wounds lampreys can inflict are shown in Figure 11.

With their sharp, rasplike, toothed tongue, they destroy trout, walleye pike, smallmouth bass, and whitefish—by attaching themselves to their victims with their sucker, rasping holes with their tongues in the body of the fish, and then sucking the blood and tissue fluids. Once the lamprey attaches itself, it holds on until the fish dies (Figure 12). A lamprey can kill a trout in 4–5 hrs. Elimination of the lamprey is exceedingly difficult because it has no known natural enemies. It is at the top of one of the Great Lake food chains.

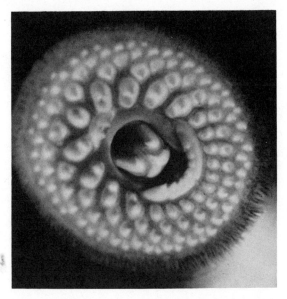

Figure 10. The lamprey's multiple rows of sharp, horny teeth outfitted for rasping through fish scales. Courtesy of the Great Lakes Fishery Commission.

Figure 11. The tenacity with which a lamprey fastens and holds onto fish is legendary. Courtesy of the Great Lakes Fishery Commission.

Figure 12. Lacerations inflicted by the lamprey are deadly. Courtesy of the Great Lakes Fishery Commission.

The lamprey could never have made it into the lakes but for the construction of the Welland Canal in 1829. This opened the inland waterways to oceangoing freighters. Although it took the lampreys close to 100 years to find their way in, once they had established themselves little could stop them. Lampreys arrived in Lake Erie as early as 1921, but they did not flourish. It was not until they arrived in Lakes Huron and Michigan that trout fishery began its downward slide. The yearly catch in Lake Huron declined from close to 2 million pounds in 1935 to fewer than 1000 pounds in 1949. Similar devastation occurred in Lake Michigan in 1935, close to 5 million pounds of trout were landed; by 1949, less than one-tenth of that was taken. The lampreys were so effective that by 1950 commercial fishing in these two lakes was literally destroyed.

After the expansion of the Soo Locks at Sault Ste. Marie in 1945, to permit the passage of huge ore tugs, the lamprey penetrated to Lake Superior. By 1958, most of the trout fisheries had collapsed under the lamprey's relentless pressure. Again, the handiwork of humans.

A. Enter the Alewife

In the absence of trout, the alewife (*Alosa pseudoharengus*) population began to explode. A very fatty, inedible saltwater denizen and cousin of the herring, the alewife has no commercial or sports fishing value. By the 1960s it represented 90% of all fish in Lake Michigan. For reasons not yet known, the alewife underwent massive population crashes every few years, which created unimaginable sanitary problems along the shores of

Lake Michigan. The stench of rotting fish made the area all but uninhabitable. Botulism, a microbial infection of the decaying fish, resulted in deaths of immense numbers of water fowl.

B. TFM

Behind the scenes, Vernon Applegate, a fisheries biologist, was investigating chemicals that would selectively kill lampreys. Over 600 compounds were tried before he found 3-trifluoromethyl-4-nitrophenol, TFM. However, before effective management, prevention, and control could be instituted, the lampreys complex life cycle had to be unraveled (5).

Adult lampreys leave the Lakes to enter spawning streams from late April through July when stream temperatures rise from 40°F (4.5°C) to 60°F (15°C). Spawning occurs in 1–2 ft of water in areas replete with sand and gravel. The female anchors herself to a stone and releases from 30,000 to 1,000,000 eggs, which are fertilized by the males. Both die from exhaustion shortly thereafter.

The fertilized eggs adhere to spaces between the stones. Hatching occurs in about 2 weeks. The $\frac{1}{4}$-in. long larvae detach from the stones and drift passively downstream, coming to rest in pools where they are attracted to decomposing organic matter. Burrowing into the soft material, they remain there for 3–14 years. This resting place proves to be their undoing. It was from this extended larval period that they transformed into parasitic predatory adults and migrated to the Lakes.

The next 2 years were spent feeding on a variety of Lake fish. Estimates indicate that each lamprey killed some 20–30 fish during the 7–10 years of its adult life.

TFM* is most effective against the larvae. The dose required to kill them is harmless to fish and most invertebrates, such as insects, snails, clams, and shrimp. In fact, the U.S. EPA requires that it have no long-term environmental effects on any but its target species. TFM's widespread use proved unusually effective. By 1961, the lamprey populations in Lake Ontario and Lake Erie crashed. By 1973, trout and other sports fish had made a remarkable recovery. But that was not to last (6).

By 1985, there were new problems. The trout were ailing because some as-yet-undiscovered toxic substances seemed to be affecting their roe. Coho and Chinook salmon (*Oncorhynchus*), introduced in the 1960s to reduce the alewife population, are thriving to the point where they have become the most prized of sports fish, but polychlorinated biphenyls (PCBs) have been found in their tissues. Warnings have been posted cautioning people not to eat the fish lest they expose themselves to the chemical. Concern about human exposure to toxic chemicals has also

---

* Although TFM is the primary lampricide, another compound, Bayer 73 (2′, 5-dichloro-4′-nitrosalicylanilide), is used at times in places where TFM is not feasible.

been heightened by the discovery of numbers of cormorants born with crossed bills. Suspicion is falling on dioxan and the dibenzofurans. Strangely enough, PCBs and toxaphene have been found in fish from lakes completely isolated from influent wastes. The presence of these chemicals seems to implicate transmission and contamination via the air.

If the lakes and we are to benefit, solutions that deal with multiple variables simultaneously should be undertaken. However, computer simulations of multiple variables may suffer from lack of basic data. It would be to our peril to forget that disruption wrought in the Great Lakes was the consequence of an engineering prowess that failed to consider the potential ontoward biological impacts.

We next consider the Aswan Dam. It typifies our tinkering with a naturally established set of interrelationships that had served the region well for thousands of years.

V. THE ASWAN HIGH DAM

The diplomats and engineers responsible for the Aswan Dam saw it as the rescuer of Egypt's agriculture, the supplier of needed electric power—in short, the savior of her poor. They were unaware of its potential impact on the sardine industry in the eastern Mediterranean Sea. At a meeting sponsored by UNESCO in Split, Yugoslavia, scientists from 13 concerned countries divulged the facts: by 1965 the Aswan High Dam had sharply reduced the sardine population; the annual catch was down by 50%. It was expected that when the dam became fully operative the catch would be reduced still further, culminating in the demise of the sardine industry—a staple of the Mediterranean economy. Figure 13a, locates the High Dam locally, while (b) indicates the area of Lake Nasser with respect to the delta of the Nile and the Mediterranean.

It is well to note that a dam on the upper reaches of the Nile can alter sardine fishing from Alexandria to Lebanon, showing again that indirect consequences, even at some distance, must be anticipated from uninformed tampering with the environment. Before the dam was built, the yearly flooding of the Nile carried millions of tons of nitrates and phosphates into the relatively mineral-deficient Mediterranean. Many of these mineral nutrients also washed down from the mountains of Ethiopia, the source of the Blue Nile.* This annual dose of fertilizer permitted the luxuriant growth of phytoplankton blooms (microscopic aquatic plants). The phytoplankton provided zooplankton (microscopic aquatic animals) with a ready source of nutrition that allowed them to flourish in great numbers. And the sardines that quickly fattened on the abundant

---

* This is the Amghar region, 14,000 ft above sea level, and the site of Lake Tonna, the highest lake in Africa.

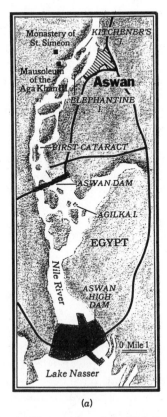
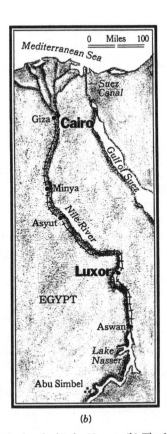

Figure 13. (a) The position of the High Dam at the head of Lake Nasser. (b) The Nile from Abu Simbel to the Mediterranean.

zooplankton filled the nets of the local fishermen. This cycle had existed longer than recorded history. Currently, the Nile no longer carries minerals and fresh water to the Mediterranean; the phytoplankton and zooplankton are drifting to more hospitable areas or dying, and with them are going the sardines. The current lack of flooding on the Nile Delta is similar in its effects to the situation in California's central valley (see Chapter 11). In both habitats excessive evaporation causes accumulation of salts in the soil, threatening the crops that are grown there.

By 1984, it was evident to the Food and Agriculture Organization of the United Nations that this dislocation was continuing. They calculated that the Aswan Dam had reduced the flow of water to the Mediterranean by 35 billion cubic meters. Nutrients had substantially decreased and plankton blooms had all but ceased. With this, sardine production had "dropped drastically." The yield at sea had declined by some 25,000 tons annually. Elimination of the sardines seems certain.

Another ill effect of the dam, also predicted prior to its erection, is the increase in the snail-borne disease schistosomiasis.* Water backing up behind the dam has extended the range of infected snails into new and hospitable habitats. Thus, this immense engineering achievement, a monument to engineering skill, has disrupted at least two major ecological niches, and they are both inimical to man's best interests.

By 1988, two seemingly contradictory consequences were being forecast for the Nile Delta. According to one set of predictions, a catastrophe may be building. With the past 7 years of drought, Lake Nasser has been falling. The level of the lake has dropped over 60 ft and behind the Aswan High Dam, the amount of stored, usable water has decreased to one-fifth of the amount recorded in 1979 (7). Thus, there is fear of a severe water shortage by the year 2000.

On the other hand, Daniel J. Stanley, Senior Oceanographer at the Smithsonian's Mediterranean Basin Project, believes that flooding could seriously endanger the Nile Delta by the year 2099. According to his studies, the sawlike action of the Mediterranean is grinding away at the soils and sediments, cutting back the shoreline by 50–100 ft/yr. This erosion could result in the shoreline being covered by up to 3 ft of seawater by the end of the 21st century (8).

Stanley believes three factors are at work simultaneously. One is the rise of the Mediterranean, another is the sinking of the ground in the area, and the third is the blocking of sediment by the Aswan High Dam. Core samples of the sediment suggest the rise and fall have been going on for nearly 8000 years—but that the loss in land mass was made up by the yearly deposit of tons of sediment washed down the Nile. These yearly deposits allowed the land to maintain its position relative to the sea. With the erection of the dam, sediment deposit halted. Without it, the shoreline has begun to wash away. Over decades the water level will of course increase. Thus, Egypt's "bread basket" could be seriously affected, even lost, creating additional food shortages for millions of people.

VI.   RFEM IN THE ARAVA

The following account indicates that ongoing impacts to ecologic systems are to be expected as new processes and new technologies are developed. Perhaps more to the point, living creates stresses that the living must continually address.

Recently, the United States and Israel agreed on the erection of a Voice of America (VOA) relay station at Idan in Israel's Arava area. The aim of this station is to penetrate the Iron Curtain to reach listeners throughout Soviet Asia and East Africa.

* Schistosomiasis is discussed in detail in Chapter 11.

Concern has been raised about the potentially disastrous effects of powerful radio emissions on birds migrating through the Arava each spring and fall on the major migratory routes between Europe, Asia, and Africa.

The 24 antennas linked to sixteen 500-kW transmitters operating at a number of shortwave frequencies will be 160–200 m high (480–600 ft). Rather than the narrow "whip antennas," these will be "net antennas" in which electronic nets will be stretched 150–400 ft (50–150 m) between pairs of towers. The station is expected to cover an area of some 6 km^2 (1 mi^2).

Public concern has focused on the millions of birds that pass through the Arava and the effect the stations' emissions might have on the birds' navigational system. Warnings have been issued that the birds' navigational ability could be disrupted or "jammed" by the powerful radio frequency electromagnetic (RFEM) radiation emitted by the installation.

Particular concern focuses on birds of prey, the eagles and buzzards. If these bird populations were significantly affected, it is feared that wheat fields of the Soviet Ukraine and other parts of Europe could be adversely affected by rodent depredation, with subsequent effect on supplies of flour.

Scientists do not yet understand the mechanism that permits birds to migrate thousands of miles between nesting places in Europe or Asia and the same tree in Africa. Experiments do suggest that the birds are guided by a geomagnetic perception.

It was only in the past 10 years that researchers have come to realize Israel's importance as a migration route for birds from a vast area stretching from central Europe to western Asia. Large predators as well as storks and pelicans, whose wing structure does not permit them to flap their way over large bodies of water, follow the narrow land bridge over Israel to winter in Africa, riding thermal currents by day and landing for the night. Although the proposed Idan site lies just outside the main migration route, a shift of the prevailing winds, or an overcast day could easily bring the birds into the VOA transmission area. To their credit, an environmental impact station assessment of the planned Arava site will include a discussion of possible damage to migrating birds of radio emissions.

The idea that radio transmitters in Israel might adversely affect bread supplies in Europe does give one pause.

VII. THE EARTH'S SOIL: A WORLD IN MINIATURE

The penchant for thinking about our world in terms of components linked into a functioning whole is nowhere more evident than in the earth's soil: We have been discussing large-scale components on a global scale, but soil is a veritable world in miniature. Let us discuss briefly a few of the systems at work in soil to show that the idea of interconnectedness extends from the molecular biochemical level to the tissue level, and that negative consequences can arise from the breakdown of these much smaller events.

The classic interdependency of roots and soil is based on the symbiotic relationship between the roots of plants, trees, and flowers and specific fungi, which together form a mycorrhiza. This interdependency or symbiosis ("life together") beneath the soil surface begins with the penetration of the roots of legumes (peas, clover, soybeans, alfalfa, and lentils) by rhizobacteria. Ordinarily, this penetration would be considered an infection, except that in this case the plants and bacteria prosper. The rhizobia obtain nourishment from the legumes and in turn supply the plants with necessary nitrogen in the form of ammonium ions (NH_4^+) (9).

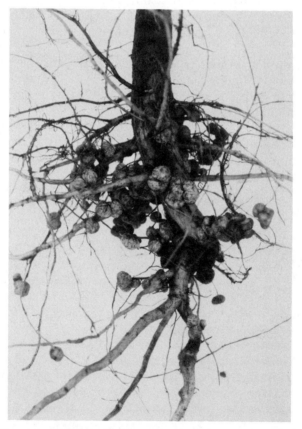

Figure 14. Well-nodulated soybean root system (*Glycine soja*). Courtesy of Dr. Donald L. Keister, Nitrogen Fixation and Soybean Genetics Lab, ARS, USDA.

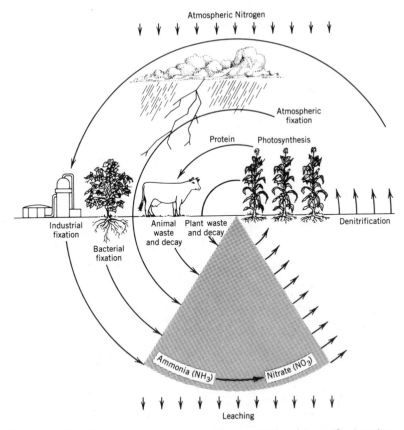

Figure 15. The nitrogen cycle. Reproduced by permission of *Scientific American*.

As a response to the bacterial penetration, the plants develop nodules of varying sizes clustered about the roots—some as small as raisins, others as large as peas or grapes. Figure 14 shows the nodule formation on soybeans. Through these nodules nitrogen fixation can take place. This is an essential life process that utilizes inert, diatomic molecular nitrogen (N_2) and converts it through nitrate (NO_3) to ammonia (NH_3) and to organic compounds, especially the peptides. It is the peptides, the precursors of proteins, that can be utilized by plants. The nitrogen cycle, a fascinating example of a symbiotic system (an interconnected web), on a global scale, is shown in Figure 15.

In addition to the rhizobia, roots provide a haven for soil fungi, which send hairlike filaments (hyphae) into the roots. The ensuing root–fungus relationship produces a mycorrhiza—an extensive network that penetrates an entire garden, forest, or cropland.

The mycorrhizal network can affect the moisture balance of plants,

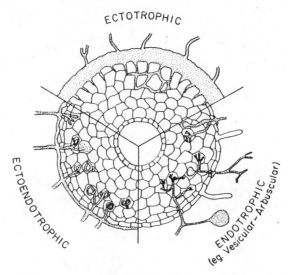

Figure 16. Types of mycorrhiza. (*a*) Ectotrophic. There is a sheath of fungal tissue and limited intercellular invasion by hyphae (the Hartig net). (*b*) Endotrophic (vesicular-arbuscular as the example). Root hair formation is not suppressed and hyphae ramify surrounding soil producing large spores. (*c*) Ectoendotrophic. Some of the coiled, fungal hyphae within the plant cells become digested (8).

especially in arid zones, by transporting to the plant water that was collected by the hyphae well beyond the reach of the plant's roots. It can also make phosphate available to plants growing in phosphate-depleted or naturally poor soils by converting the little phosphate available into soluble form and transporting it too back to the roots.

The mycorrhizal extension of the root systems of plants and trees with fungi is widespread. Some trees do it with the aid of a group of mycorrhizae referred to as endotrophs. Flowers such as orchids also profit from the endotroph system. Another group, the ectotrophs, benefit such forest trees such as birch, beech, and pine.

Figure 16 shows the major types of mycorrhiza and their structural differences. Clearly, the endotrophs have the most extensive formation of root hair. The ability of these mycorrhizal fibers to penetrate the roots borders on the phenomenal. They have been shown to extend from tree roots up beyond the forest floor to the "litter layer" above, which they tap and use as a continuous source of nutrients. Endotrophs are estimated to increase the nutrient supply and range of forest trees by a factor of 10. Endotrophs comprise the greatest mass of any group of microbes on earth: they account for some 15% of the weight of the entire world's plant roots. In fact, they are currently believed to be the only link between the biological world and the soil. A recent report by the National Academy of Sciences referred to this symbiotic relationship as "the cornerstone of

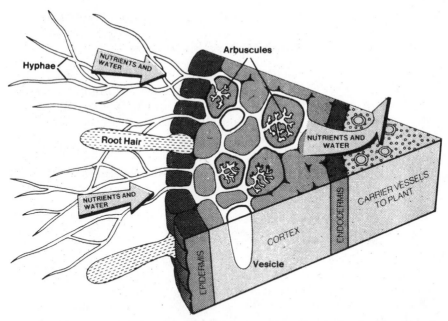

Figure 17. Enhancement of plant root systems. Components of typical endomycorrhizae in plant root are designated in white. Nutrients absorbed from soil by hyphae are transferred to the plant by arbuscules, which penetrate cells of the root cortex. In vesicles, fungal growth stores sugars it has extracted from plant. Reproduced by permission of *The New York Times*.

mineral conservation by natural tropical forests." Figure 17 presents an interpretation of the mechanism by which fungal hyphae transfer nutrients to the root cortex.

Currently, reinoculation of soils with fungi that have been suppressed or killed by pesticides is under way in nurseries and orchards. Inoculation of seedlings with fungi is also under way as a means of promoting revegetation of sites such as strip mines, where natural fungal populations have been decimated. Trees planted on mine wastes and inoculated with these fungi are far more likely to survive. More than 1000 acres of strip-mined land in Ohio and in states nearby have been regreened with the help of this fungus–root connection.

Current research is directed at selecting those species that produce more filaments faster or that can better tolerate hot, cold, dry, acidic, or other stressful conditions. The potential for increased food supplies, for example, is remarkable. The possibility of altering the genetic makeup of the fungi to further enhance nitrogen fixation is also being considered. Clearly, the mycorrhizae in conjunction with root-nodule bacteria are at the hub of an environmental system that makes life on earth unique.

VIII.   WEBS OF CAUSATION AND PERSONAL HEALTH

A. Airs, Waters, and Places

The idea that physical factors exert an influence on health is as old as humankind itself. The language and traditions of primitive peoples everywhere refer to the ill effects of certain winds, the changing of the seasons, the phases of the moon, and the influence of the sun and stars.

A cogent delineation of the unity of our habitat written some 2500 years ago, yet whose message remains vital today, was set forth by Hippocrates in his treatise "On Airs, Waters, and Places." According to this 4th century B.C. physician,

> Whoever wishes to investigate medicine properly should proceed thus: in the first place to consider the seasons of the year and what effects each of them produces. Then the winds, the hot and the cold, especially such as are common to all countries, and then such as are peculiar to each locality. In the same manner, when one comes into a city to which he is a stranger, he should consider the situation, how it lies as to the winds and the rising of the sun; for its influence is not the same whether it lies to the north or the south, to the rising or to the setting sun. One should consider most attentively the waters which the inhabitants use, whether they be marshy and soft, or hard and running from elevated and rocky situations, and then if saltish and unfit for cooking; and the ground whether it be naked and deficient in water, or wooded and well watered, and whether it lies in a hollow, confined situation, or is elevated and cold; and the mode in which the inhabitants live and what are their pursuits, whether they are fond of drinking and eating to excess, and given to indolence, or are fond of exercise and labor, and not given to excess in eating and drinking. . . . If one knows all these things, or at least the greater part of them, he cannot miss knowing when he come into a strange city, either the diseases peculiar to the place, or the particular nature of common diseases, or commit mistakes, as is likely to be the case provided one had not previously considered these matters.

Hippocrates did not have all the answers, but surely he pointed to the questions. The deeper the analysis of the web of life is pushed, the more meaningful the idea of dependence becomes. With this in mind, let us examine three diseases that illustrate this principle beautifully.

B. Heart Disease

A fundamental epidemiologic tenet is that disease does not distribute itself randomly in a population. There is a good reason why a person or group of people are ill with some condition at a given time. Such a concept considers the many predisposing factors as well as their often complex interrelations.

For example, Figure 18, not as intimidating as it may appear at first glance, suggests that a myocardial infarction* (a heart attack) may well be the consequence of a number of risk factors, that is, a complex causal "web" of multiple interactions. Looking at the problem this way can aid prevention. Instead of concentrating on one overwhelming recent cause ("What did I do today, or lately, that caused this?"), we can understand the weblike and interconnecting causes of the disease—and we can "attack" the web at a point, or, better, points, relatively distant from the occurrence of the condition itself. If we do this, it is conceivable that a heart attack need never occur.

C. Gonorrhea

Given the concept of multiple causation, it is perhaps more readily understandable why knowing that a microbe such as the gonococcus, which microbiologists consider the cause of the sexually transmitted disease gonorrhea, is insufficient to control or prevent an illness that has attained pandemic proportions.

As Figure 19 shows that knowledge of the organism and the general availability of penicillin were responsible for sharply reducing the disease immediately following World War II. In fact, by 1957, public health officials were hinting at its possible eradication. By the mid-1960s, however, it was clear that not only would that not occur, but that an epidemic was in the making. By 1975, gonorrhea had attained pandemic proportions. A detailed knowledge of the microorganism—physiological, morphological, biochemical, and clinical—obviously is inadequate to control or to prevent its wide dissemination. Estimates suggest that the actual number of cases is ten times the reported figure.

To deal with this problem, it is necessary to consider the changing patterns of sexual mores, the psychology of promiscuity, prostitution, economics, drug addition, level of education, and communications, along with microbial resistance to antibiotics—that is, all the factors that contribute to the web of causation.

According to Franz Alexander, an eminent psychoanalyst, "one of the greatest weaknesses of the human mind consists in a kind of laziness, in the urge to find either/or solutions and to explain everything by a single principle instead of considering the multiplicity of factors in their interrelationship."

This state of mind would not be applicable to the biologists who spent years unraveling the intricate ecological relationships about to be described.

---

* An infarct is a sudden loss of arterial blood to a segment of the heart, producing an area of necrosis (dead tissue) in the myocardium (the middle layer of heart muscle). With it comes intense, prolonged, deep substernal pain and changes in heart rhythm. Death often follows if appropriate management does not occur quickly.

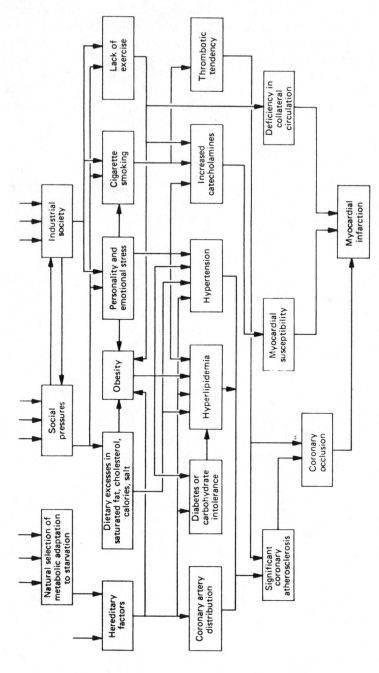

Figure 18. A web of causation for myocardial infarction. Reproduced by permission of Dr. G. D. Friedman, from *Primer of Epidemiology*, McGraw-Hill, New York, 1980.

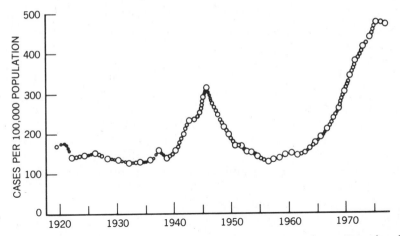

Figure 19. The fall and rise of gonorrhea in the United States. 1919–1977. Reproduced by Permission of *Morbidity and Mortality Weekly Report*, Centers for Disease Control, September 1981, Vol. 29, No. 54.

D. Sleeping Sickness

In Central and East Africa, the agent associated with the clinical symptoms of sleeping sickness is a microscopic protozoan, *Trypanosoma gambiense*. This parasite is transmitted to the human population by the bite (injection) of the tsetse fly, *Glossina palpalis*, when taking a blood meal.

The natural habitat of the tsetse fly is restricted to the dense tropical forests that border lakes and rivers. Its usual source of nourishment is the marshbuck, or Situtunga antelope, which is, in turn, the main dietary staple of leopards. Thus, there is a direct relationship between the leopard population and the incidence of human sleeping sickness. In years when leopards abound, the marshbuck population is kept low, and as a consequence the tsetse flies obtain few animal blood meals to sustain them between human blood meals. So it can be said that hunters and killers of leopards are in large measure responsible for the level of sleeping sickness.

Intensive investigations aimed at controlling sleeping sickness uncovered another unique chain of events in Zimbabwe. Control appears to depend on preventing the tsetse fly, in this instance the species *G. morsitans*, from reaching adulthood. To accomplish this, the natives in the area are prevented from burning off the grass in the treeless savannah regions.

After years of arduous searching it was found that the grass fires destroyed the ants that fed on the tsetse fly pupa (the developmental stage between larva and adult). The heavy mats of grass aided in increasing the soil moisture, which in turn encouraged the growth of the ant population, which kept the tsetse fly population in check, which ultimately meant decreased transmission of protozoan parasites and fewer cases of sleeping sickness.

There is another side to all this. Controlling predators, and thereby animal and human illness, may induce additional ecological depredation. How so? By eliminating the tsetse fly, for example, areas of Africa previously uninhabitable could be opened to development, which in the long run could be harmful. Would this be a replay of what is happening in the Sahel? Elimination of insect pests on the island country of Sri Lanka (see Chapter 14) resulted in an exponential rise in population that led to severe economic hardship.

But is this reason enough to permit prevailing adverse environmental conditions to continue when the technology exists to eliminate them? As you may perceive, solutions are neither simple nor straightforward.

E. Bioaccumulation of DDT

For years we have heard of bioaccumulation and biomagnification within ecosystems. The concept suggests that chemicals such as DDT and other organohalides are not hydrolyzed or otherwise destroyed by natural processes, but become more concentrated as they proceed up the food chain. Initially, small amounts of these substances are sprayed on soil. Species at the top of the food web levels show a much larger proportion of DDT than was originally distributed on the soil. Thus, the magnification in a lake may increase 10,000-fold, from a level of 0.02 ppm (parts per million) in the water, to 5 ppm in plankton, 40 ppm in plant-eating fish, and 2000 ppm in fish-eating birds.

The primary reason given for this accumulation is that the chemicals are soluble in fat and not in water. Had they been water soluble, they would have been excreted in urine and perspiration.

The threatening implication of this is that although the 0.02 ppm is nontoxic, the 2000-ppm dose is. People at still higher trophic levels of the food web, the final predators, will eat the birds containing these high levels and will, so the thinking goes, be at greatly increased risk of DDT intoxication. The sudden release of these toxic levels of DDT from fat tissue into our circulating blood may come about through prolonged illness or an extreme program of weight reduction. Fortunately, this type of poisoning, while perhaps reasonable, does not appear to have been observed in human populations.

From 1947, when DDT was first introduced to the U.S. environment in commercial quantities, until its banning in the 1970s, average individual exposure was on the order of 0.2 ppm. This produced an average body fat concentration of 7 ppm. Taking a worst-case scenario, Alice Ottoboni deals with this problem in her recent toxicology text via several pertinent calculations. An obese person weighing approximately 310 lb, with a fat content of 25%, would contain about 35 kg of fat (77 lb). If we use the average concentration of 7 ppm in the body fat of Americans, the 35 kg of fat would contain 245 mg of DDT. Controlled studies with human volunteers who ingested 35 mg of DDT per kilogram of body weight per day for up to 5 years produced no adverse effects. If it can be assumed that the ingested DDT is similar in effect to DDT released from fat stores, 35 mg/day would be equivalent to a body burden of DDT for the obese individual (10).

From accidents with DDT as well as experiments with volunteers, it has been learned that ingestion of up to 70 mg/day is less than an acute toxic dose. For an obese person that would be equivalent to 9800 mg of DDT. Far more would be needed to produce symptoms of acute illness.

Has anyone tried to lose 25 lb or more rapidly? Human volunteers placed on starvation diets, Dr. Ottoboni tells us, have lost as much as 0.6 lb of fat in a day (about 0.25 kg). The amount of DDT released from that much fat—containing 7 ppm of DDT—would be less than 2 mg, far less than a chronic nontoxic dose. The problem of chronic or acute poisoning as a consequence of the sudden release of DDT from fat is an unlikely, though theoretically possible scenario. Biomagnification is a compelling theory, but its adverse health effects on human populations remain to be demonstrated.

Furthermore, as Frederick Coulton, editor of the journal *Regulatory Toxicology and Pharmacology,* recently concluded, "it is now well established that DDT and its metabolites are photochemically degraded and therefore pose only a potential not actual accumulation of residues in the environment"* (11).

Furthermore, thinking in terms of ecosystems, or causal relationships, must not be carried to extremes. The anecdote of Sir Percival's cat is presented for ease of grasping the ecological concept, not for its literality. The quotation by Francis Thompson at the beginning of the chapter must not be taken at face value. Flowers can indeed be picked "without troubling of a star."

Hurricane Gilbert, the earthquake in Nepal, the floods in the Sudan, and Bangladesh are examples of natural rather than man-made disasters which have killed some 3 million people in the past 20 years and caused hundreds of millions of dollars in damage and untold suffering. Changes will need to be made, must be made. Ecological habitats will have to be reordered to prevent these depredations from continuing.

A recent editorial in *The New York Times* grappled with this idea. The editor asked: "Can Big Oil and its government regulators be trusted with the fragile environment of Alaska's Arctic Wildlife Refuge?" "Congress," it went on to say, "pressed by the Reagan administration to allow

---

* For additional details on the DDT problem see Chapter 5.

Figure 20. "Man Struggling with His Environment" by Adolf Gustave Vigeland.

exploratory drilling in what may be North America's last great oil reserve, has been wrestling with the question for years. . . The North Slope development has been America's biggest test by far of the proposition that it is possible to balance energy needs with sensitivity for the environment . . . no species is reported to be endangered, no dramatic permanent changes in ecology* (sic) are forecast. Much of the unpredicted damage has arisen because more oil has been produced than originally predicted. Even so, the total acreage affected by development represents only a fraction of 1 percent of the North Slope wilderness." They concluded

---

* Ecology is a noun. The editor should have said ecologic balance or ecologic interre-lationships. Use of the adjective ecological is also correct.

with the observation that "it is hard to see why absolutely pristine preservation of this remote wilderness should take precedence over the nation's energy needs."†

Thinking ecologically can promote intelligent planning and action. It should not be a shibboleth for inaction.

Some years ago, the Norwegian sculptor Adolf Gustave Vigeland created a statue depicting man's struggle with his environment. Manifestly, he believed people had not yet overcome—nor would in the foreseeable future—the many conjoint forces that impinge on and affect their existence. The statue "Man Struggling with His Environment" (Figure 20) can be seen in the city park of Oslo, Norway.

Vigeland saw humankind as unable to break out of confinement. I would have liked to see a small crack in the encircling ring of forces. That would have suggested gain or overcoming—which has most assuredly occurred. I would also like to add a second statue, smaller than the original, but less confining. This one would portray a human mind in chains. If we are to be finally free, it is these chains that must be rent asunder. Until we discard the unworthy notion that we are being poisoned, that we are less than healthy, we will forever be shackled, no matter what the quality of our environment.

REFERENCES

1. M. H. Glantz, Drought in Africa. *Sci. Am.* **256**(16): 34–40, 1987.
2. G. Hardin, The Limits of Altruism: An Ecologist's View of Survival. This is an expanded discussion of the thesis first presented in The Tragedy of the Commons, *Science* **162**: 1243–1248, 1968.
3. P. Lewis, Locusts Head South. *The New York Times*, Apr. 24, 1988.
4. *Fishery and Commodity Situation Outlook, 1983/84.* Infofish (FAO), Rome.
5. TFM vs. The Sea Lamprey: A Generation Later. *Great Lakes Fish. Com. Spec. Pub.* 85-6. 1985.
6. Salmonid–Sea Lamprey Management Alternatives for Lake Champlain. Prepared by Lake Champlain Salmonid/Sea Lamprey Sub-Committee for Fish. Tech. Comm., Lake Champlain Fish & Wildlife Management Coop. Feb. 1985.
7. A. Cowell, If the Nile Dries Up, Who Will Feed Egypt? *The New York Times*, Dec. 31, 1987.
8. D. J. Stanley, Subsidence in the Northeastern Nile Delta: Rapid Rates, Possible Causes and Consequences. *Science* **240**:497–500, 1988.
9. J. M. Lynch, *Soil Biotechnology. Microbial Factors in Crop Productivity* Blackwell, Oxford, 1983.
10. A. Ottoboni, *The Dose Makes the Poison.* Vincente Books, Berkeley, CA, 1984.
11. F. Coulton, Reconsideration of the Dilemma of DDT for the Establishment of an Acceptable Daily Intake. *Regul. Toxicol. Pharmacol.* **5**: 332–383, 1985.

† *The New York Times*, June 2, 1988, p. A26.

SUGGESTED READINGS

Briand, F., and Cohen, J. E. Environmental Correlates of Food Chain Length. *Science* **238:** 956–960, 1987.

McNaughton, S. J. Mineral Nutrition and Spatial Concentrations of African Ungulates. *Nature* **334:** 343–345, 1988.

2

Microbial Food Poisoning

Ideas are everywhere, but knowledge is rare.
—Thomas Sowell

Martha Jefferson,* a 34-year-old woman, complained of sporadic epigastric pain. Ms. Jefferson was a devotee of sushi, an increasingly trendy form of raw fish. With continued pain she relented and went for x-rays. Embedded in the mucosal lining of her stomach was an inch-long third-stage larva of *Anisakis simplex*. Raw salmon is considered the most likely source of this parasite.

Pomatomus saltatrix (bluefish) is not usually suspect. But for the five physicians who sat down to lunch recently at a New Hampshire inn, that was cold comfort. Although such nonscombroid fish as amberjack and mahimahi have been implicated in poisonings typical of the Scombridae family, bluefish have not. Nevertheless, within 4 h of eating they developed headache, flushing, redness of the upper body, diarrhea, abdominal pain, and pounding heart.

In New York State during an 8-month period in 1982, over 1000 people became ill with typical symptoms of gastroenteritis as a result of eating bacterially contaminated clams and oysters.

And on July 4, 1986, 41 employees of a department store in southern Taiwan were hospitalized after eating white-tipped mackerel. Dizziness, burning mouth, numbness of the lips, flushing, headache, and abdominal cramping were the clear signs of scombroid poisoning.

At Princeton University, it was nothing so sophisticated. For three consecutive years, students and nonstudents have been afflicted with

* Martha Jefferson is a fictitious name.

35

salmonellosis—one of the most common forms of bacterial food poisoning. One of the outbreaks involved 80 students and six administrators. Improper food-handling practices appeared to be the precipitating cause in all three events.

To the 1216 men, women, and children who became violently ill after eating poisoned food at two picnics in August 1959, the memory will be especially vivid.

On August 15, 1700 employees of a large pharmaceutical plant and their families gathered for a picnic lunch. Food was served beginning at about 11:30 a.m. By 1 p.m., the first cases of food poisoning (gastroenteritis) occurred. By 4 p.m., over 800 people had become ill. By 9 p.m., the Indiana Department of Health had recorded 1000 victims.

Since forewarned is usually forearmed, planners of a picnic given the following Saturday, by an electrical parts manufacturing firm, should have been alerted by the previous week's tragic events. Apparently, it was not a learning experience. August 22 was a particularly warm day: 1813 people turned out for the festivities. Food was served at 4:30 p.m. By 7 p.m., symptoms of severe gastroenteritis (nausea, vomiting, abdominal cramps, and diarrhea) were evident in 25–30 people. An additional 100 were ill by 10:30 p.m. In all, 216 cases were reported.

Indiana had its August, but other communities in the United States and around the world are not immune. Foodborne illness has little respect for boundaries, the seasons, age groups, or the sexes. Anyone is susceptible anytime, given the appropriate conditions.

I. FOOD AND ILLNESS

A. Poisoning Versus Spoilage

Bacterial food poisoning must be distinguished from food spoilage. One is not synonymous with the other. They are quite dissimilar conditions. To understand the difference, it may be helpful to recall the bacteria are unicelular organisms* that exist in three distinct forms: as rod-shaped cells, with and without flagella, as spherical cells called cocci, and as spiral or corkscrew-shaped cells. Figure 1 shows a typical rod-shaped bacterium with flagella arranged completely around the body. A curved rod (vibrio) (Figure 1a) is typical of the several organisms associated with foodborne disease. Figure 1b shows the chainlike aggregation of streptococci, 1c the grapelike clusters of staphylococci, and 1d pairs of diplococci. Spiral forms are seen in Figure 1e. They are not involved in any known foodborne disease. Figure 1f shows clostridial forms of the

---

* In 1872 Frederick Cohn classified bacteria as plants. They remained in that kingdom until 1974. Bergey's *Manual of Determinative Bacteriology*, 8th ed., has since officially placed bacteria in the kingdom Procaryotae. Whether they are plants remains problematical.

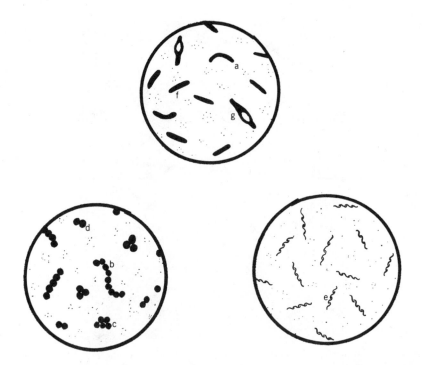

Figure 1. (a) Curved rod (vibrio); (b) typical streptococci chain; (c) grapelike clusters of staphylococci; (d) pair of diplococci; (e) typical spiral forms; (f) clostridial rods with rounded ends; (g) club-shaped spore stage.

type involved in both food spoilage and food poisoning. These cells have an unusually resistant spore state, shown in Figure 1g, which protects them against destruction by heat. This characteristic, as we shall see, is especially significant in food handling and processing.

Whereas Figures 1a–g are sketches, Figure 2 is a scanning photomicrograph of coccoid-shaped bacteria, and Figure 3 shows typical rod-shaped organisms. To help visualize the sizes of microbes, Figure 4 shows virus particles infecting a red blood cell.

Microbes are small. None can be seen without benefit of the magnifications provided by light and electron microscopes. For example, bacteria are measured in micrometers, usually one or less, and the viruses are far smaller. Recall that 1 micrometer (μm) is equal to 1/25,000th of an inch, or 1000th of a millimeter.

For an appreciation of the dimensions involved, consider Figure 4. An influenza virus is shown infecting a red blood cell. Red blood cells are between 5 and 7 μm in diameter. "Flu" viruses are approximately 0.1 μm, which means they are some 50–70 times smaller. Bacteria, especially those involved in food poisoning, are 5–7 times larger than the flu virus.

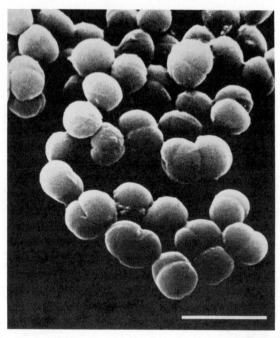

Figure 2. Scanning electron photomicrograph of a coccoid-shaped bacterium. The bar equals 2 μm. Courtesy of the American Society for Microbiology and the *Journal of Bacteriology*.

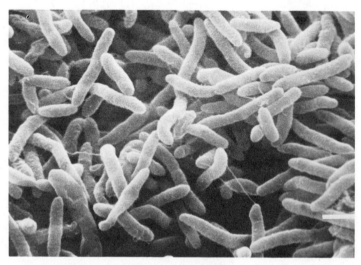

Figure 3. Scanning electron photomicrograph of a rod-shaped bacterium. The bar equals 1 μm. Courtesy of the American Society for Microbiology and the *Journal of Applied Environmental Microbiology*.

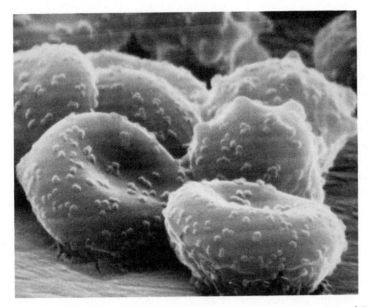

Figure 4. A red blood cell with infecting influenza virus (24,000×). Courtesy of Dr. R. A. Baker, School of Medicine, University of Southern California.

It would make a very neat package if it could be said that the cocci or grapelike forms are primarily responsible for food poisoning and the rod-shaped forms for food spoilage. Unfortunately, it is not that easy. Although cocci do induce a form of food poisoning and rods do induce food spoilage, rod-shaped organisms are often the agent of foodborne disease.

Food undergoing spoilage is usually contaminated with large numbers of rod-shaped organisms called pseudomonads. Pseudomonads of one type or another spoil cottage cheese, poultry, meats, and fish, but they do not produce toxins capable of initiating human illness. Since some rods do produce illness and others do not, it is evident that the relation of bacterial morphology—shape or form—to pathogenicity is not straightforward.

Every homemaker is familiar with the common signs of food spoilage. Many would consider fruit spoiled and unfit to eat when partially or wholly covered with mold, but while aesthetically such fruit may be unattractive, it is not poisonous. That is, it does not evoke gastric or neurologic distress when eaten. Many people pare away molded areas of fruit to enjoy the winey flavor imparted by the microbial conversion of fruit sugars to alcohol. Others simply throw the fruit away. Cottage cheese and hamburger, two highly proteinaceous foods, are readily perishable. To some, they become unfit to eat after several days of refrigeration, yet others accept them with equanimity.

To the question, "When is food fit to eat?", an appropriate reply might be that food is edible when its look and odor testify to its edibility. If this is a valid standard, then a food is inedible when the senses suggest "this is not fit to eat." Fortunately, or unfortunately, depending on individual preferences, little universal agreement exists as to what is or is not fit to eat. Some people like their meat "high." Most Americans would call such meat spoiled and not fit to eat. "High" or "gamy" meat usually has a strong odor and taste, both produced by bacterial action. Titmuck, fish buried to allow bacterial fermentation to occur, is eaten by Eskimos as a delicacy. Eskimo dogs, however, refuse to eat this semiliquid, foul-smelling delight. Most Americans would call it putrid. Yet it is putrefaction that gives Limburger cheese its gourmet qualities! These examples of "spoilage" are caused by the growth and metabolic activity of microorganisms, but food poisoning is not involved.

Food poisoning, also the result of microbial activity, is a subtly insidious process, limited to only a few types of microbe. In most cases their presence is totally unsuspected. Off odors and tastes are not apparent. Not a hint of suspicion surrounds the food. Odor, taste, color, and texture are usually unaltered.

Most foods are subject to some degree of deterioration over time. Considering the ease of spoilage, foods can be categorized as *nonperishable, semiperishable,* or *perishable.* The degree of perishability simply refers to the ability of microbes to use a food as a nutrient. Examples of relatively nonperishable foods are dry grains, sugar, flour, and dried beans. Without sufficient moisture, microbes find it difficult to spoil foods. Potatoes and apples are examples of semiperishable foods containing more available moisture, while meats, fish, poultry, and milk—the high-protein foods with large amounts of moisture—are highly perishable. Because these foods are desirable sources of nourishment for both people and microbes, it is easy to comprehend the aptness of L. V. Burton's pithy remark that "sanitation is nothing more than a race between men and the lower forms of life to determine who gets the food supply first."

B. Types of Food Poisoning: Organisms and Agents

Diseases spread by foods can conveniently be placed in three major groups: (1) foods containing a toxin as an end-product of bacterial metabolism, (2) those containing significant numbers of bacteria which on ingestion release toxins and evoke specific physiological responses, and (3) those containing a more complex form of parasite that can initiate a more complex type of illness. The following list of foodborne infections, though not exhaustive, indicates the wide range of biological forms involved. We are concerned only with those producing characteristic symptoms of food poisoning.

I. **Preformed Bacterial Toxins**
   Botulism
   Staphylococcal intoxication
   Enterotoxigenic *Escherichia coli*

II. **Bacterial, Viral, and Rickettsial Infections**
   Typhoid and paratyphoid fevers
   Salmonellosis
   Streptococcal food poisoning
   Shigellosis (bacillary dysentery)
   Hepatitis A
   Campylobacteriosis (*Campylobacter jejuni*)
   Yersiniosis (*Yersinia enterocolitica*)
   Non cholera vibrio gastroenteritis (*Vibrio parahaemolyticus* and *Vibrio vulnificus*)
   Anthrax
   Brucellosis
   Tuberculosis
   Tularemia

III. **Parasitic (Protozoal and Zooparasitic)**
   Amoebiasis (amoebic dysentery)
   Taeniasis (beef tapeworm infestation)
   Trichinosis (port roundworm infestation)
   Ascariasis
   Hydatidosis
   Anisakis simplex

IV. **Chemical/Microbial Complexes**
   A. Scombroid poisoning
   B. Paralytic shellfish poisoning

1. Reporting

Although countrywide data depend for the most part on the commitment of state and local health departments, and that is highly variable, bacterial pathogens appear responsible for the greatest number of food poisonings. Viruses, chemicals, and parasites follow in descending order. In terms of severity (i.e., deaths), bacteria and chemicals are most important. Table 1 is an example of reported outbreaks by etiologic group. Complicating the data is the large "unknown" fraction. In this instance, almost 61% of the reported outbreaks are of unknown etiology. More than likely most are of bacterial and chemical origin. Unfortunately, in too many instances neither the individual nor the suspect food are available for testing.

Outbreaks are not the same as cases. They usually involve groups of people who become ill suddenly and are in great distress. As seen in the table, these outbreaks occurred with as few as five and as many as 80. Because of the numbers involved, the suddenness of onset, and the

Table 1
Reported Outbreaks of Foodborne Disease by Etiologic Group:
United States, 1973–1984

Etiologic Group	Number of			Mean Outbreak Size	Death-to-Case Ratio[a]
	Outbreaks	Cases	Deaths		
Bacterial	1,591	73,359	152	46.1	2.1
Parasitic	106	795	4	7.5	5.0
Viral	105	8,425	2	80.2	0.2
Chemical	518	3,542	19	6.8	5.4
Unknown	3,604	84,946	25	23.6	0.3
Total	5,924	171,067	202	28.9	1.2

[a] Deaths per 1000 cases.

physiological distress, it is understandable that outbreaks will be reported and recorded more often than individual cases especially if the cases are mild.

The etiologic category can be pried open to reveal the organism involved. Table 2 displays the major contributing culprits. Between them *Salmonella* and *Staphylococcus* are responsible for some 70% of occurrences.

Although not shown in the table, new contributors have been recently identified. *Campylobacter jejuni*, poorly known only 5 years ago, has become so important that it is often grouped separately (1). And *Yersinia*

Table 2
Reported Outbreaks of Foodborne Disease Caused by Specific Bacterial Pathogens:
United States, 1973–1984

Pathogen	Number of			Mean Outbreak Size	Death-to-Case Ratio[a]
	Outbreaks	Cases	Deaths		
Salmonella	597	30,296	60	50.7	2.0
Staphylococcus aureus	358	17,332	4	48.4	0.2
Clostridium perfringens	214	12,432	12	58.1	1.0
Clostridium botulinum	183	428	37	2.3	86.5
Shigella	78	5,647	2	72.4	0.4
Bacillus cereus	53	777	0	14.7	0
Camphylobacter[b]	37	1,289	2	34.9	1.6

[a] Deaths per 1000 cases.
[b] First reported in 1978.

enterocolitica is becoming well known, as is *Vibrio parahaemolyticus.* *Aeromonas hydrophila* is among the newest and is rapidly becoming associated with diarrheal disease in children.

Some believe the figures in Table 2 are only the tip of the iceberg. From survey data, Hanschild and Bryan *estimated* the annual numbers of cases of foodborne salmonellosis in the United States. They obtained a ratio of 29.5 : 1 of illness to isolations. This ratio was designated Ri. The average number of human *Salmonella* isolates per year during the 10-year period 1969–1978 was 25,000. Applying the Ri, they obtained a total of 740,000 cases for the period (25,000 × 29.5). In 1983, the number of isolates was 38,881. Therefore, the number of cases for the year, again using Ri, would have been 1,147,000 (2).

Campylobacter infections are believed to occur 2.5 times as often as *Salmonella.* Thus, 1,147,000 × 2.5 would suggest a yearly load of 2,867,000 cases. *Shigella* isolates numbered 14,946 in 1983. If the *Salmonella* Ri is applied (14,946 × 29.5), 440,900 cases are obtained. Thus, for 1983, the three organisms could have doubled, the resulting 9 million could account for cases induced by the many other types of organism implicated in foodborne illness.

But with a factor of 50% of unknown etiology, another doubling may be in order. This brings the total of potential yearly cases to 18 million. If Ri is high or low, the number of cases could be half or double the 18 million. Clearly, foodborne illness leaves much to be desired by way of statistical rigor. On the other hand, the outcome of many of these episodes is far from life threatening: often no more than temporary discomfort—which is a good reason for the underreporting.

State Departments of Health are by law responsible for dealing with this problem. But they must ultimately rely on physicians and the public for indications of outbreaks. However, current estimates indicate that less than 30% of the total is brought to their attention.

Whether the true incidence is 100,000, 1 million, or 18 million will of course be difficult to ascertain and may in fact be irrelevant. The important point is that bacterial food poisoning is preventable. In a "high-tech" society such as ours, this type of illness need not occur. (More about this further on.)

2. Chemical/Microbial Complexes

Unusually provocative are the chemical factors that contain an assortment of synthetic and natural toxicants. In addition to the heavy metals, copper, zinc, cadmium, and tin, which in moist environments leach from piping, shelving, and other metal structures, contaminating food and beverages in the process, fish, shellfish, and mushrooms can be inherently dangerous because of the chemicals naturally present in their tissue. Mackeral, skipjack tuna, snapper, and barracuda produce ichthyosarco-toxic diseases in fairly large numbers.

The five physicians who dined on bluefish and the Chinese who fell ill on a lunch of mackeral did so after ingesting chemicals that accumulated in improperly handled fish.

Symptoms of scombroid poisoning often begin within minutes of eating. (The time frame is almost pathognomonic of chemicals compared to the hours required by bacterial toxins to exert their effects.) The headache, facial flushing, burning mouth, tingling lips, itching, and blistering resemble a histamine-like response—a far cry from the more pedestrian gastroenteric symptoms of nausea, vomiting, and diarrhea characteristically seen in bacterial food poisoning.

Acting in concert, three factors appear to be required for the typical symptoms.

1. The fish must contain free histadine in their tissue.
2. Histadine-decarboxylase-producing microbes must be present on the fish surface.
3. Conditions of time and temperature must be appropriate for microbial growth and enzymatic decarboxylation of the histadine to histamine. The reaction proceeds readily at 20–30°C (68–86°F), well above optimum refrigeration temperature.

Once the histamine is formed, boiling, baking, frying, freezing, smoking, or canning will not detoxify the meat. Since histamine taken orally is nontoxic, the presence of other synergistic chemicals has been postulated (3,4).

The Food and Drug Administration has established 50 mg/100 g (4498 μmol/L) as the as the hazard action level for histamine in tuna. The bluefish in question contained 22,492 μmol/L, clearly hazardous.

C. Considering the Problem

Food poisoning of bacterial origin is primarily a result of improper food sanitation. Food sanitation may mean many things to many people, but L. V. Burton's comment that sanitation is a race to see who gets the food supply first admirably epitomizes the subject. Food poisoning and food sanitation are often facets of the same problem; the absence of the latter may give rise to the former. There is no general method of food protection analogous to pasteurization of milk or chlorination of water supplies. Suitable protection must be achieved through accepted procedures of cleanliness and handling for each type of food. The number of illnesses that may be transmitted by food is large indeed, and the incidence of individual foodborne illnesses varies considerably throughout the world.

Further analysis reveals a recurrent theme: *inadequate refrigeration*, *poor personal hygiene, contaminated equipment, inadequate cooking, unsafe sources of food,* and *chemicals in foods* are, in descending order, the major risk factors year after year. For inadequate refrigeration to be the major risk factor in the closing decades of the 20th century strains credulity.

Unfortunately, at least for the purposes of public health officials, the role of restaurants in producing food poisonings, as opposed to homes, schools, churches, private camps, and office picnics, is poorly understood. Only about 15–20% of the total number of cases can be reliably ascribed to restaurants. Although this may be an appropriate estimate, there are nagging doubts suggesting their contribution to be far higher. One should not be misled by reports in the print or electronic media to the effect that it is safer to eat in restaurants than in homes or at private functions because health departments receive fewer reports of food poisoning from restaurants than from private gatherings. Over 100 million meals are believed to be served each day in public places in the United States. The difficulty of tracing a case of gastroenteritis to a restaurant is formidable indeed. Restaurants may well be less safe than private homes, even though fewer cases of food poisoning are reported, but further investigation and documentation are needed to establish whether this is indeed the case, and of course one should bear in mind that although a large percentage of us eat at home, much of the food we eat is prepared elsewhere.

Given our knowledge of the causes of food poisoning, why does it continue to threaten public health, and why does it persist at these high levels?

The answer to the first question has been known for some time. It has been reasonably well advertised but flagrantly ignored. Disease transmitted by food usually originates with an infected person who inadvertently inoculates the food during handling. Inadequate refrigeration and careless food-processing practices foster abundant growth of the organisms in a highly nutritious medium. The time between preparation and serving to an unsuspecting hungry population allows the multiplying bacterial population to produce sufficient toxin to infect a large number of people. Usually, the potent toxin requires the ingestion of only a morsel of food to evoke the characteristic complex of symptoms.

To answer the second question we must look to our changing environment. The rapid substitution of a technological, industrially oriented society for an agricultural one has radically changed our way of life. Intricately involved in this change are factors that abet food poisoning.

An affluent "high-tech" society encourages everyone to join the labor force. To purchase desired goods and services, the income of both husband and wife is often necessary. The homemaker of the past is now often a member of the work force. Whatever her job, she is no longer able or willing to spend long hours at home preparing meals. Domestic help, once abundantly available has also succumbed to the siren call of industry. For those women who do not have to work, free time for cultural pursuits is more compelling than kitchen drudgery. As if to punctuate the problem, the smaller homes and apartments of the urban areas have forced grandparents to live away from children. All this makes home cooking less and less common. As a consequence, we now routinely bring home ready-to-eat, precooked, prepackaged foods. We also eat out more often. And to cater to our newly educated tastes, large quantities of specialty foods are imported from areas around the world where food sanitation practices are often less than adequate. In addition, new types of food items are sold in novel packaging, whose potential for disease transmission is unknown until an illness occurs.

With regard to personal hygiene and temperature control, the points enumerated below can be primary preventives:

1. Hands must be washed with soap before preparing, handling, or serving food following use of toilet facilities. Unwashed hands, particularly those soiled with human or animal waste, can introduce food-poisoning organisms into food.

2. The nose and mouth should be covered with a handkerchief or other effective barrier when a person handling food is sneezing or coughing. After handkerchiefs containing exhaled mucus and sputum are handled, hands should be washed with soap.

3. Fingers should be kept out of glasses and dishes, and off bowls, spoons, fork tines, and knife blades.

4. In commercial food establishments, food handlers should be checked regularly for evidence of open sores, boils, sore throats, colds, and intestinal disorders. An employee who reports to work with any of these signs should not be allowed to handle food.

5. Salads made of meat, poultry, seafood, potatoes, and eggs must be held below 45°F until served. This applies as well to custards and cream-filled cakes. All these items should be eaten on the day prepared.

6. Stuffings, poultry, and stuffed meats should be heated to reach at least 165°F deep in the center of the product.

7. If a food that can be heated must be stored for 2–3 h, it should be maintained at about 145°F while stored.

8. If there is any suspicion that a canned food, particularly a home-prepared, low-acid food, is undercooked, poorly sealed, or leaking, the contents should be boiled for a minimum of 15 min before being eaten.

Although there is no substitute for proper training in food handling, adherence to these eight rules should prove helpful in preventing outbreaks of microbial food poisoning.

II. SALMONELLOSIS

Earlier, I noted that bacterial food poisoning occurs in one of two ways: eating food containing toxin and eating food containing large numbers of certain bacteria. Salmonellosis is that form of food poisoning produced by ingesting species of the rod-shaped bacteria *Salmonella* rather than preformed enterotoxin.

As with the toxin types of food poisoning, the appearance, smell, and taste of food contaminated with salmonella are unaltered. Similarly, the foods and their preparation methods offer excellent media for bacterial growth. Ten to twenty hours after eating, the first symptoms of gastroenteritis appear. The onset is sudden. Headache, chills, abdominal pain, elevated temperature, and foul-smelling diarrhea usually appear. Prostration, muscular weakness, and dehydration can be severe, leaving the patient totally debilitated. The severity of the illness depends on the number of organisms ingested and on individual sensitivity. Infants and adults over 50 tend to have severe symptoms, and considerably more fatalities occur in these two groups.

Most outbreaks occur as a result of exposure to a contaminated food by small groups at home, on a picnic, or in a restaurant. On the other hand, multistate outbreaks are now occurring more often as a consequence of cross-country distribution of products contaminated at the time of production. Powdered milk, precooked and packaged roast beef, and chocolate bars are current examples.

Of the more than 40,000 isolations reported each year for the past 5 years, salads such as turkey, chicken, macaroni, and crabmeat appeared to be the vehicle most often implicated. Not far behind were deviled eggs, hollandaise sauce, and baked goods containing egg products. These offer bacteria nutritious sources of protein, fat, and carbohydrate. We cannot change that. Prevention must therefore focus on what people do.

Interestingly enough, over the past 10 years New England and the Middle Atlantic states have witnessed a fivefold increase in the rates of isolation of *Salmonella enteritidis*. Figure 5 shows this unusual change in pattern.

In fact, *S. enteritidis* has replaced *S. typhimurium* as the single most reported serotype. Figure 6, describes its monthly distribution over a recent 15-year period. Investigation of these outbreaks has revealed a variety of transmitting sources. Scrambled eggs, liquid protein diet supplements, homemade ziti, frozen pasta, roast beef, Monte Cristo sand-

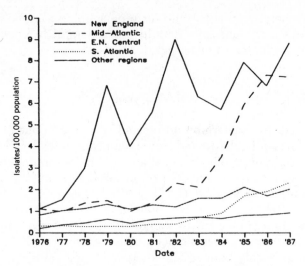

Figure 5. Isolation rates of *Salmonella enteritidis*, by region of the United States, 1976–1985.

wiches made of sliced cooked meat and cheese on bread dipped in raw egg and grilled, and Caesar salad dressing made with raw eggs have all been implicated. Obviously eggs, and products made with contaminated raw or undercooked eggs, are the single greatest infecting source. Salmonellosis associated with bulk egg products and cracked-shell eggs led to the passage of the Egg Products Inspection Act in 1970. Passage of this law led to the occurrence of fewer outbreaks. These recent outbreaks suggest a relaxing of vigilance.

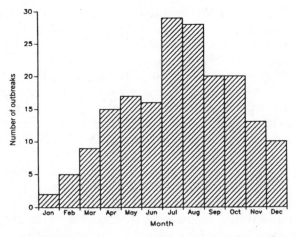

Figure 6. Cumulative distribution of reported Salmonella enteriditis outbreaks by month in the United States, 1973–1987.

III. ANTIMICROBIAL RESISTANT SALMONELLA

An entirely new route of transmission has recently been revealed. Prior to 1984, it was extremely difficult to prove that antimicrobial resistant salmonella of animal origin could provoke human illness. Recently, a team of researchers with the CDC's Enteric Disease Branch has unequivocally demonstrated the relationship (5). An outbreak of salmonella-induced gastroenteritis in 18 people in four Midwestern states was meticulously traced to a single herd in South Dakota.

Beef cattle are routinely fed subtherapeutic doses of antibiotics to achieve faster growth and disease prevention. Growing in chlortetracycline-fed calves, the microbes (S. newport) were resistent to the tetracycline, penicillin, ampicillin, and carbenicillin. Those who became ill had eaten hamburger purchased from retail outlets. The antibiotics kill a broad spectrum of microbes in a calf's system—the normal gut flora—leaving only those with genes for resistance to antibiotics.

Without the normal competition from other organisms, the resistant types flourished. Twelve of the 18 who became ill were taking penicillin for sore throats, inner ear infections, and bronchitis prior to consumption of the hamburger. It thus appears that resistant salmonella from the beef, in combination with the antibiotics taken for the infections, overwhelmed the 12. Their antibiotics wiped out their gut flora which normally would have held the salmonella in check. Simultaneously, the resistant salmonella then proliferated unabated.

Although others surely ate meat from the same herd, only those also taking antibiotics became ill.

Two things are clear: antibiotics should not be used to supplement cattle feed, and greater restriction should be placed on the use of antibiotics in treating human illness.

Prior to 1975, one of every seven salmonella isolations was traceable to contact with pet turtles. For years, microbiologists have known that pet turtles are often reseviors of this bacterium. The FDA banned both intra- and interstate distribution of small turtles (less than 4 in. long). Control seemed successful, but exportation continues. CDC's *Morbidity and Mortality Weekly Report* for March 16, 1984 indicates that the problem is still with us.

In Puerto Rico, a 4-month-old child was hospitalized with gastroenteritis after playing with a pet turtle. Following up on this, investigators found that 7 of 61 similar infections in infants most probably were turtle-related. Turtles from the United States have been implicated in outbreaks in the United Kingdom and Japan. Most of these are bred on farms in Louisiana and illicitly shipped around the country. The conclusion is straightforward: turtles are inappropriate pets for children.

The salmonella appear to have unlimited vehicles for their transmission. Between November 1981 and August 1982, 126 cases of salmonellosis occurred in Norway. Because the cases presented with an exceptionally severe gastroenteritis, and because the infecting species S. *oranienburg* is rare in Norway, a large-scale investigation was launched (6). Responses from the victims narrowed the search to mincemeat and/or minced fish products. Further investigation indicated the organism to have been spread by an ingredient in minced food, with spices the likely culprit. Final analysis implicated black pepper that was originally obtained from Brazil. How the pepper became contaminated is the subject of continuing investigation.

Gastroenteritis as a result of smoking fecally contaminated marijuana felled close to 100 people in Alabama, Georgia, Ohio, and Michigan in January and February of 1981. Fertilizing marijuana with animal feces is a well-established practice, as is adulterating it with minimal waste simply to increase bulk while diluting the essential ingredient. In either case, fingers and lips become contaminated during preparation and smoking. And, of course, with soiled fingers food can subsequently became inoculated during preparation. This can lead to secondary infections in those eating the food but not having smoked the cannabis.

One thing is certain: we are vulnerable from the most unlikely sources.

The aforementioned cases represent the typical clustering of a small group of related cases. Nothing to compare with the 18,000 plus cases that occurred between March and May 1985. Only once in our history have we seen an outbreak of this magnitude. This involved over 16,000 people in Riverside, California, in 1965, in an instance of waterborne infection.

The most recent classical foodborne epidemic occurred primarily in Illinois, but included Michigan, Indiana, Wisconsin, and Iowa where the 217 Jewel supermarkets are located.

It is believed that raw, unpasteurized milk was able to mix with pasteurized milk via a leaky valve. Typical explosive gastroenteritis occurred, with seven deaths among the very young and the very old.

The highest incidences of salmonellosis seem to be in children under 5. I say seem to be because youngsters may be tested more often than adults, and because they may also present with typical symptoms more often than their elders. On the other hand, those over 65 are at increased risk of death. The recent toll of 24 lives at a mental institution in Wakefield, England, included for the most part those of advanced age. At the height of the outbreak salmonellosis felled 381 patients and 45 members of their staff of the Stanley Royal Hospital. Most of the victims were not only elderly but suffering from senility.

Food poisoning is not the only illness caused by salmonella. Unlike other salmonella infections, the only known reservoirs of infection of typhoid fever are human carriers. Typhoid fever, also caused by the ingestion of food or water containing fecal matter from people ill with the disease, is a very different illness from food poisoning. Classical symptoms of typhoid fever include high fever, abdominal tenderness, enlarged spleen, slow pulse, a rose-colored skin rash, headache, cough, and nosebleeds—quite a different set of symptoms from those seen in salmonella food poisoning.

To the more than 500 men and women who were hospitalized with typhoid fever in Aberdeen, Scotland, in the summer of 1965, canned corned beef will always recall their narrow escape from death. On May 19, 1965, ambulances brought to the City Hospital in Aberdeen four people ill with high fevers. By midnight the following day the Health Department had identified the suspected microbe, *Salmonella typhi,* from stool and blood cultures. *Salmonella typhi* is the agent of typhoid fever.

Desultory outbreaks of typhoid occur from time to time, but in countries with modern sanitation facilities they are usually of little consequence. However, in Aberdeen, as the days passed, it became all too clear that this was no ordinary episode. By May 30, the number of confirmed cases jumped to 136 and the first death was recorded. The following day a newspaper carried the following account: "The trim streets of Aberdeen were nearly deserted tonight and in some places the air was heavy with the odor of disinfectant as the city fought to stamp out a typhoid epidemic."

By June 27 the epidemic had run its course and Queen Elizabeth sent a message indicating that she would visit the city. Over 500 people had been hospitalized and five had died. Aberdeen, a major tourist attraction, had also suffered severe financial losses from a sudden decline in tourism. Thus, the city welcomed the sign that the Queen thought Aberdeen was safe enough for her and for her subjects, safe from the havoc wrought by a can of corned beef.

During the period of the outbreak, a great deal of confusion and misunderstanding regarding the nature of the disease's communicability elicited behavior in many segments of the population more typical of the Middle Ages than the 20th century. For example, the streets of Aberdeen were sprayed with disinfectant, as though the typhoid bacterium had blanketed the environment and could be picked up and carried about on the soles of shoes. Workers in cities far to the south went so far as to refuse to unload trucks that passed through Aberdeen. Students were told to avoid public transportation, as though the bacillus could be transmitted from person to person much like influenza, a respiratory-tract infection spread by droplet nuclei from a cough or sneeze. To make matters still worse, prophecies of impending national disaster were foretold. That these tragicomic behavior patterns should have occurred in the sixth decade of the 20th century is cause for concern over our educational processes. I say *our,* for I am quite certain that similar behavior would result in the United States should a major outbreak of typhoid occur.

The cycle of events leading to the epidemic began in a meat-canning plant in Argentina. After a canned food is heated (to reduce* the bacterial population and cook the food), it is placed in cold water to lower its temperature rapidly, which prevents overcooking. Heating causes the seams to expand and offers an entrance to bacteria. The cooling water in this case was drawn from a stream that was polluted with fecal matter containing typhoid bacteria. (Chlorination, which was not used, could easily have sterilized the water and prevented the episode.) Bacteria were thus reintroduced after cooking. By the time the corned beef arrived in Scotland and was used, a significant increase in bacteria had occurred. One of the busier supermarkets on the city's main street opened one of the 6 lb tins and sliced the beef on its slicing machine. As other meats were also sliced on that machine, they in turn were inoculated with the typhoid organisms. Several of the employees became infected while handling the meat, and they spread the disease further.

Usually, the infectious organism is spread by human excreters, either those with obvious cases of the illness, those convalescing from the illness, or asymptomatic healthy carriers. Healthy carriers are our greatest problem. They do not know they are shedding salmonella organisms and outwardly appear perfectly healthy. Because of the difficulty in detecting carriers who have no symptoms, preventive measures are largely ineffective.

The problem of the carrier is particularly troublesome since the bacteria are lodged in the confines of the gallbladder, defying current forms of medication. Surgical removal of the gallbladder is the only positive means of preventive treatment, but most known carriers will not submit to it. In the gallbladder, the bacteria multiply slowly, then leave in bile and are passed out of the body in feces. These people do not appear ill and on medical examination show no signs of typhoid; they are healthy (chronic) carriers who continue for years to shed infectious organisms. These people can be the focus of explosive epidemics if allowed to work in any aspect of food handling. In New York City, for example, healthy carriers are constantly under surveillance to prevent their employment in the food industry.

Typhoid Mary is the personification of this dilemma as well as the first identified carrier of the typhoid bacillus. According to Dr. George A. Soper, who tracked her down and convincingly established the causal links in the chain of cases, Mary Mallon was a good cook but into every kitchen she entered she literally brought illness and sometimes death.

From 1900 to 1914, when she was finally caught and confined to a house on North Brother Island in New York City's East River (off 138th

---

* Although subjected to temperatures as high as 240°F for 35 min, canned food is not meant to be sterilized in the strict sense of the term. Temperatures high enough for sterilization (the complete absence of life) would produce food unfit to eat. Consequently, the heat treatment applied to various products is designed to reduce microbial populations to harmless levels.

Street), 51 confirmed cases of typhoid and three deaths were laid to her. When Health Department authorities pleaded with her to undergo chole- cystectomy (removal of the gallbladder) she refused, saying, "The Health Department just wants to use that way of murdering me." It was difficult then and continues to remain so. Our fast-paced, hi-tech society is a long way from freeing itself of this problem.

IV. STAPHYLOCOCCAL FOOD POISONING

The second most common cause of foodborne illness in the United States is staphyloccal food poisoning. On the basis of perceived estimates however, staphylococcal food poisoning is probably the most common. Of the two types of bacterial food poisoning mentioned earlier, ingesting preformed toxin or ingesting organisms containing toxin, the staphylo- cocci are examples of the former, producing enterotoxin in food as a by-product of their growth and metabolism.

Not all strains of "staph" elaborate enterotoxin, nor do the toxin- producing strains produce enterotoxin in all foods. Staphylococci are commonly found in the nose and throat, and on the skin in pimples, boils, carbuncles, and whitlows.

Such items as ham, chicken, egg salad, custard-filled cakes, meat pies, and boiled ham are the most notorious sources of food poisoning. They offer the staphylococci a nutritious, high-protein diet, and they are usually prepared in large amounts that are difficult to cool rapidly. Put the same organism on a slice of tomato and toxin production will not occur. When refrigeration is unavailable, inadequate, or simply not used, the microbe, gaining entry from a sore finger or droplet from the food handler's nose, finds the warmth and nourishment needed for growth and toxin production begins. As the hours pass, the unrefrigerated food becomes a potential bomb. At mealtime nothing is suspected. The food has not changed in appearance, odor, or taste. Two to six hours after eating, the explosion occurs. The classic triad of symptoms (nausea, vomiting, and diarrhea) occurs as the body tries to rid itself rapidly of the poison. In severe cases, blood and mucus may appear in the patient's stool and vomitus as a consequence of violent retching.

Just such conditions caused the events in Indiana in August 1959 which were described at the beginning of this chapter. The explosive outbreak of August 15 incriminated boiled ham that had become conta- minated during preparation. Toxin in potato salad and ham caused the episode on the following Saturday. Both outbreaks could have been prevented by proper refrigeration.

It has been assumed that high salt concentrations act as preservatives and prevent bacterial growth. In most cases this is true, but not with staphylococci. Enterotoxigenic strains can grow at salt concentrations as high as 20%. Consequently, ham cured with the usual 3.5% offers little challenge to the type A toxin-producing species.

On June 11, 1975, 40 hams were delivered to a small restaurant in Erwin, Tennessee. They were stored in a walk-in refrigerator at 14°C (57°F). A day later they were cooked, deboned, separated, and returned to the refrigerator for another day. Out again for slicing, then back in.

On June 14, the slices were preheated at 350°F (177°C) for $1\frac{1}{2}$–2 h—a treatment that cannot inactivate toxin—then boxed with other items and delivered to the Erwin Railroad Station.

On June 15, 627 hungry passengers on an excursion between Kentucky and South Carolina decended on the station at 12:50 p.m. for a box lunch. By 5:30 p.m., when the train reached Marion, North Carolina, on its way to Spartanburg, 93 were vomiting and diarrhetic; 36 required hospitalization. By midnight over 200 had become ill. Coagulase positive staphylococci (phage type 6/85/47/54/75/83A) were isolated from four ham samples as well as a stool sample—good evidence for a cause–effect relationship. Inadequate refrigeration had struck again.

Unusually instructive for what it suggests about food poisoning and potential disasters is the following cautionary tale. In Tokyo, 343 unsuspecting passengers boarded a Japan Air Lines 747 Jumbo Jet for a flight to Paris. On the weekend of February 1, 1975, a cook in Alaska with staphylococci-infected finger blisters had prepared some 205 portions of ham for ham and egg omelet trays.

The portions were kept at room temperature for at least 6 h prior to completing the trays. The completed trays were then stored overnight at 50°F awaiting arrival of the 747. Staphylococci were growing and toxin was being elaborated. The plane arrived and trays were loaded aboard and stored. The plane took off for Copenhagen, its next refueling stop before Paris.

Six to seven hours later, the trays were heated in 300°F ovens for 15 min (totally inadequate to inactivate the toxin). Breakfast was served. As the plane touched down in Copenhagen, nausea, vomiting, cramping, and diarrhea debilitated 145 of the passengers who had to be carried from the plane. Another 50 would be carried off in Paris.

In this episode, the crew remained unaffected. Although coincidence could be at work here, it is extremely unlikely. In Anchorage, a new crew came aboard. They had already eaten breakfast and thus had no reason to eat the prepared food. Nevertheless, in a similar situation pilot and copilot could easily become victims. The outcome would not be pleasant to contemplate. Following disclosure of this incident, Kenji Kawabara, Executive Director of Inflight Catering International, committed suicide.

Whether the 1950s or the 1980s, the problem remains the same: inadequate refrigeration and poor personal hygiene. Apparently the lessons are not being learned.

V. BOTULISM

Botulism, the deadliest form of food poisoning, results from ingestion of a neurotoxin produced by *Clostridium botulinum*. This sublimely potent chemical is the most toxic natural poison known.* Since the clostridia are normal inhabitants of the intestinal tract of animals, they are universally found in agricultural areas in close proximity to crops. It has now been established that these bacteria are also natural inhabitants of marine environments, probably as a result of soil runoff. Consequently, food fish harbor them as part of their natural flora. Unlike the staphylococci, which are introduced as a result of poor personal hygiene on the part of food handlers, clostridia are already present and not adequately removed from the produce when washed prior to canning or packaging. Personal hygiene is not the problem here. Most cases of botulism can be traced either to ingestion of under-~rocessed home- or factory-canned low- or nonacid foods, or to packaging in containers providing suitable anaerobic conditions for toxin production.

Recall Figure 1g, in which a bacterial spore was shown. Spores are generally considered to be the most heat-resistant forms of life on our planet. It is just this characteristic that food processors must guard against, especially home canners and purveyors of specialty items that recieve only light heat treatments. Foods such as stringbeans, corn, beets, asparagus, mushrooms, tuna, chicken pie, and smoked fish, known to be naturally contaminated with spores of *C. botulinum*, must be processed for at least 35 min at 240°F (115°C) to guarantee safety.

Figure 7 notes that foods with little natural acidity are most likely to support the growth of botulism organisms. Below a pH of 4 (the high acid range) this type of spoilage is not encountered. Spores of several food-spoilage (but not food-poisoning) clostridia are more heat-resistant than spores of the botulism organism. When heat processing is adequate to destroy the food-spoilage forms, food poisoning will not occur. Commercial canned food processors have developed heat treatments for each food and can size, based on the heat resistance of a test organism known to be more heat-resistant than most of the spoilage organisms. Interestingly enough, a form of built-in protection also exists. The food "spoilers" produce unbearably putrid end-products as a result of anaerobic metabolism in canned foods. Consequently, if processing has been inadequate, the insufferable odors will prevent the food from being eaten, thus protecting the consumer against the botulism toxin present along with the malodorous products.

In most cases of botulism, warning odors are not present. The foods are usually eaten at home, which accounts for the extensive family involve-

---

* Less than 1×10^{-10} g(0.0000000001 g) can kill a mouse. 10^{-10} is equivalent to half a pinch of salt in 10 tons of potato chips, or half an apple in 2,000,000 barrels.

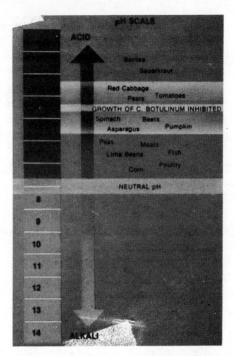

Figure 7. pH scale and inhibition of botulism.

ment. The toxin is readily absorbed by the upper intestinal tract, with the first symptoms appearing about 12–36 h later. An astounding case was that of a person who was active for 8 days before hospitalization, even attending the funerals of other victims and testifying at the coroner's inquest!

Typically, the toxin has an affinity for the cranial nerves that control the function of the head and neck. Characteristic symptoms are double vision, difficulty in swallowing, followed by slurred speech. In fatal cases the pharyngeal muscles are paralyzed as a consequence of inhibition of the release of acetylcholine at cholinergic synapses. Death results from respiratory failure. Over 70% of cases end in death, primarily because treatment with antitoxin is started too late. For antitoxin to be useful, it must be given before symptoms appear. Unless there is reason to suspect botulism, this is seldom done. In addition, symptoms of botulism are often confused with a half-dozen other neurological conditions. Thus, treatment is often delayed until information is obtained indicating that the patient had eaten foods commonly associated with botulism.

For example, around noon on December 12, a 59-year-old woman in New York City ate a meal that included eggplant, mushrooms, and peppers that she had jarred at home. That evening she had vomiting, distension of the abdomen, and abdominal pain.

December 13th—She called her family doctor, who furnished her with suppositories which produced diarrhea.

December 14th and 15th—Complaints of swallowing difficulties.

December 16th—Difficulty breathing.

December 17th—Hospitalized, with admitting diagnosis of bronchitis and respiratory failure.

Between December 17th and 22nd—Suffered a myocardial infarction and died on December 22nd.

(While in the hospital, she complained of burning eyes and was unable to close her eyes.)

After the funeral the family, 22 members, returned to her home and ate the same food. On December 24th the daughter-in-law complained of swallowing difficulties and on the 25th had difficulty talking and swallowing. Her father-in-law called the family doctor, whose impression was that her symptoms were hysterical and a placebo was prescribed. The father-in-law then called Emergency Services. Paramedics got her into the hospital on the afternoon of December 26th, where she had a myocardial infarction and respiratory failure.

Inquiring among family members, investigators found that the mushrooms, eggplant, and peppers were usually cooked in vinegar to boiling, cooled, and placed in olive oil for storage at room temperature. What happened? A reasonable explanation of the episode suggests that boiling was inadequate—not continued long enough. Spores from both the mushrooms and peppers were still viable when they were placed under the oil. The depth of the oil and the room temperature provided the necessary anaerobic conditions for the spores to germinate. Toxin production followed.

Fortunately, although botulism is the most serious and highly fatal of all the microbial foodborne gastroenterotides, it is also among the rarest. Nevertheless, it can, and does, occur in severe outbreaks. It was peppers that did it in Michigan. In the worst outbreak ever recorded (58 confirmed cases) all had eaten home-canned green peppers at Trini and Carmen's Mexican-type restaurant.

A former employee of the restaurant had canned the peppers at home and sold them to the restaurant. The cook there, wrongly believing he was out of fresh peppers, used the canned peppers.

According to CDC data, the previous record, 29 cases at one time, also occured in Michigan. That was in 1921, and spinach was the offending food. In 1963, 23 people in the Midwest, many from Michigan, became severely ill as a consequence of eating smoked white fish (7).

Given the bizarre nature of this illness, one would expect few cases to go unreported, but they are. For 1983, the CDC estimated that 200 cases would have occurred, more than 200% over the reported. It is believed that cases may be missed (attributed to other insults) because of the lack of nausea, vomiting, and diarrhea usually associated with food poisoning.

Prevention is not difficult. Botulism neurotoxins, seven in all, (A–G) have been established based on distinct antigenic differences. They are uncomplicated proteins composed solely of amino acids with molecular weights on the order of 250,000–400,000. These can be completely inactivated by the simple expedient of boiling for 10–15 min. Accordingly, prior to eating home-processed foods, boiling is called for.

As for spores, temperatures of 121°C (250°F) are necessary. This can be attained *only* in pressure cookers. Therefore, that becomes a requirement, with time dependent on the size of the batch being cooked.

Clearly, there is ample reason for the abnormally high number of home-canning related outbreaks. Commercial processing is not without blemish, but given the *billions* of cans processed each year in the United States, the canning industry must be considered a success story. Unfortunately, when an incident does occur, the nationwide distribution system places inordinate numbers of people at risk. And given the inability for expeditious diagnosis and treatment, it is not without appropriate reason that the idea of botulism still strikes terror.

Over the past 10 years, important new microbial agents of enteric infections have emerged.

VI. *CAMPYLOBACTER ENTERITIS*

Campylobacter enteritis is now recognized as a common cause of gastric illness, responsible for some 4% of outbreaks in 1982 though it was virtually unknown prior to 1975 except as an obscure veterinary pathogen. Usually, the illness is self-limiting, producing abdominal pain and diarrhea for 2–3 days. Unfortunately, pain can be severe enough to require hospitalization with, and again, unfortunately, an unnecessary appendectomy as a not uncommon consequence.

Campylobacter jejuni, the species most often isolated, are small, flagellated, endotoxin-producing gram-negative rods whose normal hosts are animals.* Indeed, campylobacter is a zoonosis (cf. Chapter 4— Zoonoses), an infection transmissible from animals to humans, and in this instance, by the fecal–oral route (8). Unpasteurized (raw) or inadequately pasteurized milk, raw clams, raw to rare hamburger, and poultry have been shown to be among the contributing vehicles.

Campylobacter appears to have its greatest incidence between May and September, and primarily among 12–30-year-olds. Among hypogamma-globulinemics (immunodeficient individuals) campylobacter infections

---

* The term endotoxin refers to the toxin released from bacterial cells as a consequence of hydrolysis within the gut. Enterotoxin or exotoxin, interchangeable terms, refers to the elaboration of toxin by bacteria, into their surrounding medium: food in this instance.

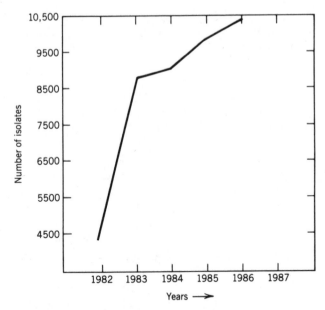

Figure 8. Human isolates of *Campylobacter*, 1982–1986, for the 50 states. Data courtesy of the CDC.

can run a chronic course, producing ulcerative colitis with bloody diarrhea mimicking Crohn's disease, which can result in highly inappropriate treatment.

As the more sophisticated microbiological techniques for cultivating and isolating these organisms gain widespread use, the incidence of *Campylobacter* enteritis should continue to rise. But the rise may only be artifactual—the cases have been present but unrecognized. Figure 8 shows this increase.

VII. *YERSINIA ENTEROCOLITICA*

Yersinia enterocolitica is another relatively new addition to the melange of organisms with a proclivity for the human gut (9). Although human yersinosis has been recognized for some 50 years, foodborne infections are new. This is a surprising circumstance considering its environmental ubiquity and close relationship to farm animals. Several germane conclusions suggest themselves: a growing tendency to consume raw milk, misuse of pasteurization procedures, and lack of attention to established hygienic milking practice. Lack of adequate sanitary milking procedures speaks to the issue of unduly large numbers of microorganisms in the milk. This would suggest that its consumption would be an inherently dangerous activity. Pasteurization makes it safe, but this pasteurization process is geared to the specific destruction of all *Mycobacterium tuberculosis* organisms and the reduction of others. Excessively larger numbers could affect that reduction. Should the process be misused, there is little doubt that with unsanitary milking conditions the product would not only contain large numbers of organisms but a wide variety of potential pathogens. As *Yersinia* does not produce an exotoxin, large numbers are usually required to produce an illness. The notion that raw milk is somehow healthier than pasteurized is not only without merit but potentially dangerous. Little if anything exists to recommend it.

A recent prospective epidemiologic study of gastroenteritis among children in Mortreal found *Y. enterocolitica* to be the second most frequent pathogen just behind salmonella. Isolates from foodborne epidemics produce a heat-stable enterotoxin that is able to penetrate guinea pig conjunctival epithelium in the Sereny test. Consequently, this species appears to exhibit both invasive and toxigenic characteristics (9).

One of the most frequent symptoms of illness with *Y. enterocolitica* is severe abdominal pain, including the lower right quadrant. This can suggest appendicitis. Considering some of the curious eating patterns gaining acceptance, we should continue to expect cases of foodborne yersinosis.

Although meals-on-wheels is an unusual vehicle for transmission of food-poisoning microbes, the delays they can involve in distribution of warm, precooked food is not. Approximately 50 elderly people in a London suburb developed abdominal pain and diarrhea 7 h (on average) after eating a meal whose main course consisted of roast chicken, stuffing, gravy, potatoes, and green beans. One person died.

On laboratory analysis, the food was later shown to have been highly contaminated with both *Bacillus cereus* and *Clostridium perfringens*. Both are well-known foodborne pathogens. *Bacillus cereus* is responsible for some 3–4% of confirmed cases, while *C. perfringens* yearly contributes between 10 and 15%. Strains of *C. perfringens* generate exotoxins of variable potency. Type A is usually mild, while type C can be quite severe.

In this instance, the chicken had been roasted the day before at 450°F (232°C) for $3\frac{1}{2}$ h and allowed to remain at room temperature for about 20 h. On the morning of the day it was to be delivered and eaten, the chicken was sliced and reheated for 30 min at 450°F. The gravy, dehydrated potatoes, and beans were reconstituted at that time.

Living as they do in soil and water, both organisms are common contaminants of much of our food. The risks inherent in inadequate heating of bulk food, especially that grown in or having contact with soil and fecal matter, are well known. Also well known is the fecund incubation period provided by storage at room temperature in an extravagantly nutritiuos medium, which permits luxurious growth of these organisms, as well as formation of their toxins.

Those responsible for meals-on-wheels, picnics, church suppers, block and business parties must be mindful of the fact that bacteria and humans, for the most part, enjoy similar food; ergo, before serving, keep the time short and the temperature high.

VIII. NON CHOLERA VIBRIO INFECTIONS

Traditionally, cholera and vibrio associated with it have been a primary concern of public health professionals. More recently, non cholera vibrio* illness initiated by V. parahaemolyticus has assumed greater importance (10).

This enterotoxin-producing organism inhabits marine environments throughout the world, the saltier the better. Given that affinity, it is little wonder that shellfish, shrimp, oysters, and crabs have been the major foods from which these organisms have been recovered. Investigation after an outbreak has frequently found these items either inadequately cooked or held at inadequate refrigeration temperatures. In either case, the organisms have multiplied sufficiently to overwhelm the body's defensive mechanisms. Accompanying the complex of nausea, vomiting, and explosive diarrhea seen within 15–20 h of ingestion are weakness and tenesmus.

This organism is responsible for about 1% of all outbreaks yearly.

IX. CHOLERA

Toxigenic Vibrio cholerae, the bacterium responsible for cholera, is known primarily as a waterborne pathogen. Consequently, the five cases of this rare disease (in the United States) produced a good deal of shock when it was learned that four of the five from in and around New Orleans had eaten crabs and shrimp, while the fifth person, a woman from California visiting in Florida, had eaten raw oysters. The Gulf Coast is currently considered a reservoir of toxin-producing cholera organisms.

X. PARALYTIC SHELLFISH POISONING

In their off-again on-again pattern, both neurotoxin and paralytic shellfish poisoning are somewhat unique. The yearly fluctuation in outbreak and cases, from none to ten or more, is related to the growth of specific aquatic dinoflagellates on which shellfish feed.

---

* Vibrio is a genus of short, curved, motile, non-spore-forming, aerobic, gram-negative rods.

These neurotoxin types of poisonings parallel the growth and "bloom" of species of the algae *Gymnodinium* and *Gonyaulax* which become so extensive and dense they produce "red tides." Oysters, clams, mussels, and scallops feed harmlessly on these doubly lashing protozoans. Nevertheless, the dinoflagellates ingested contain a formidable chemical that for human beings can be incapacitating. Curiously enough, for the *Gymnodinium* and *Gonyaulax* and for the shellfish, this otherwise venomous chemical is of no consequence.

Unfortunately for humans, nature has provided no clues as to the presence of the toxin, nor is it inactivated by the heat of cooking. Thus, unsuspecting, those with a voracious appetite for shellfish consume substantial doses. Within 15–45 min of eating there is tingling, then numbness, around the lips, mouth, arms, and legs. In its wisdom, the body attempts to throw off the toxin by vomiting and diarrhea. In severe cases, more often seen with the paralytic than the neurotoxic, loss of speech (dysphagia), paralysis, respiratory impairment, and ataxia (loss of muscular coordination of arms and legs) have been reported.

The chemical involved, saxitoxin,* is known to be of low molecular weight, thermostable, and a three-ring compound (3,4,6-triethyltetrahydropurine) with the formula $C_{10}H_{15}N_7O_2$. Considering that currently there is no known antidote and that the food yields no warning signs, it is extremely fortunate that the condition is most often self-limiting.

Generically, all illnesses produced by the ingestion of toxin contained in body parts of fish and shellfish are classified as icthyosarcotoxisms, and the condition is classified as icthyosarcotoxic.

XI. PTOMAINE POISONING

Thus far, ptomaine or ptomaine poisoning has not been mentioned, and for good reason. Ptomaine poisoning is an old idea that is hard dying. Although it has been established beyond question that ptomaines, a class of foul-smelling organisms (from the Greek *ptoma*, dead body or cadaver) are not involved in microbial food poisoning, the term persists nevertheless.

Ptomaines are aliphatic diamines formed by metabolism of proteins by bacteria. Actually, the ptomaine putrescine is formed by the decarboxylation of the amino acid ornithine, and the ptomaine cadaverine is formed by the removal of CO_2 from lysine. With the degradation of amino acids to amines and ammonia, obvious color and odor changes occur along with a

---

* So-called because it was first isolated from toxic Alaska butter clams (*Saxidomus giganteus*). It is one of the most toxic substances known.

slime formation from bacterial growth. These organoleptic alterations are typical of food spoilage, not food poisoning. Neither of these ptomaines is harmful when taken by mouth. That is, they do not produce symptoms of gastroenteritis. If a large amount were introduced directly into the bloodstream with a hypodermic syringe, illness would occur, but not food poisoning.

XII. CONCLUDING REMINDER

This brings us full circle. Whether it is the food manufacturer, a restaurant, private home, or group activity, inattention to well-established preventive practices is at the heart of the problem and can readily be circumvented. Microbial food poisoning has no place in contemporary society.

Prevention requires attention to two broad areas: temperature control and personal hygiene. Temperature is crucial for preventing microbial food poisoning. Sufficient heating and cooling are two aspects of the same coin in retardation and prevention of growth.

With a determined but hardly superhuman effort, elimination of food poisoning can be an attainable goal by the year 2000.

REFERENCES

1. R. V. Tauxe, D. A. Pegues, and N. Hargrett-Bean, Campylobacter Infections: The Emerging National Pattern. *Am. J. Public Health* **77**(9): 1219–1220, 1987.

2. A. H. W. Hanschild and F. L. Bryan, Estimate of Cases of Food and Waterborne Illness in Canada and the United States. *J. Food Protection* **43**(6): 435–440, 1980.

3. P. Etkind, M. E. Wilson, K. Gallagher, and J. Cournoyer, Bluefish-Associated Scombroid Poisoning. *J. Am. Med. Assoc.* **258**(23): 3409–3410, 1987.

4. C. Kow-Tong and M. D. Malison, Outbreak of Scombroid Fish Poisoning, Taiwan. *Am. J. Public Health* **77**(10): 1335–1336, 1987.

5. S. Holmberg, M. T. Osterholm, K. A. Singer, and M. L. Cohen, Drug Resistant Salmonella from Animals Fed Antimicrobials. *N. Engl. J. Med.* **311**(10): 617–622, 1984.

6. S. Gustavsen and O. Breen. Investigation of an Outbreak of *Salmonella oranienburg* Infections in Norway Caused by Contaminated Black Pepper. *Am. J. Epidemiol.* **119**(5): 806–812, 1984.

7. *Botulism in the U.S. 1899–1977. Handbook for Epidemiologists, Clinicians and Laboratory Workers.* Centers for Disease Control. Issued May 1979. U.S. Department of Health and Human Services, Atlanta, GA.

8. D. A. Robinson, Campylobacter Infection. *J. R. Soc. Health* **101**: 138–140, 1981.

9. E. J. Bottone (ed.), *Yersinia enterocolitica.* CRC Press, Boca Raton, FL, 1981.

10. P. A. Blake, Diseases of Humans (Other than Cholera) Caused by Vibrosis. *Annu. Rev. Microbiol.* **34**: 341–367, 1980.

SUGGESTED READINGS

Diseases Transmitted by Foods (a complete classification and 8th printing summary). U.S. Department of Health and Human Services, Pub. # 82-8237, Dec. 1979.

Sacks, J. J., Lieb, S., Baldy, L. M., Berta, S., et al. Epidemic Campylobacteriosis Associated with a Community Water Supply. *Am. J. Public Health* **76**(4): 424–428, 1986.

Spicer, A. J., "New" Enteric Infections, *The Practitioner* **224**: 789–792, 1980.

Steinfeld, A. D., and Steinfeld, H. J. Ciguatera and the Voyage of Captain Bligh. *J. Am. Med Assoc.* **228**(10): 1270–1271, 1974.

Yoshikawa, T. T., Herbert, P., and Oill, P. A., Salmonellosis—Teaching Conference, *West. J. Med.* **133**: 408–417, 1980.

3

Chemicals in Food

Loyalty to petrified opinion never broke a chain or freed a human soul.

—Mark Twain

EDB—three letters—an acronym for a chemical which 97% of the population can neither name nor pronounce (ethylene dibromide) created a crisis of massive proportions in the summer of 1983. Hysteria and panic describe the environment succinctly. To the American public EDB meant our food supply was compromised. Chemicals had struck again.

From aminotriazole,* which ruined Thanksgiving dinner for millions of American families in November 1959, to red dye #2, BHA, BHT, mercury, caffeine, cyclamate, saccharin, and of course EDB, we have been treated to what some wags refer to as "the carcinogen of the month." A disinterested observer from another planet could easily become convinced that our food supply was not safe† and that as a people we are tettering on the brink of annihilation via the chemicals we eat.

Curiously enough, this attitude is uniquely American. No other country of the Western world even approximates it. Americans traveling abroad rarely deny themselves the delights of European culinary artistry, to say

---

* Aminotriazole, a weedicide with a long history of use in corn fields, blocks formation of chlorophyll (the green pigment of plants), thus preventing carbohydrate formation and subsequent death of the weed.
† The concept of safety is discussed in detail in Chapter 15. On November 9, 1959, Arthur S. Flemming, Secretary of the Department of Health, Education and Welfare—the forerunner of our Department of Health and Human Services—announced that aminotriazole had been found in cranberries. By Thursday, the 26th, the charges, countercharges, uncertainty, and confusion had so bewildered the public that most turkeys were served without cranberries that year.

nothing of the exotic foods of Asia, the Middle East, Africa, and South America, whose foods are far less rigorously tested than our own. In some areas, food testing may be wholly nonexistent. We take our demanding food protection system for granted and assume that all countries are as rigorous as our own. Most are not.

It is a fact that nothing manufactured or produced in our society is subjected to the intense scrutiny reserved for food. Yet Americans, among the most overfed people in the world, prefer to believe ill of their food supply.

Far from being the catastrophe many people believe they are, chemicals do have appropriate places in food. But chemicals in foods can mean different things to different people, and rightly so.

I. THE CHEMICAL NATURE OF FOOD

A moments reflection will affirm that those comestibles we call food— steak, tomatoes, yogurt, herring, asparagus, coffee, jello, eggs, corn, oranges, and chicken soup, for example—are nothing more than unique collections of chemicals. Without allyl isothiocyanate there would be no bite in horseradish. And but for the presence of propionaldehyde, methyl alcohol, propyl mercaptan, hydrogen sulfide, acetaldehyde, sulfur dioxide, dipropyl disulfide, and propyl alcohol, there would be no tears in onions.

The yellow, orange, and orange-red pigments chemists call beta carotene give carrots, tomatoes, squash, sweet potatoes, red peppers, and peaches their vibrant colors. Beets, purple cabbage, black mulberries, and vinifera grapes owe their festive colors to anthocyanins, those glycosylated polyhydroxy (and polymethoxy) derivatives of 2-phenyl benzopyrilium salts. Figure 1 shows the chemical structure of four of the most

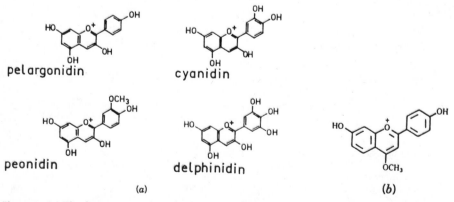

Figure 1. (a) The four most commonly encountered anthocyanidins. (b) A synthetic anthocyanidin-like red pigment.

frequently encountered anthocyanidins (the aglycones, free of their sugar moieties).

These are naturally occurring chemicals that chemists have been able to isolate and characterize. In fact, scientists have learned so much about them that they have been able to approximate their structures synthetically. Figure 1b portrays 7-hydroxy-2-(4-hydroxystyryl)benzopyrilium salt, a synthetic red anthocyanin pigment that resembles the delphinidin and cyanidin glucosides which give vinifera grapes their rich magenta color.

II. NATURALLY OCCURRING HAZARDOUS CHEMICALS

Whereas the naturally occurring chemicals noted above are for the most part benign, other naturally occurring chemicals can be inordinately injurious and lethal.

Shortly before Christmas, 1983, a group of Laotian refugees living in San Francisco went on a mushroom-hunting expedition in the Napa Valley. Being excellent woodsmen but poor botanists (mycologists would have been better yet), they gathered a delicate but deadly harvest. Sharing their colorful treasure with friends and relatives at a party, the 14 who ate the Death Angel mushrooms, the deadliest of all mushrooms, became critically ill within 8–10 h. All were hospitalized in an intensive care unit and in their comatose condition were not expected to live. There is no known antidote. Amanitin, the toxin of the mushroom *Amanita phalloides* (the Death Angel), is a complex organic ring compound that ranks in potency with the neurotoxins of botulism.

Although amateur mycophiles believe they can distinguish between safe and harmful varieties, the number of poisonings each year suggests otherwise. A dozen species contain the cyclopeptides and hydrazines which, while of no consequence to the mushroom, produce rapid cellular damage to the human liver, kidney, and heart muscles.

Mushrooms of the genera *Psilocybe* and *Panaeolus* contain powerful hallucinogens. Phosphoryl-4-hydroxy-N-dimethyltryptamine-psilocybin, an indole derivative, is the active hallucinogenic agent. For the most part, these mushrooms were used during religious rites by Indians of Mexico and the American Southwest. Unfortunately, they have been abused, with bitter experience, by people seeking relief from the stresses of daily life.

With the swing to nouvelle cuisine, away from the traditional creamy sauces, cooks and chefs have turned to the pink peppercorn for a piquant spice. But *Schinas terebenthifolius*, a delicate French import, is currently banned by the Food & Drug Administration because of the severe toxic reactions it induces.

Actually, the peppercorn is a berry, related to poison ivy. It can produce symptoms similar to ivy poisoning as well as swollen eyelids, violent headaches, shortness of breath, and chest pain.

Botanically, this exotic "spice" is the fruit of a common weed known in the United States as the Florida holly. Until the French government can assure the FDA of the peppercorn's safety, it can be banned under one of several food additives sections of The Food, Drug and Cosmetic Act.

Curious as it may be, hydrogen cyanide (HCN) is a constituent of a large number of edible plants. Cyanogens or the cyanogenetic glucosides, which on hydrolysis in the human intestine yield cyanide, are found in such garden-variety foodstuffs as lima beans, sweet potatoes, yams, sugar cane, peas, cherries, plums, and apricots. What function these cyanide-containing compounds play in the growth and metabolism of the plants is one of the intriguing questions still eluding scientists.

Although most university-level biochemistry courses cover the glucosides and frequently discuss amygdalin as an example of this class of compounds, few probably recall that on hydrolysis (splitting of the glucoside into its components) it releases HCN. Furthermore, few associate the release of cyanide with its potential for human toxicity. Interestingly enough, whereas maximum yields of amygdalin are obtained from almonds, it is also found in significant amounts in the kernels of such fruits as the cherry, plum, apricot, apple, pear, lemon, and lime.

The initial symptoms of acute cyanogen poisoning have been described as numbness in fingertips and toes and giddiness or lightheadedness. If the dose is large enough, this is followed by mental confusion, stupor, cyanosis, twitching convulsions (indication that the central nervous system is involved), and in the terminal phase, coma and death. This clinical complex is related to an oral minimum lethal dose of HCN ranging form 0.5 to 3.5 mg/kg (0.5–3.5 ppm). The lethal dose is estimated to be upward of 7 ppm (7 mg/kg). Small, nonfatal doses often produce headache, sensations of tightness in both throat and chest, perceptible heart beating (palpitation), and generalized weakness. In most cases full recovery is the norm as the body's metabolic processes eliminate the offending chemical.

For a number of years sporadic reports have hinted at the possibility of chronic cyanogen intoxication. Most recently, studies in cassava-eating areas of Nigeria and Mozambique implicated cyanogen poisoning most directly with ataxia, loss of muscular coordination, and loss of sensation (1,2).

Medical case reports have from time to time suggested the association of amblyopia (partial loss of sight and diminution of vision) with chronic cassava intoxication. Evidence does not appear to point to a relationship between a state of general nutritional deprivation and the consumption of only trace amounts of a cyanogen as necessary to elicit clinical symptoms.

The acute toxic properties of cassava, for example, have been known for hundreds of years, and there are also abundant records of acute poisoning from the ingestion of peach kernels and bitter almonds. This, in conjunction with reports of experimental animal trials, leaves little doubt that the release of HCN is the prime cause of intoxication.

The lima bean provides an illustrative example of the chemical reactions involved:

$$C_6H_{11}O_6-\underset{\underset{CH_3}{|}}{\overset{\overset{CH_3}{|}}{C}}-CN \xrightarrow{\text{glucosidase}} C_6H_{12}O_6 + HO-\underset{\underset{CH_3}{|}}{\overset{\overset{CH_3}{|}}{C}}-CN$$

$$C_6H_{12}O_6 + HO-\underset{\underset{CH_3}{|}}{\overset{\overset{CH_3}{|}}{C}}-CN \rightleftharpoons H_3C-\underset{\underset{O}{\|}}{C}-CH_3 + HCN$$

Enzymatic cleavage of the cyanohydrin, linamarin (a cyanogenetic glycoside of lima beans), yields the sugar glucose, and the cyanohydrin from which the dimethyl ketone (acetone) and free cyanide are obtained.

Analysis of such seemingly innocuous foods as bananas and various fermented cheeses reveals the presence of amines such as histamine and tyramine. These are referred to as pressor amines because of their ability to constrict blood vessels with consequent dramatic increases in blood pressure. It is not unusual to recover 100–200 mg from 100 g (approx 0.25 lb) of Camembert cheese. Normally, these amines are oxidized by the enzyme monoamine oxidase (MAO) to *para*-hydroxyphenylacetic acid and excreted in urine. However, with the current availability of tranquilizing drugs, those who eat aged cheese or drink wine or beer with high tyramine contents can experience life-threatening reactions. Certain tranquilizers can inhibit the enzyme MAO, permitting the intact nonoxidized pressor amines access to the bloodstream. Obviously the two chemicals do not mix well. Avoidance of adverse drug reactions requires careful selection of food and beverage by those on tranquilizers.

A. ALS and *Cycas circinalis*

Amyotrophic lateral sclerosis (ALS or Lou Gehrig's disease) is a fatal motor neuron condition characterized by progressive degeneration of anterior horn cells of the brain stem and spinal cord. Muscle weakness and loss usually begins in the hands and spreads to the forearms and legs. Death occurs within 2–5 years.

Scientists have long been puzzled by "rayput" laziness, a disease of the Chamorro people of Guam. During World War II, with American troops ensconced on Pacific islands, the alarming numbers of ALS seen on Guam and Rota* came to light. ALS was some 100-fold greater on Guam than in

* These islands are part of the Marianas chain of Micronesia.

the 48 continental states. As genetic and viral agents for the disease were ruled out, environmental risk factors became suspect. Suspicion fell on the lemon-sized seeds of the cycad plant, *Cycas circinalis* L. Analysis of the seeds revealed the presence of cycasin, a glycoside, and an unusual nonprotein amino acid—β-N-methylamino-L-alanine, BMAA— a known neurotoxin.

Recently Peter Spencer and co-workers at Albert Einstein Medical College investigated the effects of BMAA in Macque monkeys. They found that when these primates were fed this amino acid, ALS-like symptoms appeared within weeks of exposure (3). Although the disease has a late onset among the Chamorros, neither the millions of soldiers who passed through the Marianas or the tens of thousands of construction workers stationed there have since shown excess ALS. The illness is unique to long-term consumers of cycasin-containing products. Cassava, also grown there as a food plant, cannot be ruled out as a contributor of neurotoxin (4). It will be several years before the mechanism of food-related ALS is fully identified.

As with any chemical, those occurring naturally need not be deadly or life threatening. Often they do nothing more than cause discomfort. Psoralens are a conspicuous example. Although the specific chemicals involved have only recently been identified, the itching and rash resulting from continued contact with a variety of leguminous and other plants were well known to the Greeks and Romans who named them Psoralea, from *psora* meaning to itch.

Chemically, the psoralens are furocoumarins with a ring structure shown in Figure 2. The furocoumarins occur in celery, parsnips, dill, cloves, lime, bergamot, and figs. Biochemists and botanists believe they are naturally occurring antibiotics used by the plants as defense against fungal attack. Unfortunately, for those people who must work closely and often with them, dermatitis and erythema are often the consequences.

Supermarket employees have been among the most often affected. In a recent outbreak, 14 of 27 employees—cashiers, baggers, and produce clerks (especially produce clerks)—were observed to have well-circumscribed rashes confined to their upper arms with substantial blistering and itching. All had regular contact from April through August with fresh produce.

Clearly, naturally occurring poisons (fava beans, gossypol, lathrogens, and goitrogens from the family of crucifers—examples are endless) can be

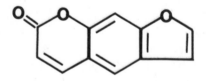

Figure 2. Psoralen, a furocoumarin.

as hazardous as anything people have developed. In fact, *few synthetic chemicals equal the potency and human toxicity of naturally occurring products.*

Commenting on this, Bruce N. Ames noted that "humans ingest about 10,000 times more of nature's pesticides than man-made pesticides. Relatively few of nature's pesticides that we are eating have been tested for carcinogenicity, but about half of the naturally occurring substances that have been tested in rats and mice are carcinogens" (5).

Nitrosamine is a outstanding example. Sodium nitrate, a precursor of nitrosamine widespread in nature and generally innocuous, has recently come under attack and scrutiny.

Vegetables, drinking water, and cured meats are the main sources of dietary nitrate. Mean daily intake in the United States, with amounts varying considerably from country to country, is estimated to be approximately 100 mg,* 80% of which comes from vegetables as a consequence of their contact with soil. Beets, celery, lettuce, spinach, radishes, and rhubarb contain about 100 mg per 100 g. Accumulation of nitrates from soil is characteristic of food plants. Less than 10% comes from water. Cured meats can contribute another 10%. The Japanese consume three times the amount of nitrate consumed in the United States, while the English consume 60–80 mg/day.

Most of this dietary nitrate passes to the stomach and on to the small intestine for absorption into the bloodstream and tissues and is rapidly excreted. Some 25% is secreted in saliva. The proportion of salivary nitrate reduced to nitrite varies markedly between individuals, primarily as a consequence of normal microbial flora resident in the mouth. The concentrations of nitrates and nitrites in saliva are often far higher than amounts consumed in cured meats.

Nitrates and nitrites can undergo nitrosation to form nitrosamines. It is these newly formed compounds that are believed to be carcinogenic. Sander (6) reported the action of specific strains of bacteria in the formation of nitrosamines. The importance of his finding was the fact that the reaction proceeded at pH levels close to 7, that is, at neutrality. For people, the implication meant that such a reaction could occur at any site within the body where secondary amines, nitrate, and an appropriate microbial population could coexist. Within 4 years of Sander's report, G. M. Hawksworth and M. J. Hill of St. Mary's Hospital Medical School, London, showed that such a set of conditions does not exist in healthy people. Again, data from animal studies must be interpreted with extreme caution (7).

---

* 1000 milligrams (mg) = 1 gram (g).
  454 g = 1 lb.
  100 g = ¼ lb.
  100 mg = 0.1 g.

The general formula for the series of compounds known as N-nitrosamines is

$$
\begin{array}{c}
R_1 \\
\diagdown \\
N\!-\!N\!=\!O \\
\diagup \\
R_2
\end{array}
$$

The nitroso moiety is common to all these compounds. It is the R_1 and R_2 substitutions, which may be alkyl, aryl, or alicyclic, that confer the widely differing properties of each compound. Over 100 are known. The most commonly isolated are N-nitrosodimethylamine (DMN), N-nitrosodiethylamine (DEN), N-nitrosopyrrolidine (PYR), and N-nitrosopiperidine (PIP).

In acute toxicity studies, large doses of dialkylnitrosamines fed to animals have shown that the liver and esophagus are the primary organs affected. However, acute toxicity is of little real interest since most foods will contain quantities in the microgram per kilogram level, that is, in parts per billion.

Studies of the occurrence of nitrosamines in foods have shown that cured meats, especially fried bacon, followed by fish and cheese, are the major sources of these compounds in the diet. DMN and DEN have been shown to be present in luncheon-type meats such as bacon, ham, and pickled meat in the United States, Canada, the United Kingdom, and The Netherlands. Although animal testing has shown a number of nitrosamines capable of inducing tumors in animals, as of 1985 cancer in human populations as a consequence of consumption of nitrosamines has not been demonstrated. Curing of meat is not another example of the excesses of a "hi-tech" society. Although the origins of curing are lost in antiquity, it is well known that the Romans cured pork and fish extensively and that they learned the technique from the Greeks. By Homer's time, 900 B.C., curing with salt, with nitrate as an impurity, was well established. The relationship of the salts and nitrate to the red color of the meat was also understood. It was not until 1899 that two German chemists, K. B. Lehman and K. Kisskalt, demonstrated that the typical color of cured meat was due to nitrite rather than nitrate. By 1975, nitrate was reduced or eliminated completely in some products. Given this long history of use and consumption, it is not surprising that human illness associated with low doses is difficult to demonstrate (8).

If N-nitroso compounds are important in human cancer, populations ingesting larger amounts of nitrate or nitrite should have a higher incidence of cancer of the liver, stomach, and perhaps the esophagus. In the United States, for example, both liver and esophageal cancer are relatively rare and gastric cancer has been declining since the late 1930s, while the consumption of cured meats has risen steadily. If the consumption of nitrate in food is related to cancer of any body site, it has yet to be observed. Nevertheless, strong pressures exist for regulatory action to reduce the amount of nitrate added in the curing process.

Recently, a research team from the Imperial Cancer Research Fund, Oxford, England, conducted an epidemiologic study, taking advantage of the well-known regional differences in gastric mortality in England. They compared populations from consistently high areas with those from consistently low areas to determine if nitrate exposure was a risk factor.

To estimate exposure, they measured nitrate/nitrite levels in saliva. To their surprise, levels of both ions were significantly higher in the low-risk group. This inverse relationship suggests that "nitrates or something associated with them exert a protective effect" and is inconsistent with the idea that nitrate exposure is a risk factor for cancer of the stomach (9).

Although the accumulated evidence does not support the opprobrium heaped upon it, the public and apocalyptic view is that nitrate is a "clear and present danger."

At the 1985 meeting of the American Chemical Society in Honolulu, Michael W. Pariza of the University of Wisconsin's Food Research Institute reported on the extraction from beef (hamburger) of a "mutagenesis modulator" that inhibits the effects of carcinogens produced by frying.

Since they were first demonstrated in 1929 by Isaac Berenblum, currently of the Weizman Institute of Science in Israel, anticarcinogens have become well-established entities found in a wide spectrum of chemical classes, ranging from simple inorganic compounds to complex organic molecules.

Pariza dissolved this compound in acetone and painted it on the backs of 60 mice. Five minutes later, having allowed it to dry on the shaved skin, he applied a known carcinogen, DMBA (7,12-dimethylbenz[a]anthracene). Control mice were treated with only acetone prior to receiving applications of DMBA.

The modulator-treated mice developed far fewer papillomas (wartlike tumors of the skin and mucous membranes) than did the control mice. Although the protective compound had not then been isolated or identified, he believes it is present in both cooked and uncooked beef and inhibits carcinogens formed during cooking.

We human beings are constantly exposed to carcinogens and anticarcinogens. Some of us may be getting more of one than the other, which could account for differences in cancer incidence between groups. This, however, remains to be verified. If shown to be true, prevention could take a quantum leap.

There is another twist to this carcinogen/anticarcinogen problem for policy considerations. We know, for example, that vitamin A is essential to health. We know too that at high levels it can be toxic. It may even be carcinogenic. Since it is essential to health, and as it may be carcinogenic, should it be banned under the Delaney Amendment to the Food, Drug and

Cosmetic Act? Is vitamin A protective against cancer, or is it a cancer producer? Can it be both at different concentrations? This is not a frivolous question, but one that can be asked of hundreds of substances. On which side do the decisionmakers come down?

Foods are chemicals. Because they are, they can cause illness in some people. Given the proper conditions, these foods/chemicals can react to yield new chemicals. And as many provide nutrients in the form of chemicals we call proteins, carbohydrates, fats, and vitamins, there is very often a race between humans and all manner of microbes, insects, rodents, and mammals to get to the supply first. To give us an edge in this costly race, chemicals are added.

Unfortunately, hazardous* substances can enter foods unintentionally as a consequence of attempts by people to protect their food supply from the above-named pests. Pesticides, an umbrella term for weedicides, fungicides, herbicides, insecticides, and rodenticides, are the primary weapons. The relevant question is: Are our foods, and by extension our health, imperiled by their use? What certain knowledge do we have about the levels of these contaminants?

B. Market Basket Studies

Conceived in 1961 to determine whether fallout from atmospheric nuclear tests resulted in elevated levels of radionuclides (cf. Chapter 7) in foods, total diet studies to monitor chemical contaminants were begun by the Food & Drug Administration in 1965. Since then, there has been an ongoing testing program to determine the levels of pesticides and biphenyls (PCBs), chlorophenoxy acids, heavy metals, and radionuclides, such as strontium-90.

The objective of these market basket or total diet studies is the collection and analysis of foods obtained from retail markets in 20 widespread urban areas, for the express purpose of ascertaining the levels of pesticide residues, polychlorinated biphenyls (PCBs), trace metals, and other chemical contaminants in the diet of adults, infants, and toddlers.

Originally, the study was directed at the type and amount of food consumed in a single day by an average male teenager, who is considered to have the most voracious appetite of any age group and of either sex.

The cities surveyed for adult food consumption in fiscal year 1979 are listed in Table 1. Cities sampled for infant and toddler diets, are shown in Table 2.

Perusal of the lists suggests adequate geographical representation. Cities appear to have been chosen from among the more typical rather than the major population centers.

---

* Hazardous and toxic are not synonymous (cf. Chapter 13).

Table 1
FY 79 Total Diet Sample Collections

Market Basket	Sample Collection Area	Geographic Region	Date of Collection
1	Los Angeles, CA	W	10/02/78
2	Miami, FL	S	10/10/78
3	Chicago, IL	NC	10/23/78
4	Portland, OR	W	10/30/78
5	Springfield, MA	NE	12/04/78
6	Columbus, OH	NC	01/08/79
7	Biloxi, MS	S	01/22/79
8	Harrisburg, PA	NE	02/20/79
9	San Francisco, CA	W	02/26/79
10	Wichita, KS	NC	03/06/79
11	Lakeland, FL	S	04/09/79
12	Portland, ME	NE	04/16/79
13	Salt Lake City, UT	W	04/30/79
14	Nashville, TN	S	05/07/79
15	Indianapolis, IN	NC	06/18/79
16	Phoenix, AZ	W	06/25/79
17	Providence, RI	NE	07/09/79
18	Sioux Falls, SD	NC	08/06/79
19	Little Rock, AR	S	08/20/79
20	Albany, NY	NE	08/27/79

Table 3 lists the 12 food groups included in this study, as well as their proportions with respect to the total consumed. These 12 food groups include 129 different items of which grains and cereals, fruits, meat, fish, and poultry represent almost 50% of the total. Consider for a moment, the range of food items you have consumed over the last 7 days, or the next seven. How close to 129 are you?

Table 2
Total Diet Sample Collections

Market Basket	Sample Collection Area	Geographic Region	Date of Collection
1	Green Bay, WI	NC	10/16/78
2	New York, NY	NE	11/17/78
3	Columbia, SC	S	01/15/79
4	Boise, ID	W	02/20/79
5	Grand Rapids, MI	NC	03/12/79
6	Hartford, CT	NE	04/23/79
7	Des Moines, IA	NC	06/11/79
8	Manchester, NH	NE	07/02/79
9	Houston, TX	S	08/13/79
10	Albuquerque, NM	W	09/04/79

Table 3
Average Daily Composition of FY79 Adult Total Diet by Food Group and Weight

Food Group (Composite)	Average[a] (g/day)	Total Diet (% by weight)
I. Dairy	756	26.3
II. Meat, fish, and poultry	256	8.9
III. Grains and cereals	404	14.0
IV. Potatoes	159	5.5
V. Leafy vegetables	55	1.9
VI. Legume vegetables	73	2.5
VII. Root vegetables	30	1.0
VIII. Garden fruits	74	2.6
IX. Fruits	220	7.7
X. Oils and fats	73	2.5
XI. Sugars and adjuncts	82	2.9
XII. Beverages (including drinking water)	697	24.2
Total	2879	100.0

[a] Average of four regional diets.

Source: U.S. Food & Drug Administration, Washington, DC.

After purchase, the items are prepared for consumption in the way they are usually eaten. The foods in each group are then combined: the 22 fruits, for example, or the nine leafy vegetables, or the eight garden fruits, to yield a homogeneous mixture. Each of these composite samples is then analyzed.

Given the known quantity of consumption of each food, as well as the level of contaminants, an estimate of daily intake of each chemical can be made. Thus, the daily intake of six major pesticides for the years 1980–1984 is shown in Table 4. The consistently low levels should be reassuring, as market basket studies relate to foods at point of purchase. These are not laboratory "experiments" or artifacts. Consequently, the pesticide residues reported effectively demonstrate the condition of the

Table 4
Daily Intakes, Teenage Male (ng/kg per day)

Pesticide	ADI	80	81	82/84
DDT, total	20,000	34	34	42
Malathion	20,000	203	243	107
Diazinon	2,000	4	7	13
Heptachlor	500	7	7	3
Dieldrin	100	22	16	8
HCB	—	4	2	2

Data reproduced by permission of Division of Contaminants Chemistry, Center for Food Safety and Applied Nutrition, FDA, Washington, DC.

food supply. Comparison with the Acceptable Daily Intake, ADI, an internationally agreed upon standard, assists in making a judgment of overall safety.

III. THE ADI CONCEPT

One of the major problems with food additives both intentional and unintentional has been the appropriate control of their safe use. The essential basis for evaluation has been the use of long-term experimental studies in laboratory animals. Table 5 indicates the total intake of the six metals tested over the period 1976–1979 (10).

Underlying all toxicologic studies is the hypothesis that the degree or severity of an adverse effect is directly related to the dose or concentration of the chemical to which test animals are exposed. For most chemicals, there is a dose below which no effect occurs. This dose is referred to as the no observable effect level—the NOEL. The NOEL is the point of departure for establishing an Acceptable Daily Intake. It is the daily dose of a chemical without appreciable risk. "Without appreciable risk" means the practical certainty that injury will not result even after a lifetime of exposure (11).

Figure 3 suggests the complexity of the evaluation process. It shows that specific toxicologic procedures (1) lead to the design of suitable studies (2) from which adequate information is obtained (3) which, with appropriate interpretation (4), can assist in the formulation of recommen-

Table 5
Trends in Daily Intake (μg/person per day) of Selected Elements

Element	Acceptable Limit	FY76	FY77	FY78	FY79
Arsenic (As_2O_3)	—	66.5	7.16	59.1	61.5
Cadmium	57–72[c]	32.9	36.9	30.9	32.6
Lead	429[b]	71.1	79.3	95.1	81.7
Mercury	42.9[c]	6.5	6.3	3.4	5.2
Selenium	50–200[e]	135.6	110.7	156.2	152.3
Zinc[a]	15[d]	19.1	18.0	16.8	17.8

[a] Zinc is expressed in milligrams per day.
[b] Maximum limit proposed by the FDA Bureau of Foods for lead from all sources.
[c] Provisional Tolerable Daily Intake (PTDI) calculated from the Provisional Tolerable Weekly Intake (PTWI) proposed by the FAO/WHO.
[d] Recommended Dietary Allowance (RDA) proposed by the Food and Nutrition Board of the National Academy of Sciences.
[e] Estimated safe and adequate daily intake (ESADI) proposed by the Food and Nutrition Board of the National Academy of Sciences.

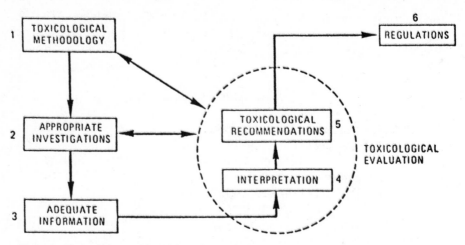

Figure 3. Flow diagram identifying the critical points and objectives of toxicological assessments of intentional food additives.

dations (5) which should provide a rational basis for regulation of the chemical being tested (6).

From the tables the intakes are seen to be consistently well below established limits. Dieldrin, however, was "pushing" its ADI far more than the others. As a result, in 1975 the EPA canceled all of its registered uses. As noted in the table, by 1982 it had decreased substantially.

This type of study and validity of the data should help to reassure the public of the safety of the food supply. Unfortunately, the media, both print and electronic, find this information less than exciting. Consequently, instead of having a positive impact on the public's fund of knowledge and decisionmaking, it has little effect. With the aid of hindsight, and a data base consisting of over 20 years of analytic results, it can be demonstrated that levels of contamination were neither as high nor as threatening as some would have us believe.

Furthermore, on an international level, reports of pesticide residues form 11 countries (Figure 4) show that the dietary intakes are well within established guidelines (12).

Interestingly enough, although 30 pesticides are regularly included in the analytic studies, ethylene dibromide (EDB) has not been. Two reasons are advanced. Tests for EDB in table-ready foods had regularly found small to nonexistent levels, and EDB requires a totally different type of chemical analysis, which would have raised the analytical procedures to a far higher level of complexity.

Testing was done, however, and of approximately 500 samples of EDB tested between 1983 and 1984, 12 had levels of just above 1 ppm (between 1.2 and 1.3), a fraction of their ADIs. This seems to have been overlooked

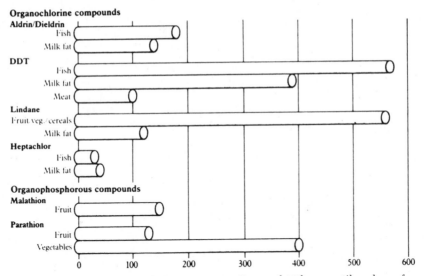

Figure 4. Pesticide residues in foods, 1980–1983. Range of 90th percentile values of concentrations (μg/kg) in participating countries.

in the frenzy of the EDB "scare." (See Table 6 for the relationships between the various units used in measurement.)

Because of the special sensitivity of infants and toddlers, their consumption of chemicals is of considerable interest. Table 7 indicates the food classes included, as well as the relative composition of each to the total. Lead, cadmium, and mercury, as noted in Table 8, were well below current acceptable limits. Zinc, a required nutrient, in some instances exceeded the Recommended Dietary Allowance (RDA) but not enough to pose a hazard to health. Arsenic, for which no WHO acceptable daily

Table 6
Relationship Between ppm, ppb, and ppt

mg/kg = milligrams/kilogram = parts per million = ppm
μg/kg = micrograms/kilogram = parts per billion = ppb
ng/kg = nanogram/kilogram = parts per trillion = ppt

ppb is 1000 times less than ppm
ppt is 1000 times less than ppb
ppt is also 1,000,000 times less than ppm

Ordinarily, a trillion is greater than a million, but in context of toxicity testing, the reverse is true. One part in a trillion or a billion is far less than one part per million. Consequently, caution must be exercised when interpreting data. For example, 1 ppm = 1000 ppb = 1,000,000 ppt. Of course, 1000 ppb looks much more dreadful than 1 ppm.

Table 7
FY 79 Relative Composition of Diet by Food Class[a]

		Infant Diet		Toddler Diet	
Food Class		Average (g/day)	Percentage of Total	Average (g/day)	Percentage of Total
I.	Drinking water	208	15.4	336	21.5
II.	Whole milk, fresh	672	49.7	515	33.0
III.	Other dairy and dairy substitutes	112	8.3	70	4.5
IV.	Meat, fish, and poultry	44	3.2	127	8.1
V.	Grains and cereal products	35	2.6	117	7.5
VI.	Potatoes	6	0.4	36	2.3
VII.	Vegetables	100	7.4	72	4.6
VIII.	Fruits and fruit juices	135	10.0	151	9.7
IX.	Oils and fats	6	0.4	14	0.9
X.	Sugars and adjuncts	12	0.9	30	1.9
XI.	Beverages	23	1.7	93	6.0
		1353	100.0	1561	100.0

[a] Average of four regional diets. Past results of the total diet studies (infant, toddler, and adult) have shown that pesticide residues and the levels of selected elements in the diet are well below limits established by the joint Food and Agriculture Organization/World Health Organization (FAO/WHO) Expert Committee on Food Additives or Pesticide Residues, that is, acceptable daily intakes (ADIs) and provisional tolerable weekly intakes (PTWIs).

Table 8
Daily Intake of Selected Elements (ug/day)

Element	Acceptable Limit	Infant			Toddler		
		FY77	FY78	FY79	FY77	FY78	FY79
Lead	100–150[a]	22	25	36	28	35	46
Cadmium	57–72[b]	6	6	4	8	11	9
Zinc[c]	3–10[d]	4	5	5	8	9	8
Arsenic (As_2O_3)	—	5	2	5	25	18	23
Selenium	10–80[e]	15	18	15	46	52	45
Mercury	43[b]	0.9	0.2	0.03	1.0	0.7	0.4

[a] Maximum limit proposed by the FDA Bureau of Foods for lead from all sources.
[b] Provisional Tolerable Daily Intake (PTDI) calculated from the Provisional Tolerable Weekly Intake (PTWI) proposed by the FAO/WHO.
[c] Zinc is expressed in milligrams per day.
[d] Recommended Dietary Allowance (RDA) proposed by the Food and Nutrition Board of the National Academy of Sciences.
[e] Estimated safe and adequate daily intake (ESADI) proposed by the Food and Nutrition Board of the National Academy of Sciences.

intakes have been established, was detected at the lower ranges, similar to findings for adult samples. This helps to validate the analysis.

Pesticide and industrial chemical residues were essentially unchanged for the past 4 years. Some 40 different chemicals, including the six elements, have been detected each year in infant and toddler diets, but all well within acceptable low levels. In 1979, PCBs were not detected at all. This is down from the 0.09 μg/day for infants and 1.35 μg/day for toddlers found in 1978. Here, we are approaching the lower limits of the parts per billion range. Clearly, from both the adult and infant samples, the food supply has to be given a vote of confidence.

A. Animal Testing

The FDA does not have the field to itself. Since 1967, the USDA has also been monitoring one food supply. Its National Residue Program focuses on animals coming to slaughter. Three hundred random samples are collected yearly from healthy animals at time of slaughter. Tests are geared to the detection of levels of some 50 plus chemicals. The number can be increased should tests suggest the need. Both the USDA and the FDA contribute residue data for development of regulatory policy by the EPA. USDA analysis of pesticide residues over the past decade reveals a declining trend. Figure 5, for dieldrin, and Figure 6, for DDT, show dramatic declines in tissues of the three major animal food sources.

Other chlorinated hydrocarbons were not detected as residues more than 30% of the time through 1979. Since then, detection levels have fallen off to less than 10%. The data clearly show that the animal food chain responds quickly to elimination or control of sources of food contamination.

In 1982, a new program geared to prevention was introduced. The Total Residue Avoidance Program (TRAP) emphasizes the integration of residue prevention into basic husbandry practices. Among its unique features are:

- On-farm assistance and provision of educational materials.
- Development of simplified, accurate residue tests for farm use.
- Opportunity for farmers to discuss management practices onsite.

Animal exposure to chemicals can occur at any time and from a number of sources, including food, housing, and disease control efforts. Chemicals and drugs so used can accumulate in tissues and organs. Consequently, the TRAP relies heavily on the integration of residue prevention into fundamental animal management practices. While this down-on-the-farm approach is novel, it may indeed be an appropriate place to effect change. It should have been started years ago.

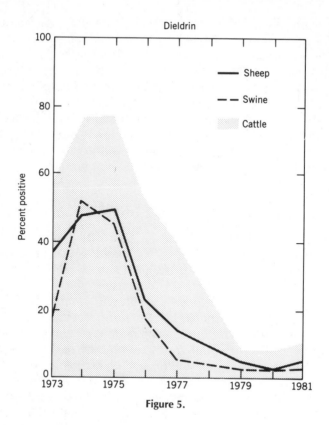

Figure 5.

Most of the pork and much of the beef and poultry produced in the United States is raised on feed treated with low levels of such antibiotics as penicillin and tetracycline. This addition promotes growth, increases efficiency of feed conversion to animal tissue, and helps prevent a number of microbial diseases. Unfortunately, these antibiotics are also used to treat human illness.

Feeding antibiotics to cattle, pigs, and chickens poses a risk of increasing the prevalence of antibiotic-resistant bacteria that can cause human illness, which then may not be treated successfully with these same antibiotics.

A ban on antibiotic addition to feed was suggested by the FDA in 1977, but it was blocked in Congress by farm-state legislators at the urging of the livestock and pharmaceutical industries. Toward the end of 1984, an array of new evidence linking human illness to drug-resistant bacteria in meat had been developed by a team of investigators at the Centers for Disease Control (13).

Given the new evidence, the FDA will again attempt to seek restrictions on the use of these drugs. However, even if Congress allows that agency to move ahead with a proposed ban, the process of hearings and appeals

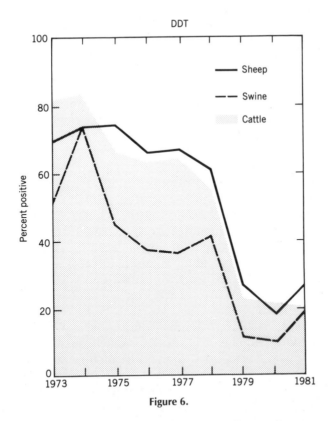

Figure 6.

would normally take more than 3 years before the rules could take effect. Even if a ban were to be imposed, the problem of resistant bacteria would barely be affected because of the widespread use of antibiotics to combat infections in both animals and humans. No one said that we were dealing with simple problems or equally simple solutions.

Additives often get a "bad press." Leavening agents such as sodium aluminum sulfate or potassium acid tartrate did not, but BHA and BHT did. And so did DES.

Butylated hydroxyanisole (BHA) and butylated hydroxytoluene (BHT) are antioxidants used primarily in "cold" cereals, cakes, and pastry products to protect against the adverse effects of rancidity and reversion which produce off odors and tastes in foods with high oil or fat content.

For years people believed these antioxidants were responsible for birth defects and cancer. Currently, each has an established FAO/WHO ADI and both are approved under Section 409 of The Act.

The belief among many people that BHT and BHA were responsible for birth defects and cancer was hard dying. Not only was this not so, but they have now been shown to counter oxygen radical formation by sequestering oxygen radicals as they form. Although yet to be firmly established, it is being suggested that dietary control of oxygen radical damage could add, on average, five or more years to our lives.

B. DES

In terms of complexity, addition of antibiotics to animal feed is not in a class by itself. Even more difficult to unravel has been the adverse health effects of diethylstilbestrol, DES.

From the late 1940s to the early 1970s, a synthetic estrogen, DES was prescribed for pregnant women whose physicians believed the possibility of miscarriage existed. It was *proscribed* in 1971 when medical reports indicated that daughters of mothers treated with DES were at increased risk of developing an unusual cancer of the vagina.

DES was also used in animal feed to spur growth and thereby shorten time to market. Was there a relationship between the use of DES in feeds and vaginal cancer for daughters of pregnant mothers who consumed meat? Some thought so. Its use was banned in 1979. Did the evidence warrant such action? It may well have. The evidence seems divisible into three parts.

Diethylstilbestrol (DES), a white, odorless, crystalline powder, is a nonsteroidal synthetic chemical, exhibiting female sex hormone (estrogenic) activity. It was used for a time in the poultry industry to produce caponettes. (A capon is a male bird that has been castrated surgically to fatten it. Castration by chemical means produces a caponette.) Although the Food and Drug Administration authorized the use of DES in 1947, tests on laboratory animals later showed the chemical to be toxic for them and thus potentially hazardous for human consumption. As a result, it was withdrawn from use in poultry. It remained certified, for use in cattle and sheep. Here, too, it is used to achieve rapid growth. An animal fed DES can often gain as much as 300–400 lb in a matter of months; far more weight, and in less time, than when fed standard fodder without DES.

It is reasonably well established that DES is excreted in the feces unchanged chemically, which means that it does not remain in the tissue as residue, nor does it become metabolized to other chemically reactive substances. In addition, certification for its use was granted to feeders by the FDA on the condition that it not be fed for a least 7 days prior to slaughter.

Until 1971, pressure to remove DES from the food supply was almost nonexistent. Then on April 22, 1971, a study with startling and profound implications was published by the *New England Journal of Medicine* (10). Physicians from St. Vincent's Memorial Hospital in Boston described a striking association between the use of DES during the first trimester of pregnancy (for protection against abortion and premature delivery, miscarriage) and the development, 15–20 years later, of vaginal cancer in daughters born of these pregnancies. Although cancer of the vagina rarely occurs in women under 50, between 1966 and 1969, these physicians diagnosed adenocarcinoma in seven 15–22-year-old women. Prior to 1966, this type of tumor, especially the clear cell, had not been seen at their hospital.

Although seven cases may not seem like many, when they occur suddenly and in circumstances where they have not been previously seen, that number could constitute an outbreak of epidemic proportions.

Given the number of women treated by this, then standard, procedure, additional follow-up studies ensued. Data from both a study done by the Mayo Clinic and a study by the University of Chicago failed to show tumors in daughters of DES-treated mothers. They concluded that the relationship was rare: not nonexistent, but rare.

The question that follows is: What if anything do these findings mean for DES fed to cattle? Would women consuming meat from cattle and sheep fed with DES be at increased risk?

Of primary importance is the question of dose or amount consumed. Since DES is implanted as pellets in beef cattle, the possibility exists that daily consumption of beef liver could amount to 2 ng (1 mg = 1,000,000 ng). The dosage of DES used to treat potential miscarriage in pregnant women was 5–150 mg/day, on average approximately 60 mg/day for a total of some 10–15 g over the pregnancy period, a difference of 30,000,000 : 1 between the two. As Jukes has cogently demonstrated (14) the content of DES in beef liver from implanted cattle has been found to be on the order of 0.12 ppb, thus 12.3 g would be present in 100,000 tons. He wrote: "The total annual production of beef liver in the U.S.A. is about 140,000 tons."

According to Jukes, given the 1.5 million female births per year, and given an ingestion dose by mothers of 1.9 ng/day, one could anticipate a DES-induced vaginal cancer rate of one every 133 years. Even assuming an error of 50% in his calculations, the risk seems infinitesimal.

In his reversal of the administrative law judge's opinion* that that the Delaney Clause did not apply in this instance, the then FDA Comissioner Donald Kennedy considered a study of DES fed to C3H mice—the most susceptible to breast cancer of any mice strains. This study, done in 1964, had a number of problems associated with it, including the fact that most of the data were destroyed in a fire. What did survive was information that at a dose level of 6.25 ppb, tumors occurred, but at a level of 12.5 ppb, they did not. This study was not repeated.

A third problem for the commissioner was the test for DES. To detect the unusually low levels in beef liver and muscle tissue, a two-step procedure was used. The mouse uterine/paper chromatographic method could detect DES as low as 2 ppb. According to the commissioner, that

---

* The judge who presided at the 1978 hearings concluded that the Delaney Clause did not apply because DES residues were not found in edible tissue of beef or sheep.

was not sensitive enough since there was no knowledge as to whether DES in doses as low as one molecule could cause cancer.

Given the 6.25-ppb level of tumor production in mice, the close proximity of this number to the 2-ppb detectable levels, and the fact of DES's relationship to vaginal cancer in women, DES was banned on June 29, 1979 (14).

With the findings of Herbst at St. Vincent's Hospital, DES received an extremely bad press. That other follow-up studies failed to support the relationship made little difference. The fact that exposure levels in beef bordered on the miniscule also seemed to make little difference. Emotion was running high. Perhaps DES has no place in food, but the evidence supporting the decision to ban seemed thin indeed.

This is not a plea for the reinstatement of DES; rather, it is concern for the process by which the safety of our food supply is maintained. Surely a decision could go the other way, permitting a chemical to remain when in fact it should be withdrawn. If the decisionmaking process is to be reliable and meaningful, it must follow from the scientific data. If data are ignored, there is no process. Can this be prudence?

Although abetting the growth of cattle with DES has been banned, chemical fattening continues. Several hundred cases of premature breast development have been reported among Puerto Rican girls for each of the three years 1982, 1983, and 1984.

The Centers for Disease Control found the chemical Zeranol in 21% of blood samples taken from girls aged 1–8 years old. Zeranol, $C_{18} H_{26} O_5$ (dihydroxyundecyl betaresorcyclic acid) also known as Ralone and Ralabol, is an anabolic steroid fed to cattle to stimulate growth. It has not been known to affect people.

Premature breast development has been reported all over the world from areas free of chemically treated cattle. Obviously, it has more than one cause. Nevertheless, Department of Agriculture officials began testing meat and poultry in October 1982 for "estrogenic substances" which might be linked with abnormal development in children. Of the more than 100 samples tested, none was found contaminated. Whether premature breast development in Puerto Rico is a fortuitous event unrelated to chemical treatment of cattle remains to be determined.

C. Toxicity Testing

Whether a chemical additive is natural, synthetic, intentional, or incidental, the question invariably asked is: Is it poisonous? Before a reliable answer can be given, it is necessary to know what a poison is. The ancient adage, "One man's poison is another man's porridge," reminds us that a poison—a chemical—has variable consequences depending on the individual involved. This should perhaps suggest that the unavoidable question, "Is it poisonous?" is not readily answered by yes or no.

A fish dealer in Philadelphia wanted to vend a batch of spoiled fish. In order to do so, it was necessary to mask the "off" odor and color to make the fish seem fresh. He liberally dosed the fish with sodium nitrate. Sodium nitrate, or Chile saltpeter, is one of the most abundant naturally occurring nitrates. When used properly, it is a rather innocuous substance. It is also well known that nitrates in sufficient quantity can cause a rapid drop in blood pressure through vasodilation (enlargement of the blood vessels).

In this instance, the fish was freely laced with the nitrate and shipped out to markets along the Atlantic coast. Within a short time 150 people became ill with nitrate poisoning; many were hospitalized and a 3-year-old child died. When this substance is used in properly prescribed amounts, no problems arise. When greater than prescribed doses are administered, some adults become mildly ill, others so ill as to require hospitalization, and others may not even be uncomfortable. On the other hand, babies often die. Thus, any definition of a potentially toxic chemical must consider who is to use it and how much is consumed: the dose to response relationship at work. In addition, such factors as age, sex, nutritional status, and general state of health exert much influence on the type and degree of response.

Paracelsus was right: everything and nothing can be poisonous depending on a constellation of factors. Common table salt is essential to life; yet an "overdose" would probably kill if taken over a short time. Many metals such as copper, manganese, zinc, and cobalt are essential to health in trace amounts but can become toxic in larger quantities. Both vitamins A and D, when taken in doses larger than prescribed, are known to cause severe adverse health effects. Thus, to answer the original question, a poison can be any substance that may cause death or illness when taken in sufficient quantity.

This idea was first expressed by Theophrastus Bombast von Hohenheim, who referred to himself as Paracelsus. It is easy to see why. In his *Epistula dedicatora St. Veit/Kärnten* published in August 1538, he noted that "anything may be poisonous and nothing may be without poisoning property; it is the dose which makes a thing non-poisonous" (*dosis sola facet veneum*—what makes a poison is dose). That was approximately 450 years ago, but the message continues to elude most of us. Perhaps the reverse is true. Do most of us refuse to believe it?

It had to await the experimental studies of Frederich Flurry and Wolfgang Heubner* of the Kaiser Wilhelm Institute for Physical and Electrochemistry, to demonstrate unequivocally, in 1919, the existence of a dose that does not induce harmful effects, that is, nontoxic exposure levels. They used cyanide, hydrocyanic acid, to establish a dose–response relationship.

---

* *Biochem. Zeit.* **95:** 249–256, 1919. Uber Wirkung und Entgiftung eingeatmeter Blausäure.

The dose–response relationship is the central and unifying principle of modern toxicology. The greater the dose, the greater the effect; the smaller the dose, the smaller the effect. Two aspirin every 4 h for several days is a therapeutic dose. One hundred aspirin at one time can be deadly. An ounce of whiskey (a "shot") is warm, even friendly. But a "fifth" (750 mL) at one time is often fatal. In between, effects range from giddiness, loss of balance, slurred speech, and unconsciousness. We are all aware of this continuum. Are we unable to translate the idea to other chemicals— hazardous wastes, air pollutants, food additives? Are these so different from alcohol or aspirin? A chemical, is a chemical, is a chemical??

Within the dose–response relationship, there is an essential element—the threshold. This is a point on a dose–response curve above which measurable effects occur, and below which they do not. Again, the mere presence of a substance, a chemical, does not, should not imply harm. If it did, homo erectus would barely have advanced much beyond Peking man.

A further distinction that must be made is that between *hazard* and *toxicity*. While these terms are used synonymously and interchangeably, in reality they connote totally different ideas. Toxicity is the capacity of a substance to produce injury. Hazard, on the other hand, is the probability that injury will result from the use of the substance, if continued in the manner and amount proposed. Whereas toxicity implies a known fact, hazard is a presumption based on the statistical likelihood derived from a number of previous events or experiences. In short, it is speculative.

To determine the safety of an additive—or any chemical—for human consumption, two major types of test are employed: short-term bioassays and animal feeding studies.

1. Short-Term Bioassays

Over the past 15 years, a variety of short-term tests or systems were developed to assess the potential of chemicals to induce cancer. Impetus for this derives from the tens of thousands of substances that remain to be evaluated for potential hazard, as well as the sky-rocketing costs of currently acceptable laboratory animal assays.

The more than 100 available short-term systems can be classified as one of four types in terms of their biological end-points. These are gene mutation, chromosomal effects, general DNA damage, and neoplastic transformation.

a. Gene Mutation. Thus far the system used to test the largest number of potentially carcinogenic chemicals—over 5000—is the salmonella/ microsome assay, widely known as the Ames test, after its primary discoverer, Bruce N. Ames of the Department of Biochemistry, University of California, Berkeley. This is an example gene mutation.*

Mutagenesis (the end-point) is detected by mixing an extract of rat liver (or human, if available, to supply a semblance of mammalian metabolic activity) with the tester bacteria, in a medium containing appropriate nutrients and agar. As a consequence of prior modification, the microbial cells cannot grow without the addition of histidine to the medium. A spot of chemical to be tested is placed on a filter paper disk, which in turn is placed on the solid agar surface.

The plates are incubated at 37°C (98.6°F) for 48–72 h. During this period most of the histidine-deficient organisms die off for lack of appropriate nutrients. However, around the chemical "spot" there may be a luxuriant growth of bacterial colonies. The chemical, diffusing from the filter paper to the agar, can induce damage to the bacterial DNA, giving rise to mutants.

These mutants have the ability to produce histidine and can therefore grow. Depending on the number of colonies—compared to control plates with no added chemical—the substance can be assigned a value that indicates its level of carcinogenic potential. This "spot" test is primarily qualitative and is useful for water-soluble, diffusible chemicals.

Given the simplicity of the test, its relative low cost, and extreme short term, it has gained worldwide acceptance as a screening test, a first appraisal of potential carcinogenicity.

Other microbial systems employ the common intestinal bacteria E. coli, which has been modified to require the amino acid tryptophan for growth. In addition, cultured mammalian somatic cells (which are zoologically closer to human cells than are bacteria), such as mouse lymphoma or Chinese hamster ovary cells, are used to detect genetic mutations induced by either chemicals or radiation.

b. Chromosome Effects. Short-term tests have demonstrated that a number of chemicals can produce structural alterations of chromosomes in bone marrow of exposed animals or in cultured cells derived from a variety of tissues. However, neither chromosome effects, DNA damage tests, nor cellular transformation tests have attained the level of acceptance of the gene mutation test. All have the inherent shortcoming of

---

* It is no small concern that the terms mutation and carcinogen are used interchangeably. Although a positive response on these tests is the development of mutation, the results are interpreted as though the chemical had been tested directly for carcinogenic potential. It has not. This interpretation has led to a significant scientific controversy. As yet, evidence for such an assumption is incomplete.

Conceptually, the Ames test is based on the idea that a chemical with the ability to induce mutations will stimulate growth of salmonella bacteria by causing a mutation that restores an enzyme needed for synthesis of the amino acid histidine.

attempting to translate effects in isolated cell systems to potential effects in complex, intact human beings. To attain information more closely approximating the human system, laboratory animals are a more suitable model for study.

2. Animal Tests

In performing animal studies an appropriate species must be chosen, as no one species is a satisfactory substitute for people, humankind.

Long-term feeding (chronic toxicity) studies are carried out on the premise that the effects of lifetime ingestion of an additive by humans cannot be predicted from experiments less rigorous than the total life span of a short-lived animal (2–3 years). In chronic toxicity studies, large doses are used, so the argument goes, to try to not miss weak carcinogens, to compensate for the limited number of animals in any experiment, and for the shorter life span of the animals. Limits to the highest doses depend on what the species can tolerate without the test substance itself becoming lethal. This level is referred to as the MTD, the maximum tolerated dose. Two additional doses are also used: one-half or one-quarter of the MTD, and none or zero, the control group. Actually, there are usually six groups, three male and three female, each group consisting of 50 animals for a total of 600.

Feeding is started as soon as possible after weaning and is continued for at least 2 years. During the 2 years of the tests of qualitative and quantitative observations are made; these include total food consumption, growth, weight, mortality, chemical analysis of blood and urine, behavior, and reproductive capacity.

An unusually instructive example of the way researchers develop information on a potentially harmful additive is exemplified by the report of Löfroth and Gejvall (15). Their experiments showed that diethyl pyrocarbamate (DEP), Baycovia, an antimicrobial food additive used primarily in beer, wine, and citrus juices, itself a nontoxic substance, can on interaction with ammonia (normally present in a number of foods as a result of the natural conversion of amino acids) produce ethyl carbamate (urethane), a known carcinogen.

A most interesting feature of this work is its bringing to light of new data on a substance that had previously been tested without finding the potentially carcinogenic by-product. It is particularly noteworthy that the new findings had to await the advancement of our technical capability. The unusually sensitive isotope dilution technique used by Löfroth and Gejvall had not yet been developed for use in this area when the substance was first tested. From studies such as this one, additional recommendations can be made on the amounts of DEP that can safely be used in foods.

3. HUMTRN

But what of the future? Will we continue to rely on laboratory animals or short-term microbial assays as predictors of harm or safety to people? I think not. There is a better way. There's a HUMTRN in your future.

HUMTRN, the creation of a team of scientists at the Los Alamos National Laboratory is the "research rat of the future," a computer program that duplicates the body's complex systems. HUMTRN, short for human transport, is a vast data bank that gives simultaneous access to 10 million pieces of information on what occurs in the body when a chemical is ingested—or what can be expected of that chemical from either acute or chronic ingestion.

Since the human system is constantly changing, so does HUMTRN. It is programmed "to eat, breathe, work, perspire, eliminate waste, grow, develop sexually, age and die." Thus, it allows experimentation without manipulating people or animals. Moral and ethical prohibitions are not involved; and with continued refinement of inputs into one of the worlds most sophisticated computer systems, it is certainly possible that by 1995 computerized models will predict the effects of every new chemical prior to its reaching commercial production.

Such a system will be extremely rapid, cost relatively little, require little space, and of course dispense with the hundreds of thousands of animals used annually.

It may just be capable of substituting for people and provide data that need no translation, safety factors, or risk assessments based on mathematical models (see Chapter 15).

At this moment it is a hope for the 21st century. Toxicity is then usually determined on both a short-term and a long-term basis. An estimate of the acute toxicity of a chemical may be obtained relatively rapidly (within a few days to a few weeks) by determining how much of the chemical in a single dose is necessary to kill 50% of the test animals. This is the LD_{50}. The purpose of the acute toxicity test is to obtain a first approximation of the inherent pharmacological character of the compound. If the proposed level of use of the substance approximates the lethal dose, the tests can be concluded and the chemical quickly disapproved.

Experiments to *estimate* the effects analogous to lifetime ingestion can be carried out over a period of 90 days. These subacute toxicity tests can serve as guides for selecting feeding levels for the chronic toxicity studies. Figures 7 and 8 demonstrate methods of introducing test substances into animals.

4. Transgenic Mammals

In April 1988, The U.S. Patent and Trademark Office issued patent No. 4,736,866 for "transgenic non-human mammals" to Philip Leder of

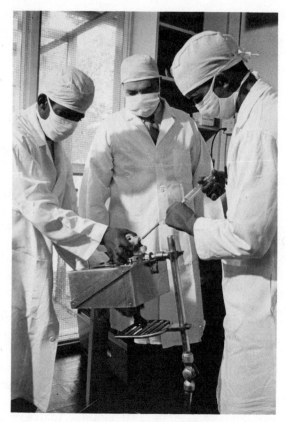

Figure 7. Insertion of a chemical additive. The monkey is in a restraining cage. FDA photo.

Harvard Medical School and Timothy A. Stewart of Generntech Inc., a biotechnology company in San Francisco. This was a historic event.

Leder and Stewart isolated a gene that induces cancer in a number of mammals and injected it into fertilized mouse eggs, thereby developing a new breed of mice. Because half the females will (normally) develop cancer in their lifetime, this genetically altered species can serve as an effective model for studying how genes contribute to the development of cancer. However, these mice can also be seen as a new, highly complex biological system for testing whether chemicals used in foods, for example, are toxic. According to Leder, "while half of the untreated transgenic animals will develop cancer within lifetimes, a greater proportion should develop a malignancy when treated with a carcinogen. The increased sensitivity of the reaction takes into account the fact that cancer is a multi-step process that may involve the recruitment or mutation of additional oncogenes" (16). Down the road then, a means of generating additional information about the safety or potentially deleterious effects of food additives should be available to scientists.

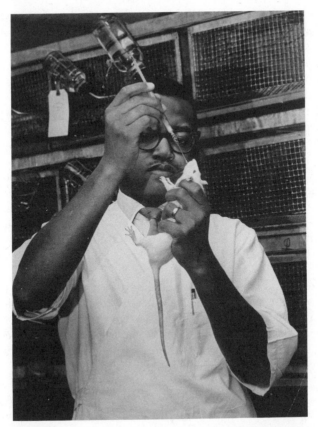

Figure 8. Testing of food additive in an albino rat. FDA photo.

The final decision as to whether a proposed chemical is safe for human consumption can be arrived at only after evaluation of several factors: the maximum dietary level that produced no untoward response in test species, experience with other chemicals of similar structure, estimation of the potential human consumption of all foods containing the additive, and perhaps most important of all, epidemiologic studies. Epidemiology can be defined as the study of the risk factors responsible for human illness. Because its subjects are people, species to species translations are unnecessary.

D. Epidemiologic Studies

Earlier, I noted that some 8700 different chemicals (additives) have been approved for use in foods. Most are unfamiliar to the general public, and the fact that they are largely unpronounceable increases the dislike. Adding to this the erroneous belief that synthetic chemicals must some-how be different from and more hazardous than natural chemicals, as well as the very idea of intentionally putting strange chemicals in foods, has produced an aversion to chemicals in a large number of people.

An aversion, a dislike for the concept, is understandable, but what of their safety? What can we glean from epidemiologic and toxicologic investigations over the past 20 years, especially with respect to cancer, the most feared disease in the United States today?

For the most part, exposure to cancer-inducing chemicals (oncogens) usually occurs 20–30 years prior to the appearance of frank clinical indications. Consequently, the cancers we see today are the result of exposures in the 1950s and 1960s.

With increasing knowledge, there is now convincing reason to believe that diet has indeed contributed to current cancer incidence and death. However, it must be clear that what has been found are associations between a suspected oncogen and cancer, not causal relationships. A major distinction exists between association and cause. That cautionary note must be borne in mind. Nevertheless, the relationship goes beyond suspicion.

In 1981, Doll and Peto (17) published the results of their penetrating epidemiologic evaluation of the relationship between environmental risk factors and cancer. This work was commissioned by the Office of Technology Assessment of the U.S. Congress. They concluded that diet could account for as little as 10% and as much as 70% of all cancers, and that diet was more than likely the most important risk factor for human cancer.

The ink hardly dried on their report, when the National Academy of Sciences report *Diet, Nutrition and Cancer* (18) was published. Two of their most prominent recommendations for preventing cancer in the future were the reduction of dietary fat from the current 40–45% to 30% of total caloric intake, and the need for "including fruits, vegetables and whole grain cereal products in the daily diet."

Evidence of the importance of diet as a factor in cancer initially came from epidemiologic studies of migrant populations. For example, the Japanese in Japan have lower levels of cancer of several anatomic sites than those Japanese who migrated to Hawaii. Japanese who migrated to California from Hawaii have far higher cancer rates than the Japanese in Hawaii or Japan. In California, their rates have approached those of the indigenous white population. Similar types of changes have been seen with eastern and southern Europeans who left their homelands for Canada, the United States, and Australia. More recent studies with Amerindians (19), whose cancer rates were far lower than the general U.S. population, show their cancer rates are now catching up. Certainly, genetic changes cannot account for the higher rates; nor can pollutants in air or water. Diet is a likely risk factor.

However, although diet and cancer appear related, food additives are neither a major nor a minor risk factor. According to Doll and Peto, food additives may have been protective. They estimated that additives contributed from less than 1% to −5%. Evidence suggests that additives may protect against the deleterious effects of oxygen radicals in tissue cells.

It does appear, opinion to the contrary not withstanding, that with respect to our food supply and the health of the American people, food additives have been a success story.

The women's liberation movement has propelled women from the kitchen to the executive suite. As a consequence, both men and women have little time or inclination for food preparation. Families and households of all descriptions depend more than ever on precooked, prepackaged, ready-to-eat foods and meals. In addition, "eating out" has become a necessary adjunct of contemporary living. Often eating out is less expensive than "home cooking." In 1988, the average person ate out 4.5 times per week. Clearly, the bulk of all meals currently eaten in the United States are not prepared from raw ingredients at home. For this to occur, a variety of added chemicals are necessary. This is yet another pathway for chemicals to enter the food supply.

Foods are among the most difficult products to package successfully. Such considerations as protection from microbial invasion, penetration by atmospheric oxygen, loss of volatile substances through evaporation, and choice of a container that will not itself adversely affect the food are important if the item is to come to the consumer in an edible state the year round. Chemicals help make this possible.

Synthetic chemicals may be incorporated into a food item during its growth or during its storage and processing. Such substances are described as *additives*. It is generally understood that an additive is any substance used in food, directly or indirectly, that affects its characteristics.

After much debate, Congress arrived at a definition of food additives which includes all substances whose intended use can reasonably be expected to result in their becoming (directly or indirectly) components of food.* By this definition, Section 409 applies not only to direct additives but also to indirect ones such as packaging materials whose components may migrate into food or animal drugs that may leave detectable residues in food-producing animals. The term "food additive" is not limited to synthetic chemicals but also includes agricultural products. Thus, a potato, for example, would be an additive when it is one of several ingredients of a canned stew. Even though a potato would be Generally Recognized As Safe [GRAS, Section 201(s)], it would also come under Section 409.

* In Congressional language, "a food additive means any substance the intended use of which results or may reasonably be expected to result, directly or indirectly, in its becoming a component of or otherwise affecting the characteristics of any food, if such substance is not generally recognized, among experts qualified by scientific training and experience to evaluate its safety, as having been adequately shown by scientific procedures to be safe under the conditions of its intended use."

When introduced to preserve or improve the quality of a product it is known as an *intentional* additive because it is purposely added to serve a specific purpose. Such materials as artificial coloring, synthetic flavors, sweeteners, vitamins, microbial inhibitors, antioxidants, emulsifiers, and minerals are all intentional additives. All are added during preparation to the food product in controlled amounts. Common table salt, sodium chloride, is one of the oldest intentional additives known; aspartyl phenylalanine is one of the newest; and cyclopentanoperhydrophenanthrene,* used to enhance the nutritive quality of milk or bread, is simply the chemist's name for vitamin D! Chemicals in foods are not new; what is new is the means of adding the correctly measured chemical in pure form, which results in a uniform product.

When primitive humans first learned to preserve a portion of their food supply by smoking it in a fire, they were utilizing chemicals unknowingly. It was (and still is) the production of formaldehyde, phenol, and related compounds, coupled with the drying action of heat, that retarded microbial spoilage. Chemicals in the wood smoke also impart unique and desirable flavors.

New discoveries, the result of nutritional investigations during the past 50 years, have measurably enhanced our food supply. Most of us are aware of the importance to general health of vitamins, minerals, proteins, fats, and carbohydrates. Many of these chemicals can be manufactured so that foods can be "fortified" or "enriched." Potassium iodide, for example, added to common table salt, is almost solely responsible for the near elimination of simple goiter. Enrichment of bread and milk with chemicals such as the B vitamins (thiamin, niacin, and riboflavin) has almost eliminated pellagra from the United States. The minerals, vitamins, and other synthetic nutrients so common in many foods today—the list would read like an inventory of a chemist's laboratory—have undoubtedly contributed to the robust health of our population. Use of chemicals in foods, combined with other forms of preservation, permits stockpiling to cope with emergencies. Food technologists and chemists are in fact modern-day Josephs, evening out the fat and the lean years.

Although some foods keep better than others, all are subject to microbial attack. Because of their very low moisture content cereal grains (wheat, barley, rye), nuts, and seeds, for example, require little intervention for their preservation. Such items as potatoes, carrots, flour, dried fruits, and butter, because of their relatively low moisture content, have a fairly long storage life. Foods with high moisture content, such as meats, fish, and poultry, have poor keeping characteristics and thus require effective means of preservation if they are to survive microbial onslaughts. Of the many methods of food preservation (heating, freezing,

---

* Cyclopentanoperhydrophenanthrene is actually the skeletal framework for a group of compounds such as steroids, hormones, natural pigments, and vitamins. Each of these differs by specific additions of carbon and hydrogen atoms to the basic skeleton.

dehydration, smoking, pickling) chemical methods assume increased importance.

Certain chemicals are used to color foods; others add flavor; some impart firmness, while others soften; still others thicken, moisten, dry, or acidify, while others retard or hasten chemical reactions. Recall that what are called "foods" are in reality plant and animal tissues composed of proteins, carbohydrates, fats, minerals, water, and a host of other organic and inorganic substances, all of which are chemicals that can enter into many chemical reactions. The browning of sliced apples, peaches, and potatoes is an example of the chemical change that occurs when the oxygen in the air contacts the naturally present chemicals released by the cut tissue. To prevent or retard these chemical reactions, such chemicals as sodium bisulfite or ascorbic acid (vitamin C) are added to processed foods.

Approximately 8700 substances have been certified for use in foods. It is appropriate to ask why so many are needed. The use of these additives can be justified by serving several essential purposes:

- Maintaining a food's nutritional quality.
- Enhancing keeping quality and/or stability resulting in decreased losses.
- Enhancing acceptability without deception.
- Facilitation of food preparation and processing.

1. Sweeteners

Low-calorie, nonnutritive sweeteners, among the most thoroughly studied of all food additives, are being reevaluated in the light of new research.

While working at Johns Hopkins University during the summer of 1879, Constantin Fahlberg, a German organic chemist, accidentally discovered saccharin. In the course of oxidizing derivatives of toluene, some of the material spattered from a flask onto his fingers. Contrary to good laboratory practice, he licked them and was surprised by the sweetness. The Princess of Serendip could not have been more delighted. Fahlberg had discovered saccharin. The controversy surrounding it, continues unabated.

Nevertheless, Americans have been consuming saccharin since 1900, when it first became commercially available. Ingestion on a grand scale began some 25 years ago, during the late 1950s and early 1960s, with the remarkable popularity of bottled soft drinks.

In 1972, responding to published reports suggesting that saccharin produced bladder tumors in rats, the FDA removed it from the GRAS* list

* Generally Recognized As Safe—GRAS.

of chemicals. Removal from this list meant that saccharin was now subject to the Delaney Clause of The Food, Drug and Cosmetic Act. That clause states in part that "no additive shall be deemed to be safe if it is found to induce cancer when ingested by man or animal, or if it is found, after tests which are appropriate for the evaluation of the safety of food additives, to induce cancer in man or animal."

In 1977, a study performed by the Food and Drug Directorate of Canada found a similar relationship. The FDA moved to ban saccharin under the requirements of the Delaney Clause. Given the controversial aspects of these studies, and the public desire for a noncaloric sweetener, Congress passed the Saccharin Study and Labeling Act before the ban could go into effect. The moratorium has since been extended three times and was due to expire on April 22, 1985. An additional extension is clearly in sight. Studies have now confirmed that saccharin is a weak carcinogen with a very small risk.

On the basis of data produced by the International Research and Development Corporation, mathematical models suggest that exposure at a level of 0.01% of the total diet (approximately 2.3 12-oz. cans of diet soda per day over a lifetime) could be expected to increase the risk of developing bladder cancer by about one in two million.

By contrast, cyclamate, once the most feared of sweeteners, and banned as a carcinogen in 1970, may shortly be reinstated. In 1937, Michael Sveda, a graduate student at the University of Illinois, was investigating the ability of sulfamide derivatives to inhibit bacterial activity. Legend has it that, being an inveterate smoker, he brushed a ribbon of tobacco from his lips and tasted the sweetness of the chemicals with which he was working. Sveda discovered cyclohexylsulfamic acid. Sucaryl (calcium cyclamate) became commercially available in 1950.

Cyclamates were considered safe until tests revealed gross abnormalities in chick embryos that developed from eggs into which cyclamates had been injected. In other experiments, rats fed massive doses of cyclamates developed bladder cancers. On October 21, 1969, the FDA ordered all production of general-purpose products containing cyclamates to cease. It also required that by January 1, 1970, all cyclamate-containing beverages be removed from markets. Although evidence of human bladder cancer was not involved nor were data from studies with rats translatable to human cancer, the FDA acted prudently on the authority of the Delaney Clause.

However, cancer is not the only ill effect ascribed to cyclamate. Both the FDA and the NRC will be required to evaluate evidence indicating the production of chromosome breakage with consequent mutagenic hazard, as well as the induction of testicular atrophy in laboratory animals.

If ever a chemical received thorough testing prior to approval, aspartame would surely be its name. Aspartame, the official chemical designation and major ingredient in Equal and NutraSweet, the artificial sweeteners produced by G. D. Searle & Co., is the methylester of aspartylphenylalanine. While a synthetic linking of the amino acids phenylalanine and aspartic acid, both occur naturally in a variety of plants and animals.

Aspartame's compelling features are its natural origins coupled with the palatable fact of little or no unpleasant bitter aftertaste. It is, however, far more expensive than the other nonnutritive sweeteners and tends to deteriorate on prolonged storage or when heated, sharply limiting its range of uses.

In 1974, the FDA approved aspartame for use but stayed the regulation because of objections and concern over its safety. After further analysis of the data, the stay was lifted and aspartame was finally approved in 1983.

In a recent review of safety issues, the American Medical Association's Council on Scientific Issues noted that evidence indicates that consumption of aspartame is safe except by individuals with homozygous phenylketonuria or other individuals needing to control their phenylalanine intake (20).

Responding to concern about headaches after aspartame consumption, Schiffman and co-workers at Duke University Medical Center designed and conducted a randomized, double-blind, placebo-controlled crossover study of people who claimed headaches within 24 h of eating aspartame-containing products. They concluded that "aspartame was no more likely to produce headache than a placebo" (21).

In a decision that could effectively end the debate over the safety of aspartame, the U.S. Court of Appeals for the District of Columbia said in September 1985 that consumer groups that had questioned aspartame's safety failed to show it was not adequately tested by the FDA. According to Judge Abner Mikva, "the FDA could properly exercise its discretion," as far as ascertaining safety, "as long as the total amount ingested per day did not exceed the level shown to be safe."

Perhaps the most unabashedly outrageous statement about artificial sweeteners was recently enunciated by President Reagan. In response to a question by a student at the University of Tennessee, the President said he had recently quit using sugar substitutes because "we don't know what's in them." Is there any wonder that people remain confused?

Currently, most diet drinks contain a blend of saccharin and aspartame with saccharin being the primary sweetener. Apparently, this blend is sweeter than the sum of the two. This combination enables producers to combine the cheapness and heat stability of saccharin with the more acceptable taste and safer reputation of aspartame.

None of these sweeteners has sugar's ability to impart density or "mouthfeel" to soft drinks. Consequently, bodying agents must be added to beverages made from synthetics to avoid a thin or "watery" taste. To obtain the appropriate density, one of a variety of gums, such as cellulose, carrageenan, arabic, algin, or carboxymethyl cellulose, are used. Another reason for additives.

2. A New Addition, Sunette/Acesulfame K

Considering that the nonnutritive sweetener market is estimated at a billion dollars annually, it is passing strange that in so competitive an economy as ours, aspartame alone has challenged saccharin's hegemony.

Before 1989 gets underway, Sunette—known also by its generic name acesulfame K—pronounced a-see-sul-fame-kay—developed in West Germany by Hoechst AG, will be available in tablet and powder form. Developed in 1967 and introduced into England in 1983, Sunette has had a long period of testing outside the United States. Nevertheless, approval by the U.S. FDA required 6 years.

Sunette is a white, odorless crystalline powder, with an intensely sweet taste. Compared to a 3–4% sucrose solution, the international standard of sweetness intensity, it is 180–200 times sweeter: about equal to aspartame and half as sweet as saccharin. As with other sweeteners, its sweetness decreases with increasing concentration. The relationship is shown in Figure 9.

Chemically, Sunette is the potassium (the "K" in acesulfame K) salt of 6-methyl-1,2,3-oxathiazine-4(3H)one-2,2-dioxide, or 3,4-dihydro-6-methyl-1,2,3-oxathiazine-4-one-2,2-dioxide, and can be considered a derivative of acetoacetic acid. Empirically, it is $C_4H_4NO_4KS$. Figure 10 shows its molecular structure.

In toxicologic animal-feeding studies, Sunette was found to be excreted intact. It is neither metabolized nor converted to other forms. Consequently, it is noncaloric. And given its lack of carcinogenicity, WHO/FAO found it safe for use in foods and recommended an acceptable daily intake (ADI) of 0–9 mg/kg. An ADI of 9 mg/kg of body weight corresponds to the consumption of 80–125 g of sugar per day by a person weighing 70 kg

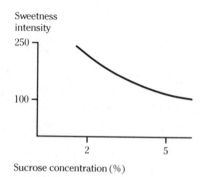

Sucrose concentration (%)

Figure 9. Sweetness intensity curve.

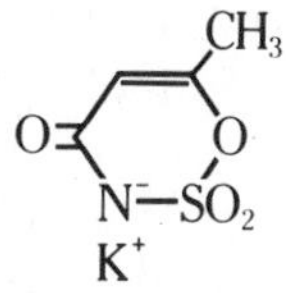

Figure 10. Sunette's molecular structure.

(approximately 155 lb). In addition, it is not metabolized by bacteria in the mouth; thus, it cannot contribute to caries formation.

Sunette manifests synergistic effects with the nutritive sweeteners sorbitol and fructose. It exhibits little synergism with saccharin but shows a pronounced increase with aspartame—as much as 30%. Considering that blends of sweeteners are often more desirable than a single sweetener, in terms of quality of sweetness, its optimum use may be in combination with others.

Current FDA approval permits its use in dry mixes for beverages, instant coffee and tea, gelatins, puddings, nondairy creamers, and chewing gum.

3. Sulfites

Sulfites are another widely used additive. They have been added to foods for centuries. The earliest known use was the preservation of wine by sulfur dioxide, an antibacterial agent. The primary use of sulfites has been as an antioxidant during food processing to prevent browning of sliced apples and instant potatoes. Sulfites can exert a general preservative effect by protecting against losses of carotene and ascorbic acid in a wide variety of foods.

Dehydrated apples, apricots, raisins, pears, and peaches usually contain the highest amounts of 1000–3000 ppm, on a dry-weight basis. Dehydrated soups and vegetables can contain from 200 to 2000 ppm. Although estimates of total human intake are difficult to determine, approximations are in the range of 1.5–3 mg (ppm) of SO_2 per person per day. The FAO/WHO ADI for sulfite is equivalent to 50 mg.

Since 1959, the six sulfating agents sulfur dioxide, sodium sulfite, sodium and potassium bisulfite, and sodium and potassium metabisulfate have been listed by the FDA as GRAS, that is, for use in foods.

With the advent and widespread use of salad bars, a previously unrecognized reaction to sulfite appeared. Given the sulfite characteristic of preventing or forestalling the browning of leafy and other vegetables, restauranteurs had taken to the liberal sprinkling of sulfite on aged products rather than discarding them.

Some 100 instances of adverse reactions to sulfites, including one death, have been reported to the FDA. Nausea, diarrhea, and severe allergic reactions (respiratory distress, skin eruptions, abdominal cramps) have not been unusual, especially as the majority of these cases, but not all, occurred in asthmatic individuals. More than likely these cases represent only the "tip of the iceberg," only a fraction of those actually occurring.

The FDA response has been to require all interstate carriers, buses, trains, and planes, to clearly notify passengers if sulfites have been used in catered foods which will be eaten raw. In their monitoring of food establishments, many state and local departments of health have been requiring the conspicuous posting of signs warning of sulfites added to raw foods. Given the use of sulfites as a "cover-up," little comfort can be taken that such advertising will in fact occur. Complete banning of the use of sulfite is in the consumer's best interest. After 2 years of mulling, the FDA decided (June 1988) that broad restrictions on the use of sulfites in foods was unwarrented. They estimated that sulfites are dangerous to 8–10% of the 10 million people with asthma, and to a smaller number of nonasthmatics. They would have to work out means for their individual protection.

A separate action to extend a 1986 ban on the use of sulfites in fresh fruits and vegetables, salad bar items, remains pending.

4. Colorants

Before biting into any food we almost unconsciously note its color to judge whether it falls within an acceptable range. Instantaneously, the brain returns the message—eat or reject.

When in a test bread and rolls were baked from flour containing a substantial amount of algal protein, the resulting green loaves and rolls were flatly rejected by test participants. Green bread is too far removed from our image of "breadness," which has a well-defined color characteristic. Similarly, yellow meat or blue potatoes would find few takers.

Coloring agents (colorants) are added to a host of foods to bring them to a familiar color range. However, every color used in food must be certified by the FDA prior to use. The undeclared addition of any color is a flagrant violation of both state and federal laws. In 1960, Congress placed natural as well as synthetic colors under a single law, Section 6 of the Food, Drug and Cosmetic Act. Prior to that, only synthetic colors had been subject to a federal program of pretesting and certification of individual batches of dye. Nevertheless, a chemical is still a chemical.

Cochineal, a rich crimson or scarlet dye widely used in confectionary, baked goods, beverages, and spices, is still obtained from coccus cactil, an insect, primarily females, which lives on the Opuntia cactus in the Spanish Canary Islands and South America. Although the dye is obtained by squeezing the insect's abdomen, it is actually the alcohol extracted (carminic acid-methyl tetra-O-methyl carminate) that is responsible for the coloring quality.

The natural versus synthetic wrangle has its humorous and baffling side. For example, the chemical that gives tomatoes and carrots their natural red-orange color is carotene. Carotene is of course widely distributed in nature. Not too many years ago, chemists succeeded in duplicating carotene's molecular structure in the laboratory. The natural and currently synthetic are so alike, much like single ovum twins, that they cannot be told apart. In the United States, if a food processor wants to add carotene to a food but prefers to use the synthetic twin, which is cheaper, more readily available, and always of the same potency, the FDA requires that the food label carry the designation "artificial color." Across the border in Canada, the same colorant is labeled "natural color."

As noted earlier, under the Delaney Clause of the Food, Drug and Cosmetic Act, any ingredient that produces cancer in laboratory animals, in any amount, cannot be used in food. The ban is absolute. As a consequence of this stringent criterion, only eight synthetic colors have been certified for use in food. Table 9 lists the colors along with their

Table 9
Current Status of FD&C Certified Colors

Color	Status		Provisional Listing Expires	Comments
	Dye	Lake		
Blue 1	Permanent[a,b]	Provisional	Indefinite	
Blue 2	Provisional	Provisional	Indefinite, pending hearing	Order permanently listing the dye was stayed. Hearing requested by HRG held Apr. 16–27, 1985.
Green 3	Permanent[b]	Provisional	Indefinite	
Yellow 5	Permanent[a]	Provisional	Oct. 2, 1984[d]	
Red 3	Permanent	Provisional	Oct. 2, 1984[d]	Await results of three current studies in process.
Yellow 10	Permanent	Provisional	Indefinite	Not currently approved for food use. CCMA plans future petition.
Yellow 6	Provisional	Provisional	Oct. 2, 1984[d]	
Red 40	Permanent[c]	Permanent[c]	—	

[a] Permanently listed 1969.
[b] Permanently listed or listing reaffirmed following review of recent chronic studies.
[c] Permanently listed because not in use prior to 1960.
[d] Extension of provisional period anticipated.

Table 10
Natural Coloring Ingredients

Annatto	Iron oxide
Beet powder (dehydrated)	Paprika and oleoresin
Beta-apo-8-carotenol	Riboflavin
Carthaxanthin	Saffron
Carminic acid	Tagetes (Aztec marigold) meal and extract
Corn endosperm oil	Titanium
Carrot oil	Toasted cottonseed flour
Ferrous gluconate	Tumeric
Grape skin extract	Vegetable juice

permanent or provisional status. Is it possible for eight colors to satisfy the needs of the thousands of commercially available foods? Hardly. Obviously more than eight colors are used. Most are drawn from the more than 20-odd natural extracts that have intense coloring ability and are generally recognized as safe (GRAS). See Table 10.

E. Additive Safety

Until the enactment of the Food Additives Amendment of 1958, the burden of proving that a substance was unsafe was upon the FDA. The amendment shifted the responsibility to the more appropriate sector, the manufacturer. It would henceforth be their responsibility to show and convince the FDA that a proposed ingredient or chemical was in fact safe. In addition, the amendment established in Section 409 a premarketing approval process. This meant that prior to use, a substance for use in food would have to be shown to be safe.

1. Fraudulent Cover Up

Food additives do not include chemicals added by unscrupulous businesspeople to conceal off odors and tastes, or to enhance taste and/or odor where lacking.

An example of an attempt to enhance taste and odor, and thereby avoid dumping an unfit product, recently created a scandal of international proportions. In July 1985, West German authorities revealed that wine imported from Austria was contaminated with diethylene glycol, a chemical normally used in automobile antifreeze. Some wines contained lethal levels. Diethylene glycol is known to induce nerve, brain, and kidney damage in people who have consumed it at various levels.

Apparently, a large group of Austrian wine producers had surreptitiously added the antifreeze to their wines to increase sweetness, which was lacking in the natural product. By the end of July, the wines had also been found in Japan and the United States. Arrest warrants for ten people accused of adding the unauthorized chemical had been issued.

Under great international pressure, the Austrian Minister of Agriculture, Guenther Haiden, indicated that his ministry was preparing new rules for testing and labeling wine. "The Austrian wine law will be the strictest in the world," he said. There would be a ban on adding sweetener of any kind.

It is unfortunate that we remain at the mercy of charlatans. The world remains an imperfect place where vigilence must continually be practiced. Nevertheless, these unlawful *additions* are not *additives*.

The trial of more than 60 jailed defendants began October 16, 1985. They face up to $50,000 in fines for violating the existing wine law, and possibly 5–15-year jail terms for fraud.

F. Hyperactivity in Children

Food additives have been accused of initiating allergies and hyperactivity in children. In 1975, Benjamin Feingold, a pediatrician, fired the first salvo. His book, *Why Is Your Child Hyperactive?*, categorically stated that food colorings, flavorings, and salicylates produce hyperactivity as well as learning disabilities in children. The battle was joined. Parents bought his book and went to great lengths to prepare food free of the offensive chemicals.

At the same time, the scientific community, discomforted by the data Feingold offered, as well as the cause–effect relationship claimed for the diet and its mollifying effects on hyperactivity, mounted studies to test the proposition. During the period 1976–1983, a number of studies were published indicating that the Feingold diet was of little value, that other family factors were a more likely cause. Unfortunately, the studies lacked sharp end-points. Evaluation of hyperactivity depended on subjective responses of both parents and teachers, both with vested interests. A randomized double-blind trial was needed. It was provided in 1984 by Judith L. Rapoport, Chief of Child Psychiatry at the National Institute of Mental Health.

"Parents," she noted, "have been telling us for years that sugar upsets their children, so we decided to demonstrate in a double-blind fashion whether or not their perceptions were correct." In a double-blind study, interviewers or others responsible for determining behavior, do not know whether a child is in the test group (receiving the substance under study) or control group (receiving a placebo).

Her study on sugar sensitivity involved testing children whose parents responded to an ad calling for grade-school children with a history of adverse reactions to sugar consumption. Many parents claimed their children were hyperactive, temperamental, easily upset, and argumentative because of their consumption of sugar. These were children who habitually loaded their cereal with sugar and often ate large amounts each day.

In the double-blind sugar challenge study, the 21 prepubertal boys enrolled were given 1.75 g of glucose, sucrose, or placebo* in random succession on separate occasions. Rapoport's conclusion was that there was no evidence of any behavioral effects from sugar. Results of glucose tolerance tests were also normal.

A caffeine challenge was also undertaken. Forty-one 6–12-year-olds were selected from among 800 school children. The test group consisted of 19 whose daily diets included more than 300 mg of caffeine, equivalent to six cans of soft drink and usually considered sufficient to induce "jitters." In the control group were 22 whose diet was almost caffeine-free and with no signs of hyperactivity.

This study was a double-blind crossover challenge with a 2-week high-caffeine diet and a 2-week low-caffeine diet for all participants. The children were observed for 4 weeks by both parents and teachers.

Those in the control group experienced adverse effects from 2 weeks of a high-caffeine diet. But those who ordinarily consumed a high-caffeine diet experienced no ill effects from 300 mg daily. And on a caffeine-free diet they became more active rather than less. Clearly, these challenges with sugar and caffeine fail to support a link between food and behavior.

Childhood hyperactivity is currently referred to as attention-deficit hyperactivity disorder and may affect as many as one in five school children. However, on a worldwide basis, 3% seems to be a more realistic number. It also appears to be the single most prevalent disability reported by elementary schools.

Today, research interests have shifted from food additives to genetic influences, as well as indications that reduced concentrations of homovanillic acid in the spinal fluid of affected children suggest abnormalities in central dopaminergic systems. Danish researchers have reported that positron emission tomography (PET) scans of the brain of ADHD children showed metabolic differences in areas of the brain rich in dopamine (22).

The 15 years spent chasing after food additives may not have been wasted. Much has been learned about their chemistry and activity that helps establish them as safe for human consumption.

Speaking to the New York Academy of Science's Conference on Public Control of Environmental Health Hazards, held in New York in 1978, Peter Barton Hutt, former General Council for the Food & Drug Administration noted that "it is undoubtedly true, moreover, that these risks† are much lower than the risks faced by our ancestors. The risks faced by the first settlers of this country in crossing the Atlantic, or in enduring winter hardship, or in exploring the wilderness, or indeed in the food supply, were far greater than those that any of us face today. Modern technology has reduced our individual and societal risks, not increased them."

---

* In effect, a placebo does not contribute to the outcome of a test. It is a dummy pill or substance made up of inert constituents such as water or milk.
† The risks he referred to included such common everyday events as crossing the street, driving, and mountain climbing.

REFERENCES

1. B. O. Osuntokun, Cassava Diet, Chronic Cyanide Intoxication and Neuropathy in Nigerian Africans. *World Rev. Nutr. Diet.* **36:** 141–173, 1981.

2. Ministry of Health, Mozambique. Mantakassa: An Epidemic of Spastic Paraparesis Associated with Chronic Intoxication in a Cassava Staple Area of Mozambique. 1. Epidemiological and Clinical and Laboratory Findings in Patients. *Bull. WHO* **63:**477–484, 1984.

3. P. S. Spencer, P. B. Nunn, J. Hugou, A. C. Ludolph, S. M. Ross, D. N. Roy, and R. C. Robertson, Guam Amyotrophic Lateral Sclerosis–Parkinsonism Dementia Linked to a Plant Excitant Neurotoxin. *Science* **237:** 517–522, 1987.

4. P. S. Spencer, Guam ALS/Parkinsonism-Dementia: A Long-Latency Neurotoxic Disorder Caused by "Slow Toxins" in Food? *Can. J. Neurol. Sci.* **14:**347–357, 1987.

5. B. N. Ames, Paleolithic Diet, Evolution and Carcinogens. *Science* **238:** 1634, 1987.

6. J. Sander, Nitrosaminosynthese durch Bakterien. *Hoppe-Seyler's Z. Physiol. Chem.* **349:** 429–432, 1968.

7. G. M. Hawksworth and M. J. Hill, Bacteria and the N-nitrosation of Secondary Amines. *Br. J. Cancer* **25:** 520–526, 1971.

8. *The Health Effects of Nitrate, Nitrite, and N-nitroso Compounds.* Committee on Nitrite and Alternative Curing Agents in Food, Assembly of Life Sciences, National Academy of Science. National Academy Press, Washington, DC, 1981.

9. D. Forman, S. Al-Duggagh, and R. Doll, Nitrates, Nitrites and Gastric Cancer in Great Britain. *Nature (London)* **313:**620–625, 1985.

10. Compliance Program Report of Findings FY79 Total Diet Studies—Adult. U.S. Food and Drug Administration, Washington, DC, 28 June, 1982. U.S. Department of Commerce, National Technical Information Service PB83-112722.

11. G. Vettorazzi, Advances in Safety Evaluation of Food Additives. *Food Add. Contaminants* **4**(4): 331–356, 1987.

12. Global Pollution and Health Sentinel, *Health & Environ. Int.* **4**(2): 6, 1987.

13. S. Holmberg, M. T. Osterholm, K. A. Singer, and M. L. Cohen, Drug Resistant Salmonella from Animals Fed Antimicrobials. *N. Engl. J. Med.* **311**(10): 617–622, 1984.

14. T. H. Jukes, Diethylstilbestrol in Beef Production: What Is the Risk to Consumers. *Prev. Med.* **5:** 438–453, 1976.

15. G. Löfroth and X. Gejvall, Diethyl Pyrocarbamate: Formation of Urethane in Treated Beverages. *Science* **174:** 1278, 1971.

16. P. Leder, Transgenic Non-human Mammals. Personnal communication. Harvard University, School of Medicine, April 29, 1988.

17. R. Doll and R. Peto, The Causes of Cancer: Quantitative Estimates of Avoidable Risks of Cancer in the U.S. Today. *J. Natl. Cancer Inst.* **66:** 1191–1308, 1981.

18. U.S. National Academy of Sciences, *Diet, Nutrition and Cancer.* National Academy Press, Washington, DC, 1982.

19. T. K. Young and T. W. Frank, Cancer Surveillance in a Remote Indian Population in Northwestern Ontario. *Am. J. Public Health* **73**(5): 565–520, 1983.

20. Council on Scientific Affairs, Aspartame: Review of Safety Issues. *J. Am. Med. Assoc.* **254**(3): 400–402, 1985.

21. S. S. Schiffman, C. E. Buckley, H. A. Sampson, E. W. Massey, J. N. Baraniuk, J. V. Follett, and Z. S. Warwick, Aspartame and Susceptibility to Headache. *N. Engl. J. Med.* **317**(19): 1181–1185, 1987.

22. V. S. Cowart, Attention Deficit Hyperactivity Disorders: *J. Am. Med. Assoc.* **259**(18): 2647–2652, 1988.

SUGGESTED READINGS

Alternatives to the Current Use of Nitrite in Food. Part 2 of a 2-Part Study. Assembly of Life Science, NAS/NRC, Washington, DC, 1982.

Ames, B. N. Dietary Carcinogens and Anticarcinogens. *Science* **221:** 1256–1264, 1983; **224:** 659–6701, 757–760, 1984.

Benarde, M. A. *The Chemicals We Eat.* McGraw-Hill, New York, 1975.

Byers, T., and Graham, S. The Epidemiology of Diet and Cancer. *Adv. Cancer Res.* **41:**1–69, 1984.

Cohan, L. A. Diet and Cancer. *Sci. Am.* **257**(5): 42–48, 1987.

Conway, R. G., and Pretty, J. N. Fertilizer Risks in the Developing Countries. *Nature* **334:** 207–208, 1988.

Crosby, N. T. Nitrosamines: A Review of Their Chemistry, Biological Properties, and Occurrence in the Environment. In *CRC Handbook of Naturally Occurring Food Toxicants,* CRC Press, Boca Raton, FL, 1981.

Fennema, O. R. Food Additives—An Unending Controversy. *Am. J. Clin. Nutr.* **46:** 201–203, 1987.

Fraser, P., Chilvers, C. Beral, V., and Hill, J. J. Nitrate and Human Cancer: A Review of the Evidence. *Int. J. Epidemiol.* **9**(1): 3–9, 1980.

Kinlin, J. J. Meat and Fat Consumption and Cancer Mortality: A Study of Strict Religious Orders in Britain. *Lancet* **1**(8278): 946–949, 1982.

Tannenbaum, S. R. A Policy Perspective on Safety: Nitrite and Nitrate. Public Policy Planning Department, Hoffmann–LaRoche Inc., Nutley, NJ, 1984.

Toxicants Occurring Naturally in Foods, 2nd ed. National Academy of Sciences, Washington DC, 1973. This is a general overview of the entire range of naturally occurring chemicals.

Willett, W. C., and MacMahon, B. Diet and Cancer: An Overview. 2 Parts. *N. Engl. J. Med.* **310**(10): 633–637, 1984; **310**(11): 697–670, 1984.

4

Zoonoses

*As a rule we disbelieve all facts for which we have
no use.*

—William James

I. HUMAN–ANIMAL RELATIONSHIPS

Twenty-four-year-old Helen Wilson* was hospitalized in Rochester, New
York, with encephalomyelomeningitis. Her 16-year-old brother had painful orchitis, and her mother and father experienced severe muscle pain,
fever, and headache. The painful myalgia encompassed their necks,
shoulders, backs, and legs, and especially their eyes each time they
moved.

Just before Christmas, the family had purchased a hamster. By late
January, a total of 19 cases of lymphocytic choriomeningitis (LCM) had
occurred in Rochester. By April, a total of 57 LCM cases had been
recorded. All were associated with pet hamsters (1). Five hundred
hamsters appeared to have been sold between December and January. If
this number is representative of the 100,000 hamsters sold nationwide
during that same period, some 4000 cases may well have occurred.

Over a period of 3 months, two physicians in the Orlando, Florida area
became extremely frustrated with their inability to eliminate recurrent
pharyngitis in a family of four: mother, father, and two young children.
Each time they were treated by antibiotic for the complex of sore throat,
fever, beefy red pharynx, and tonsillar exudate. The cycle of infection
seemed to be broken, only to reappear within 2–3 weeks.

---

* Not her real name.

From persistent questioning, it was discovered that the family had a pet dog. Brought to the physician's office, the animal was microbiologically cultured and found to be positive for group A beta-hemolytic streptococci (2). Treating the dog with antibiotic cured the family's pharyngitis, and cured the dog of his carrier state.

Animals have long been responsible for human illness. Whether the reverse is true is not known with any degree of certainty.

Animals and humans have lived in close association for thousands of years. People have used animals for food, clothing, and shelter and have taken animals into their homes to warn of danger, control rodents, and serve as family companions. The friendship of people with animals, especially dogs and horses, has been immortalized in song and story. The breeding and selling of domestic animals and the commercial processing of animal products are fundamental to the economy of many countries.

Nevertheless, it is through these close associations that we expose ourselves to their diseases. When animals are brought into living quarters, exposure becomes even more thorough. Members of the household play, hug, kiss, and occasionally share their food and bed. As increased leisure time permits more people to camp out, additional contact with wild animals is made, and as we push our communities farther into sylvan areas, still more contacts are made. Living in close proximity, it is not surprising that people contract illnesses from the many surrounding animals.

The animal world must be considered a constant reservoir of a variety of potentially infectious diseases. Tabe 1 lists seven occupational categories whose workers have frequent contact with animals and who are therefore at increased risk of contracting specific diseases. Although Table

Table 1

Occupational Groups at Risk of Acquiring Specific Zoonotic Diseases

Group	Infections
Agricultural	Anthrax, brucellosis, campylobacteriosis, erysipeloid, glanders, leptospirosis, meliodosis, salmonellosis, tetanus, tick-borne relapsing fever, tuberculosis, tularemia, yersiniosis.
	North Asian tick-borne rickettsiosis, psittacosis/ornithosis, Q fever, Rocky Mountain spotted fever, scrub typhus.
	Argentine haemorrhagic fever, Bolivian haemorrhagic fever, bovine papular stomatitis, California, encephalitis, Colorado tick fever, contagious ecthyma, cowpox, European tick-borne encephalitis, Japanese encephalitis, Korean haemorrhagic fever, Kyasanur forest disease, louping ill, lymphocytic choriomeningitis, Murray Valley encephalitis, Newcastle disease, pseudocowpox, rabies, Rift Valley fever, Russian spring–summer encephalitis, St. Louis

Table 1 (*Continued*)

Group	Infections
	encephalitis, vaccinia virus infection, vesicular stomatitis (=40).
Animal product manufacture	Anthrax, brucellosis, campylobacteriosis, erysipeloid, glanders, leptospirosis, salmonellosis, tetanus, tick-borne relapsing fever, tuberculosis, tularemia, yersiniosis. Psittacosis/ornithosis, Q fever. Bovine papular stomatitis, contagious ecthyma, cowpox, louping ill, Newcastle disease, pseudocowpox, Rift Valley fever, vaccinia virus infection (=22).
Sylvan and campestral	Brucellosis, campylobacteriosis, erysipeloid, glanders, leptospirosis, pasteurellosis, plague, salmonellosis, tetanus, vibriosis, yersiniosis. North Asian tick-borne rickettsiosis, psittacosis/ornithosis, Rocky Mountain spotted fever, scrub typhus. Bolivian haemorrhagic fever, California encephalitis, Crimean–Congo haemorrhagic fever, European tick-borne encephalitis, Japanese encephalitis, Korean haemorrhagic fever, Russian spring–summer encephalitis, yellow fever (=23).
Recreational	Campylobacteriosis, glanders, leptospirosis, pasteurellosis, plague, salmonellosis, tetanus, tick-borne relapsing fever, tuberculosis, yersiniosis. Psittacosis/ornithosis. Herpes simiae infection, lymphocyctic choriomeningitis, Newcastle disease, rabies (=15).
Laboratory	Anthrax, brucellosis, campylobacteriosis, glanders, leptospirosis, plague, salmonellosis, tetanus, tuberculosis, tularemia, yersiniosis. Psittacosis/ornithosis, Q fever. Argentine haemorrhagic fever, Bolivian haemorrhagic fever, European tick-borne encephalitis, herpes simiae infection, Japanese encephalitis, Lassa fever, lymphocytic choriomeningitis, Marburg/Ebola infection, Newcastle disease, Omsk haemorrhagic fever, Rift Valley fever, Russian spring–summer encephalitis, vaccinia virus infection, vesicular stomatitis, yellow fever (=28).
Epidemiological/ clinical	Campylobacteriosis, salmonellosis. Psittacosis/ornithosis. Japanese encephalitis, Lassa fever, rabies, yellow fever (=7).
Emergency	Plague. Flea-borne typhus fever. Korean haemorrhagic fever, rabies (=4).

Source: Reproduced by permission of the World Health Organization.

2 lists the same seven major categories, it also provides a detailed breakdown of the types of professional at risk.

Veterinarians and other public health scientists refer to diseases of animals transmissible to humans as zoonoses. Close to 200 are known. Table 3 lists a selected number as well as their known routes of transmission, and Table 4 groups them by type of causal agent.

Not only has human welfare been endangered by epidemics of animal diseases, but development of many areas has been restricted by our inability to control certain zoonoses. A dramatic example of this exists in Central Africa, where sleeping sickness (Gambian fever) has prevented

Table 2

Specific Occupational Groups at Risk of Acquiring Zoonotic Infections

Group	Components
Agricultural	Farmers, other agricultural workers, veterinarians, livestock inspectors, transporters of livestock, often their families
Animal product manufacture	Butchers, slaughterhouse workers, abattoir and freezing plant workers; processors and handlers of meat, milk, eggs, hides, furs, and other animal products; processors and handlers of animal by-products and wastes and dead animals
Sylvan and campestral	Wildlife workers, foresters, hunters, trappers, taxidermists, fishermen, naturalists, ecological researchers, surveyors, resource explorers and developers (e.g., petroleum and minerals), project construction workers (e.g., dams, highways, pipelines), campers, tourists
Recreational	Pet dealers (in domesticated and wild animals); pet owners, their families, and visitors; zoological garden and wildlife parks employees and visitors; veterinarians
Laboratory	Health professionals, scientists, technicians, laboratory animal caretakers, other handlers of animals or animal tissues (e.g., for diagnosis of human and animal diseases, research, biologicals manufacture, product safety assay)
Epidemiological/ clinical	Physicians, nurses, veterinarians, other health professionals and paramedical personnel in contact with sick animals or people or highly contaminated surroundings during performance of epidemiological field investigations or hospital management of cases
Emergency	Refugees, disaster victims, participants in major pilgrimages, other congregations of persons living under temporarily crowded and stressful conditions or in absence of usual feeding, housing, sanitary, or other amenities

Source: Reproduced by permission of the World Health Organization.

Table 3
Selected Animal Diseases that Commonly Affect People

Disease	Route of Transmission
Anthrax	Exposure, ingestion of infected livestock
Bacterial food poisoning	Ingestion of contaminated meat
Brucellosis	Ingestion of contaminated meat or dairy products
Glanders	Contact with infected horses
Plague	Bites from infected wild animal fleas; also airborne
Relapsing fever	Bites from infected rodent fleas or lice
Shigella (dysentery)	Contact with infected primates
Ringworm	Contact with infected mammals or birds
Cryptococcosis	Contact with infected animals, especially pigeons
Toxoplasmosis	Contact with infected birds or mammals, especially cats
Swimmer's itch	Exposure to water contaminated by infected birds or rodents
Hydatid disease	Ingestion of eggs from infected livestock or wild animals
Rabies	Bite of infected mammal, especially bats and carnivores
Yellow fever	Bite of mosquito that has fed from infected monkey
Encephalitis, several types	Bite of mosquito or tick that has fed from infected bird or mammal
Cowpox	Contact with infected cattle
Lassa fever	Contact with infected rodents; ingestion; inhalation
Cat scratch fever	Wounds, scratches inflicted by infected cat or dog

Table 4
Selected Major Zoonoses by Type of Infectious Agent

VIRAL

Rabies
Encephalitis
Cat scratch fever
Ornithosis/psittacosis

RICKETTSIAL

Q fever
Spotted fever
Typhus
Rickettsial pox

PROTOZOAL

Toxoplasmosis
American leishmaniasis (Chiclero
   ulcer, espundia)
Trypanosomiasis (Chagas' disease)

HELMINTHIC

Hydatidosis
Taeniasis (beef tapeworm)
Trichinosis (pork tapeworm)
Hookworm

BACTERIAL

Anthrax (wool sorter's disease)
Brucellosis (undulant fever)
Leptospirosis
Tularemia (rabbit fever)
Tuberculosis
Diphtheria
Salmonellosis
Listeriosis

FUNGAL

Ringworm
   tinea corporis—body
   tinea capitis—scalp
   tinea pedis—foot
Histoplasmosis
Coccidioidomycosis (San Joaquin
   fever

the use of hundreds of thousands of acres of arable land (cf. Chapter 1). Not only is this the tsetse fly belt, it is also the yellow fever, lymphoma, and malaria region.

During World War II, the pressing need for South American rubber galvanized efforts to clean out the hitherto uninhabitable yellow fever mosquito areas, in order to plant and cultivate additional rubber trees. In this instance, cost was not a consideration.

In other circumstances, cost might well have been considered prohibitive, and the effort would not have been made.

This interaction is not without its benefits. The fact that certain illnesses are shared by animals and people makes it possible to use animals in experimental biochemical research and to produce vaccines used to immunize both animals and people against a host of communicable diseases.

Even though we have moved from an agricultural to a technological society, the health of animals and the health and well-being of people remain closely linked. This may be seen in the numbers of domestic animals that play an intimate part in our lives. Table 5 indicates the number and type of household pets in the United States. Table 6 indicates the number of domestic animals. Clearly, there are far more domestic animals than people. If to these are added the great number and variety of wild animals, such as opossum, rats, foxes, raccoons, wolves, deer, rabbits, and bats, it becomes clear that animals must play a significant role in our lives.

Zoonoses do not occur with equal frequency around the country. Of the 189 known, 49 occur in the southern states, 46 in Texas alone. Although brucellosis (undulant fever) occurs chiefly in Illinois, Iowa, Nebraska, and Kansas, where raw milk and dairy products abound and large numbers of carcasses are handled, it is not restricted to rural areas; 35–40% of reported cases occur in urban areas. On the other hand, tularemia (although often called rabbit fever, it occurs in a wide variety of lagomorphs and rodents) is found primarily in the rural areas of Missouri, Arkansas, and the Gulf Coast states, where squirrels, rabbits, and other rodents are hunted and trapped. The virus of rabies, once thought to be transmitted solely by the bite of rabid dogs, is now known to be

Table 5
Domestic Animals in the United States—January 1984

Cattle (all types)	114,040,000
Hogs and pigs	55,819,000
Sheep and lambs	11,411,000
Horses	3,400,000
Goats	1,330,000
Mules and donkeys	64,500
Approximate total	186,000,000

Table 6

Household Pets in the United States—November 1984

Dogs	56,700,000[a]
Cats	33,100,000[a]
Parakeets	
Canaries, finches, and all other birds	28,000,000
Gerbils, hamsters guinea pigs, turtles, monkeys, and skunks	100,000,000[b]
Approximate total	217,800,000

[a] Figures for dogs and cats, based on several surveys, are fairly firm. Figures for other pets are approximations, based on data available from the Pet Information Bureau in New York City.
[b] Although included under the rubric of household pets, the 340,000,000 tropical fish believed to be in homes throughout the United States are not included in this accounting.

disseminated by bats, foxes, skunks, and raccoons. These four species are concentrated from the East Coast westward to the central states.

Of particular importance for those concerned with the prevention and control of zoonoses is the observation that many exhibit marked seasonal fluctuations. For example, the several types of arthropod-borne encephalitis (St. Louis encephalitis, western equine encephalitis, eastern equine encephalitis) principally occur from July to October, whereas the incidence of brucellosis climbs sharply between April and August. Leptospirosis appears to begin its rise in May, peaks in July, and fades as fall approaches.

For convenience, zoonoses are divided into two major categories: those occupationally induced (see also Chapter 8, Occupational Health) and those transmitted via other routes. Of course microbes, being rather perverse creatures, do not read books and consequently are not aware of what is expected of them; thus, there are overlaps in which a disease agent normally of occupational origin infects an individual having nothing to do with that industry or occupation.

A. In Other Circumstances Animal Quarantine

Several states are implementing new laws aimed at protecting the public from zoonotic illness by imposing strict quarantines on wild animals brought into the state. These laws cover as many as 145 species of animals including a wide variety of primates, such as spider monkeys and wildcats. The laws are expected to improve the chances of purchasing a disease-free animal. They are also expected to provide a mechanism for ascertaining the incidence of disease in imported animals. The laws spring from a warning issued by the National Academy of Sciences, which noted that imported primates "should be viewed as possible vectors of a number of diseases transmissible to man, such as tuberculosis, the dysenteries (salmonellosis, shigellosis, amebiasis), yellow fever, and infectious hepatitis."

II. LEPTOSPIROSIS

Leptospirosis is a common zoonotic disease of livestock, pet animals, and wildlife, causing mastitis and abortion. In human beings, it occurs for the most part in sewage-plant operators, miners, loggers, and agricultural workers raising such crops as rice and sugar cane by irrigation. Children may become infected when swimming in stagnant or slow-moving ponds, often in farm areas. The infecting organism is passed into water via cattle, rodent, or dog urine. The corkscrew-shaped microbe, shown in Figure 1, enters the body through abraded skin and exposed mucous membranes and conjunctiva, nasal or oral. The infection is characterized by fever, abdominal and muscular pain, jaundice, kidney injury, and a tendency to hemorrhage.

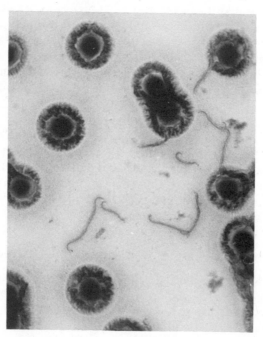

Figure 1. Leptospira in hamster blood. Photomicrograph courtesy of Dr. Donald Blenden, School of Veterinary Medicine, University of Missouri.

Leptospirosis is a worldwide problem, especially important among dairy farm workers. Thousands of cases occur among them annually. Until milking becomes totally automated, milkers will continue to be at high risk from urine splash. Total elimination of the organism from cattle herds by vaccination must await appropriate procedures.

In many areas of the world, the rat is the primary natural carrier of the spirochete *Leptospira icterohaemorrhagiae*, the predominant bacterial strain. However, the NCDC routinely employs about 23 serogroups in their agglutination tests for organism identification. In the United States, the raccoon, *Procyon lotor*, is the most important natural reservoir. Other firmly implicated feral hosts are the opossum, gray fox, skunk, armadillo, and muskrat.

The precipitous rise of leptospirosis in May and June corresponds to the period of use of the "old swimming hole" into which much animal urine has been passed.

For the 48 contiguous United States, the yearly level of total cases is no more than 0.05 cases/100,000 population. Hawaii, with an annual incidence of 1.5/100,000, is 30 times higher as a consequence of its heavy rat infestation which so contaminates water at recreational sites. The island of Kauai with a rate of 20/100,000 is nothing less than a hotbed of leptospirosis. Both the local population and tourists availing themselves of scenic swimming sites are at severely increased risk.

III. BUBONIC PLAGUE

Henry Turner* of Valencia County, New Mexico, gave one of his 40 cats to his grandson, Roger. When Roger and his new feline friend arrived home, the cat scampered up a nearby tree. Scrambling up after him, Roger was bitten on his left hand and scratched on both hands and forearms (3).

Three days later, both Roger and the cat became ill. Because of his 104°F fever, chills, vomiting, and underarm pain, Roger was quickly hospitalized. *Yersinia pestis*, a gram-negative bacterium, was isolated from a swollen lymph note (bubo). Roger had bubonic plague, which he had received from the cat.

Cats rarely convey plague, but their predatory activities among wild rodents makes it a distinct possibility. Of the approximately 1100 cases that have occurred in the United States since 1900, most have resulted from contact with wild rodents and their *Yersinia*-containing fleas. And in the decade 1970–1979, there was a decided increase in the number of new cases. Eleven percent of the total occurred during that time. Between 1925 and 1965, the incidence was one case per year. During the subsequent 17 years (1966–1983), the incidence rose to 15–16 cases per year.

* Not his real name.

But it was in the summer of 1984 with 28 confirmed cases and six presumed that the United States witnessed the worst outbreak since 1925. This sudden upsurge may have been the result of an unusually cool, moist spring. Fleas are more abundant in moist weather and survive for longer periods.

Were it not for our collective memories, plague would be of little concern. Treated early, with antibiotics, as in Roger Turner's case, it is completely curable. If diagnosed late, the death rate can be as high as 60%. Because of its rarity, many physicians do not recognize it.

Plague only appeared in the United States in 1899. It was brought from the Orient to San Francisco by shipboard rats. Although a rare disease, it is still associated with the "black death"* that ravaged Europe between 1363–1369. During those 8 years, 20–25% of the population died. Estimates range from 20 to 30 million dead.

To be sure, plague known as far back as biblical times decimated Europe until the middle of the 18th century. Before it subsided, it had wrought profound changes on the entire social structure, primarily as a consequence of the lack of people to carry on the traditional culture.

Although not firmly established, it is believed that the disease could not maintain itself because of the disappearance of the smaller, less aggressive black rat (*Rattus rattus*) driven out by the hardier brown rat (*Rattus norvegicus*) whose fleas are less effective in transmitting illness to people.

In the United States, wild rodents, ground squirrels (*Citellus*), chipmunks (*Eutamias*), prairie dogs (*Cynomys*), rabbits (*Sylvilagus*), and mice (*Peromyscus*) have been the primary carriers. People visiting or living in known plague areas can acquire the infection via the bite of infected fleas (*Xenopsylla, Pulex, Ctenocephalides*) or contact with animal carcasses while handling and/or skinning. Figure 2 depicts the typical routes of transmission.

From its initial foci in California, plague has moved slowly but inexorably eastward. Over the past 80 years, it has moved northward to Washington and Idaho, southeast to Arizona and New Mexico (the two most plague-ridden areas of the United States), and eastward into Nevada, Utah, and Colorado. It has not yet reached the 100th meridian, a line passing through the center of the country. This eastward march is not expected to continue because of the absence of large rat and rodent populations (with their accompanying fleas) in the Midwest, thus offering fewer harborages.

Although five documented cases have occurred east of the 100th meridian, all were initially exposed in enzootic areas—those in which plague normally occurs—such as the southwestern part of the United States (four cases) and Vietnam (one case).

---

* It is called black death because of the extremely dark color of the blood beneath the skin and draining from the buboes, the lymph glands in the groin.

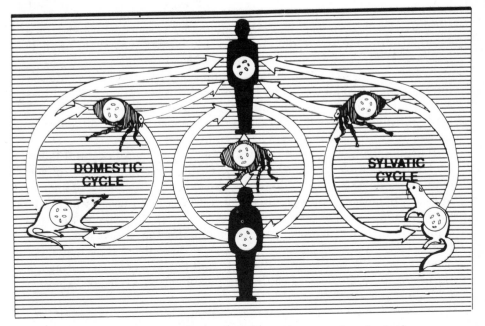

Figure 2. Cycles of transmission of plague between people and animals.

In the countries of the Americas, Brazil and Peru appear to have the highest number of cases of plague, at least in so far as they are reported. Between 1980 and August 1984, 407 cases were reported from Brazil and 344 from Peru. For Brazil, that represented a decrease of 31% over the prior 5-year period. Peru, on the other hand, registered an increase of 340%. This unfortunately suggests a lack of attention to a serious and rising problem.

IV. RABIES

Rabies is another disease with a rising incidence. Between 1960 and 1984, 45 cases of human rabies occurred in the United States. Robert Agerson's was the second of two that year. Twelve-year-old Robert was in good health until the morning of September 14th, when he complained of a runny nose and sore throat. In the afternoon, he refused to eat and was unable to swallow, not even an aspirin. He developed muscle spasms on the 17th and had difficulty breathing. He was admitted to a hospital that afternoon. No one had any idea that rabies was the problem, until the following day when he refused all liquids (hydrophobia) and had periods of violent screaming and facial grimaces. A skin biopsy taken on the 20th was positive for rabies virus. All measures to save him failed, and he died on September 29th. Although Pennsylvania, Robert's home state, is included among the four mid-Atlantic states (Maryland, Virginia, West Virginia, Pennsylvania, and Washington, DC) currently involved in a raccoon rabies epizootic (rampant among wild raccoon), rabid animals have not been found in his home county since 1978.

In Decemer 1987, a 13-year-old Filipino boy living in San Francisco died of rabies. He and his family had come to this country in 1981 and, as far as could be discovered, had no contact with wild or rabid animals. His case was diagnosed after his death. It demonstrates the difficulty of diagnosis lacking accompanying animal involvement.

During the almost 3-week interval from the onset of his painful lower back spasms on November 26th, to his death on December 15th, physicians believed he had appendicitis, pyelonephritis, psychiatric problems, and arthropod-borne encephalitis. Not until his brain tissue was examined by fluorescent antibody staining methods were Negri bodies (intracytoplasmic inclusion bodies), pathognomonic for the presence of neurotropic rabies virus, discovered and a definitive diagnosis established. Then it was necessary to quickly identify all possible contacts during his hospitalization. Over 100 health care personnel received postexposure antirabies vaccine.

Many of the boy's early symptoms—fever, chills, and back pain—are nonspecific, occurring in many common infections. Unfortunately, the source and time of his exposure remain unknown.

A most curious case of rabies recently occurred in Australia. If confirmed, it would be the first laboratory confirmed case of human rabies ever reported from Australia. It is one of the few rabies-free countries.

On July 16, 1987, a 10-year old boy who had traveled with his mother and friends through Asia between February and October 1986 complained of headache. Vomiting, chills, and fever followed. Over the next week he developed pain in his right arm, had double vision, and progressive weakness in his legs. Following hospitalization he became paralyzed, sunk into a coma, and died 23 days after his initial complaint. The diagnosis was encephalitis.

According to a friend who was along on the trip, he had been bitten on the finger by a wild monkey in a marketplace in India 16 months prior to the onset of illness. The bite had not been mentioned to his mother. Postexposure vaccine was administered to his mother, friends, and hospital personnel.

Monkey-transmitted rabies is rare, and its incubation period extremely long. If this is one of those cases, this is not of Australia, but imported. And if those receiving vaccine are properly protected, then Australia rabies defenses have not been breached.

Although rabies has been well known throughout recorded history, it is only recently that reservoirs of rabies other than the dog have become apparent. Rabies as a community health problem was considered well controlled as long as dogs were the only reservoir of infection. Now that it has been firmly established that wild animals such as the raccoon, fox, skunk, and bat are additional natural reservoirs, a serious public health problem has developed. Figure 3 shows the typical expression in a dog with rabies.

Robert's early symptoms followed a well-established pattern: restlessness, excessive salivation, tightness of pharyngeal muscles, and refusal to swallow, with consequent dehydration. Figures 4a and 4b portray the path taken by the rabies virus from point of penetration upward along nerve fibers to the brain.

Robert's case was only the tip of the rabies iceberg. By 1986 it was evident that the worst rabies epidemic since 1946 had occurred among

Figure 3. This canine, part shepard, husky and wolf, exhibits the expression typical of the furious form of rabies. Courtesy of Roberto Suarez, Jr. Merieux Institute, Inc., Miami, FL.

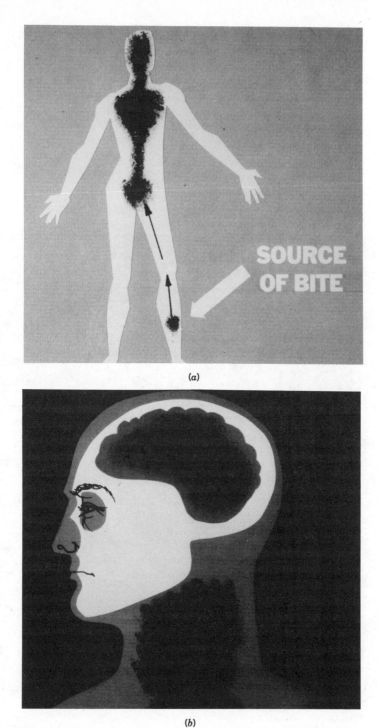

(a)

(b)

Figure 4. Pathway of rabies virus from point of penetration to the brain.

wild animals in the mid-Atlantic and southeastern states. In 1986, 5551 animal rabies cases were confirmed: twice the number reported 10 years earlier.

This recent outbreak is the result of raccoon infecting cats, dogs, cows, horses, and donkeys. Curiously enough, hundreds of people have been bitten by or been exposed to rabid animals in this most recent outbreak, yet not a single case of human rabies has occurred.

This outbreak crossed the Susquehanna River from the west and moved toward Baltimore and Philadelphia. Animal rabies is now widespread in Maryland and Pennsylvania and rampant in South Carolina, Georgia, and Florida. At the same time, an outbreak of skunk-borne rabies is spreading the infection from Texas to Minnesota, and from Kentucky to North Dakota and South Dakota. This means that as more people opt for the woods as their means of recreation, their risk of contact with rabid animals increases measurably.

A. Prevention

Veterinarians and epidemiologists at the NCDC believe this latest outbreak actually began in 1977–1978, when hunters released some dozen or more South Carolina raccoons between Virginia and West Virginia in an effort to revive the game hunting in that area. The released raccoons were rabid and the disease began spreading slowly north and east at about 30 mi/yr (4, 5).

In Robert's case, as in others, it is disturbing that without clear evidence of animal bite, human rabies apparently cannot be diagnosed until its last stages, when it is beyond help. This being so, prevention is essential. However, with so few cases occurring countrywide, vaccination of everyone, with booster doses at regular intervals, is unwarranted. Vaccination is appropriate for high-risk groups. And indeed, as of 1984, the vaccine has been administered to over 150,000 people, primarily veterinarians, letter carriers, and others whose work brings them into contact with animals, as well as those requiring postexposure treatment. Mitigating against wide preexposure vaccination is the cost—$350–450 for the series of six inoculations.

Currently, appropriate prevention would be at the secondary level, before symptoms appear. In the case of rabies, postexposure immunization and prompt local wound treatment are necessary. That cannot be stressed enough, especially as cases have occurred 6 months after an initial bite or scratch. Within 72 h is the recommended period. It is worth remembering that not a case of rabies has occured in a person who received the recommended postexposure treatment.

In addition to the difficult biological problems, religious considerations can impede control. In Arab and Asian countries, dogs are considered unclean. Accordingly, few Moslems are willing to handle them for the routine vacinnation procedures. In those countries of Southeast Asia where Buddhism is the dominant religion, stray dogs are considered an integral part of the cycle of life and must not be killed. Thus, cultural practices can impede prevention and control measures.

In 1983, the National Association of State Public Health Veterinarians (NASPHV) set forth a series of recommendations to augment community rabies control programs. In place of the yearly rabies vaccine, a 3-year vaccine would be available. They proposed a standardized system of identifying tags for vaccinated pets as follows:

Calendar Year	Color	Shape
1989	Blue	Rosette
1990	Orange	Fireplug
1991	Green	Bell

Since more cases of rabies are now reported in cats than dogs, immunization of cats should be required. Laboratory-confirmed cases in dogs have been reduced from 6949 in 1947 to 94 in 1986.

As noted in Table 7, dogs and cats account for less than 10% of all rabid animals. Nevertheless, they accounted for over 90% of all animal bites. Consequently, evaluation of rabies in dogs and cats is vitally important. Figure 5 demonstates an algorithm for evaluation.

The fact of feral animal types replacing domestic animals as the major vehicle of rabies transmission in the United States is shown in Figure 6. This can be compared with the Mexican experience, Figure 7. There, stray dogs are the outstanding problem. Figure 8 shows the abrupt increase in total numbers of animal rabies cases, as well as the proportion contributed by feral animals since 1975.

Yet another preventive measure would be the removal of all stray animals from a community. The NASPHV recommends that if a stray

Table 7
Reported Rabies Cases in the United States[a] by Type of Animal—1975–1983

Year	Dogs	Cats	Farm Animals	Foxes	Skunks	Bats	Raccoons	Other Animals	Humans	Total
1975	129	104	200	276	1226	514	192	31	3	2675
1976	116	106	198	187	1468	737	277	55	2	3146
1977	120	108	217	122	1631	637	281	65	1	3182
1978	119	96	254	148	1657	567	404	49	4	3298
1979	196	156	284	145	3031	756	543	34	5	5150
1980	247	214	499	213	4096	726	394	92	0	6481
1981	216	285	581	196	4480	858	481	111	2	7210
1982	153	209	381	222	3088	975	1156	94	0	6278
1983	132	169	282	111	2285	909	1906	66	2	5862

[a] Includes Guam, Puerto Rico, and Virgin Islands.

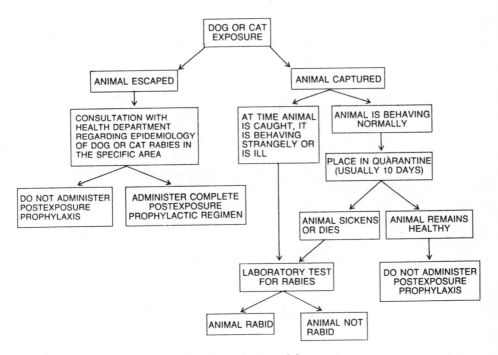

Figure 5. Algorithm for evaluation of dog or cat exposure.

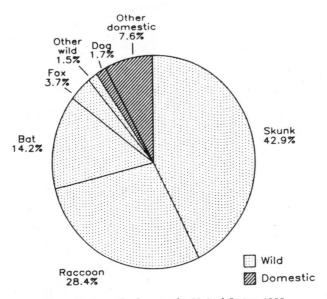

Figure 6. Animal rabies in the United States, 1986.

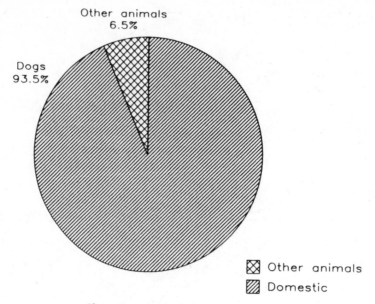

Figure 7. Animal rabies in Mexico, 1986.

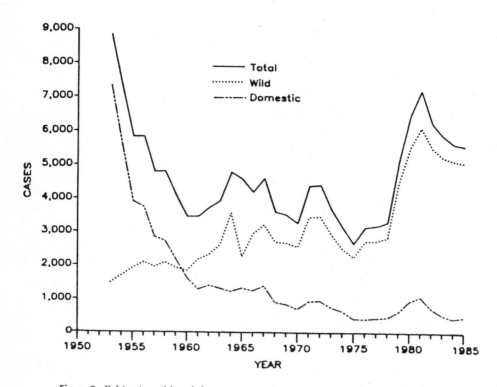

Figure 8. Rabies in wild and domestic animals in the United States, 1953–1985.

animal bites, it should be killed immediately. And following their guidelines, its head should be removed (by a person professionally qualified to do so) and shipped, under refrigeration, to a laboratory designated by the State Health Department.

B. Rabies Virus Areas

On the basis of nucleocapsid reactivity to a range of monoclonal antibodies, five antigenically distinct groups of rabies virus can be formed from isolates collected from the major wildlife rabies areas of the United States. While Figure 9 describes the boundaries, specifically they are as follows:

1. Skunk rabies areas of California and the north central United States and gray fox rabies areas of central Texas.
2. Skunk rabies areas of the south central United States.
3. Raccoon rabies areas of the mid-Atlantic and southeastern United States.
4. Red fox rabies areas of the northeast United States.
5. Gray fox rabies areas of Arizona.

C. Western Europe

The United States is not alone in its developing rabies problem among feral animals. Since 1950, foxes have been spreading rabies from Poland, Czechoslovakia, and East Germany toward western Europe. Obviously, they know nothing about an Iron Curtain.

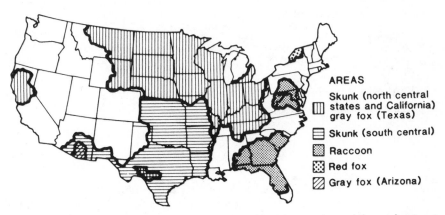

Figure 9. Distribution of five antigenically distinct rabies virus strains and the predominant wildlife species affected in the contiguous United States.

After crossing the Elbe River in 1950, foxes spread their infection by biting cows, pigs, dogs, and deer. By 1960, rabies had reached the Rhine; by 1967 it was in Switzerland and Germany, and by 1968, France.

The migrating fox population is so large and the idea of rabies so disturbing to the Germans and Swiss that Operation Chickenhead was launched to immunize feral foxes against rabies. The first approach was to use an encapsulated American-made vaccine secluded in chicken heads and dispersed by helicopter in the Rhone Valley. More recently, the one-quarter million member German Hunting Protection Association volunteered to decorate the pristine sylvatic landscape of Bavaria, Hesse, and Baden-Wurttenberg with vaccine-containing chicken heads. That worked, but chicken heads were not all that plentiful.

A "chocolate-brownie" made of foul-smelling fats and fish meal was devised. The little foxes loved it. By the fall of 1986, the epidemic was checked and some stretches of Bavaria were declared rabies free. However, the fundamental problem remains: how western European nations deal with rabies rampant in the eastern bloc counties, in order to prevent its movement westward, is high on the agenda of many western countries. Why it is not in Poland and East Germany is unclear.

D. A Genetically Engineered Vaccine

The Wistar Institute of Philadelphia may have solved a number of rabies problems. Researchers at Wistar have developed an animal vaccine by removing one of five genes from the rabies virus, splicing it into the genetic pattern of the vaccinia virus—ordinarily used to make smallpox vaccine.

The gene removed from the rabies virus was a harmless one which cannot induce rabies, but it can stimulate the immune system to produce antibodies to the rabies virus, thereby protecting animals receiving it.

Plans calls for conducting tests on wild raccoons on islands off the coast of Virginia or South Carolina. The raccoons are expected to bite into the bait and absorb the vaccine through the lining of the mouth. This has already occurred in Belgium, where the vaccine has been successfully field tested on foxes.

Should the next develement be a human oral vaccine, injections would become a thing of the past.

V.   BRUCELLOSIS

Brucellosis, or undulant fever as it is often called (suggesting the cyclic, wavelike rise and fall of fever during a seizure), is a chronic illness of long duration with periods of fever and pain between periods of apparent health. Brucellosis is predominantly an occupational disease of those working with infected animals or their tissues, especially farm workers, veterinarians, and abattoir employees. Consequently, it is more frequent among men than women. Sporadic cases and outbreaks occur among consumers of unpasteurized milk or milk products (especially cheese) obtained from infected cows, sheep, or goats.

Since 1927, when brucellosis was first recognized as a quite common cause of human illness, it has been steadily diminishing in incidence. Figure 10 clearly shows the reductions attained since 1955. Ten and 20 years earlier, 10,000 cases per year were not uncommon.

Between 1960 and 1972, the majority of abattoir-associated cases were related to contact with hogs and pigs. Since 1973, the greatest number of cases have come from infected cattle. The reason for this shift is not readily apparent. What is evident is that the microorganisms involved, *Brucella abortus* (cattle), *Brucella suis* (swine), and *B. melitensis* (sheep and goats) can gain entrance to the body via several routes: skin contact, aerosol exposure, contact with the conjunctiva, and ingestion of raw milk and milk products. For the past several years the number of reported cases has remained below 200. This may be an irreducible minimum so long as infected animals continue to exist and come to slaughter.

VI. LISTERIOSIS

Between January and August 1985, 83 people in southern California died of listeriosis—a bacterial disease—traced to a manufacturer of Mexican-style soft cheeses.

Although a human case of listeriosis was first described in 1929 (6), the ecological aspects of this disease remain poorly understood. Initiated by the gram-positive bacillus *Listeria monocytogenes*, listeriosis occurs most commonly in infants, pregnant women, and immunosuppressed individuals. In fact, neonatal listeriosis accounts for the largest number of infected individuals (7). In the California episode, mother–infant pairs proved to be the group at highest risk.

In pregnant women the disease exhibits flulike symptoms that often result in fetal infections and interrupted pregnancies—abortions. Infants may also be stillborn or develop meningitis within a month of birth. The true incidence of listeriosis in the United States is not known. Estimates suggest one case per million population per year. Obviously, this is a rare condition, but when it does occur, it is formidable.

Infected domestic and wild mammals and fowl are known to be reservoirs of the microbe, and transmission in the Los Angeles area was via contaminated cheese. Both Cotija and Queso Fresco, soft cheeses made by the Jalisco Mexican Products Company, contained the *Listeria*. These organisms can be passed from mother to unborn infant in utero or during the baby's passage through an infected birth canal.

Although not yet determined, the large number of cases in this outbreak indicates cheese made from contaminated raw milk. This would also

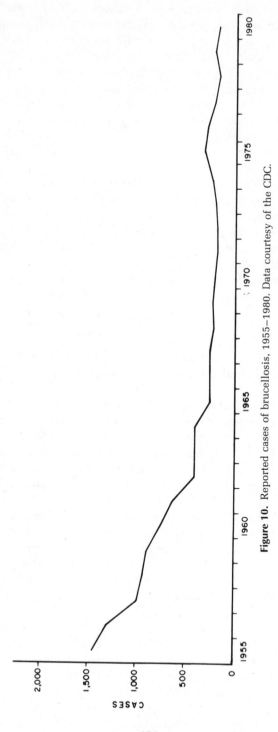

Figure 10. Reported cases of brucellosis, 1955–1980. Data courtesy of the CDC.

suggest heavy infection of the cattle on the farms. With heavily con-taminated milk, pasteurization may not kill all the organisms. Even with only a few remaining, the numbers increase rapidly given the geometric pattern of microbial growth.

Although listeriosis was not linked to food contamination until the 1950s, current concern appears to have begun with an outbreak in Massachusetts where 43 people acquired listeriosis during the summer of 1983 as a consequence of drinking milk contaminated with the organism (8,9).

After the many deaths and illnesses in the Los Angeles area, the Food & Drug Administration (FDA) and the NCDC concluded that milk and milk products were the most likely vehicle of microbial transmission. Accord-ingly, the FDA has since placed ice cream manufacture under close scrutiny. Their tight surveillance resulted in the impounding and dump-ing of hundreds of thousands of gallons of contaminated ice cream products between 1985 and 1988.

VII. ARTHROPOD-BORNE ENCEPHALITIS

The arboviral encephalitides, eastern equine encephalitis (EEE), western equine encephalitis (WEE), St. Louis encephalitis (SLE), and California encephalitis (CE) are a group of acute inflammatory diseases that involve the brain, spinal cord, and meninges. Generally, the five types of encepha-litis present with similar signs and symptoms but vary in severity and rate of progression. Mild cases often occur as meningitis. Severe infections are marked by headache, drowsiness, fever, vomiting, stiff neck, disorien-tation, tremors, paralysis of the arms and legs, and occasionally convul-sions.

Each of these is transmitted by the bite of infective mosquitoes. On puncture, virus particles contained in mosquito saliva are injected di-rectly into the human bloodstream. For the most part, mosquitoes acquire the virus from other wild birds, rodents, pigs, or horses. Figure 11 indicates the usual cycle of transmission. Birds, insects, rodents, and mammals were involved in this cycle long before humans stumbled into it. Human blood is unnecessary for any part of the life cycle. Given a choice, however, of bird, rodent, mammal, or human blood, the female mosquito seems to choose human every time.

Typically, as noted in Figure 12, cases begin to occur in April and May, building to a peak in mid-August and September, then crashing quickly with the onset of cooler weather. This, of course, parallels the decline in mosquito populations as well as the migration of birds. The buildup in the spring and summer is the reverse, with the gathering of bird populations, emergence of mosquitoes, and the extended periods of daylight and warm weather which allow both children and adults longer periods outdoors. Although many types of wild bird are natural reservoirs of the virus, they

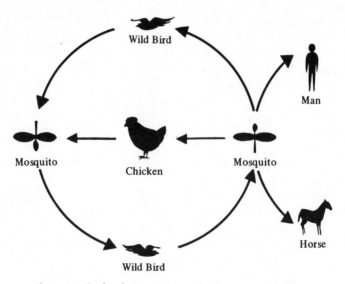

Figure 11. Cycle of transmission of St. Louis encephalitis.

do not appear to be affected. Female mosquitoes in the vicinity obtain a blood meal from these birds, at which time they pick up the virus. As the bite is made and blood is drawn up, saliva and virus are delivered into the wound; the saliva contains an enzyme that prevents the blood from clotting too rapidly, thus allowing the mosquito to imbibe at its leisure.

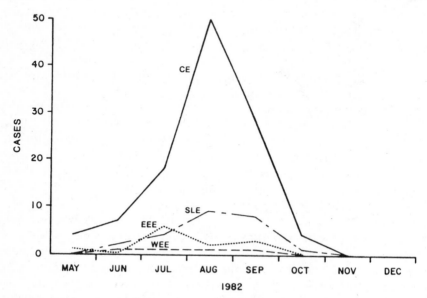

Figure 12. Arboviral encephalitis cases in the United States for 1982 by month of onset. Data courtesy of the CDC.

Figure 13 depicts the recurring encephalitis cycle between 1975 and 1982. With the exception of 1975, when a major outbreak of SLE occurred in several major population centers, the trends are predictable.

Between 1976 and 1983, an average of 150 cases have been reported each year, which is far below the 2113 that occurred during the epizootic of 1975.

In 1983, southeastern Massachusetts experienced a sudden increase in the number of cases. The area involved is a marshy, wooded swamp, dotted with ponds and lakes, inordinately suitable for unrestricted growth of mosquitoes. In this area of the state, the danger of EEE outbreak was clearly related to two successive years of excessive rainfall.

To obtain early warning of potential arbovirus pressure, many states maintain sentinel flocks of chickens and horses in strategic locations. Mosquito traps are also set out. Together these indicate the size of the current mosquito population, as well as their infection rate.

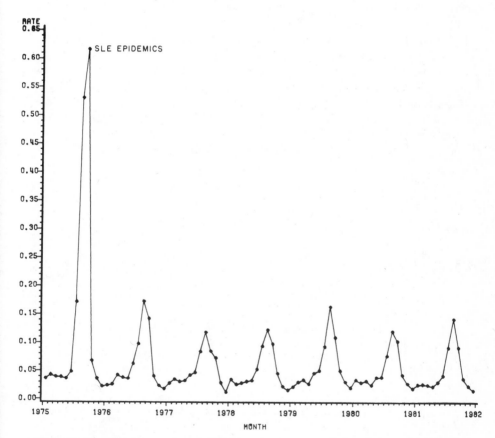

Figure 13. Reported cases of encephalitis per 100,000 population by month of onset, 1975–1981. Data courtesy of the CDC.

In Minnesota and North and South Dakota, just such a constellation of events indicated that WEE was building in the spring of 1983. By the end of summer, eight cases had been confirmed. It is likely that additional cases would have occurred if preventive measures to reduce and repel the mosquito population had not been taken. Clearly, regular, planned control of mosquito and bird populations is essential to prevent the transmission of encephalitis to humans.

VIII. TRICHINOSIS

Trichinosis occurs from the ingestion of a specific parasitic worm. *Trichinella spiralis* larvae can be found in inadequately cooked or heat-processed pork, pork products, wild boar, walrus, bear, and other carnivores.

The number of cases reported yearly has declined sharply since 1947 when the NCDC first began collecting these statistics. In the period 1977–1981, the average annual number of cases reported was 137. In the following five years, 1982–1986, the numbers decreased to approximately 57 per year. Figure 14 shows the generally downward trend of reported cases over the past 50 years. However, the number of infected and unreported as well as asymptomatic cases may be as high as a million. Reporting leaves a good deal to be desired.

Interestingly enough, Alaska persists as the state with the largest number of cases. This may be related to the Eskimos predilection for eating raw or minimally cooked walrus and bear meat.

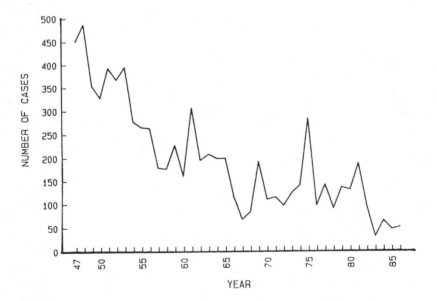

Figure 14. Reported cases of trinchinosis in the United States, 1947–1986.

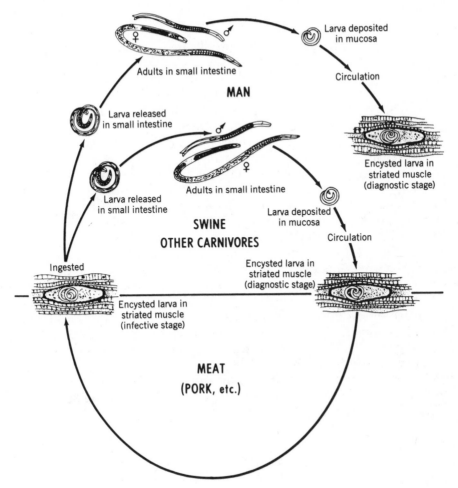

Figure 15. Life cycle of *Trichinella spiralis*.

Following up identified cases has established that in fully 26% of cases meat was eaten raw. In another 43%, the meat was boiled, while it was fried in another 14%. Obviously, further reductions in trichinosis could be expected if consumption of raw meat was strongly disavowed. That boiling and frying were responsible for another 48% indicates that adequate heat penetration failed to occur, suggesting that the size of the portion being heated was too large for the cooking time, and in the case of frying, only the surface received sufficient heat.

Over the past several years, the normally consistent pattern of trichinosis which showed peak incidences in December and January—related to the consumption of homemade pork sausage during the Christmas holidays—has not been seen. Figure 15 shows the life cycle of the parasite.

A. France, 1985

Trichinosis was rare in France—until 1985. Between August and October, two large outbreaks occurred: 1073 cases were identified, six of whom subsequently died. The first outbreak occurred in Paris in the 14th District, and the second in Melun, 30 mi (48 km) southeast of Paris (10).

The patients in both areas reported eating horsemeat during the weeks prior to the onset of illness. Symptoms rarely appear before 7–14 days, and edema of the upper eyelids is one of the earliest and most characteristic signs. This proved true in both Paris and Melun.

Over 98% of those who ate horsemeat (purchased from horsemeat butchers who usually sell no other meats) ate it raw or rare. The most frequent method of preparation was minced and raw as "steak tartare," or minced and dunked raw in soup, or lightly grilled as steaks or roast, but hardly sufficient to kill the parasitic worms.

Although all the patients were French, and all became ill in France, the horsemeat was imported from the United States and West Germany, where trichinosis is endemic. But the mystery remains. As natural infection of horses has never been reported, how were horses, which are not meat eaters, infected? Think about it. Meanwhile, these outbreaks demonstrate that horsemeat can be a resevoir of human trichinosis.

In France, where the national custom is to eat horsemeat fresh and rare, advising the public to freeze and thoroughly cook the meat—as is done with pork in the United States—may be tantamount to asking them to foreswear horsemeat. For the French that would be intolerable, a fate worse than the disease.

IX.   SWINE INFLUENZA

One of the most interesting examples of the interrelatedness of human and animal health is the relationship of influenza virus to both humans and animals. It is well known that a wide range of viruses of group A influenza occur not only in humans and pigs but in birds and horses as well. Recent studies have shown that human influenza A_2 (Asian) can cause natural, inapparent infections in horses and swine. The reverse is also true. Swine strains can produce human illness. The great pandemic of 1918, which was responsible for the death of more than 20 million people worldwide, was also responsible for the outbreak in swine that followed.

Laboratory findings suggest a disturbing possibility concerning the human and avian (bird) influenza viruses. It has been found that an avian and a human strain can be hybridized. It is thus conceivable that a strain could emerge with the virulence of fowl plague (not normally pathogenic for humans) and the host specificity of a human strain. Fowl plague has a mortality rate of nearly 100%. While this unique combination of traits may be unlikely in nature, the possibility of the emergence of entirely new strains as a result of genetic recombination remains a serious threat.

Nevertheless, microbes from animals can protect humans against illness. Swine influenza, noted above, is a case in point. A virus for a vaccine to protect humans against human influenza has been found in the snouts of pigs from Taiwan. The virus appears to be a weakened version of A_2, which originated in Hong Kong and continues to evoke "flu" in a number of countries.

When susceptible volunteers were injected with samples of the Taiwan virus, antibodies against the Hong Kong virus appeared in their blood, but no colds occurred. The importance of this discovery is that the Chinese swine flu virus seems to have crossed the "species barrier" between humans and other animals. It is therefore highly likely that other pigs and even birds are presently incubating the viruses suitable for use in preparing human vaccines at an early stage of future outbreaks. A search for these ready-made vaccines has already started among animals in the Far East as so many pandemics appear to originate there.

There is yet another aspect of the swine–human influenza relationship which is both controversial and political. Shortly after New Year's Day, 1976, several thousand new recruits arrived at Fort Dix, New Jersey, to begin their basic training. As is common at any military post, respiratory infections are anticipated within 10–14 days within the new crop of susceptible troops. Fort Dix was no exception. Infections occurred. Several required hospitalization. One soldier, who should have been hospitalized, refused. Instead, he joined his company for an overnight hike and died.

Influenza virus, especially the then predominant A-Victoria strain, was isolated from nasal and throat cultures of several of the hospitalized men. From four others, including the fatal case, swine influenza virus was recovered. Virologists were concerned on several counts. More than likely contact with pigs was remote to nonexistent. Therefore, person-to-person spread was the probable route of transmission.

Considering that swine flu had not been found in the human population since the World War I period, only the population well over 50 would have antibody protection. It was also recalled that during the 1918 pandemic, of the half million men and women who died in the United States, most were young people in the 20–30-year age group. Concern swirled around another facet of the problem. The A-Victoria strain and the swine flu strain had totally different surface proteins (antigens).* This

---

* The influenza virus is classified as a myxovirus, from the Greek *myxa* meaning mucus. Its surface is convered by projections called spikes or rods, and its core contains the genetic material which is a nucleoprotein composed of ribose nucleic acid (RNA).

These two components, the core RNA and the spikes, have antigenic properties that serve as the basis for its classification. First, it can be divided into three main types A, B, or C on

difference would vitiate any resistance established by exposure to the other. But did these four cases foreshadow a potentially massive outbreak in a largely susceptible population? Scientists were divided, both on the possibility and appropriate procedures for protecting the public. The media fueled speculation of a natural disaster of the magnitude of the 1918 pandemic.

Several meetings of eminent scientists were held by the National Centers for Disease Control to review available data with the Advisory Committee on Immunization Practices. The preeminent questions were: Should a nationwide immunization program be mounted? And, if so, which vaccine should be used—Victoria or swine flu?

It was then mid-March. Little time was left for protracted discussion, vaccine manufacture, and immunization of 200 million people. Fourteen days were needed for adequate development of immunity. The task was staggering. President Ford faced a dilemma: order production and be considered a bumbler if a pandemic failed to materialize; be a bumbler if adequate preparations had not been made and an epidemic occurred. For him the issue seemed clear. Politics had no part in it. His strong belief was to gamble on the side of caution. Unfortunately, there was yet another problem. Serious side effects were possible with the swine vaccine.

On March 24th, the President called for an appropriation of $135 million. Both the Senate and House ratified the supplemental bill. It was signed on April 15th. Between October 1st and December 16th, more than 40 million people were immunized. Late in November, the first case of paralysis was reported. By December, hundreds of cases of Guillain–Barré syndrome had occurred. Swine flu never appeared.

Guillain–Barré syndrome, rare in the population, is an acute, rapidly progressive form of multiple nerve involvement characterized by muscular weakness and sensory loss at finger and toe tips that can ascend to legs, arms, and face. In severe cases, respiratory function can be fatally impaired.

On December 16, 1976, the swine flu program was suspended. By the winter of 1984, the federal government was still attempting to ascertain which cases of Guillain–Barré syndrome were vaccine-induced and which were of other etiologic consequences. However, this was no more

---

the basis of the reaction of the nucleoprotein. The A viruses have emerged as the most important in terms of human health. These A viruses contain the two different antigenic glycoprotein spikes, which are referred to as the H and N spikes, for hemagglutinin and neuraminidase, respectively. The H portion enables the virus to attach itself to the surface of a cell so that it can enter and infect. The neuraminidase (N) is an enzyme that permits the virus to detach itself, thus enabling it to infect another cell.

Thus far, analysis of these structures has revealed five major H (H_0, H_1, H_2, H_3, and H_{sw1}) and two N components (N_1 and N_2). It is from these H's and N's that all group A viruses are identified. So, for example, the Hong Kong flu of 1968 was known as $H_3 N_2$, while the flu of 1977–1981 was $H_1 N_1$.

than a scientific exercise as the government, through the agency of the Department of Justice, had decided to compensate all cases that could in any way be associated with the period in which the vaccine was used.

Since 1977, the predominant forms of influenza in the United States have been A (H_3N_2 and H_1N_1) and B, with neither attaining more than normally expected illness frequencies. Swine flu has not been present.

X. COMPARATIVE MEDICINE

Animals and humans are in reality so kindred that knowledge of a certain disease in one can often be applied to the other. This area of research is known as comparative medicine. Cardiovascular, degenerative, nervous, and rheumatic diseases occur in animals as well as humans. Because domestic animals are slaughtered for food or other useful products, they seldom live on to old age. To try to obtain the necessary data, research centers have been established in which animals are studied as they live out their full potential of years.

Atherosclerosis occurs in primates (including people), swine, chickens, turkeys, and pigeons. Cerebrovascular symptoms appear to be common in aging pigs. Viral leukemia is well established in chickens and mice and may also occur in dogs, cats, and cattle. Since this virus can infect a large number of animal species, researchers are concerned that humans can be infected from the animals around them. As yet, the viral etiology of human cancer remains controversial. Much more research is needed to establish this relationship.

The existence of degenerative diseases among animals raises the possibility that animal research studies could be as helpful in understanding and controlling chronic degenerative human ailments as they were in understanding communicable diseases. One of the most beneficial aspects of the animal relationship is the far shorter life span of most animals, which conveniently compresses into a few years events that are spread over a half-century in humans.

As we continue to obtain information about our environment, it becomes increasingly evident that human health is intimately associated with that of the many animals occupying the same habitat. Thus, real comprehension of many community health problems must encompass knowledge of both animal and human diseases.

REFERENCES

1. R. J. Biggar, J. P. Woodall, P. D. Walter, and G. E. Haughie, Lymphocytic Choriomeningitis Outbreak Associated with Pet Hamsters. *J. Am. Med. Assoc.* **232**(5): 494–500, 1975.
2. G. Mayer and S. Van Ore, Recurrent Pharyngitis in Family of Four. *Postgrad. Med.* **74**(1): 277–279, 1983.

3. J. M. Mann, G. P. Schmid, P. A. Stoesz, M. D. Skinner, and A. F. Kaufman, Peripatetic Plague. *J. Am. Med. Assoc.* **247**(1): 47–48, 1982.

4. V. F. Nettles, J. H. Shaddock, R. K. Sikes, C. R. Reyes, Rabies in Translocated Raccoons. *Am. J. Public Health* **69**: 601–602, 1979.

5. Rabies Surveillance, 1986. *MMWR Suppl.* **36**:35, 1987, CDC/USDHHS.

6. A. Nyfeldt, Etiologie de la mononucliose Infectieuse. *C. R. Soc. Biol.* **101**: 590–592, 1929.

7. W. L. Albritton, S. L. Cochi, and J. C. Feeley, Overview of Neonatal Listeriosis. *Clin Invest. Med.* **7**(4): 311–314, 1984.

8. D. W. Fleming, et al. Pasteurized Milk as a Vehicle of Infection in an Outbreak of Listeriosis. *N. Engl. J. Med.* **312**(7): 404–407, 1985.

9. W. F. Schlech, et al. Epidemic Listeriosis—Evidence for Transmission by Food. *N. Engl. J. Med.* **308**: 203–206, 1983.

10. T. Ancelle, et al. Two Outbreaks of Trichinosis Caused by Horsemeat in France in 1985. *Am. J. Epidemiol.* **127**(6): 1302–1311, 1988.

SUGGESTED READINGS

Acha, P. N., and Szyfres, B. Zoonoses and Communicable Diseases Common to Man and Animals. Scientific publication No. 354, Pan-American Health Organization, Washington DC, 1980.

Anderson, L. J., Nicholson, K. G. Tauxe, R. V., and Winkler, W. G. Human Rabies in the United States, 1960–1979: Epidemiology, Diagnosis, and Prevention. *Ann. Intern. Med.* **100**: 728–735, 1984.

Hanson, L. E. Leptospirosis in Domestic Animals: The Public Health perspective. *J. Am. Vet. Med. Assoc.* **181**(12): 1505–1509, 1982.

Malkin, H. Louis Pasteur and "La Rage"—100 years ago. *Perspec. Biol. Med.* **30**(1): 40–46, 1986.

Neustadt, R. E. and Fineberg, H. V. *The Swine Flu Affair: Decisionmaking on a Slippery Disease:* U.S. Department of Health, Education and Welfare, Washington DC, 1978.

5

Pesticides

And when it was morning, the east wind brought the locusts.
And the locusts went up over all the land of Egypt. . . . They
covered the face of the earth, so that the land was darkened;
and they did eat every herb and every fruit . . . and there
remained not any green thing.

— Exodus 10:14–15

Taona Zina (Year of the Abundant Harvest) is a remarkable achievement. In its first year, 1983, this finely tuned pest control project on the world's fourth largest island increased rice yields by tonnages sufficient to feed 36,000 families for a year, and that was only the first season. The objective of the cooperative undertaking took shape in 1982: to increase rice yields over an area of 124,000 acres by 500 lb/acre. It could not have been accomplished without the use of a new synthetic chemical pesticide.

In the Malagasy Republic,* as elsewhere, a complex of insect pests that sharply curtails potential crop yields posed a major problem. The entomological designations of these pests are *Malirapha separatella, Sesamia calamistes, Trichispa sericea,* and *Hispogestroi.* More familiarly the first two are the white and red stem borers; the other two are leaf-chewing beetles. As their name implies, borers tunnel rice stem, producing dead "hearts." They are also the cause of "empty ears," in which heads fail to produce normal grains.

Since 1960, the Malagasy Ministry of Agricultural Productions and Agrarian Reform (MPARA) has been amassing data on combatting these crop destroyers which infest the rice-growing areas of the world. Early in 1982, working with local extension services and the farming community in the Lake Alaotra region 270 mi north of Tananarive, Malagasy's capital,

---

* The Malagasy Republic, another name for the Democratic Republic of Madagascar, lies in the Indian Ocean off the southeast coast of Africa.

they instituted Taona Zina, a project geared toward wresting their crops from the grip of the pests. In 1983, The Year of the Abundant Harvest, they brought in the first bumper crop in the country's history.

The citrus-producing areas of Florida were less fortunate. Citrus canker disease threatens to wipe out Florida's $2.5 billion orange, grapefruit, and tangerine industry. The canker* is being spread by a previously unknown strain of bacteria that horticulturists and botanists believe may be one of the most aggressive and virulent microbes ever to imperil citrus crops. The year 1984 will not soon be forgotten in Collier, Hendry, Highlands, and Polk counties—Florida's prime citrus-growing areas. Pesticides capable of stopping the canker are unavailable. The campaign to destroy the organism was being waged by fire in burning millions of orange and grapefruit trees as the last resort.

I. THE ENTOMOLOGICAL BACKGROUND

Unfortunately, nature has no favorite species. From an ecological perspective, the human population is but one of many populations inhabiting planet Earth, and we must vie with all other species interested, in these instances, in the same nourishing food supply. Fortunately, we humans are endowed with the intelligence to moderate and counterbalance hostile environmental forces. However, it appears an endless struggle, and our management of the recurring problem of insect depredations may suggest just how intelligent we really are.

Insects entered "the game" with a great advantage; they have evolved over some 350 million years, compared to our 1 million years. This seniority has led to some remarkable survival characteristics that have attuned them almost perfectly to their ecological niche. Their ability to adapt to a hostile environment was seen in the resistance they displayed to chemical agents they had never before encountered. How resourceful we humans will be remains to be seen.

Insects pose a major threat to our food, fiber, health, and comfort. Despite modern agricultural practices, losses caused by a variety of "pests" approximate 23% of potential production, or $20 billion annually. As a consequence, chemical pesticides have become an integral part of agriculture practice for protection of crops. Table 1 indicates the annual usage during the period 1977–1982.

Insects and other arthropods† are all around us. There is scarcely a

---

* Although the word canker derives from the Anglo-Saxon, meaning cancer, to botanists a variety of lesions and discolorations are referred to as cankers.

† Phylum Arthropoda

Class Crustacea: crabs, shrimps, barnacles, water fleas.
   Insect: ants, bees, flies, mosquitoes, beetles, wasps, lice, roaches, aphids, moths, fleas, termites, and locusts (to note a few).
 Arachnida: spiders, scorpions, ticks, mites.
 Myriapoda: centipedes, millipedes.

Table 1
Pesticide Use[a]

Year	Total	Herbicide	Insecticide	Fungicide
1977	1388	674	570	143
1978	1416	664	605	148
1979	1429	657	617	155
1980	1468	806	506	156
1981	1430	839	448	143
1982	1113	623	379	111

[a] In millions of pounds.

place on the planet that is not home to at least one insect. They have been found in deep underground caves, and termites have been trapped at altitudes of 20,000 ft. Some 40 varieties live in the Antarctic region; mosquitoes and other biting insects penetrate the polar regions as far north as the warm-blooded animals on which they feed. Insects are abundant in desert areas and in rushing waters. Different species are adapted for life in the air, on land, in soil, and in fresh water, brackish water (1% salt), or salt water (3% salt). Wherever they live, they seem to be indestructible. People have never eradicated a single insect pest from the earth. In fact, a number of them exhibit increased resistance to chemical agents designed specifically for their destruction. Insects have been frozen solid at −35°C and have still lived. Others inhabit hot springs where temperatures reach 50°C (120–125°F). Petroleum flies spend part of their lives in pools of crude oil around well heads. Many insects can endure long periods without water; they possess fuel reserves and can get the water they require by metabolizing carbohydrates to carbon dioxide and water.

Insects have an enormous size range, probably greater than any other major animal group. The smallest, approximately 250 μm, are smaller than some single-celled protozoa; the largest are larger than mice. The Atlas moth of India measures 12 in. from wing tip to wing tip.

Why are insects so successful? As a group their success seems to be due to at least six major assets:

1. *Adaptability.* As previously noted, they can live within a wide range of environmental conditions. They can eat anything—corks, mummies, tobacco, cotton, paper. They appear also to possess biochemical mechanisms that permit metabolism of chemicals with which they have had no previous experience. In the contemporary world this may be the most sophisticated of all their adaptive and survival capabilities.

2. *External Skeleton.** Their cylindrical shape offers the strongest

* The exoskeleton or cuticle contains a polysaccharide, chitin, which is made up of a series of glucosamine units, joined by beta-linkages, which imparts structural integrity.

possible construction for a given amount of material. Being wax-coated, it is resistant to drying from inside and outside.

3. *Small Size.* As a consequence of their usually small size, their food and water needs are comparatively quite small.

4. *Ability to Fly.* Wings allow wide distribution and thus greater choice of food and environment and escape from predators.

5. *Metamorphosis.* This is the gradual change in form or structure that occurs during the insect's developmental period. Three types of developmental changes are recognized. Some, such as flies, mosquitoes, and moths, undergo complete metamorphosis, consisting of four stages—egg, larva, pupa, and adult. In those insects that undergo incomplete metamorphosis, with egg, nymph, and adult, several nymphal stages can occur in which the insect simply grows larger. Cockroaches, grasshoppers, and body lice are examples of this type. The third type exhibits no metamorphic development. The young possess all the features of the adult; they simply increase in size. The silverfish, a primitive type of insect, is an example of this type of development. These various types of metamorphosis *confer enormous survival value,* as each stage is often spent in a different environment. Thus, the insect is not dependent on a single food supply or on one set of environmental conditions. (While these types of metamorphosis serve the insects, they also permit a variety of methods for controlling insect populations.)

6. *Specialized System of Reproduction.* After mating, the female can often delay fertilization until the proper food supply and environmental conditions are located. When mating occurs, the sperm from the male is stored in a special sac; when conditions are favorable, the sperm is released for fertilization. This mechanism has extraordinary survival value.

Thus, it can be seen that after 250 million years insects have evolved their own bag of tricks to frustrate our attempts at preventing them from taking our food supply or making us ill.

Although the number of insect species is between 670,000 and 1,250,000, the number of insect pests* is relatively small; fewer than 2000 types attack people, animals, and plants. Of the 2000, by far the largest number are agricultural pests. Upward of 30% of the world's food supply is destroyed or rendered unfit to eat either during growing, harvesting, or storage. Until the advent of chemical pesticides, it was evident that the battle for the food supply between humans and insects was being won by the insects. Crops, like people, are susceptible to disease and injury, with either destroy them completely or reduce their food-yielding capac-

---

* Bear in mind that an insect becomes a "pest" only as we call it so. It is a pest when it annoys us in some way.

ity. Because plants cannot as yet be "vaccinated" against disease (this may be possible by the year 2010), pesticides must be used to ensure their health and productivity.

Trees and plants are susceptible to over 1500 diseases. Together, the loss in food crops and trees is placed at $10–15 billion annually. In addition, carpet beetles, silverfish, and moths destroy another $200 million worth of property. The losses from termite destruction are far higher.

Before chemical insecticides were introduced, hundreds of thousands of people died each year from a host of insect-borne diseases. Table 2 lists examples of plant and human diseases transmitted by arthropods. Figure 1 indicates the dramatic effectiveness of DDT against typhus, a disease of rats transmitted to humans by the bite of the rat flea. Clearly, its application markedly reduced the number of cases of typhus. Similar charts could be presented for relapsing fever, yellow fever, and bubonic plague.

If we are to eat, remain relatively disease free, and reduce the losses from a host of pest-induced damages, the activity of "pests" must be sharply curtailed. However, insects are not the sole object of this effort. Crop and animal losses and human illness occur as a consequence of the

Table 2
Examples of Diseases Transmitted by Arthropods

	Disease	Transmitting Arthropod (Vector)	Pathogenic Agent
Plant	Dutch elm Fungus	Bark beetle	
	Cucumber wilt	Cucumber beetle	Bacterium
	Curly top of sugar beet	Beet leafhopper	Virus
Carrot	Aster yellows	Leafhopper	Virus
Potato	Leaf roll	Aphid	Virus
Peach	Mosaic	Mite	Virus
Cotton	Boll rot	Weevil	Fungus
Human	Yellow fever	Mosquito	Virus
	Bubonic plague	Flea	Bacterium
	Malaria	Mosquito	Protozoan
	Typhus	Louse and flea	Rickettsia
	Tularemia	Fly and tick	Bacterium
	Encephalitis	Mosquito	Virus
	Filariasis	Mosquito	Nematode worm
Animal	Rocky Mountain spotted fever	Tick	Rickettsia
	Texas cattle fever	Tick	Protozoan
	Chagas' disease (South American sleeping sickness)	Giant bedbug	Protozoan
	Dog tapeworm	Louse and flea	Worm

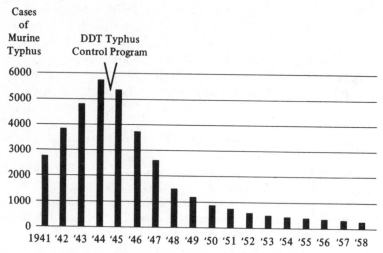

Figure 1. Effect of DDT on the incidence of typhus.

activity of rodents, weeds, worms, fungi, and bacteria. All are "pests" in the broadest sense of the term, and each requires a specific approach.

While "pesticide" is an umbrella term meaning killer of pests—from the Latin *cida* to kill—scientists have developed a range of agents specific for the target species. Table 3 lists the most common of these "cidal" agents.

The classes of agents noted in the table have generally referred to a wide variety of organic and inorganic chemicals, especially prior to 1975, when biological methods of control did not have the importance they currently do.

II. TRADITIONAL CLASSES OF CHEMICAL PESTICIDES

A. Naturally Occurring Plant Products—Botanicals

Plants have provided relief from insects for hundreds of years. Flowers, leaves, and roots have provided naturally occurring toxic ingredients long before analytic procedures were able to isolate, purify, and characterize

Table 3
Types of Pesticide

Type of Agent	Destroys
Bactericide	Bacteria
Fungicide	Fungi molds
Larvicide/grubicide	Insect larvae
Acaricide/miticide	Mites
Molloscicide	Snails, slugs
Nematocide	Worms
Rodenticide	Rats, mice

them. Accordingly, pyrethrin, derris, hellebore, quassia, tobacco, camphor, and turpentine were used around the world, for as much protection as they might afford.

Extracts, dusts, and smokes made from the dried heads of the pyrethrin-containing chrysanthemum have long been used as insecticides. Pyrethrin is a mixture of four active chemicals: pyrethrin I ($C_{21}H_{28}O_3$) and II ($C_{22}H_{28}O_5$) and cinerin I ($C_{20}H_{28}O_3$) and II ($C_{21}H_{26}O_5$). Pyrethrin sprays have a reputation for rapidly paralyzing flying insects.

Nicotine, a plant alkaloid obtained from tobacco leaves, has a typical nitrogen-containing heterocyclic ring structure. Its toxic effects occur in the ganglia of the insect's central nervous system.

Rotenone or actually rotenoids are obtained from the roots of the East Indian leguminous plant derris. Its insecticidal activity occurs as a consequence of its being a powerful inhibitor of mitochondrial electron transport.

B. Inorganics

These are compounds containing the heavy metals lead, copper, zinc, arsenic, and mercury as the cidal ingredient. In the 1870s, *Paris green*, an arsenical (actually a copper acetoarsenite, with the copper conferring its green color), was used to control the codling moth in apple and pear orchards. Bordeaux mixture, containing copper as its active agent, was used in the 1880s to control *Phylloxera vastatrix*, a plant louse of the aphid family that was destroying the vineyards of Europe. Ultimately, the only successful method was grafting of resistant American vines. Consequently, all European vineyards now grow on American roots.

Currently, inorganic, metal-based pesticides are of little commercial importance.

C. Halogenated Organics

Although the halogens include chlorine, bromine, fluorine, and iodine, bromine and chlorine have been the most successful against a wide variety of pests.

Currently, the most widely known is ethylene dibromide (EDB). It has been used as a fumigant on citrus fruits, grains, and papayas for over 40 years. Insects chew through the protective outer seed coat which surrounds kernels of grain. Without the coating mold spores grow luxuriantly on the nutrient-containing seed, ruining it for human food use. EDB had effectively prevented the loss of the seed coat. In September 1983, the U.S. EPA suspended its use as a consequence of detecting EDB in groundwater and declaring it an "imminent hazard."

Although its insecticidal properties were not discovered until World War II (the 1940s), benzenehexachloride (BHC) is relatively ancient, having been first prepared in 1825. The γ-isomer (one of five) is the only one exhibiting insecticidal activity. Unfortunately, its use on crops is limited as a consequence of the musty odor and taste it can impart.

DDT (dichlorodiphenyltrichloroethane) was until the 1970s the pesticide of choice. The modern era of chemical pesticide control began with the commercial introduction of DDT in 1946; it had been used successfully during World War II to combat louse-borne typhus fever in military and civilian populations.

It was found to be soluble in fatty (adipose) tissue and resistant to metabolism, which suggested accumulation in body tissue. It was banned in 1976. Other members of this group are the cyclodienes, dieldrin, chlordane, and heptachlor. Because of their stability in soil, all four have been successfully used against root-boring insects and termites. Thus far, their safety has not been questioned.

D. Organophosphorus Compounds

The substances in this group are rapidly replacing the halogenated compounds because of their instability—their lack of persistence.

Prior to World War II, the development of specific pesticides and the development of nerve gases for chemical warfare had been almost inseparable. Tabun, the nerve gas recently identified by a United Nations team in Iran from samples of battlefield soil and liquid, was first made by a German chemist, Gerhard Schrader, in 1936, during the course of commercial insecticide research for I. G. Farben. Nerve gases and the organophosphate insecticides work by disrupting the electrochemical action of nerve cells, specifically inhibiting the enzyme cholinesterase.

Parathion, malathion, and amiton are members of this group. Most commercial pesticides are used in extremely dilute concentrations. When additives are included to help their entrance into the body, they become dangerous substances. They can also become dangerous when used inappropriately, such as preparing in concentrations well beyond label directions.

On August 6, 1984, the Mississippi State Department of Health was notified of the death of an 11-year-old girl and the hospitalization of her six brothers and sisters. Apparently, an adult member of the household (who remained unharmed) had sprayed the interior of the house with methylparathion, using a concentration three times that used for outdoor agricultural use. The air, water, and food in the house were found to be highly contaminated. Although parathion has a long and successful history of effective use, it should not be available for home use.

When California's citrus industry was threatened by the Mediterranean fruit fly in 1981, malathion was used to save the $14 billion industry. The decision to spray brought a great deal of controversy about whether or not malathion posed a health hazard. Fortunately, experience was available from the state of Florida. A massive aerial spraying had been done in south Florida to eradicate the medfly in 1956, as well as another heavy application in 1962–1963, over three counties as well as continual use for mosquito control.

E. Petroleum Oil Fractions

These oils are generally used alone or in combination with one of the other classes of insecticides. They have proved most useful as mosquito larvicides. The oils are applied as a thin film on the surface of water of mosquito breeding sites. The larvae coming to the surface to feed or breathe contact the oil film, which penetrates the breathing tube (trachea) and kills by suffocation or poisoning.

III. BIOLOGICAL CONTROL

A. Resistance

Pesticide use has been increasing. Worldwide sales have climbed from $8.1 billion in 1972 to $13 billion in 1983 in constant dollars. At the same time, farmers the world over have reported that they must use heavier concentrations and often multiple chemicals to achieve control. Over the years pests have developed resistance to the chemicals.

According to a recent report (1) between 1970 and 1980, the number of resistant arthropods has doubled from 224 to 428. Some 25 species of beetles, caterpillars, mites, and other insects that attack cotton in 36 countries are now resistant to pesticides. The problem of resistance not only affects agriculture but public health as well. *Anopheles*, the genus of mosquito that transmits malaria, currently the world's number one disease, is now resistant to several of the major chemical insecticides in six countries. The number of new cases of malaria doubled between 1972 and 1976. In addition, fungi and rodents have also developed resistance to chemicals that formerly controlled them. Figure 2 shows this increasing pattern of resistance, while Table 4 lists the number of resistant species known to be immune to specific types of pesticides.

In the United States, costs of added chemical applications are estimated at $130 million, which farmers of course pass along to consumers. It is further estimated that this figure can be tripled if added costs to combat plant pathogens, weeds, worms, and rodents are included.

Resistance is a natural process, and it may take (or require) other natural processes to overcome it. Resistance is nothing more than an evolutionary survival mechanism. Those species, or individuals within species, genetically able to withstand a chemical, or those surviving as a consequence of a protective random mutation, begin to multiply as their

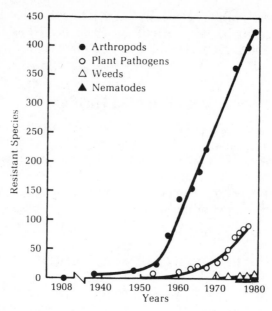

Figure 2. Increases in the number of pests resistant to pesticides. Courtesy of World Resources Institute, November 1984.

nonresistant members die off. As a result, populations of resistant species are developed. They have in fact been "selected out" by the pesticides meant to control them. How then can these populations, as well as those not yet resistant, be managed or controlled? One way is biological or natural control.

Table 4
Increases in Arthropod Resistance, 1970–1980

	Number of Resistant Species	
	1970	1980
Number of species (irrespective of number of chemicals resisted)	224	428
Number of species resisting		
DDT	98	229
Cyclodienes	140	269
Organophosphates	54	200
Carbamates	3	51
Pyrethroids	3	22
Fumigants	3	17
Other	12	41
Total for all groups	313	829

Source: World Resources Institute, November 1984.

B. The Idea of Biological Control

Recent research has focused on a variety of alternatives to broadcasting chemical pesticides over the landscape. Biological control has as its objective the reduction of pest populations to tolerable levels by manipulation of the insects' own physiological and social systems and their ecological niche.

Biological control emerged almost a century ago when the citrus groves of California were threatened with destruction by the cottony cushion scale. When the available means of control failed, investigators went to its original home in Australia to find the mechanism that kept it in check there. The check proved to be a predator—the Vedalia lady beetle. The successful introduction of the vedalia beetle into California resulted in the complete control of the scale in one year, at low cost and with no untoward side effects. The idea of using specific predators, pathogens, and parasites remains a part of the armamentarium of pest control, but new dimensions have been developed.

Many insects locate their mating partner, food supplies, and favorable egg-laying sites via potent physiologically active chemical stimuli. Among these are sex attractants, food attractants, and oviposition lures.

C. Pheromones

Insects appear to communicate with chemicals. They release infinitesimally small quantities of highly specific organic compounds that vaporize and disperse quickly and that are detected by insects of the same species. One of these is the pheromone,* a sex attractant that scientists hope will disorient the insects' sex lives. If this can be achieved, there will be no reproduction; no reproduction, no pest; no pest, no crop damage.

Disorientation of mating partners is just the type of nonproductive concept needed. If a little attracts, a little too much may just repel or neutralize. Either way there's no sexual contact. It is hard to conceive of a better way of keeping populations low or nonexistent.

1. Periplanone-B

In September 1984, Stuart L. Schreiber of Yale University reported on his successful production of a synthetic cockroach aphrodisiac. This will offer the waiting world—waiting for as long as humans have been on earth—sure-fire means of ridding its habitat of *Periplaneta americana*, the American cockroach. Figure 3 shows the usually slender, larger American cockroach, compared with the German (Figure 4) and Oriental (Figure 5).

---

* In 1959, Karlson and Luescher defined pheromes as "substances which are secreted by an individual and received by a second individual of the same species, in which they release a specific action, for example, a definite behavior or development process." They coined the term from the Greek *herein*, to transfer and, *hormon*, to excite (2).

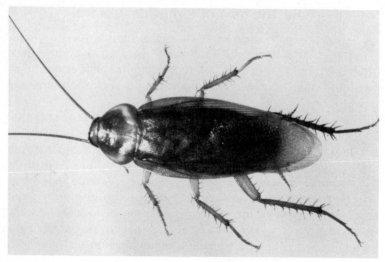

Figure 3. The American cockroach, *Periplanata americana*. Courtesy of the CDC.

These three make up the predominant types of cockroach that infest most cities today. Figure 6 is a close-up of a German female cockroach extruding an egg capsule.

The synthetic substance Schreiber calls periplanone-B, a complex ten-member ring compound that took years to synthesize, is so potent that 2–3 femtograms (several quadrillionths of a gram) can send half-a-dozen male roaches into sexual frenzies from which they barely recover. Traps

Figure 4. German cockroaches (*Blatella germanica*): male at left, female at right. Courtesy of the CDC.

Figure 5. Oriental cockroaches (*Blatta orientalis*). Courtesy of the CDC.

using the periplanone attractant may well become available commercially by 1989.

The pheromone of the spined soldier bug was recently discovered, patented, and should become commercially available by 1991. The soldier bug is different. It is not a pest. The soldier bug preys on the tomato hornworm and the Mexican bean weevil. One soldier bug can

Figure 6. German cockroach with egg capsule (ootheca). Courtesy of the CDC.

consume as many as 50 eggs/day of these two pests. Its pheromone could readily lure beneficial bugs into a field where by dieting on pests they could protect crops against predation. It would be easy and inexpensive.

2. Gypsy Moths

Sexual lures are also being tried on gypsy moths, a scourge that has defoliated 13 million acres of prime woodland across the Northeast, primarily in New Hampshire, Massachusetts, Connecticut, New York, New Jersey, and Pennsylvania. Figure 7 graphically depicts gypsy moth depredation as seen from a NASA U-2 aircraft flying at 65,000 ft. Dark areas around the reservoir and roads have been treated for gypsy moth and retain their foliage. Gray areas have been completely denuded by the moth.

The gypsy moth, *Porthetria dispar*, is one of the most despised pests in the country. Scientists have devoted their careers to the moth and its extermination with little success. The moths owe their persistence to two principal factors: their remarkably effective survival equipment and the inability of entomologists to understand and thereby control the moths' "boom and bust" population cycles. They have an extraordinary capacity for reproduction. A single egg mass can produce more than a thousand larvae, and they have an awesome appetite.

Figure 7. Effects of gypsy moth seen from NASA U-2 aircraft flying at 65,000 ft. Dark areas around reservoir have been chemically treated and retain their foliage. Gray areas are those completely denuded by the caterpiller. Courtesy of the USDA Forest Service.

Chemical insecticides, often effective, are seldom used because their potential toxicity, or perceived toxicity, makes them politically unpalatable. With DDT, eradication of the moth seemed imminent. When it was banned the moths staged a spectacular comeback. Sevin (carbaryl, a methyl carbamate), considered better than DDT because of its lack of persistence in soil, is also avoided because it too is looked on dubiously by a public fearing contamination of water supplies. Meanwhile, biological control of the gypsy moth has not matured, though several approaches are being explored.

The resistance to chemical sprays has forced the New York Department of Environmental Conservation (DEC) to abandon its gypsy moth control program. This decision was made because of the difficulty the state had in obtaining a consensus on spraying from towns, counties, and private landowners. Prior to 1985, the DEC conducted aerial and ground surveys and certified pesticides. After 3 years of decline, the gypsy moth population of the Northeast rose sharply in 1986, causing almost double the damage done in 1984. Substantial defoliation was reported in New York, New Jersey, and Connecticut, and the moths were spreading south and west into Maryland, West Virginia, and Pennsylvania.

They eat so voraciously that their damage is recorded in acres defoliated. Trees weakened by defoliation often die. Oaks, birch, willows, and maple are the moths favorite varieties. Two states, Rhode Island and Delaware, had far less defoliation because of their aerial spraying programs—in spite of public opposition. For the foreseeable future it will be up to local communities to deal with their own woodlands, make the surveys, contract with private applicators, and pay the full cost of control. Funds will not be available from either the states or the federal government. Given this fragmented approach, the gypsy moth will have unfettered access to the trees for years to come.

D. Microbial Pesticides

Although diseases of insects were found, almost simultaneously with human diseases, to be the result of microbial pathogens, circa 1868, the idea of turning them loose on insect pests was slow in coming. *Bacillus thuringiensis Berliner*, originally isolated from a diseased flour moth in Thuringen, Germany, in 1911, has the distinction of being the first microbial pesticide. Thuricide became commercially available in the 1970s and was certified by the FDA for use on crops.

Microbial pesticides have an inherent advantage. They can be broadcast over an area, in a field, much the same as conventional chemicals; however, they can initiate disease among susceptible species and reduce them to tolerable numbers quickly. Because they reproduce, they can remain viable in an area as a constant check or control on a pest population. They become part of the ecological niche.

An outstanding example of pest control as a consequence of a microbial insect pathogen is "milky disease" of the Japanese beetle (*Popillia japonica*). The bacterium *Bacillus popilliae*, ingested by the beetle grub (larva), multiplies in the insect's gut. Shortly thereafter, the larva ceases activity and dies. Its blood becomes milky-white, hence the name of the disease.

In the commercial preparation, spores of *Bacillus popilliae* are mixed with powder to be broadcast over soil. The larvae of beetles or other susceptible insects ingest the spores and are destroyed within 2 weeks by milky disease. Applied to turf in the eastern part of the United States, the preparation has provided excellent control of the Japanese beetle. The spores of *Bacillus popilliae* persist for long periods in soil and achieve long-term control of the beetles.

Thuricide, also known as BT, works in another way. Its spores produce a proteinaceous crystal that can destroy the gut lining of insects. However, BT has proved short-lived and thus requires multiple applications. A team of biologists at Temple University, Philadelphia, may have solved the BT problem. William R. Harvey and his colleagues have isolated a crystal BT peptide 20 times as potent for insects, yet harmless to people.

Harvey's approach, however, is different. Perhaps it should be called ingenious. Rather than spray crops as has traditionally been done, they are using gene-splicing techniques to implant the gene for this peptide into the seeds themselves. Imagine, seeds of the various food crops containing a gene for resistance to a variety of insect pests. Planted, they would develop unmolested. Yields would be unaffected, and no pesticides would be used in the field. This is Nobel-type thinking.

Currently, his efforts are being directed against the gypsy moth, corn borer, and tobacco and tomato horn worms. With respect to the gypsy moth, seeds of trees resistant to the moths—trees the moths would avoid—are being "engineered." Translated, that should mean that the caterpillars would starve for lack of a suitable food supply. This is an idea whose time has come. The ability to do it is at hand: and corporate America knows it. A race is on to get in on what will surely be a huge, lucrative industry. The stakes are high.

Grasshoppers are a major pest around the country and around the world. In 1985–1986, some 20 million acres of rangeland were infested to some degree by grasshoppers. One of the problems with chemical insecticides used to control them is that by the time fields are treated, the crop is lost.

In Broomfield, Colorado, Evans Bio Control, Inc. hatches 40,000 grasshoppers weekly, infecting them with a microscopic protozoan *Nosema locustae* contained on a wheat bran bait. This one-celled parasite attacks the midget—especially fat bodies—competing for energy reserves. In so doing, it disrupts circulation, the excretory and reproductive systems. As the infection builds, insect feeding declines and death ensues

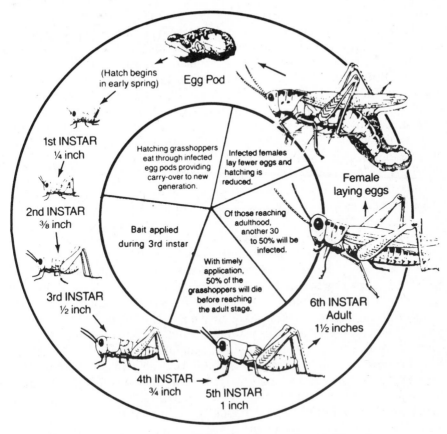

Figure 8. How the NOLO bait affects the grasshopper life cycle.

within 3–4 weeks. By releasing infected insects, the disease spreads through the population via egg pods to their offspring.

Nosema appears not to harm domestic pets, plants, birds, livestock, or other insects. Figure 8 shows the grasshopper life cycle and the point of application of the *Nosema* bait.

E. Gene-Splicing Biotechnology

Gene splicing, or recombinant DNA techniques, is opening totally new horizons. Consider that a gene in reality is a region of DNA and can therefore code for the production of the protein. What is needed then is the ability, the technology, to slip or splice a gene known to possess a specific code into the organism's genetic material. Given the proper signals, the microbe, usually a bacterium but it can be others, will produce the new protein as part of its "normal" metabolism. And it will continue to do so as long as it is given appropriate sustenance.

First, the need is to identify the gene that codes for a specific product or metabolite. Having identified it or its position on a chromosome, there is an excellent possibility that it can be isolated and recovered.

Genes are molecules of DNA carried neatly packaged in the chromosomes within the nucleus of each cell. DNA directs the assembly of amino acids into the proteins essential for normal functioning. Using recently developed analytic techniques, it is entirely possible to remove DNA from cells and, using restriction enzymes, cut the DNA strands into segments and inset new and foreign genes. In addition to their single, large-ringed chromosomes, cells contain much smaller closed loops of DNA called plasmids, an extra bit of DNA that consists of only a few genes.

Once the restriction enzymes slice DNA and the new gene is inserted, a ligase enzyme, acting as a genetic "glue," reattaches the cut segments and the splicing is complete. Thus, bacteria that divide and double every 30 min are virtual chemical manufacturing plants. These engineered microorganisms can be made to produce or metabolize almost anything.

In a test tube, using detergentlike chemicals, the cell membrane is dissolved, allowing the cell contents, including the small plasmids, to spill out. These are the most easily modified. The segregated plasmids are mixed with restriction enzymes that cleave it at various points, opening it up and "stretching" it out. Conveniently, sticky ends are produced. The solution also contains the ligase which cements the sticky ends together and the new gene into place. The result is a new plasmid loop. The complete process is shown in Figure 9. New plasmids are placed in a solution of cold calcium chloride, which contains normal untreated bacteria. This solution is heat shocked. As a result, the membranes become permeable, allowing the new plasmids to pass through and become part of the microbes' new genetic constitution. When the bacteria reproduce, they create "carbon copies" of themselves, including the new plasmids.

Harvey has another project. He would like to eradicate malaria—quite obviously a worthy goal. At this moment in history, however, the long war being waged against the mosquito is going badly. Around the world mosquitoes are among the many insects becoming resistant to every class of chemical pesticide. This suggests multiple resistances.

Harvey's approach is to create algae that are toxic to mosquito larvae. Breeding as mosquitoes do in water, and feeding on algae as their means of nourishment, algae with a built-in BT gene will be lethal in larvae that ingest them. Since the new gene is now a part of the algae, it will be there forever, and mosquito control should become part of the natural ecological interrelationships of an area. Ridding the world of malaria would be worth several Nobel prizes. The theoretical approach is sound. Whether it will work "in the field" remains to be tried and seen.

A variation of the bacterial pesticide concept is currently being field tested by scientists of the Monsanto Company of St. Louis. In an attempt

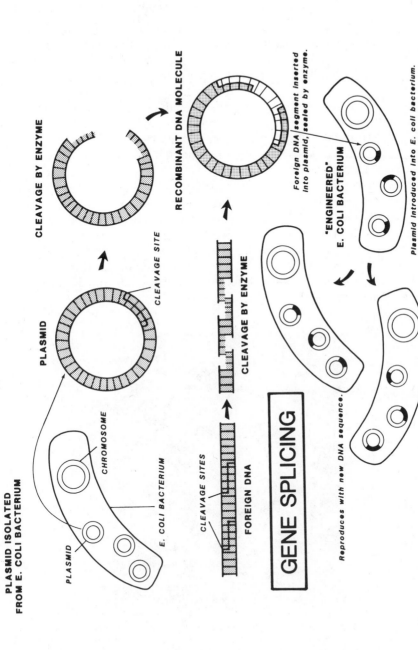

Figure 9. The gene splicing process.

159

to control the effects of the black cut worm, which destroys the roots of corn and soybean plants, they have engineered (spliced into) a common soil bacterium, *Pseudomonas fluorescens*, with a gene taken from *Bacillus thuringiensis*. This gene governs (codes for) the production of a chemical toxic to the cut worm. Corn and soybean seeds will be *coated* with the bacteria and planted. By the end of the growing season, data from the test and control plots will be compared for differences in crop yields. This will be the first field test of a genetically engineered biopesticide. Tests were concluded in September 1985. Should these tests, along with safety tests, prove reliable, actual commercial products may be available to farmers by 1989 or 1990. The photomicrograph in Figure 10 shows the bacteria on the surface of corn seeds.

Tobacco farmers in hot and humid areas such as Mississippi may benefit handsomely if recent field trials by scientists of the Rohm & Haas Company are successful. They have spliced a single gene, again from *B. thuringiensis*, into tobacco plants which are the prey of hornworms and bud worms. This gene induces the production of sharp crystal formation in tobacco plant leaves. When a caterpillar bites into the leaf it ingests these crystals which rip its gut and kill it. Hybrid seeds may be ready for marketing by 1992.

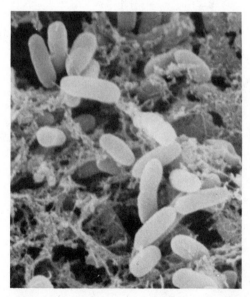

Figure 10. Isolates of a common nonpathogenic soil bacterium, *Pseudomonas fluorescens*, have been genetically engineered to contain the pesticidal, lepidopteran specific δ-endo-toxin gene from *Bacillus thuringiensis* sub-sp. *kurstaki*. This scanning electron micrograph shows the organisms on the surface of corn roots. Courtesy of Dr. Lidia S. Watrud, Research Manager, Monsanto, September 1985.

F. Frostban

An altered strain of *Pseudomonas* is also currently being field tested to determine its ability to reduce frost-induced crop damage. Such damage costs farmers some $1.6 billion a year in crop losses—which of course increases costs to consumers.

Experiments performed several years ago showed Dr. Steven Lindow that spraying a field of corn with the antibiotic streptomycin reduced frost damage. Frost damage is not generally seen as a microbially induced condition. Nevertheless, the presence of certain bacteria on plant leaves fosters ice formation. Streptomycin, by removing the organisms, controlled ice formation. In the absence of bacteria on leaf surfaces, water supercools but does not freeze when temperature drop below 0°C. Widespread use of an antibiotic is not an environmentally sound practice.

The use of an indigenous soil bacterium can be. Strains of the gram-negative bacterium *Pseudomonas syringae* produce a protein that inhibits formation of ice crystals.

Although the freezing point is 32°F (0°C), impurities in water generally prevent freezing until the temperature drops several degrees lower. Frost can form at higher temperatures because of the presence of *Pseudomonas*. Lindow's group engineered a mutant organism that had a single gene deleted. This ice-minus strain is no longer capable of producing the ice-nucleation protein but in all other respects is identical to the naturally occurring organisms.

Working with scientists at Advanced Genetic Sciences, Inc., Lindow and his co-workers at the University of California have developed a protective product called Frostban.

By appropriately timing the application of Frostban, it is possible to prevent the growth of ice-seeding leaf and blossom bacteria, which produce much of the frost injury at temperatures between 32 and 23°F (0 and −9°C). Frostban reduces the freezing temperature of the plant tissue.

This treatment can replace or reduce the need to use chemicals, smudge pots, burning tires, wind fans, and sprinklers to achieve frost protection. Farmers will also benefit by using Frostban to control the periodic occurrence of spring frost that damages blossoms and small fruits of Rosaceous plants—strawberries, almonds, pears, peaches, cherries, and apricots, as well as frost-sensitive seedlings such as tomatoes and peppers. Expectations are that the public would also benefit; with reduced losses to growers, savings should be passed on to consumers.

After 6 years of delay, due primarily to legal challenges, Frostban appeared to have been successfully tested in June 1987. Sixty percent of a crop of strawberries were protected by the application of Frostban, which reduced freezing an average of two degrees. Of singular importance was the observation that these mutant bacteria were controlled with the experimental plots. When tests are completed and all objections satisfied, AGS hopes to obtain marketing approvals by 1990.

The idea of genetically altered microorganisms has fanned the fears of people who believe they might behave unpredictably, setting off ecological catastrophes or at least disrupt local ecosystems. To allay the public's exaggerated fears of an "Andromeda strain," tests are being conducted at Clemsen University, using another *Pseudomonas, Pseudomonas fluorescens,* that has been altered by the addition of two genes taken from the common intestinal organism, *E. coli.* When grown on a nutrient medium containing lactose and a chemical called X-Gal, this newly created *Pseudomonas* has the ability to convert X-Gal to bright blue, which makes the organism easily detectable. This ability to turn blue will be a marker for tracking the spread of the cells. This should be seen as a major advance and should go a long way toward satisfying the public's "need to know" where new organisms are used. If this system proves out, then organisms altered for whatever purposes can also contain the marker for "blueness" or other color that may be introduced. Tracking and identification will no longer be a problem. This means that scientists can add safety to abundance. Time should work in favor of the public.

G. Arcelin

A naturally occurring insecticide, Arcelin-1, a seed protein of the Mexican wild bean Phaseolus vulgaris, has toxic effects on the bean weevil *Zabrotes subfaciates.* Larvae of susceptible insects that feed on these seeds die at the first instar stage.

Addition of purified Arcelin to seeds confers a high level of resistance to the weevil. Similarly, breeding the gene into cultivated bean plants increases resistance.

The weevil is a major threat to stored beans. Unless the beans are fumigated with chemical insecticides, weevils can consume 15% of the stored crop. Dry beans provide much of the world's population with as much as 50% of their protein needs. Use of naturally resistant plants would be an inexpensive and safe way of protecting the bean supply. Credit for discovery of the protective bean goes to Thomas C. Osborn and Frederick A. Bliss of the Departments of Agronomy and Horticulture, University of Wisconsin (3).

H. Glyphosphate

The ingenuity of scientists may provide yet another potential weapon against pestilential depradations. A problem currently limiting the usefulness of certain herbicides is their inability to distinguish friend from foe, the crop from the weed. Consequently, the search is on for herbicides with far greater specificity, which will selectively kill specific weeds but not nearby crops with similar biochemical constitutions.

The Calgene Company of Davis, California, has introduced into tobacco plants a bacterial gene "aroA" which encodes for an amino-acid-modified enzyme. This enzyme is resistant to the herbicide glyphosphate (a phosphomethyl glycine). In the tobacco leaves the transplanted gene provides fully one-quarter of the total activity of the enzyme normally inhibited by glyphosphate. As glyphosphate is used to protect a wide variety of crops, Calgene is experimenting with the aroA gene in soybeans, tomato, corn, and oilseed rape* to produce glyphosphate-*tolerant* varieties unaffected by the pesticide. Such plants would permit more liberal use of herbicides to kill weeds without injuring the crops.

I. Phototoxins

Among the most promising forms of the biopesticides are the light-activated phototoxins. These are chemicals that become highly reactive when absorbing energy from sunlight. Some of the reactions can injure and destroy tissue. And insects that feed on plants containing them can be killed when the sun comes up.

At the University of Illinois, Constantin A. Rebeiz and co-workers found that the synthetic pathway used by plants to make chlorophyll-a is a series of branched pathways that lead to the synthesis of several chlorophyll species, each with different photosynthetic functions. The phototoxic herbicides take advantage of this.

The Illinois group developed a family of herbicides containing a biodegradable amino acid, δ-aminolevulinic acid (ALA)—Figure 11— along with a series of modulators. These dozen modulators when used singly or in combination can kill a number of plant species while leaving others in close proximity unharmed. In recent field tests they have demonstrated that their ALA–modulator combinations can kill the ten

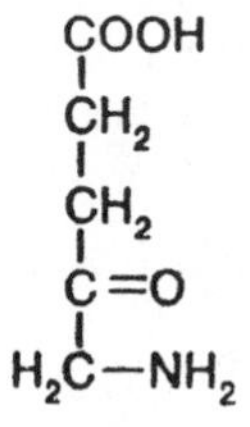

$$
\begin{array}{c}
COOH \\
| \\
CH_2 \\
| \\
CH_2 \\
| \\
C = O \\
| \\
H_2C - NH_2
\end{array}
$$

Figure 11. δ-Aminolevulinic acid.

* Plants of the Brassica family, which include mustard grown as food for sheep, as well as its seed from which oil is expressed.

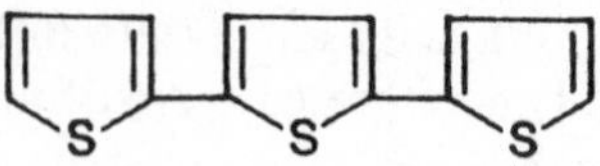

Figure 12. α-Terthienyl.

most common weeds among corn and soybean plants, while leaving those crops intact.

The approach is to treat the plants with ALA. In the dark, ALA is converted to magnesium tetrapyrolle—a precursor of chlorophyll-a. Magnesium tetrapyrolle is light sensitive and is used to form singlet oxygen, a powerful oxidizing agent which can trigger free-radical reactions that can ultimately destroy nucleic acid, enzymes, other proteins, and cell membranes. Thus, if plants produce more magnesium tetrapyrolle than they can utilize, the accumulated chemical, in the presence of sunlight, will kill weeds.

Two naturally occurring chemicals, α-terthienyl and furanoacetylene, isolated from asters have been shown to be phototoxic to mosquito larvae. Both of these are light absorbing and use light energy to produce singlet oxygen. Note too (Figure 12) that phototoxic activity can be achieved by a broad spectrum of molecular configurations.

The singular idea embodied in all the preceding examples is that novel approaches to eradication and/or control of pests spring from detailed knowledge of plant and insect metabolic pathways. It is from such fundamental knowledge that humanity is served. Basic scientific research has often been decried and attacked as wasteful expenditure of tax dollars, when in fact it is a prerequisite for innovative pest control measures.

J. Enzyme Conversion

Another imaginative innovation turns the insect's metabolism on itself. Researchers at the State University of New York at Stony Brook employ otherwise harmless chemicals which are converted to poisons by enzymes present only in insects. For example, they have chemically attached fluorine atoms to natural plant phytosterols. These compounds are chemically related to such human steroids as vitamin D, sex hormones and drugs such as cortisone. By adding fluorine to the phytosterols, they produce 29-fluorophytosterols. Insects possess enzymes that can split the molecule, producing a new compound, fluoroacetate, deadly to tobacco and tomato horn worms. The 29 fluorophytosterols are innocuous to plants, animals, and people who do not possess the chemical "key" to open the ring. This then, is a safe chemical pesticide and highly specific.

If there is one idea that these examples underscore, it is that we are moving from treatment to prevention. Plants are being bred with innate resistance to predators. No longer will it be necessary to broadcast chemicals over a field of crops. This must be seen as a remarkable development. But it had to await the slow, steady accretion of knowledge of underlying plant and parasite biochemical and physiological mechanisms. Unfortunately, acquisition of these basic biologic and chemical prerequisites requires time, talent and funding—a triad not regularly present in appropriate proportions.

K. Natural Predation

Enlisting nature's help in combatting pests has much to recommend it. A virus successfully controlled the rabbit population in Australia. The Mexican bean beetle is being controlled by a parasitic wasp. The Arabs of Yemen, for example, have for hundreds of years used predaceous ants to protect date palms from other harmful ants. Each year the date growers of Yemen bring colonies of the beneficial ants down from the mountains and place them in palm trees, where they eat the ants that feed on the palms. Snails, fish, amphibians, birds, and mammals have been used from time to time to control economic pests. Insects that prey on other insects (entomophagous species) and microbes that initiate infection and disease have also been used.

In 1973, I noted that the state of Florida imported several manatees, sea cows, to check the growth of aquatic plants that had made boating almost impossible in Florida's inland waterways. Figure 13 shows a 10- and a 12-ft-long manatee gliding along the bottom of a waterway. The smaller specimens weigh from 1000 to 1500 lb, while the larger ones weigh as much as 3500 lb. Manatees belong to the zoological order Sirenia, and

Figure 13. Manatees gliding along the bottom of one of Florida's inland waterways.

Figure 14. A manatee close up. Obviously this was once a land animal. Courtesy of Patrick Rose, Florida Department of Natural Resources, and the Save the Manatee Club.

their latin designation is *Trichechus manatus.* Figure 14 presents a close-up view of this friendly, helpful creature.

Although the manatee is a voracious consumer of aquatic plants, and their ponderous bodies require prodigious amounts of plants each day they reproduce too slowly to keep up with plant growth, unless, of course, manatees are frequently imported.

This was a splendid example of man and beast working together for the benefit of one another: an unalloyed ecological interrelationship. Unfortunately, it was not to be. By 1984, the manatee population had been reduced to 1000 survivors by the burgeoning boat population and their reckless drivers (cf. Chapter 6, Accidents).

Manatees have been mauled and cut to death by the propellers of speed boats criss-crossing the waters, totally unmindful of the sea cow's presence. Of course, aquatic weeds continue to outpace the application of synthetic herbicides which must be used if the waterways are not to become totally impassible.

In New Jersey, the alfalfa weevil (*Hypera postica*) is being successfully controlled by a fly imported from India. The thistle Cardurus, a widespread weed, is being effectively controlled by two beetles. Examples are legion (4).

Nevertheless, the gypsy moth eludes all attempts at such control. This moth was brought to the United States from Europe in 1869 to start a silk-producing industry in New England. As is often the case, several moths escaped captivity. By 1890, with none of its natural enemies to

Figure 15. Female gypsy moths laying eggs. Courtesy of the USDA Forest Service.

control its growth, the gypsy moth had become a menace, through the destruction of trees by the larvae, and has remained so ever since.

The female moth, which cannot fly, deposits buff-colored egg masses on the bark of trees. They emit a potent lure which attracts male moths, who are strong fliers. Eggs hatch within 6 weeks, and the young caterpiller larvae have voracious appetites. Figure 15 shows a cluster of females laying eggs, and as shown in Figure 16 the female is lighter and larger than the male. The caterpiller larvae seen in Figure 17 seem always to be chewing away on a leaf.

Figure 16. Male gypsy moth (left) and female (right).

Figure 17. Caterpillers chewing away leaf. Courtesy of the USDA Forest Service.

It is at the caterpillar stage that the gypsy moth appears most vulnerable to biological control. Just about every insect and animal feeds on them.

As an example, scarlet tangers, nuthatches, downy woodpeckers, and black-billed cuckoos thrive on them as they pick them off leaves and twigs. A wasp, Apanteles, parasitizes young caterpillars. The female wasp lands on the back of the caterpillar, inserts her ovipositer, and deposits an egg. The growing maggot destroys vital organs and ultimately kills the caterpillar.

Predacious ground beetles also devour caterpillars. Calosoma, a large, green European import, is the primary beetle predator. In addition, spiders attack them, white-footed mice dine on them, and the tachinid fly Compsillura uses the caterpillar as an incubator for its eggs. It lands on a caterpillar's back and quickly cuts a slit in its skin and lays an egg inside. It happens so fast the caterpillar hardly knows it has happened. Yet, more of these, or any combination thereof, have been adequate to make even a dent in the number of gypsy moths.

Given the current level of development of new means of biological control, it will be at least 5 years and 5 million additional ravaged acres of forest and woodland before the damage is reversed—testimony to prevailing unreasoned fear of chemical pesticides in many communities. That too requires study.

However, biological control is being waged on a broad front. The tools are numerous, and the ideas creative. Opportunity exists as never before. Nevertheless, the time for cheering is not at hand. Given the pests'

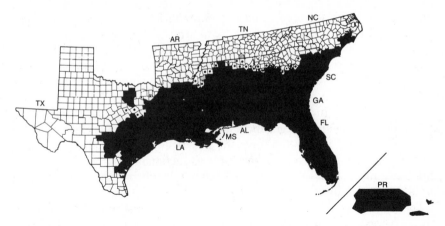

Figure 18. Area of distribution of imported fire ant quarantines. Courtesy of the USDA lab in Gainesville, FL.

penchant for survival, and their millions of years of experience,* it is too soon to raise the victory sign.

L. Imported Fire Ant

If the gypsy moth is despised, *Solenopsis invicta Buren*, the imported fire ant (IFA) is detested and feared. This nasty, fiercely stinging, omnivorous, red beast, along with its less extensively dispersed black relative, *Solenopsis richteri Forel*, have made life miserable for farmers all the way from Texas to South Carolina and Puerto Rico. The extent of their depredations and distribution are shown in Figure 18.

The IFA arrived in Mobile, Alabama, in the 1930s. South America is its natural habitat, but whether it comes from Uruguay, Paraguay, Argentina, or Brazil is not clear. Nor is it clear how they arrived, and still being speculated upon was the means of their wide and rapid dispersion from Mobile, both east and west. What is clear is that these ants have been spectacularly successful in establishing themselves, given the freedom from their natural enemies.

Particularly distressing is their venom. Five distinct alkaloids have been isolated from invicta and richteri; all 2,6 substituted piperidines: solenopsin A (*trans*-2-methyl-6-*n*-undecyl piperidine); solenopsin B (*trans*-2-methyl-6-*n*-tridecyl piperidine); solenopsin C (*trans*-2-methyl-6-

---

* It was during the Upper Cretaceous period, the age of the great coal forests, that cockroaches made their appearance. That was 320 million years ago. As the earth entered the Age of Reptiles, some 230 million years ago, many of the insect types known today—grasshoppers, crickets, mayflies, dragonflies, leafhoppers, cicadas and others—became established. When the flowering plants arose some 130 million years ago, the insects that pollinate them came along; for example, moths, bees, wasps, flies. Indeed, they have all had appreciable time to learn much of earthly adventures.

n-pentadecyl piperidine); dehydrosolenopsin B; and dehydrosolenopsin C. All are C_{11} to C_{15} ring compounds, natural and toxic.

The venom is used to kill or immobilize prey and has been shown to possess insecticidal, bactericidal, and fungicidal activity (5). It is also an extremely effective weapon for preventing large animals from disturbing their mounds.

Unfortunately, human beings also disturb their mounds and in the course of farming simply get in their way. Figure 19 shows one result of this interaction. Figure 20 shows the beasts at close range as unusually formidable creatures. The worker ant uses its tenacious mandibles to bite and hold fast as it lowers its abdomen to swiftly insert the venom-containing "needle."

The initial reaction is a shockingly burning sensation followed by a wheal whose diameter can attain 4 in. Pustule formation follows with scarring. Secondary infection often occurs if the pustule is broken, as is the case with laborers and farm hands who do not obtain adequate protection for the exposed sites. In addition to the piperidine alkaloids, the venom contains hyperallergenic proteins. Susceptible individuals can

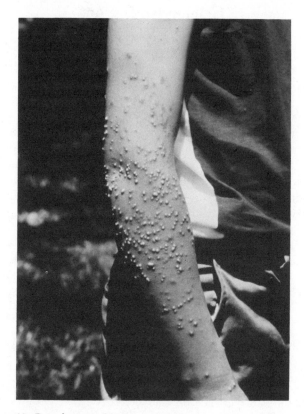

Figure 19. Pustules raised by bites of fire ants. Courtesy of the USDA.

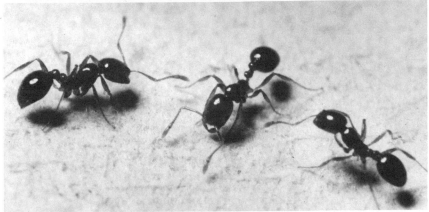

Figure 20. Close-ups of imported fire ants. Courtesy of the USDA.

anticipate respiratory distress akin to severe asthma. Deaths have been reported. More often the responses are nausea, vomiting, dizziness, and cyanosis. Lofgren and Adams of the USDA's Imported Fire Ant Research Laboratory at Gainesville, Florida, estimated the *minimum* yearly incidence of allergic systemic reactions at 1500 cases (6).

David Williams of the same facility commented that "anyone who has inadvertently stepped in a fire ant mound has a problem. For a one-time occurrence, it is an unforgettable experience. To be subjected to it frequently because of a large number of mounds around your home and work place is not only unbearable, but should not have to be tolerated."

M. Biological Control

For a number of years the IFA was reasonably controlled by the fat-soluble, polychlorinated hydrocarbon Mirex, $C_{10}Cl_{10}O$, almost identical in structure to Kepone (Chlordecone). However, in 1978, after years of controversy, it was banned and its registrations canceled by the EPA. Here again, evidence for human toxicity was based on carcinogenicity in laboratory rats. With its cancellation, there was a resurgence in fire ant populations and activity. Therefore, biological controls are currently being sought. Several pheromones are primary candidates. For example, it was found that the ants leave a chemical trail, dropping a pheromone from their stinger which others can follow.

A queen ant pheromone is also in the experimental stage. This chemical draws worker ants to it in great numbers. In addition, a brood pheromone is being studied. It appears to be the means of individual recognition. Obviously, any one of these could be translated into a bait for ant traps. However, little can be expected by way of real control before 1990.

Accordingly, it is anticipated that there will be a severe economic impact from large agricultural losses to soybeans, corn, peanuts, beans, potatoes, and cabbage, which the ants relish. But the effects do not stop there. Livestock are stung, young animals and ground-nesting birds are killed, and damage to farm machinery occurs as a result of striking mounds. Control is needed, but research is hampered because it is considered a low priority item (7).

Biological control has a long way to go before it becomes effective. Chemical control should not be dismissed out-of-hand because laboratory animal data suggest hazard to people.

IV. HUMAN HEALTH EFFECTS

Pesticides, for the most part synthetic chemicals, are purposefully constructed to achieve an end—the destruction of living things: microbes, insects, plants, and animals. And they are constructed or chosen to do this via the most efficient route for the pest in question. They do not all act in the same way, and for good reason: insects do not all feed in the same way, nor do they attack the same parts of plants, and their habits are markedly different. For these reasons, insecticides must be developed to perform a specific function. Thus, the five groups of insecticides are also categorized by the way they get into the insect's body. Stomach poisons must be swallowed or absorbed through the gut; this restricts their use to insects whose mouth parts are suitable for biting, sucking, or lapping food. Contact poisons enter through the cuticle or skin, and thus must be fat-soluble, while fumigants enter the respiratory system through the spiracles, the external openings of the trachea.

Although insecticides have been categorized as to what types of substance they contain and how they enter the insect's body, neither of these categories should be confused with the mechanism by which the insecticide kills or stuns. For example, pyrethrin penetrates through the cuticle, but it causes paralysis by blocking nerve impulses in the insect's central nervous system. Rotenone appears to cause paralysis of the breathing mechanism. Paris green, a stomach poison, is considered a general protoplasmic poison that disrupts several enzyme systems. Lindane, a chlorinated organic, is a fumigant, entering via the spiracles, while DDT is classed as a contact insecticide; both appear to affect the peripheral nervous system and muscle tissue of insects. The organophosphorus compounds are powerful contact insecticides that inhibit the action of cholinesterase, thereby preventing the normal mechanism of nerve-impulse transmission.

Given the similarity of basic life processes and chemicals (DNA, for example, in all species is composed of two polynucleotide strands in a helical configuration) across the zoological spectrum, it is reasonable to assume that pesticides used to destroy beetles may also harm people. It is this reasonableness that underlies use of all biological test systems, making use of the similarities between animals, bacteria, and tissue cultures.

Nevertheless, "similarity" between species cannot be translated as meaning "the same." Accordingly, data obtained from animal studies must be treated cautiously when evaluating potential harm to human populations. Four of the six people who had received the Jarvik-7 artificial heart implants suffered strokes and injury to the brain due to imparied blood flow. Yet in the more than 100 animals that have received similar implants during experiments at the University of Utah, only three developed strokes, and these were the result of mechanical blockage by bacterial clumps which had broken away from infected implants, not the result of blood clots.

Fifteen calves received implants. Not one developed a cerebrovascular accident. Human blood differs markedly from animal blood—sheep, cows, goats—all mammals. Rats are certainly far removed zoologically, and their hearts beat far faster. The rabbit heart beats even faster, at several hundred times per minute. We are not rabbits, we are not calves, nor are we rats, mice, or goldfish. It is time we understood that animals can be of only limited value in relieving the human condition. Decisionmaking and standard setting regulations must be based on criteria more nearly approximating the human animal itself. This is even more the case with data from microbial or tissue culture experiments.

Unfortunately, knowledge of human toxic effects of most pesticides is almost completely lacking. Based on available laboratory animal data, however, these chemicals appear to be among the most toxic of substances. To make interpretations all the more difficult and engender still greater uncertainty, harmful effects to similar dose levels vary considerably by type of animal used. Doses of a chemical inducing similar effects or death can vary by a factor of 4000–5000.

Direct human experiments with these chemicals are both too dangerous and unethical. If that is so, how then can human data be obtained? The answer is inadvertently, through carelessness, and via accidents. Recall that much, if not all, of our knowledge of acute respiratory effects of air pollutants came via this route: Donorra, Pennsylvania, London, and Belgium's Meuse Valley (cf. Chapter 10, Air Pollution).

Frederick Coulston, editor of *Regulatory Toxicology and Pharmacology*, recently commented on this. "No more," he said, "should we listen to those who preach give the highest dose to the most sensitive species and this will predict what happens in man without any regard whatsoever to the metabolic fate of the chemical in the particular species to be studied. What good does it do to study a chemical for 2 years in the dog when the dog does not metabolize the chemical as it would be metabolized in man" (8).

And Joseph D. Rosen, Department of Food Science, Rutgers University wrote the following:

> In 1977, Weisberger reported that EDB caused a high incidence of tumors in rats and mice. The doses used in these experiments were about 40 mg/kg/day in rats and 107 and 62 mg/kg/day in mice. [Two groups of rats (one male and one female) were tested at about 80 mg/kg/day, but this dose had to be cut back considerably because of the very high toxicity.] In order to err on the side of safety, EPA used the 40 mg/kg/day rat dose for dietary assessment because it was lower than the dose administered to mice. From both analytical and human grain consumption data, the agency was able to calculate an average EDB daily intake for humans of 5.5 mg/kg/day a dose *approximatly 7 orders of magnitude lower* than the dose administered to rats. Somehow, EPA was able to calculate an increased dietary risk of 1 per 100 due to EDB use on the basis of the consumption data, *one rat data point*, and the Wiebull mathematical model. This calculation has little underlying science to back it up. Unfortunately, numbers lend credence to regulatory decisions, and we will be given numbers no matter how these numbers are obtained. (9)

Accidents, in their scale, can often provide far more information than any planned experiment, but a salient omission, or missing link in an accidental episode, or inadvertent contact is lack of appropriate exposure data. The events may either occur so rapidly or so insidiously or under such careless conditions that no one is prepared to gather that essential information. Without it, we must fall back on estimates. But then again, estimates are what we get from animal data.

Several trenchant "case histories" follow, from which illuminating information can be drawn. This information suggests that animal data can be misleading for people.

A. Dioxin

2,3,7,8-Tetrachlorodibenzo-p-dioxin is a colorless, crystalline solid at room temperature. With eight ring positions that could be occupied by the four chlorine atoms of a tetrachloroisomer,

Chlorinated dibenzodioxin

22 tetrachloroisomers are possible. However, chemists have isolated and identified 75 dibenzo-p-dioxins containing chlorine atoms. The 2,3,7,8-isomer appears to be the most toxic to animals, but the lethal dose varies considerable from species to species. In 1982, Poland and Knutson reported on the differential lethalities (9). Table 5 lists these in order of increasing dose. Clearly, the ordering is haphazard. There is no indication of relationship of dose to zoological complexity.

Note that the monkey, a primate, is far more sensitive than the amphibian, yet the guinea pig, a rodent, is far more sensitive than the monkey. And the hamster is remarkably more resistant than the guinea pig. Yet both the guinea pig and the hamster are rodents. But the mouse and rat (some would include the rabbit, a lagomorph) are also rodents. So there is a great diversity of response to dioxin just within the Rodentia. Doubtless there are significant physiological (biochemical) differences in enzyme systems or metabolic pathways between species. Certainly size is unrelated to dioxin activity. It would be useless even to try to predict the effects on either elephants or shrews.

The tantalizing question is which species is most predictive of human health effects. Or are human effects unpredictable from these animal species?

There is another question. Does this table contain the entire range of dose levels? Do all animals fall between 1 and 5000 ppm? Are some

Table 5
Dioxins' Lethal Dose Varies Between Species

Animal	LD_{50} (mg/kg body weight)[a]
Guinea pig	1
Rat (male)	22
Rat (female)	45
Monkey	70
Rabbit	115
Mouse	114
Dog	>300
Bullfrog	>500
Hamster	5000

[a] The LD_{50} is the toxicologist's way of indicating the lethal dose to 50% of a group of animals being tested.

animals sensitive to less than one or more than 5000? Obviously, the choice of species is central, if not crucial, to any decision making strategy.

It is true that for the most part human exposure has been to high levels of dioxin for relatively brief periods, with the exception of the "Ranch Hands," who applied the defoliants in Vietnam and had 3-month tours of duty, and the agricultural workers in New Zealand, who regularly sprayed 2,4,5-T with its low levels of TCDD as a contaminant (10).

Consequently, we are without significant long-term low-level human exposure effects. Much the same situation currently pertains to air pollution and human effects. This is worth much reflection. Animal data, at least in this instance, are clearly at variance with human experience. Species specificity requires more attention than it is currently receiving.

The words of Stephen M. Ayres, Chairman of the Department of Internal Medicine at the St. Louis University School of Medicine, are instructive. "Society," he said, "has probably been undeservedly fortunate in the matter of dioxin. Synthesized by German scientists in the mid-nineteen fifties and known since then to produce an acne-like lesion when applied to rabbits' ears, it has been shown to produce all sorts of problems in experimental animals, including cancer. Extremely large exposures in man however, have not produced many lasting illnesses probably because the insoluble material rarely reached high tissue concentrations in man or gets into the human food chain" (11).

This is not to say that data obtained from animal experimentation are useless. The foregoing is offered as a means of bringing a perspective that appears to be lacking in current discussions. Data from animal and other biological studies are presented to the public as though there is a one-to-one correlation between animals and humans. There is not. On the other hand, the available human data appear to be treated as though they do not exist. Not only is there a great deal of it, but it is well reasoned and appropriately documented. Public discussion and decision making would benefit measurably from its inclusion.

In fact, the first of a dozen accidental releases of TCDD and other dioxins, as a consequence of the manufacture of 2,4,5-trichlorophenol (TCP) or 2,4,5-trichlorophenoxyacetic acid occurred in Nitro, West Virginia, in 1949 (12). At least 11 other industrial accidents or exposures have occurred since then. By 1985, an estimated 600 people are known to have been heavily exposed—including the 160 employees at the ICMESA plant at Meda, Italy. From these episodes, it appears that human health effects simply do not follow those suggested or predicted by laboratory animal data.

In a lecture at the School of Veterinary Medicine, University of Pennsylvania, in April 1985, Philip Dorn* presented data clearly showing

---

* Philip Dorn is Professor and Chairman, Department of Veterinary Preventive Medicine, Ohio State University.

this marked species difference. From investigations of the effects of airborne lead in the Ozark area of Missouri, he found that horses were acutely susceptible, while cows and human populations were unaffected by similar blood and bone lead levels. These striking differences can no longer be ignored. Public discussion and decisionmaking at all levels require its inclusion.

B. The Meda Incident

Saturday morning, July 10, 1976, was not unlike other Saturday mornings in Meda. Fifteen miles north of Milan, Meda with a population of less than 20,000 is a quiet, rural community. One hundred sixty of its citizens worked for ICMESA (Industrie Chemiche Meda Societa Anonima), a subsidiary of Givaudin, whose parent company is Hoffmann–LaRoche, a Swiss pharmaceutical firm headquartered in Basel.

The plant at Meda produced trichlorophenol (TCP) for use in making hexachlorophene. Plant workers knew that the safety valve of the 10,000-L reactor vessel had not been provided with protective devices, but few gave it a second thought; production had proceeded smoothly for years.

Between noon and 2 p.m. a runaway reaction began in the reactor (Figure 21). Although the plant was shut down for the weekend, the

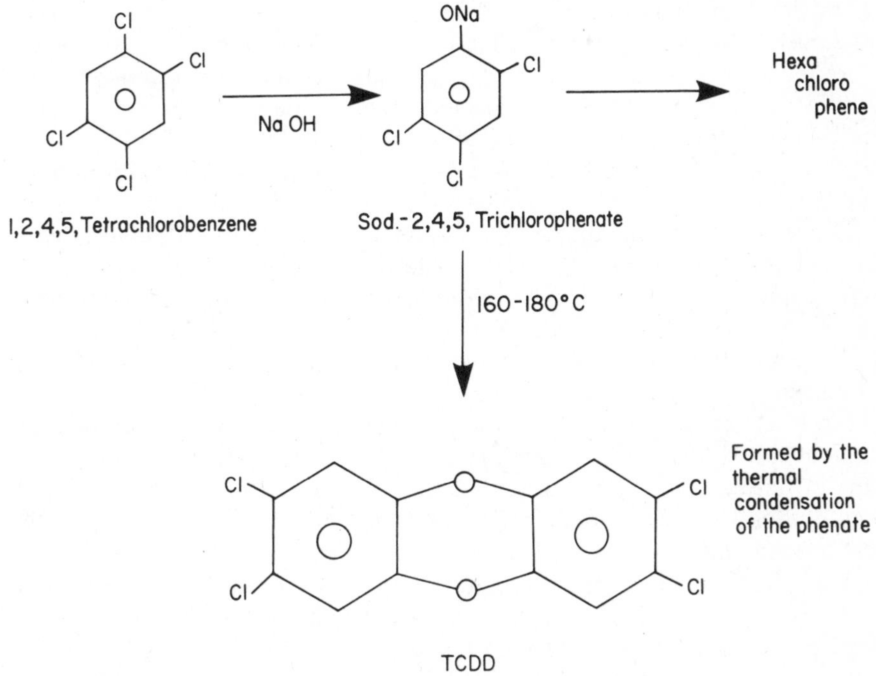

Figure 21.

reactor had been charged with ingredients necessary for production of a new batch of TCP. An unusual amount of pressure developed within the reactor. Without the protective units, the rupture disk burst and hot gases, vapor, and particulates exploded into the air above the plant. Dense white clouds formed, and the prevailing winds carried the chemical mist to the southwest. Although the incident occurred in Meda, Seveso became the "epicenter" of the toxic rain. By Monday, children began to complain of red, burning skin lesions. Rabbits, cats, birds, poultry, and frogs were dying. But it was not until the following Monday, July 19th, that ICMESA officials acknowledged that 2,3,7,8-tetrachlorodibenzo-p-dioxin (TCDD), the most potent and toxic of the dioxin isomers, had been expelled along with TCP and sodium hydroxide. In the reactor vessel the reaction proceeds as indicated in Figure 21.

Between July 10th and 20th, no one was alerted nor was medical attention provided anyone to control for possible TCDD damage. Given an almost complete lack of public health facilities and organization, the inhabitants were exposed for approximately 3 weeks before the first group of children and pregnant women were evacuated (13).

For at least 30 days after the release, the only clinically observable signs noted were the florid skin lesions, showing approximately second and third degree burns. Two months later chloracne began to appear. New cases occurred well into the spring of 1977. For some children chloracne persisted well into 1979.

One of the most pronounced effects of several of the chlorinated dioxins is its tendency to induce blackheads (comedones) which are characteristically distributed on the cheeks—particularly on the prominences beneath the eyes, spreading to the chin, behind the ears and back of the neck. It was first described in 1899 and its cause was identified in 1917 as contact with certain chlorinated hydrocarbons.

Because data from laboratory animals indicated that birth defects were a prominent sequelae of TCDD, obstetricians and pediatricians in the entire province of Lombardy were alerted. Interestingly enough, under Italian law reporting of birth defects is mandatory. Nevertheless, between 1972 and 1978, one to two defects were reported in a population of approximately 1800 births per year within the four most exposed towns. This was an unusually low rate. However, Italy is well known for underreporting birth defects. During 1977, 38 birth defects were reported in 11 cities; 7 in Seveso and 16 in Meda. Of the three zones of contamination established (high, medium, and low), Seveso was in the high zone and Meda in the medium. Obviously, with so large a surge in level of birth defects, something untoward appeared to have occurred. But the data are curious. Meda with less than half the exposure had more than twice the number of defects. In addition, doublings of defects were reported from Bergamo, Como, Brescia, and Cremona in the northeastern reaches of Lombardy, in the opposite direction of the path of the drifting chemical cloud.

Unfortunately, though of inestimable value, because of the fear of birth defects that the publicity generated, 30 known medically induced abortions were performed, four among women of Seveso. None of the embryos had any type of malformation. Is TCDD fetotoxic? Animal data are clear on this point. It is. Laboratory data are also clear. Tumors (cancers) were to be expected. By 1984, 8 years after the accident, examinations showed that birth defects, spontaneous abortions, deaths, and tumors were not adversely affected or induced by the heavy "rain" of TCDD and TCP (14,15). It is especially pertinent to consider that levels of TCDD averaged 20 ppm in Seveso. That is equivalent to 20,000 ppb, an astronomical concentration. However, 8 years may be too early to tell about tumor development.

C. Times Beach, Missouri

Moscow Mills, Times Beach, and Environs. Dusty roads and horse arenas have been oiled down for years as a common practice. On May 26, 1971, a new ingredient was added. Russel Bliss, an independent oil salvage contractor, stopped at Shenandoah Farms near Moscow Mills, southwest of St. Louis, to apply a coat of oil to their horse arena. The oil was in fact undiluted still bottoms* containing TCDD at a concentration of 350 ppm—350,000 ppb!! By May 29th, horses became ill and dead birds were found on the arena floor. By the middle of June, insects, worms, mice, rats by the bucketful, hundreds of birds, a dozen cats, and four dogs were found dead in or near the arena. Of 85 horses that had used the arena, 43 had died. Twenty-six pregnant mares aborted and foals exposed in utero died at birth or shortly after (2). Not a species within the principal animal phyla was spared, except one: people. Children and adults regularly used the Shenandoah arena. For one 6-year-old girl the arena was her "sandbox." Three younger children and a number of adults were regularly in and out. Other than headaches, bloody noses, hematuria, diarrhea, and dermatitis, all had uneventful recoveries (15).

In another arena similarly sprayed and contaminated, two 3-year-old boys were heavily exposed during that summer. They exhibited clear cases of chloracne that persisted for over a year. Today, 17 years later, those children are all healthy adults.

Russell Bliss also played a major role in contaminating Times Beach, Missouri. The roads in that area were also oiled with TCDD-contaminated still bottoms. Times Beach sits astride the flood plain of the Meramec River. In 1982 and 1983, flood years, the Meramec overflowed its banks and inundated Times Beach. The U.S. EPA estimated that 100 times more TCDD was sprayed in Missouri than was released from Meda's ICMESA plant and fell on Seveso. It has been estimated that the TCDD level in the still bottoms was upwards of 350 ppm—350,000 ppb. Remember too that

---

* Still bottoms are the waste materials remaining from chemical distillation processes.

TCDD is completely insoluble in water. This raises the possibility that TCDD was not spread evenly and equally throughout Times Beach. Nevertheless, the U.S. EPA ruled that Times Beach was a health hazard and required that its inhabitants leave. Times Beach was abandoned. It no longer exists (cf. Chapter 15). Whether it is in fact an actual health hazard remains to be determined.

D. Vietnam, Defoliation, and Agent Orange

Early exposure to dioxin resulted from production or use of TCDD-contaminated herbicides. The dioxins have never been purposefully or deliberately manufactured. They are generated as inadvertent by-products during the production of chlorinated phenols such as 2,4,5-trichlorophe-noxyacetic acid (2,4,5-T). These herbicides were widely used because they had been shown to be effective against broadleaf weeds and plants. 2,4,5-T was one of the most common defoliants, the best known of which is Agent Orange.

The U.S. Air Force began using 2,4,5-T in Vietnam in 1962. At that time, the defoliant was "coded" as Agent Green. The drums containing the chemical had a green band painted around them. It was not until 1965 that "Orange" came into use, following Agents Pink, Purple, Blue, and White.

Agent Green contained 2,3,7,8-TCDD as a contaminant. Estimates indicate the level at 66 ppm. At that time, however, the number of U.S. troops was still relatively small. Agent Purple, a newer product, contained 33 ppm TCDD, and from 1965 to 1970, when the spraying was halted, Agent Orange was used. It contained about 2 ppm TCDD.

In Vietnam, airborne crews were used to broadcast the defoliant over the landscape. Some of the crew received daily exposures of 2 ppm. Consequently, they were enrolled in Project Ranch Hand, an epidemiologic study initiated by the U.S. AF to ascertain long-term human effects from exposure to TCDD (16).

In this study, 1274 exposed military personnel were compared to 6171 unexposed men who also served in Vietnam at the same time. The study found that for these groups, the overall mortality rates were almost identical. Neither chloracne nor phorphyria cutanea tarda were observed in the "test" group. Porphyria cutanea tarda is characterized by fluid-filled vesicles (blisters) on the skin, excessive growth of hair, and excessive urinary excretion of uroporphyrin, a pigment produced by the liver. Twenty years after contact with Agent Green, and 15 years after Agent Orange, the lack of tumors (cancers) between the groups is a definite plus. A retrospective study completed in 1984 by Erickson and his colleagues at the Centers for Disease Control attempted to ascertain the risk among Vietnam veterans of fathering babies with birth defects. They reported that "this study does not provide support to the notion that those men who may have been exposed to Agent Orange in Vietnam have had an increased risk of fathering babies with most types of defects." They went on to remark that "this study provides strong evidence that Vietnam veterans in general are not at increased risk. This suggests that if there is any increased risk related to exposure to Agent Orange, either the risk must be small, must be limited to select groups of Vietnam veterans, or the increased risk must be limited to specific types of defects" (17).

A follow-up study was reported in 1988: telephone interviews with some 8000 enlisted men who had done a tour of Vietnam and a similar number who had served elsewhere during the same period. A group of more than 4000 received thorough physical exams and psychological testing.

The mental tests showed that the Vietnam vets suffered more depressive episodes and were prone to drinking problems. But there was no evidence that exposure to Agent Orange affected them or their children. Hospital records showed that the rates of birth defects were similar for the children of men in both groups (18).

This study has not ended. The enrolled men will be examined again in 1992, 1997, and finally, during the concluding year of the study in 2002. By 2002, 37 years will have passed from initial exposure. Given the 20–30-year latency period for tumor development, excess incidence should be apparent.

E. Constant State of Crisis: Receiving the Wrong Messages

The possibilities and potential of biological control notwithstanding, synthetic chemical pesticides need not be summarily banished. There is unfortunately an antichemical atmosphere abroad in the land. It is difficult to be unaware of the controversy between those who are certain that synthetic pesticides are a proven health hazard, and those who are of the view that little real evidence supports such a claim.

The difficulties and uncertainties of scientific research, as well as the treatment accorded scientific reports by the various media, make such controversy inevitable. Two views stand out in stark relief: the scientific with its shades of gray, and the public's with its black or white "is it or isn't it?" mentality.

For scientists an issue or problem is rarely fully solved. Constant study is required to elucidate the fullest understanding. And, as all investigations are not equally reliable or comparable, it is understandable that confusion can arise. But it is not a requirement. Writers and broadcasters of the electronic and print media often appear to prefer crisis and confusion to enlightenment. Certainly they do not appear prepared or willing to discuss the whys of seeming contradictions. For example, differences can arise because one investigator uses mice while others use dogs, chickens, guinea pigs, or goldfish. Some researchers may use too

"What it comes down to is you have to find out what reaction they're
looking for, and you give them that reaction."

Figure 22.

few animals to draw valid statistical inferences, while others may include
only male animals or only females in their tests, or only animals of one
age group. Some researchers may keep an experiment going for 3 months,
while others feel that suitable data can only be collected after a minimum
of 1 year. Differences may also arise because a pesticide was administered
in different concentrations or in a different manner. Was it painted on the
exposed skin of the animal's belly? Or was it placed in the eye? Or was it
inoculated into the muscle or directly into the bloodstream? Or was it
surgically implanted into the bladder? Furthermore, most important and
all too often lacking in many studies that rush into public print, is use of
controls, untreated as well as treated animals. Were the controls properly
selected and maintained, and are they comparable to the treated group?
Figure 22 suggests additional uncertainties and complexities.

It would be surprising if the public was not confused. It takes a great
deal of discussion, conferences, meetings, and symposia for scientists
themselves to finally agree, and that is rarely ever total. "There must be a

better way" to impart information to an interested and concerned public, better than the current fragmented often adversarial approach.

REFERENCES

1. M. Dover and B. Croft, Getting Tough: *Public Policy and the Management of Pesticide Resistance, Study I.* World Resources Institute, Washington, DC, November 1984.

2. P. Karlson and M. Luscher, "Pheremones," A New Term for a Class of Biologically Active Substances. *Nature (London)* **183:** 51–56, 1959.

3. T. C. Osborn, D. C. Alexander, S. S. M. Sun, C. Cardona, and F. A. Bliss, Insecticidal Activity and Lectin Homology of Arcelin Seed Protein. *Science* **240:** 207–209, 1988.

5. C. S. Lofgren, W. A. Banks, and B. M. Glancey, Biology and Control of Imported Fire Ants. *Annu. Rev. Entomology* **20:** 1–30, 1975.

6. C. S. Lofgren and C. T. Adams, *Economic Aspects of the Imported Fire Ant in the United States. The Biology of Social Insects.* International Union for the Study of Social Insects, xxx, 1982, p. 124–128.

7. D. P. Jouvenaz, C. S. Lofgren, and W. A. Banks, Biological Control of Imported Fire Ants: A Review of Current Knowledge. *Entomol. Soc. Am. Bull.* **27** (3): 203–208, 1981.

8. F. Coulston, Reconsideration of the Dilemma of DDT for the Establishment of an Acceptable Daily Intake. *Regul. Toxicol. Pharmacol.* **5:** 332–383, 1985.

9. J. D. Rosen and F. M. Gretch, Analytical Chemistry of Pesticides. Evolution and Impact. *Silent Spring Revisited,* G. J. Marco, R. M. Hullingsworth, and W. Durham, (Eds.), American Chemical Society, Washington, DC, 1987, Chap. 8.

10. A. H. Smith, D. O. Fisher, N. Pearce, and C. J. Chapman, Congenital Defects and Miscarriages Among New Zealand, 2,4,5-T Sprayers. *Arch. Environ. Health* **37** (4): 197–200, 1982.

11. S. M. Ayres, The Prospects for Chemical Darwinism. *Bull. Environ. Contam. Toxicol.* **33:** 631–635, 1984.

12. R. R. Suskind and V. S. Hertzberg, Human Health Effects of 2,4,5-T and its Toxic Contaminants. *J. Am. Med. Assoc.* **251 (18):** 2372–2380, 1984.

13. G. Reggiani, Anatomy of a TCDD Spill: The Seveso Accident. In *Hazard Assessment of Chemicals: Current Developments,* Vol. 2 Jitendra Sexena (Ed.), Academic Press, New York, 1983, pp. 269–342.

14. F. Pocchiari, V. Silano, and G. Zapponi, The Seveso Accident and Its Aftermath. In *Insuring and Managing Hazardous Risks: From Seveso to Bhopal and Beyond,* P. R. Kleindorfer and H. C. Kunreuther (Eds.), Springer-Verlag, New York, and International Institute for Applied Systems Analysis, Laxenburg, Austria, 1987, Chap. 3, pp. 60–78.

15. R. D. Kimbrough, C. D. Carter, J. A. Liddle, R. E. Cline, P. E. Phillips, Epidemiology and Pathology of a Tetrachlorodibenzodioxan poisoning episode. *Arch. Environ. Health* **32:** 77–85, 1977.

16. G. D. Lathrop, W. H. Wolfe, R. A. Albanese, and P. M. Moynahan, *An Epidemiologic Investigation of Health Effects in Air Force Personnel Following Exposure to Herbicides: Baseline Morbidity Study Results.* USAF School of Aerospace Medicine, Brooks Air Force Base, Texas, 1984, National Technical Information Service Document No. AD A 138340.

17. J. D. Erickson, J. Mulinare, P. W. McClain, T. G. Fitch, L. M. James, A. B. McClearn, and M. J. Adams, Jr., Vietnam Veterans' Risks for Fathering Babies with Birth Defects. *J. Am. Med. Assoc.* **252** (7): 903–912, 1984.

18. The Centers for Disease Control Vietnam Experience Study. Health Status of Vietnam Veterans. III. Reproductive Outcomes and Child Health. *J. Am. Med. Assoc.* **259** (18): 2715–2719, 1988.

19. A. Poland and J. C. Knutson, 2,3,7,8-TCDD and Related Halogenated Aromatic Hydrocarbons: Examination of the Mechanism of Toxicity. *Annu. Rev. Pharmacol. Toxicol.* **22:** 517–554, 1982.

SUGGESTED READINGS

Brachmachary, R. L. Ecology and Chemistry of Mammalian Pheromones. *Endeavor* **10** (2): 65–68, 1986.

Crow, K. D. Chloracne—An Up to Date Assessment. *Ann. Occup. Hygiene* **21:** 297–298, 1978.

Introduced Parasites and Predators of Arthropods Pests and Weeds. A World Review. Agric. Handbook No. 480, ARS USDA, Washington, DC, 1978.

Korgeski, G. P., and Leon, G. R. Correlates of Self-Reported and Objectively Determined Exposure to Agent Orange. *Am. J. Psychiatry* **140** (11): 1443–1449, 1983.

Prestwich, G. D. Chemistry of Pheromone and Hormone Metabolism in Insects. *Science* **237:** 999–1006, 1987.

Public Health Risks of the Dioxins, William W. Lowrance (Ed.). Proceedings of a Symposium held in New York City on October 19–20, 1983, by the Life Sciences and Public Policy Program of the Rockefeller University. Distributed by William Kaufman, Inc., 95 First Street, Los Altos, CA 94022.

Rappe, C. Analysis of Polychlorinated Dioxins and Furans: A Critical Review. *Environ. Sci. Technol.* **18** (3): 78A–89A, 1984.

Kogan, W. J., Gladen, B. C., Hung, K., Koong, S., et al. Congenital Poisoning by Polychlorinated Biphenyls and Their Contaminants in Taiwan. *Science* **241:** 334–336, 1988.

6

Accidents

On a crisp September day in 1899, H. H. Bliss stepped from a tolley car in New York City and was struck down by a horseless carriage. He was the first person to be killed by an automobile. By 1951, more than one million people had died in traffic accidents. With the close of 1974, a second million deaths had been recorded. Unless there is a substantial change in rate of occurrence, a third million will have died by 1990. That is 450% more than the number killed in all our wars.

I. ACCIDENTS AND OTHER TRAUMA

Fifty-two years elapsed for the first million, 23 years for the second, and only 16 for the third. Clearly, people are being killed at a significantly increased rate. Simply calculating an average indicates that at least 33,000 men, women, and children are killed every year. In fact, since 1949, it has been closer to 47,000 deaths per year. Only heart disease, cancer, and stroke claim more lives.

If an infectious disease took that many lives, there would be national panic. Consider that between 1981 and May 1988, acquired immune deficiency syndrome (AIDS), a fairly new disease, was responsible for

Table 1
Fatalities from Major Accidental Causes[a]

Type	1986
Motor vehicle	47,900
Falls	11,000
Workplace	11,200
Drowning	5,600
Fire	4,800
Poisoning	4,900
Totals	85,400[b]

[a] The term "accident" should not be viewed as the dictionary definition meaning chance or fortune. When looked at closely, most of these can be seen as purposeful events. More about this further on.
[b] For 1986, the number of accidental deaths from all causes was 94,100.

some 62,763 cases and 35,202 deaths: less than 5000 deaths per year.* Yet the vehement public outcry forced Congress to appropriate additional millions of dollars of research into this problem. No such action has ever occurred with motor vehicle accidents: not even in 1972, when the death toll exceeded 56,000.

This, of course, does not include deaths from other accidental causes. To do so would double the yearly death rate. Currently, accidents constitute a greater threat to the health of our communities than do all communicable diseases combined. In economic terms, the financial costs are astronomical: $59 billion in 1986, $24.9 billion in lost wages, $12.9 billion in medical costs, and $21.2 billion in property damage and lost production of goods and services.

But that was not to last. From 1984 to 1988 highway deaths began a rise that appears headed for 50,000—numbers we have not seen since the carnage of the 1960s and 1970s.

The first year the number killed on our highways dipped below 45,000 was 1983. Table 1 indicates the number of fatalities for the six major classes of accidents for 1986. Figure 1 shows the trend lines for public accidental deaths, all accident deaths, along with those for motor vehicle, the workplace, and the home. The trends are clearly downward, which may be comforting. In terms of absolute numbers involved, however, the problem begs for more rapid solution.

In addition to death, accidents also produce a profusion of disabling injuries. The numbers above suggest the problem must rank as the nation's most grave and unremitting environmental conundrum. Accord-

---

* During that same period, 1981–1988, motor vehicle accidents were responsible for 377,100 deaths: an average of 47,137 deaths per year, and not quite ten times that of AIDS.

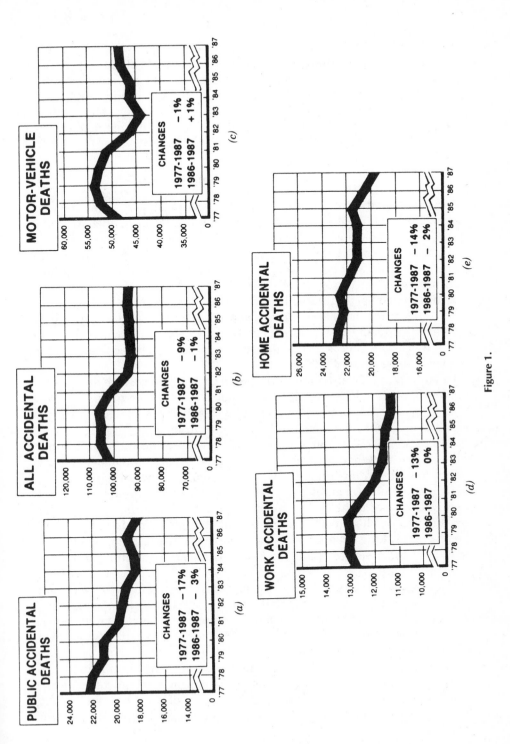

Figure 1.

ing to the National Safety Council, accidents of all types were responsible for approximately 9 million disabling injuries. A glance at the most recent census figures reveals that 9 million is more than the combined population of Delaware, New Hampshire, Vermont, Rhode Island, North and South Dakota, Montana, Idaho, Nevada, Utah, Wyoming, and Alaska. Rather than being seen, and dealt with, as the formidable problem it is, it appears to be accepted with an undeserved equanimity.

A. Causes of Accidents

Injury and death may be accidental. For the most part, they are not. "Accident," which according to most dictionaries is defined in such terms as unintentional, unplanned, fortuitous, and unexpected, is misleading in concept. Current estimates set the rate of solely fortuitous events (the truly accidental occurrences), which the involved individual had no hand in causing, at no more than 10–15%. Eighty-five to 90% appear to have been initiated by the victims themselves.

A trailer truck, traveling at night, goes off the road and turns over in a riverbed. The accident report reads: "Driver asleep at the wheel."

An 8-year-old boy is struck down by a city bus. The accident report reads: "Ran in front of vehicle."

A lawyer, homeward bound from a late party, misses a turn and rams a culvert buttress. The accident report reads: "Driver inattention."

A garage worker pins another employee against a wall while testing a car. The accident report reads: "Shifted the wrong gear."

Let us look at the four accidents just cited. In the first one, further probing revealed that the dead driver, in order to earn more money to pay debts, had driven for three nights without sleep and was keeping awake on amphetamine. An autopsy on the 8-year-old boy disclosed a tumor that deprived him of sight on the side on which he was struck. The lawyer recalled that he had reached for the cigarette lighter in his new car and had turned out the headlights instead. The garage worker admitted that he had been under severe emotional stress after learning that his child was mentally retarded.

Researchers seeking to isolate all the tangled components that make up any single accident have suggested a useful tool: *the accident syndrome.* A syndrome is usually understood to be a group of related symptoms that together characterize a disease. The components of the accident syndrome as put forth by Schulzinger (1) are depicted in Figure 2.

Another way of viewing this is seen in Figure 3. The top line indicates seven stages between initiation of an event and the final outcome, death. The stages are expressed generically in the top of each box, and in the bottom box in terms of an auto accident. The stages are linked by causal pathways denoted by triangles.

Six control stages are linked to pathways between hazard states by the

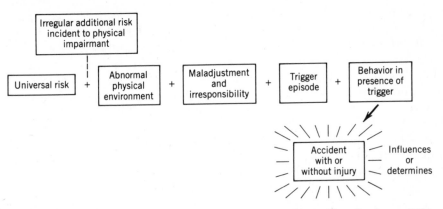

Figure 2. The accident syndrome. From M. S. Schulzinger, *The Accident Syndrome*, 1956. Courtesy of Charles C. Thomas, Publisher, Springfield, IL.

vertical arrows. Each is described generically as well as by specific control actions. Thus, control stage 2 would read: "You can modify technology choice by substituting public transit for automobile use and thus block the further evolution of the motor vehicle accident sequence arising out of automobile use."

The *universal risk* is considered a constant factor. From the day of birth an individual lives in peril. Injury and death can occur at any time. For the most part, the universal risk becomes part of the syndrome because of fear of injury. Many accidents are brought about because of a state of mind induced by knowledge that danger exists; the dread of an accident may well be a factor in its cause. A nervous, jumpy, nagging passenger may so upset an otherwise reliable driver that an accident occurs, when ordinarily none would have.

Irregular additional risk is set above the other components of the syndrome as it does not always enter into the equation. It can enter the

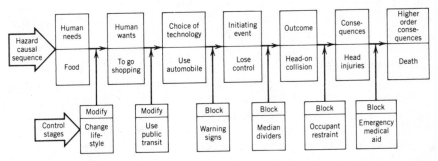

Figure 3. Illustration of events in a causal chain of the evolution of an automobile accident. Adapted from Bick, Hohenemser, and Kates, Target: Highway Risks. *Environment* **21**(2): 7–15, 29–38, 1979.

syndrome as an individual's physical defect or handicap. The fact that a person is a diabetic, which would be an irregular additional risk, would enter the syndrome only if insulin deficiency initiated a series of metabolic events resulting in coma. Similarly, epilepsy and cardiovascular disease can increase the risk of accident. However, many physically handicapped persons compensate for impairments and may in fact have fewer accidents than people without handicaps.

An instructive investigation was recently carried out in Vermont. During the period 1975–1980, individuals who had been involved in a motor vehicle accident and who had a prior history of hospitalization for coronary insufficiency were studied. Levels of heart disease severity were assigned based on an estimation of the extent to which the disease might affect the ability to respond appropriately to the multiplicity of demands of driving. To determine if the crash experience of those with cardiovascular disease was the same or different from that of a similar group matched for age, sex, and place of residence, two comparison groups were assembled: an age, sex, and residence group and a sex–residence matched group (2).

The results of analyses of yearly crash rates for the heart disease group and their matched "healthy" opposite numbers revealed a consistent pattern of underrepresentation in the accident records of those with a history of heart disease. Overall differences in crash rates as a consequence of severity of heart disease did not appear. In no case was the presence of coexisting medical complication (diabetes, lung disease, hypertension) suggestive of increased risk of crash. Again, in no case did the crash rates for those with previous coronary complications exceed the rates of the two comparison groups, nor did they exceed the crash rate for the average Vermont driver.

Not so obvious are those without "true" handicaps. An individual of 6'3" or over may not have the line of vision needed for ready observation of overhead traffic lights. In many automobiles, windshields are tinted more heavily at the top; this restricts the vision of very tall people at night. On the other end of the scale, short people have difficulty reaching automobile pedals. Individuals at the extremes in stature also have difficulty with doorways, furniture, machinery, tools, and household appliances.

Abnormal physical environment, the third factor in the syndrome, is not at all sharply defined. Nevertheless, certain points can be noted. Extremes in weather, such as heavy downpours, can obstruct vision and make conditions ripe for skidding. Poor ventilation, upsetting odors, sudden startling noises, and excessively hot weather, as well as sustained periods of cloudiness and rain, may contribute to accidents.

Although abnormal physical environment remains hard to pin down, it is clear that *maladjustment and irresponsibility*, which lead to the blunting of good judgment, are at the core of the accident syndrome. The main psychological and physiological elements that sharply increase the probability of accidents in maladjusted persons are anxiety, fear, hostility, worry, fatigue, frustration, and a host of other states resulting from pressure in the home, on the job, and in the community. Some 250 factors contributing to the blunting of good judgment and hence to accidents have been suggested. Experts in accident prevention look to this area as the most promising for reducing accidents. Mechanical devices and educational programs appear to be achieving less and less.

The most readily observable portion of the syndrome is the *trigger episode*—the nail that is stepped on, the blown tire, the fire, and the bullet entering the body. To students of accident prevention, this is the least important part of the syndrome, since the prospective victim will almost invariably reach some accident trigger. The maladjusted person ready for an "accident" may find a trigger anywhere: at home, at work, on the street. The trigger is only the detonator of a far more powerful cause or combination of causes. Consider for a moment how you would appraise a blown tire in the light of the syndrome theory. Check your ideas with Figure 2.

A particularly important aspect of the syndrome is the individual's *behavior in the presence of the trigger*. An individual's pattern of behavior when confronted with a sudden decision or with danger will frequently determine whether the forces set in motion will result in a near miss or an "accident." In addition, behavior may affect the severity of the accident. For example, if clothing is ignited, one person may try to smother the flames by wrapping himself in a coat or blanket. Another person, in stark panic, may run aimlessly, and as a result may die or suffer severe injury. Behavior in such a case is determined, at least in part, not by the immediate situation, but by long previous adjustment—one's life situation—that paves the way for disastrous mishaps or survival. Perhaps in Figure 2 this factor should be related directly to maladjustment rather than linked to the trigger episode.

Even if all the factors making up the syndrome are present, it does not follow that an *accident with or without injury* is inevitable. In the main, however, it does occur. Viewed in this manner, accidents appear to be set in motion by a constellation of events. And it may well be that this type of theorizing offers means to break the links at various points in the accident chain. The accident syndrome theory may be a tool that will lead to significant reductions in "accidents."

II. WHERE ACCIDENTS OCCUR

A. Accidents in the Home

Motor vehicles were responsible for 1.8 million severe injuries in 1987. The workplace produced another 1.8 million, and the home 3 million. Unlike motor vehicle and occupational injuries, the likelihood of report-

Table 2
Nursery Accidents (1982)

	Injuries	Deaths
Cribs	14,500	129
Strollers, carriages	12,500	1
High chairs	9,900	9
Baby walkers	16,200	1
Playpens	4,000	8
Baby gates	1,500	1
Other	32,100	28
Totals	90,700	177

Source: Consumer Product Safety Commission, October 1984.

ing home accidents is extremely low. Six to 9 million would be a more realistic figure. The home may be one's castle, but it is an extremely hazardous one at that. In addition to injuries incurred as a consequence of falling in slippery tubs and showers, lacerating fingers with high-speed electric tools, and being burned from the assortment of fires that occur every 50 s of the day and night, there are others which are entirely needless.

Table 2 points out the wide variety of objects that place infants and young children at undue risk. Poor product design and negligence are major contributors to these accidents. Similarly, the U.S. Consumer Product Safety Commission estimated that 594,100 toy-related injuries to children were treated in hospital emergency rooms in 1983. Table 3 lists the number of injuries and the type of toy involved. Most injuries occurred from impacts with toys. Choking from ingestion of small toys, or parts of toys, was the second most frequently reported incident. Half of the deaths involved children who choked on balloons, rode tricycles into pools, or were struck by automobiles while riding their bikes.

Children under five are also the most vulnerable to bathtub-related electrocutions. From 1979 to 1982, approximately 100 people were

Table 3
Estimated Injuries Among Children Under 15 years old
Treated in Hospital Emergency Rooms—United States,
1983

Cause of Injury	Number of Injuries
Toys	118,000
Bicycles	387,300
Sleds	16,600
Skates	61,900
Skateboards	10,300
Total	594,100

electrocuted while bathing in tubs. Although the range is broad, from under 1 year to 75, children under five had the greatest mortality. Interestingly enough, the much maligned telephone is not the villain it is made out to be. Not a single electric-current bathroom-related incident involved a telephone. The real culprit is the hair dryer. Sixty percent of the deaths can be directly attributed to it. The remaining 40% are due to TV, frayed or inadequate wiring, lamps, and electric heaters (3). None of the fatalities as self-inflicted. Suicide is not at issue here.

More than 2 million Americans are burned each year. Some 70,000 require hospitalization, and approximately 11.5%—8000—die of their injuries. Some 2500 of these deaths can be laid to smoking in bed. Cigarettes contain additives in both the paper and tobacco which allow the cigarette to burn for approximately $\frac{1}{2}$ h. If these additives were omitted, an unpuffed cigarette would burn out in less than 4 min. Mattresses and other upholstered furniture require more than 4 min of exposure to the heat of a cigarette for ignition to occur. Obviously, omitting the additives would make smoking a great deal safer, and the taste of the cigarette would not change. Nevertheless, resistance to such action remains insurmountable.

Some of the most disabling injuries have occurred, and continue to occur, as a result of the use or abuse of toothpicks. From 1979 to 1982, an estimated 8176 toothpick-related injuries were recorded (4). The rate of injury was greatest among 5–14-year-olds. But children under 5 had eyeball and ear injuries 20 times greater than any other age group. Although toothpicks are presumably commonplace and innocuous, they have been responsible for performation of the gastrointestinal tract, penetration of heart muscles, pleurisy, and liver abscesses. Death is not uncommon. Whereas a 51-year-old Alabaman recently died of cardiopulmonary arrest (the immediate cause), the underlying cause of death, determined at autopsy, was perforation of the colon by a toothpick. In California, another person died after eating stuffed cabbage held together with toothpicks.

B. Alcohol and Motor Vehicle Accidents

Alcohol has a profound effect on behavior. Consequently, it is to be expected that it will have a significant impact on driving ability, injury, and death. From 1977 to 1981, data from the National Highway Traffic Safety Administration's Fatal Accident Reporting System (FARS) show that the proportion of drivers with measurable blood alcohol concentrations (BAC*) increased steadily. The percentage of 16–19-year-old drivers tested who had positive BACs rose 8% in those 5 years. Comparable

---

* A BAC of 0.10% is the level of legal intoxication in most states.

increases occurred among the 20–24-year age group, as well as those over 25.

In 1981, BAC results showed that 21% of the 8790 teenage drivers involved in fatal crashes had been drinking alcoholic beverages. Most young intoxicated drivers become involved in single vehicle accidents though not exclusively, and five times as many men as women are involved in these crashes. As might be expected, most accidents occurred on weekends. Figure 4 shows the disproportionate hourly crash activity that occurs so frequently between midnight and 3 a.m.

The announcement by the Department of Agriculture in November 1984 of the increasing consumption of alcohol, more beer, wine, and liquor, to the tune of 38.2% increase between 1962 and 1982, may be of more than pedestrian interest.

Recent FARS data indicate a decrease in the number of fatal accidents in 1982. This has been interpreted as an outcome of the economic recession which prevented frequent purchases of gasoline and fewer alcoholic beverages. If it is true that economic factors influence driving frequency, behavior, and fatality rates, with economic recovery, these can be expected to rise again.

According to the U.S. Department of Transportation, 250,000 people lost their lives in alcohol-related accidents between 1974 and 1983. That's

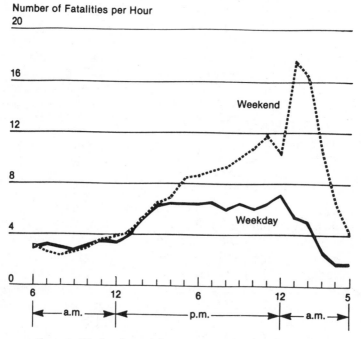

Figure 4. Weekend–weekday contrast in fatality rates, 1981.

25,000 per year or one death every 20 min. Alcohol-related injuries occurred at the rate of 45 every second—700,000 between 1982 and 1983.

House and co-workers (5), of the University of North Carolina's Highway Safety Research Center, have shown that alcohol-impaired drivers were more apt to be seriously injured or killed independent of such common risk factors as age, sex, model of car, speed prior to crash, type of crash, and use of safety belt.

They found that alcohol consumption increased a person's vulnerability. Of particular significance was the discovery that as BAC increased, the speed of a crash decreased but that the severity of injury increased. It is possible to speculate that a highly intoxicated person may have depressed recuperative powers.

Evidence supporting the theory comes from a number of studies. For example, Benveniste and Thut (6) suggested that biochemical data indicated that wound healing is depressed by chronic alcohol consumption. They subjected a group of male mice to a diet of standard Purina Laboratory Chow, but whose sole source of liquid was a 10% ethanol solution. A control group received Chow and water. The study lasted 6 months.

They found that long-term alcohol consumption retarded the wound-healing process as reflected in granuloma formation in polyvinyl sponges imbedded subcutaneously at a variety of sites. This finding parallels clinical indications that alcoholics exhibit poor or slow wound healing. House and her colleagues reported another striking observation. In their study, "more of the drinking drivers died, even though their injuries were no worse than those of sober drivers."

The distinct possibility that small amounts of alcohol, enough to produce mild intoxication, may radically increase the potential for injury and death following motor vehicle crashes, was tested by James Liedtke and William DeMuth (7). At the Milton S. Hershey Medical Center, Hershey, Pennsylvania, they investigated the effect of a standardized nonpenetrating blow to the chest of dogs pretreated with alcohol. To be certain of the relationship between chest trauma and alcohol consumption, their study included three groups of dogs. Nine anesthetized animals in one group received a blow to the chest. A second group of five anesthetized dogs received an infusion of alcohol, while a third group of 12 anesthetized dogs received both alcohol and a blow to the chest at approximately the fifth intercostal space. The trauma delivery system consisted of a Cash-X captive bolt pistol device which generated a spike force of 238 kg and produced a selective injury to the heart "with little additional trauma to other organ systems."

Of the nine animals receiving impact alone, eight survived. Several manifested disorders in cardiac rhythm and conduction. By the end of the study period, the eight dogs had reverted to their pretrauma normal heart rhythm. None of the five animals receiving alcohol alone was statistically different from pretreatment electrocardiographic and hemodynamic values. In group III, of the 12 receiving both alcohol and impact, only one survived. Major disorders in rhythm and conduction were evident. The investigators concluded that the effect of alcohol on cardiac injury is profound. "Alcohol," they noted, "can affect catastrophic reductions in mechanical performance, in the presence of otherwise non-fatal injury." They went on to say that the "effect emphasizes the potential dangers of alcoholic excess on electrical and mechanical functions of the heart and may have a clinical analog in the sudden death of intoxicated drivers soon after *seemingly minor* cardiac injuries" (italics mine).

Evidence from a variety of animal species consistently indicates a similar correlation. According to House et al. (5), the results of many new studies "contradict the time-honored folklore that maintains that the intoxicated person is more relaxed and hence more likely to emerge unharmed from a given crash."

Approaching the issue from a different direction, driver simulation studies have generally yielded similar results. Using interactive simulators, which allow "the driver" control of both steering and speed on a video-projected two-lane highway, behavior and performance during a 10-mile drive can be measured and evaluated.

In a recent study, funded by the National Highway Traffic Safety Administration, comparing blood alcohol levels of 0 and 0.10% and marijuana (in the form of δ-9-tetrahydrocannabinol) at 0.50 + 100 μg per kilogram of body weight (ppm), a number of significant observations emerged.

At a BAC of 0.1%, steering and speed control as well as response speed and accuracy deteriorated markedly. Increased numbers of accidents occurred, and increased speeds were observed. This is in marked contrast to the findings of House and co-workers working with real accidents, which indicated speed *decreased* with increased blood alcohol.

Marijuana at 50 and 100 ppm apparently did not contribute significantly to driver impairment in any of the criteria in this particular test. The combination of alcohol and marijuana, as far as obstacle avoidance, passing, and maintaining speed are concerned, was no different from the effects of each separately.

Efforts to influence the public's attitude toward drinking and driving have generally met with little success. Recently, however, Mothers Against Drunk Driving (MADD) has achieved a modicum of success. Their efforts have moved state legislatures to enact laws that will impose more severe penalties for "driving under the influence."

1. Increased Risk Among the Young

Listed in Table 6 in Chapter 14 are the current leading causes of death in the United States. This listing combines all ages and both sexes. For each event, it displays the average figure for the entire population. Accidents of all types are in fourth position, a rank unchanged for more than 20 years. Concealed within this average are several surprising yet candidly revealing facts.

For example, if rates for specific age groups are teased out, we find that for both men and women in the age group 1–34, accidents are in fact not the fourth, but the first leading cause of death.

With even greater specificity, the 25–44-year age group emerges as the population at greatest risk. In 1983, they contributed 27% of all accident-related deaths. The 15–24-year-olds followed closely with responsibility for another 22%.

In *Healthy People, The Surgeon General's Report on Health Promotion and Disease Prevention* (8) the devastating point is made that between 1960 and 1979 every age group in our population had shown an increase in life expectancy with the exception of those aged 15–24. They had in fact exhibited a steady decrease, with the major cause being deaths from motor vehicle crashes.

The reason is not difficult to divine. Analysis of data from FARS revealed that 48,000 people lost their lives in motor vehicle accidents in 1987. Alcohol was responsible for 49%. Thirty-three percent of all drivers in fatal crashes were between 16 and 24 years old.

This group represents only 20% of all licensed drivers, yet they are involved in most alcohol-related fatal crashes.*

In 1986, 2820 inebriated teenagers aged 15–19, an increase of 17% over the previous year, were killed in single vehicle night crashes. With the new increased speed limits now in effect, this number can be expected to rise along with it. But teenagers are not the only ones who drink and drive.

Twenty-seven counts of capitol murder were recently brought against Larry Mahoney, a 34-year-old Kentuckian who drove his pickup truck the wrong way down a highway into a school bus. In the crash and fire that followed, 27 teenagers were killed. Mahoney, a repeat offender, had a BAC of 0.24, three times the level designated as impaired and twice the level most states set as drunk.

There is a curious and macabre twist to the tougher anti-drunk-driver laws that have become effective. In Florida, for example, fatalities involving drunken bicycle riders have doubled since 1981. The increase has been traced to drunken drivers who lose their licenses and take to riding bicycles—while drunk.

* To this are added deaths from homicide (murder) and suicide, so that trauma becomes the leading cause of death for the 15–29-year age group. While "trauma" is not a specific rubric within the International Classification of Diseases, the fact is plain enough; for these people death is the obvious result of violence. After age 45, motor vehicle and other nontransport accidents assume relatively minor importance. This type of information can be useful in directing appropriate types of control and prevention. A shotgun approach is not only meaningless, it is economically wasteful. Young drivers are patently overinvolved. Lack of attention to this calamitous environmental issue may stem from a feeling of indestructability. Nothing could be further from the deadly truth.

Attention to metabolic wisdom may be useful. The primary influence on BAC is the rapidity with which ethanol, the alcohol of whiskey, wine, and beer, moves from the stomach to the small intestine and into the liver where it is ultimately metabolized by the enzyme alcohol dehydrogenase. The movement of alcohol out of the stomach depends on what—or what is not—in the stomach. Drinking on a full stomach or drinking along with eating is the optimum way to slow alcohol down—allowing it to leave in smaller amounts over an extended period. Drinking on an empty stomach is tantamount to "driving under the influence," if driving is the next activity.

2. Preventing Automobile Accidents

Nevertheless, it appears that engineering approaches may provide the surest forms of prevention. For example, the General Motors Corporation has developed a Critical Tracking Test* (CTT) in the form of an electronic device mounted in the instrument panel designed to prevent or discourage an impaired driver from operating the vehicle.

As shown in Figure 5, when the ignition is turned on, the driver must turn the steering wheel to keep a fluctuating needle on an instrument panel mounted display within two red lines. During this time, the needle moves in progressively wider swings requiring quicker movements of the wheel to keep it centered. If the driver fails to keep the needle centered during the allotted time, a red "reset" button lights up, indicating failure of the test. Two more opportunities are provided "to pass" after pressing the reset button. A "pass" is given when the needle is kept within the center area, photo (c) in Figure 5.

On passing, the car can be driven without activating a beeping horn and flashing lights, which occurs if the test is failed. The beeping and flashing mechanism is activated when the car's velocity exceeds 10 mi/h, ostensibly to warn other drivers and alert enforcement officials.

The CTT device is currently undergoing final trials in California by the NHTSA. The test utilizes multiple drunk-driving offenders. Courts involved in the study have assigned cars with CTT to drivers convicted a second time for DUI. These drivers ordinarily would have had their licenses revoked, but they chose instead to participate in the program.

The CTT can also be modified to detect drowsiness while driving. When a steering wheel sensor and microcomputer are added to the CTT, a change in steering action can be detected and a warning signal activated.

Experience with the ignition interlock system used in the early 1970s, requiring occupants to buckle up before a car could be started, suggests that forcing drivers to pass a test would meet with stiff and widespread buyer resistance.

---

* Although the CTT was developed by GM for use in automobiles, the original concept derived from a test employed by NASA to evaluate stress on astronaut performance.

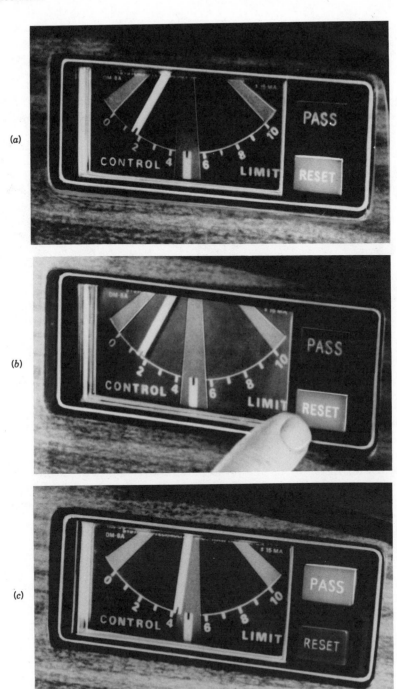

Figure 5. The critical tracking test. (*a*) Driver must turn steering wheel to keep fluctuating needle within two red lines. (*b*) Failing test, driver must press red reset button to take test again. (*c*) "Pass" is given when needle is kept centered.

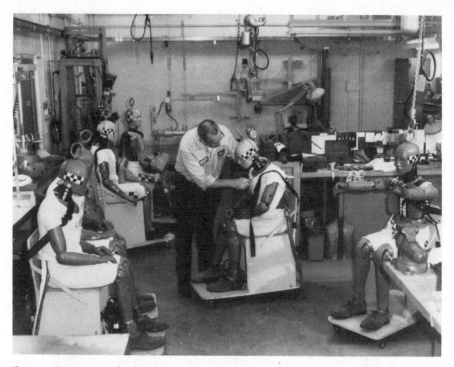

Figure 6. Preparing Hybrid III dummy for a crash test at GM Proving Ground Safety Research and Development Laboratory.

Other countermeasures depend less on the driver and more on original design and construction—the so-called "built-in" safety.

Figure 6 shows a series of dummies in various states of readiness for impact testing. Instruments similar to those used in space flight programs are installed in the dummies to allow engineers to record and later evaluate the dummies' reactions during impact sled or barrier crash tests. Shown in the figure are Hybrid III dummies, the latest in a series of crash test dummies.

A 30-mi/h barrier impact crash, similar to that seen in Figure 7, is equivalent to hitting a parked car at 60 mi/h. In a 30 mi/h front end collision, a 165-lb man hits the windshield with a force of 3 tons. A 10-mi/h increase in impact speed from 30 to 40 mi/h means that 79% more energy must be absorbed. Design criteria must build in this capability to provide the necessary margin of safety.

Figure 8 demonstrates the thorax certification test, in which a ballistic pendulum strikes the dummy's thorax while records of speed and force imposed on the chest area are made. Hybrid III, the third generation of test dummy, has been designed to respond much like a human chest under this condition of impact. As seen in Figure 9, a variety of dummies are used in impact testing. Each, of course, provides different information.

Figure 7. This 1982 Chevrolet Celebrity has just slammed into a barrier at GM's proving ground.

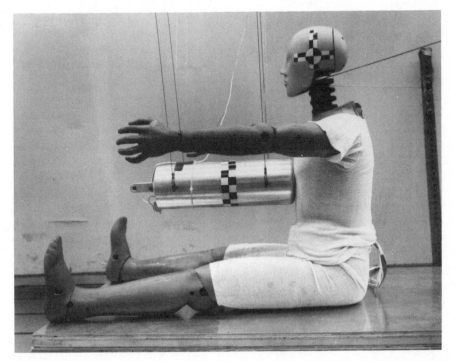

Figure 8. A Hybrid III dummy being subjected to the thorax certification test at GM's proving ground.

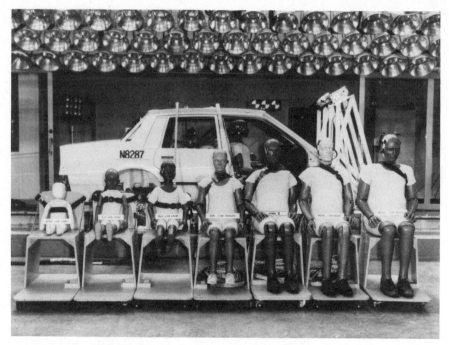

Figure 9. Dummy family waiting for its turn to "ride."

The dummies represent different ages, weights, and varying degrees of "human" sophistication. In this figure, they range from a "6-month-old rag doll to the highly complex Hybrid III."

Sixth from left, Hybrid II, represents 50th percentile males. Their weights and heights place them at the midpoint of the total male population. Fourth from left is a 5th percentile female, and at the extreme right, a 95th percentile male. Each represents an extreme segment of the population—small for females, large for males.

Structural engineers make use of computer crash simulators to design greater protection into automobiles. Figure 10 shows a crash through the "eyes" of a computer. The model consists of "springs" representing the stiffness of the structural members and masses representing objects such as the engine. During a crash, the model can be used to estimate the magnitude of forces likely to be imposed on the passenger compartment, the distribution of forces among the various structural members, the duration of a given vehicle impact, and the effect of engine size and engine compartment options on the other three.

Seat belts are an enigma. Fewer than 15% of all American drivers use seat belts when they drive. In Great Britain the number is upward of 90%. We know that the chances of being killed in a crash increase 25 times if an occupant flies from a vehicle. We know that restraint devices virtually eliminate the risk of death for children 4 years and under. Nonetheless,

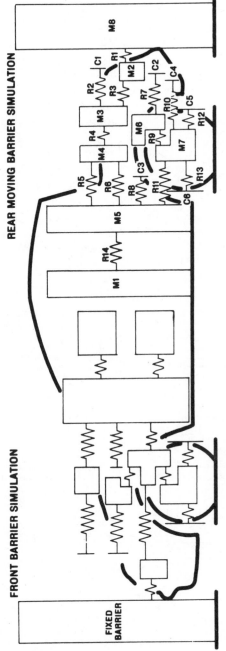

Figure 10. A crash through the "eyes" of a computer.

203

Americans simply refuse to use seat belts. Why won't motorists buckle up? The reasons are numerous and mistaken. Psychologists researching the problem find that people hold false perceptions of their vulnerability, have poor assessments of accident risks, and believe in the false notion that being belted is dangerous in the aftermath of a crash.

Elizabeth Hanford Dole, former Secretary of Transportation, believes that without legislation requiring mandatory use, buckling-up will never get beyond a 25% compliance rate—if that. Each time a driver negotiates a trip uneventfully there is "positive reinforcement" for the notion that seat belts are unnecessary. Psychologists are unanimous in believing that if a room full of people were asked to rate their driving ability, they would all give themselves a 10. In addition, for many motorists, the decision not to wear a belt is a way of denying the risks of driving. For too many people the presence of seat belts destroys the illusion. Belts are constant reminders of reality in a world where fantasy is widely prevalent.

If three-point belts are not mandated by most states, all cars will be equipped with air bags by 1990. Air bags are another built-in safety measure that does not depend on drivers—sober or inebriated—to be activated. As shown in Figure 11, on impact, a bag inflates coming

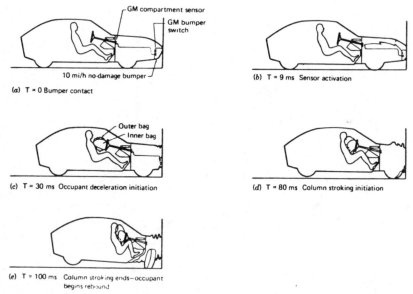

(a) T = 0 Bumper contact

(b) T = 9 ms Sensor activation

(c) T = 30 ms Occupant deceleration initiation

(d) T = 80 ms Column stroking initiation

(e) T = 100 ms Column stroking ends—occupant begins rebound

Figure 11. Operation of an air bag during a frontal barrier crash. The top picture illustrates that nothing happens when the collision speed is 10 mi/hr: the car front end is not significantly crushed nor is the airbag deployed. The next four pictures show successive "snapshots" of the vehicle as it is crumpled in a 50-mi/h frontal barrier crash. The time of each picture is given in milliseconds. The air bag is inflated by a pressure-sensitive device mounted in the bumper. The essential function of the airbag is to prevent the driver from traveling freely forward as the passenger compartment slows down. Courtesy of the Department of Transportation, Washington, DC.

between the driver and the steering column and windshield, preventing impact trauma. The bag deflates quickly, permitting the driver to escape.

William Hadden, president of the Insurance Institute for Highway Safety, believes that air bags should not replace seat belts, rather they should complement them. Given the chilling statistics, and the inability while drunk to perform simple tasks such as buckling up, air bags are a sensible expedient—to protect people against themselves and others.

Chrysler Corporation's decision to install air bags as standard equipment on all domestically made cars signals the end of a nearly 20-year battle over their use. Chrysler's decision follows federal regulations calling for additional passive restraints that require no action by passengers—such as buckling a belt. It is estimated that 10,000 lives a year may be saved by these devices. Unfortunately, most will be installed only on the driver's side—which is not the most dangerous seat in the car.

a. How Air Bags Work. An air bag protects by sensing that a crash is taking place and inflates in approximately 1/25 of a second—half the time it takes to blink an eye—to cushion the head and chest from contact with the steering wheel and windshield. After contact the bag deflates in about 2 s.

On the drivers side the folded bag and chemical gas generator are mounted inside the hub of the steering wheel. If any of the sensors are jolted hard enough to close an electrical contact—a trigger—the chemical charge (sodium azide, molybdenum disulfide, and sulfur) generates nitrogen gas and the bag bursts through the foil seal under the steering wheel cover. These neoprene-coated nylon bags are about 26 in. in diameter 6–8 in. deep, with a volume of 2 ft^3 when inflated. A coating of talcum powder covers the inside to help expansion when fully activated. Figures 12 and 13 show the arrangement of the bag system.

b. Anticollision Systems. Collision warning systems, standard airplane equipment, may soon help car and truck drivers avoid crashes.

The anticollision devices use an array of electronic circuitry to determine if a potential hazard is imminent. With the Rashid Radar Safety Brake, drivers are alerted by signal lights and buzzers that a slow-moving vehicle is nearby. Brakes would automatically be applied should the driver fail to do so.

Left Coast Technologies, Inc., a California firm, has developed a sonar unit called The Assistant, which is mounted on either the front or back of a car. A companion unit mounted on the dashboard displays the distance to an obstacle. Both audio and visual alarms are triggered when the car comes within a foot or two of an object.

Should these devices not be financially out of reach of most drivers, they could reduce the current burden of crashes.

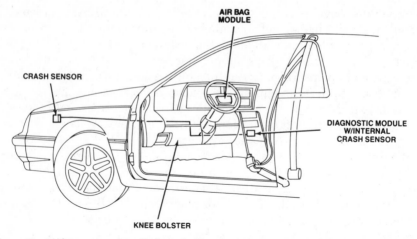

Figure 12. Air bag system installed on six Chrysler Motors' 1988 models consists of a crash sensor, knee bolster, air bag module, and a diagnostic module with an internal crash sensor. The air bag supplements the conventional safety belt systems on the Chrysler LeBaron coupe and convertible, Dodge Daytona, Chrysler Fifth Avenue, Dodge Diplomat, and Plymouth Gran Fury. Courtesy of Chrysler Motors Sales and Marketing Public Relations, Detroit, MI.

c. Accident Reconstructions and Simulations. Computers can contribute in another unusually beneficial way. Teasing apart the components of an accident or staging crashes with dummy occupants may lead to accident-prevention measures or accident-prevention behavior,

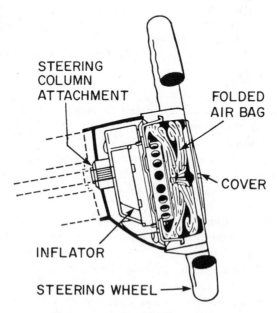

Figure 13. Air bag folded in place.

but reconstructing an accident after it has actually occurred is a higher order of sleuthing.

By using powerful new software programs such as EDSMAC (Engineering Dynamics Simulation Model of Automobile Collisions) and ED-CRASH (Engineering Dynamics Corporation Reconstruction of Accident Speeds on the Highway) computer-aided drafting tools developed by the Engineering Dynamics Corporation of Lake Oswego, Oregon, computers can recreate an automobile accident, for example, showing who hit whom, at what speed and direction, and how driver and passenger(s) were affected by the crash. After accidents, crash investigators—police, accident scientists, insurance officials—can obtain all pertinent data gathered at the crash site and place it in their programs to deduce the force of impact from the severity of damage, or its speed from the length of the skid marks. The incident and the events immediately preceding it can be displayed as a "slow motion" replay of how the "accident" occurred. Such reconstructions can be used to improve road and car safety as well as aiding in development of regulatory guidelines.

Figure 14a and b show an EDSMAC trajectory simulation of a front-end collision. Vehicle 1 was struck while turning, by vehicle 2, which struck it with a forward velocity of 35 mi/h. The final positions of both cars and passenger are displayed. A trajectory simulation uses the position versus time history and vehicle exterior dimensions to show the vehicle moving from the start until the end of the simulation.

EDSMACH is an IBM-compatible personal computer program developed from the Simulation Model of Automobile Collisions (SMAC) mainframe computer program. As a simulation program an investigator supplies conditions believed to have preceded the crash. It then uses vehicle properties, physics, and numerical integration to create the vehicle paths and damage profiles which would be expected to result from those initial conditions.

EDSMACH analyzes a crash in precisely the opposite way that a reconstruction program such as EDCRASH does. EDCRASH, which estimates the initial conditions, creates an EDSMAC input file. The anatomy of an accident is shown in Figure 15, an EDCRASH reconstruction of a two-car collision.

EDCRASH is an interactive program requiring the user to respond to questions requested at the terminal. This prompts the user to consider a wide range of potentially contributing factors. Reconstructions can show for example, where and how forcibly the heads of pedestrians hit car hoods. Data such as these resulted in a recommendation for increased space between engine hood so that pedestrians struck by cars have a greater chance of escaping serious injury.

d. Yet Another Approach. Reasonably, one would expect the majority of motor vehicle accidents and deaths to occur in areas of heaviest vehicle

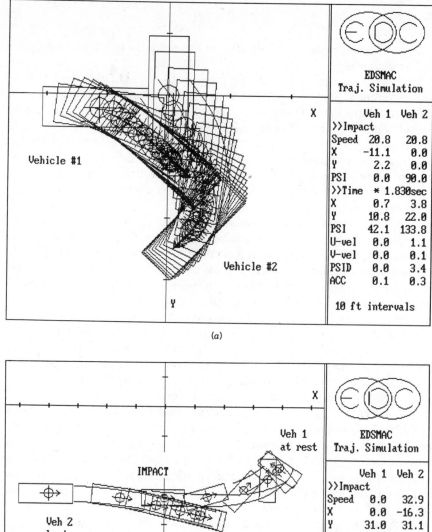

(a)

(b)

Figure 14. (*a* and *b*) EDSMAC trjectory simulations of collisions of vehicle 1 with vehicle 2. The location and speed of impact are displayed. At the right of the screen are the simulated results in numerical form.

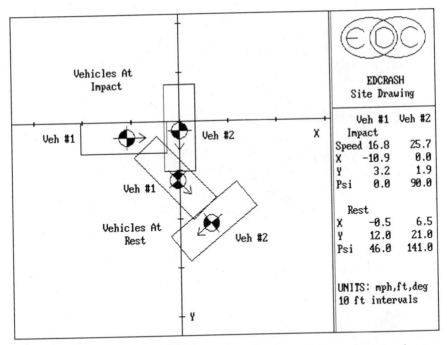

Figure 15. Reconstruction of a crash scene, showing the positions at impact and rest.

density. Actually quite the opposite is true. According to a recent study, drivers are much more likely to die in crashes in Nevada or Colorado than on the concrete urban expressways. Only a glance at Figure 16 is needed to grasp the inverse relationship.

Researchers from John's Hopkins University and the Insurance Institute for Highway Safety mapped the 1979–1981 auto accident death rates and the 1980 population density for each of the 3500 counties of the continental United States. They found geographic hot spots, but not where you would suspect. New York, Philadelphia, and Los Angeles, for example, were lower than the rates in the most sparsely populated areas of the West. Manhattan had a 3-year death rate of 2.5/100,000 compared with 558/100,000 for Esmeralda County, Nevada (9). It is worth speculating on the possible reasons for the differences in magnitude. Is driving in great open space conducive to accidents? Are the roads there of poorer quality? Are the people less safety minded? Are the vehicles out West less safe? Is there stricter law enforcement in high population density areas? Much needs to be learned if we are to reduce the slaughter on our highways. Dealing with motor vehicle safety must be at least as important as any disease in the United States today.

Whitfield and Fife of the Insurance Institute for Highway Safety offer data from their recent study that should give us pause. They found that overall motor vehicle mortality rates varied little between 1940 (26

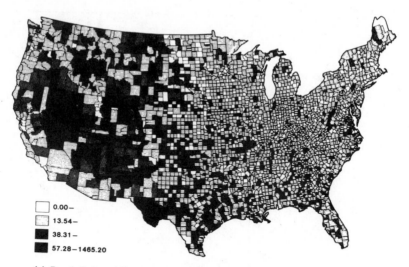

(a) Death Rates of Occupants of Motor Vehicles per 100,000 Population According to County, 1979 through 1981.

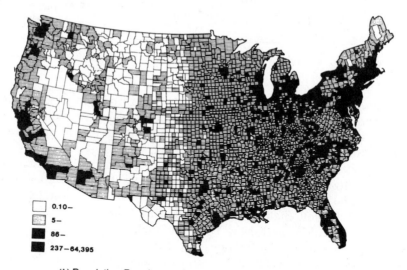

(b) Population Density per Square Mile According to County, 1980.

Figure 16.

deaths/100,000 per year) and 1980 (23/100,000 per year) (10). The distribution, however, is worth considering. Among the elderly there has been a substantial decrease in rates while among those below age 34, the increase has been considerable, so that the number of years of life lost to motor vehicle crashes has increased 84% between 1940 and 1980. If we looked only at the overall deaths for 1940 and 1980, one might be comforted by the fact of a 10% decline. One can take no comfort from that, given the unprecedented years of life lost. If this is not a national calamity, it should be.

C. Boating Accidents

Unfortunately, the accident–alcohol synergy is not limited to our highways. We also have troubled waters. Boating accidents around the country have increased significantly. Between 1960 and 1980, the number of privately owned boats more than trebled. In Florida the numbers climbed from 200,000 to 600,000. This does not include the 200,000–300,000 from other states and nations that use Florida's waters. Nor does it include another 500,000 that are exempt because of their size. And today's craft are far faster, thus the potential for collision has increased prodigiously.

Figure 17 depicts the trend in accidents and fatalities reported in Florida over the 5 years from 1980 to 1984. The Coast Guard estimates these numbers to be only 10% of all accidents. They maintain that accidents are not reported because owners do not have sufficient insurance to cover damages or believe reporting serves no purpose.

Three major reasons for accidents are lack of seamanship (rules of the road are not known or are disregarded), lack of courtesy, and drunkeness. Figure 18 shows the number of collisions in the five major boating states. Many, if not most, of these are alcohol related.

Almost 90% of all recreational boating deaths result from drowning, with the remaining 10% attributed to falls, burns, and other causes. In 1985, 1116 drownings occurred among boaters. Unlike automobile operators, boaters are not generally required to be licensed. Many have had no formal training in operation and safety procedures. Of all fatalities involving recreational boaters, one-third to two-thirds involve alcohol.

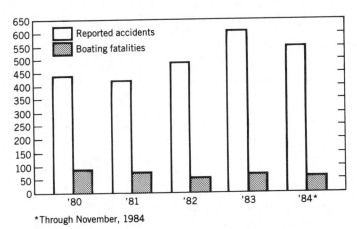

Figure 17. Accidents and fatalities in Florida.

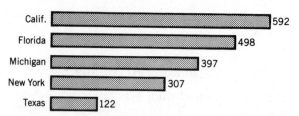

Figure 18. Number of reported collisions in 1983. Data courtesy of the U.S. Coast Guard and Florida Marine Patrol.

As good as he is, Willie Nelson is sending the wrong message:

> Whiskey river take my mind,
> Whiskey river don't run dry.

Alcohol is the single most important risk factor for accidental death; it is time it ran dry.

D. Trauma

Trauma is defined as personal physical injury. Together, accidents, suicide, and homicide contribute more lost years of life than heart disease and cancer. Despite this frightful loss, the specialized care of major injury has not yet received adequate attention and support. If there is real community concern for saving lives, trauma is the problem that must be addressed.

Violence and mayhem are as much a part of our lives in "peacetime" as in war. To the traditional contributions of accidents, homicide, and suicide, we can add the casualties of such natural disasters as flood, earthquakes, volcanic eruptions, and tornadoes, as well as contributions from such newer risks of contemporary life as terrorist bombings, chemical leaks and explosions, mob violence, air crashes, and the risk of nuclear accident. The recent leak of methyl isocyanate from the Union Carbide plant in Bhopal, India, put emergency resources to the ultimate test. An "accident" of such magnitude in the United States or any other Western world country would create no less of a test. Our communities are unprepared for emergencies of this magnitude.

War can be considered as prolonged disaster. That being the case, casualties are expected and anticipated, and the military has moved efficaciously to deal with them. Many of the most successful advances in emergency medical treatment, including rapid evacuation by helicopter, have come from experience gained in wartime. Table 4 shows the notable reductions in death rates with each successive conflict.

For the most part, these reductions reflect decreased time between injury and appropriate medical care. MASH units, Mobile Army Surgical

Table 4
Reduction in Death Rates as a Consequence of Rapid Medical Care

Conflict	Time Delay Between Injury and Care (h)	Mortality Rate (%)
World War I	12–18	8.5
World War II	6–12	5.8
Korean War	2–04	2.4
Vietnam	0.03–1.1	1.9

Hospitals, proved their effectiveness in wartime. RTCs, Regional Trauma Centers, can be just as effective in peacetime. Theirs is an idea whose time has come, but it would require a national effort to be put in place.

We are not without precedent in this. Excellent regional centers currently exist in West Germany. There is a curious twist to the establishment of these centers. Germans love speed. It may be a national characteristic. It certainly is a cause for nationwide trauma. Severe motor vehicle crashes are a common and daily occurrence along the Autobahns—the express motorways. Here, too, the German people accept these as a way of life. Nevertheless, during the Vietnam War—in which they did not participate—German observers were impressed by the med-evac care provided by our armed forces. In 1979, a decision was made to adopt and apply similar procedures throughout West Germany, especially along the Autobahn. Currently, 90% of the population, as shown in Figure 19, are within 15 min of a trauma center.

Figure 20 indicates the flight times and distances (in kilometers) for the ADAC's (the German Automobile Club) emergency heliocopter.

Strangely enough, having developed these proven prehospital procedures for our own troops, attempts to translate them for civilian use seems never to have been made.

Another interesting wrinkle has been the classification of all German hospitals according to their ability to provide specific levels of trauma care. The institution of this overall scheme in West Germany is credited directly with a 25% reduction in the motor vehicle mortality rate.*

From purely economic considerations, establishment of a network of trauma centers would repay the country handsomely. If the 25% reduction in mortality rate could be achieved here, and all estimates suggest it can—and more—and if we use an earning power of only $10,000 per year per individual able to return to the work force, the gross national product would be increased by more than $2 billion per year, and the federal government would collect an additional $550 million per year. The return on investment would more than please the most conservative entrepre-

---

* Approximately 20% of the people killed in road accidents died between the time the accident occurred and the time they arrived at a hospital. The object is to reduce the "therapy-free interval."

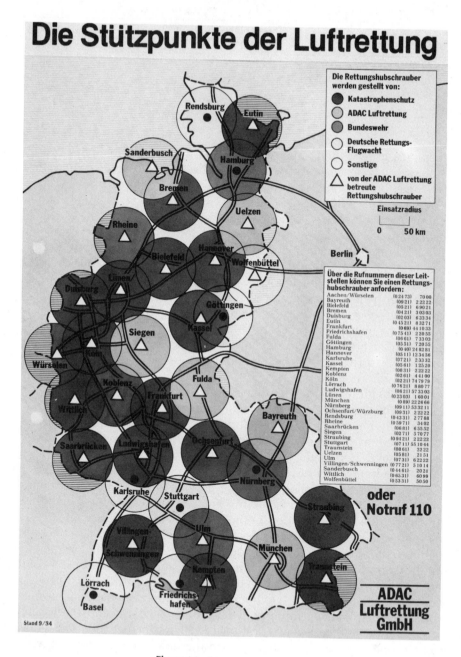

Figure 19. Air rescue centers.

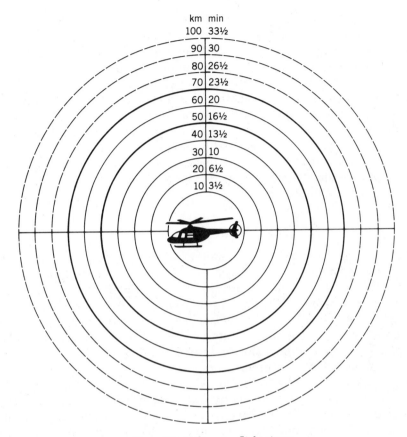

km	min
100	33½
90	30
80	26½
70	23½
60	20
50	16½
40	13½
30	10
20	6½
10	3½

Figure 20. Helicopter flight times.

neur. What will it take to move state governments? If state governments are not ready to depart from conventional thinking and action, the Center for Environmental Health of the Centers for Disease Control is.

With the passage of the Emergency Medical Services Systems Act of 1973, the United States was divided into 304 Emergency Medical Service regions in an attempt to coordinate services. But that was only a beginning.

According to R. Adams Cowley of the Maryland Institute for Emergency Medical Services Systems in Baltimore (11), hospitals need to be designated as to the level of care they can provide. This will permit trauma victims to be taken quickly to the facility best equipped to deal with the specific problem.

Tragic experience has shown that upward of 80% of victims can be adequately cared for on a local level, while 10% require intensive care facilities. For another 10%, survival depends on medical services available only at advanced life support trauma centers. Cowley believes that in order to attain these tripartite levels of care, every state "should have a major trauma center as well as satellite area wide centers. There should be at least 150 dedicated trauma centers in the United States."

The concept of regional trauma centers depends on a smoothly functioning transport system. To achieve this, the emergency medical professions must receive specialized training in the field evaluation, rapid stabilization, and transport of injured individuals, especially children.

A basic component of a regional trauma center is a resuscitation unit. An effective communication system permits a sophisticated multidisciplined trauma team to form in the resuscitation area prepared for prompt action when the injured arrive.

E. Accidents in Complex Systems

Accidents can be viewed from another dimension. The world of high technology provides unique possibilities for accident occurrence.

At what point the United States and western Europe became high-tech societies is debatable. What is not debatable is that all of us now live alongside technologies whose levels of sophistication, or perhaps more to the point levels of complexity, are beyond the comprehension of most people. Lack of comprehension and understanding results in fear, which is to be expected. On the other hand, it may well be that comprehension and understanding will also breed a degree of fear because of the simple fact that disasters such as occurred at Bhopal, Chernobyl, Three Mile Island, or North Sea Platform Bravo can affect everyone regardless of one's level of comprehension. Currently, innocent bystanders can be placed in jeopardy by such necessary high-tech industries as nuclear power and petrochemicals. A new and more comprehensive approach to community safety is necessary.

Engineers can describe how complex machines and systems are supposed to function. Individuals and communities learn how they actually work—or fail to work—when a serious accident occurs. The accident at Three Mile Island (near Harrisburg, Pennsylvania), the sinking of the uranium-laden French freighter Mont Louis after colliding with a Belgian freighter in the North Sea, the release of dioxin into the surrounding community at Seveso, Italy, the explosion and fire at the Pemex gasoline storage facility near Mexico City, the release of methyl isocyanate at the Union Carbide plant in Bhopal, India, and the fire at Utah Power and Light's Wilberg Mine share common characteristics. They are not the result of engineering failure, operator error, or faulty equipment. Nor are they random events. Rather, they share in common a failure of the system.

It is irrelevant to blame these accidents on any single part or person, yet that is exactly what happens. Training and disposition cause us to "see" what it is we are trained to see. Engineers see things their way, sociologists another way, and lawyers yet another. In our litigious society, damages will be sought and all parties involved will seek to disavow responsibility. Little thought is given to solutions that will prevent similar occurrences: similar only in that the end result—a gas leak—occurs. The mechanism next time may be via an entirely different route than the previous one. Then again, it could be similar.

1. Bravo and TMI

At 10 p.m. on April 22, 1977, an oil well went out of control in the North Sea off Norway. By the time it was capped off 8 days later, the blowout at platform Bravo had spewed some 13,000 tons of oil over the surface of the sea. "It was the first blowout of an off shore rig in the North Sea, and it came as a shock to the North Sea oil industry and to the governments of several countries in the vicinity. All had been assured that such an event was impossible" (12).

On March 28, 1979, a series of events culminated in a nuclear plant accident at Three Mile Island, near Harrisburg, Pennsylvania. Within a short period the plant's uranium core become seriously overheated. By the time the initial danger had subsided a week later, radiation had been released into the environment, however small. It was the first major nuclear accident in the United States, and it raised the issue that nuclear accidents caused by human error were not something that happened only in the movies. Prior to TMI, such an accident was considered almost impossible.

David W. Fisher, of the Institute for Industrial Economics in Bergen, Norway, looked at lessons to be learned from a comparison of organizational responses to these two accidents. He found that these accidents were not as unique as some believe.

Table 5 suggests quite dramatically that the organizational similarities between the two accidents indicate there are significant generalizable lessons to be learned about large-scale accidents (12). Fisher's aim was to identify common defects in accident planning as a means of reducing the risk of their occurring again.

Charles Perrow, professor of sociology at Yale, believes with good cause that accidents such as those noted above are normal in a high-tech society. Normal and inevitable because these technologies possess two system characteristics which he calls "interactive complexity" and "tight coupling."

"Linear interactions," according to Perrow, are those interactions of one component in the system with one or more components that precede or follow it in the production sequence. "Complex interactions" are those in which one component can interact with one or more other components

Table 5
The Obvious Similarities

	Bravo[a]	TMI[b]
Type of accident	Oil blowout, first in the North Sea	Nuclear core overheat, first in a U.S. commerical reactor
Indirect cause	Maintenance program	Same
Early warning	Leaking valve	Same
Initiating event[c]	Valve stuck open	Same
Backup event	Not ready to install backup valve	Not aware of need to close backup valve
Shortcomings of crew	Warnings were not acted on	Same
	Information was not shared between shifts	Same
	Lack of formal engineering education	Same
	Lack of experience with events	Same
Shortcomings of supervisors	Lack of theoretical knowledge	Same
	Lack of experience with events	Same
Basic cause of accident	Weak organizational and administrative systems	Same
Safety program	Lacking in details	Same
Maintenance program	Poorly organized	Same
	Neither detailed nor approved	Not detailed but approved
Communication among actors	Internal contact poor	Same
	No plan for informing public	Same
Off-site damages	Not significant	Same
Political consequences	Temporary moratorium on oil drilling in North Sea	Temporary moratorium on construction of new nuclear plants in United States

[a] The blowout of North Sea oil production platform Bravo that occurred off Norway on April 22, 1977.
[b] The nuclear power plant accident resulting in an overheated fuel core that occurred at Three Mile Island, Pennsylvania, on March 28, 1979.
[c] The event that marked a departure from normal recovery from early warning, that is the fault that should have been controlled.

outside the normal production sequence, intentional or not. They can involve unfamiliar, unplanned, and unexpected sequences and they are either invisible or not immediately comprehensible.

"Tight coupling," a term borrowed from mechanical engineering, refers to the lack of give or slack between two elements. Thus, what happens in one element directly affects what happens in the other. "Loose coupling,"

Table 6
Types of Interaction

Type of Coupling	Linear	Complex
Tight	Dams Marine and rail Transport (1)	Nuclear Power accidents Nuclear weapons DNA Research (?) (2)
Loose	(3) Assembly-line production Most manufacturing	(4) Mining

of course, is the opposite. Table 6 shows the possible interactions (13). Each of the four cells contain examples of an appropriate system. Clearly, cell 2 with tight, linear coupling and complex interaction contains the most sensitive systems.

Perrow maintains that "we need two or more failures among components that interact in some unexpected way. No one dreamed that when X failed, Y would also be out of order and the two failures would interact so as to both start a fire and silence the fire alarm. Furthermore, no one can figure out the interaction at the time and thus know what to do. The problem is just something that never occurred to the designers. Next time they will put in an extra alarm system and a fire suppressor, but who knows, that might just allow three more unexpected interactions among inevitable failures. This interacting tendency is a characteristic of a system, not of a part or an operator; we call it the 'interactive complexity' of the system."

Fortunately, this type of interaction has enough give or slack designed into it to allow preventive measures to be instituted to keep the reaction from spreading. If the system is "tightly coupled," with processes occurring very quickly and unable to be turned off or closed down, then spread will occur and the process will be out of control for some time. Multiple failures will have taken place. The outside world hears of it in the form of an accident, a widespread gas leak or fire and explosion covering 30 acres.

Accidents can be the result of failure of any combination of six major components: design, equipment, procedures, operators, supplies and materials, and environment. Different parts of a system can be quite dependent on one another or they can be independent.

As an example, with the development of the nuclear energy industry, designs have become so complex that it may be difficult to impossible to anticipate all the possible interactions which may result in failure. Could a Three Mile Island occur again elsewhere? Of course. Because of the complex nature of the system it was to be expected and can be expected to occur again. Does this mean it has to occur? No. The problem is that although we have reached a point where we have the ability to design and build highly complex (high-risk) processing systems, we do not yet think and design in terms of "catastrophe potential." We design for the specific process, not for what a failure within the process can do to the world around it.

To deal effectively with this level of problem we need to think in terms of systems rather than in the conventional and traditional terms of individual components. However, thinking in terms of systems is in itself neither the solution nor a panacea. Each time an additional component is added, the reliability of the entire system is correspondingly diminished. The problem is not simple, nor will it disappear if ignored.

The recent spate of accidents hold messages for us. The burning question is, how will we respond to these messages?

III.   GOALS FOR THE 1990s

In 1979 and 1980, the Government Printing Office issued two reports which set the tone for the decade. *Healthy People: The Surgeon General's Report on Health Promotion and Disease Prevention* made the essential point that Americans—for all their intense concern about their health— are a healthy people. The following year, with little fanfare, the Department of Health and Human Services (DHHS) issued its manifesto— *Promoting Health/Preventing Disease: Objectives for the Nation.*

Healthy People identified 15 highest priority needs requiring pursuit if health of the American people was to be progressively improved. It established broad national goals for desired and attainable reductions in death rates and/or disability.

In *Promoting Health/Preventing Disease* specific goals were set forth and quantified. Among them were the following:

By 1990, reduce motor vehicle mortality rates to 18/100,000 (1979 baseline = 23.6).

By 1990, reduce motor vehicle fatality rate for children under 15 to 5.5/100,000 (1978 = 9.0).

By 1990, reduce home injury mortality rate for children under 15 to 5/100,000 (1978 = 6.0).

By 1990, 75% of communities with a population over 10,000 should have the capability for ambulance response and transport within 20 min of a call.

By 1990, everyone should have access to regionalized systems of trauma, burn, and spinal cord injury units.

Fully comprehending the forces necessary to reduce their ideas to action, the DHHS noted that if these objectives were to be attained, they required the energy and resources of the states, private industry, and voluntary agencies, as well as the support of a variety of federal agencies and individual legislators.

Although it is to their everlasting credit to have set forth these priorities, it is also clear that short of war, national agreement on any issue is an illusion. Even regional agreement may not be possible. That brings us full circle to the separate and individual states. If these worthy objectives are to be achieved, it will ultimately fall to people, groups of people willing to take the time and persistent effort to convince their elected representatives that enacting the necessary legislation is in everyone's best interests. The year 1990 is a useful milestone; 2000 may be more realistic. The Center for Environmental Health will be the catalyst for keeping these objectives in the public's eye.

REFERENCES

1. M. Schulzinger, *The Accident Syndrome*. C. C. Thomas, Springfield, IL, 1956.

2. J. A. Walter and T. J. Naughton, Driving and Crash Experience and Ischemic Heart Disease. Implication for Counter Measures. Presented at the Annual Meeting of the American Association for Automotive Medicine San Antonio, TX, Oct. 3–6, 1983.

3. L. D. Budnick, Bathtub-Related Electrocutions in the United States, 1979 to 1982. *J. Am. Med. Assoc.* **252** (7): 918–920, 1984.

4. L. D. Budnick, Toothpick-Related Injuries in the United States, 1979 through 1982. *J. Am. Med. Assoc.* **252** (6): 796–797, 1984.

5. E. G. House, P. F. Waller, and J. R. Stewart, Blood Alcohol Level and Injury in Traffic Crashes. University of North Carolina Highway Safety Research Center, Chapel Hill, NC. Presented at the Annual Meeting of the American Association for Automotive Medicine, Ottawa, Ontario, Canada, Oct. 1982.

6. K. Benveniste and P. Thut, The Effect of Chronic Alcoholism on Wound Healing. *Proc. Soc. Exp. Biol. Med.* **166**: 568–575, 1981.

7. J. A. Liedtke and W. E. DeMuth, Effects of Alcohol on Cardiovascular Performance After Experimental Non-penetrating Chest Trauma. *Am. J. Cardiol.* **55**: 243–249, 1975.

8. *Healthy People: The Surgeon General's Report on Health Promotion and Disease Prevention.* DHEW Pub. No. (PHS) 79-55071, U.S. Government Printing Office, Washington, DC, 1979.

9. S. P. Baker, R. A. Whitfield, and B. O'Neil, Geographic Variations in Mortality from Motor Vehicle Crashes. *N. Engl. J. Med.* **316**: 1384–1387, 1987.

10. R. A. Whitfield and I. Fife, Changing Patterns in Motor Vehicle Crash Mortality: 1940–1980. *Accident Anal. Prev.* **19** (4): 261–269, 1987.

11. A. R. Cowley, Accidental Death and Disability: The Neglected Disease of Modern Society—Where Is the Fifth Component? *Ann. Emerg. Med.* **11**: 582–585, 1982.

12. D. W. Fisher, Lessons from Major Accidents: A Comparison of the Three Mile Island Nuclear Core Overheat and The North Sea Platform Bravo Blowout. Executive Report 6, International Institute of Applied System Analysis, Laxenburg, Austria, 1981.

13. C. Perrow, *Normal Accidents: Living with High Risk Technologies*. Basic Books, New York, 1984.

SUGGESTED READINGS

Accident Facts. National Safety Council, Chicago, 1983.

Baker, S. P., O'Neil, B., and Karpf, R. S. *Injury Fact Book.* Lexington Books/Heath, Lexington, MA, 1984.

Davis, H. F., Schety, A. V., Ing, R. T., and Weisner, P. J. The 1990 Objectives for the Nation for Injury Prevention: A Progress Review. *Public Health Rep.* **99** (1): 10–23, 1984.

Department of Health and Human Services, *Promoting Health/Preventing Disease: Objectives for the Nation.* U.S. Government Printing Office, Washington, DC, 1980.

Office of the Assistant Secretary for Health and Surgeon General, *The Surgeon General's Report on Health Promotion and Disease Prevention.* DHEW Pub. No. (PHS) 79-55071, U.S. Government Printing Office, Washington, DC, 1979.

7

Noise Pollution

Humanity either makes, or breeds, or tolerates all its afflictions, great or small.

—H. G. Wells

During the spring and summer of 1983, Sweden's security was sorely tried. Russian submarines tried to penetrate restricted naval installations in the Hors Bay area along Sweden's northeast coast. According to initial reports by officials of the Naval Ministry, underwater sound-detecting gear functioned poorly making it difficult to pinpoint the location of the subs.

It was later determined that the problem lay elsewhere. The young Swedish sailors manning the monitors had in fact developed hearing losses as a consequence of excessive exposure to rock music. On learning this, other governments moved quickly to ascertain the hearing acuity of their own personnel responsible for monitoring a variety of sound detectors. Consequently, the problem of exposure to sound levels detrimental to hearing takes on new and more serious dimensions.

Yet it has been argued that because noise produces no dramatic ill effects, the public has been largely uninterested in its suppression. It may be more to the point to say that the degree of annoyance and discomfort that people will endure is astonishing.

Although noise is an integral part of civilization, it would appear that unless some definite steps are taken to reduce the present inordinate levels in both industry and the community generally, more and more people will become auditory cripples.

Fortunately, in the past few years large segments of the population have expressed an increasing determination to revolt against noise. A measure

223

of this is seen in the number of communities moving forcefully to reduce
noise levels.

I.   WHAT IS NOISE

Noise can be defined in terms of particle displacement, variations in
pressure, and particle velocity in an elastic medium, or we can accept the
concept by the British Committee on the Problems of Noise, which with
characteristic British pithiness defined noise as "unwanted sound." It has
also been defined as any sound regarded as a nuisance. While wholly
satisfactory, these latter definitions share a measure of subjectivity: that is,
different people are annoyed by different sounds, depending on individual thresholds. Generally, however, nuisance increases with frequency
and pitch.

Sound (noise) in industry, in the home, or in traffic is a by-product of
the conversion of energy. No process using power is completely efficient;
some energy is inevitably wasted. Most is converted into heat; some is
converted into sound, as when surfaces vibrate or turbulence is set up
in air.

Noise includes two inseparable elements: a subjective, judgmental
preference and a physical phenomenon, which makes control necessary
and exceedingly difficult, if not frustrating in a democratic society. The
key or operational difference between these two elements is the "unwanted." At acute levels disagreement over offensiveness diminishes.
The question currently being argued is how high does unwanted sound
have to go before the annoyance factor becomes a maddening factor and
encompasses an entire community.

Sound is transmitted by pressure waves traveling through air from a
vibrating source to the human ear. Pressure waves can be produced in
almost any material that has both mass and elasticity—glass, metal, wood,
water, dry leaves, and leather, as heels striking tile, wood, or concrete
testify.

A plucked guitar string or jackhammer boring through pavement
compresses air molecules, which in turn push against other molecules
thereby causing a wave to move through air, not unlike a row of falling
dominoes (see Figure 1).

Figure 1. Each domino pushes the next one over and a pressure wave is generated.

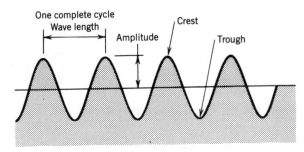

Figure 2. Wave motion.

As the wave (see Figure 2), passes across the ear hair cells that line the cochlea convert the waves into electrical impulses which the auditory nerve conveys to the temporal lobe of the brain. Here the wave is registered as sound—music, jackhammers, speech. But the strings continue to vibrate, producing a series of repeated waves.

Accordingly, what we call sound consists of a series of repeated waves impinging on the ear.

Of course, rarely are we in a setting with single vibrating source. More often than not, someone is talking, a radio is playing, and a tea kettle on the stove in the kitchen may be whistling, just as a friend pulls into your driveway and lets fly with several blasts of the horn, as the neighbor's teenager streaks past the house on a moped. Is it any wonder you did not hear the telephone ring? Each of these sources has a different vibrating level and thus different pressure wave, and all are impacting on your ear at different moments. The auditory system and brain must deal with each simultaneously. Clearly, there is room for error, such as not hearing the phone or misunderstanding what is being said. Given noise-induced hearing losses, the errors increase significantly, generating as they do a set of personal and social problems.

The pattern of wave motion shown in Figure 3 describes a neutral sound, neither high nor low. Figure 4 indicates the changes in wave amplitude responsible for loud or soft sounds. Thus, the intensity of sound depends on the amplitude of its waves, while more rapid vibrations with decreased wavelengths produce higher pitched sounds.

Higher pitch can also be understood in terms of expansion of the vibrating source as in Figure 5. When a bell is struck, particles of air are pushed out until they bump against their neighbor. This creates a steadily

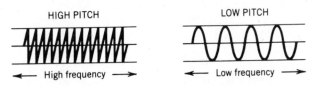

Figure 3. Frequency change can alter pitch.

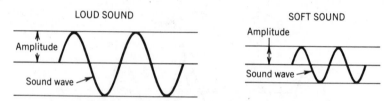

Figure 4. Sound as a function of amplitude.

moving region of collision, called a compression. As the clacker moves back from the bell a rarefaction is created—a region of emptiness into which the particles rebound. When the large bell rings, the compression and rarefaction around it are greater than that of the particles around the smaller bell in Figure 6, and a louder sound is produced.

Figure 7 demonstrates how interactions of waves can affect the loudness of sound, a phenomenon known as interference.

In Figure 7a, two sounds 1 and 2 are in phase; that is, their compressions coincide. The waves' intensities, shown by their heights above the line, are reinforced to make a stronger sound, 3. Figure 7b shows waves half a wavelength out of phase. When they are of varying intensity their total strength is diminished; should they have the same intensity, as in Figure 7c, no sound is heard at all. This is a rare phenomenon. Well, perhaps not so rare if it is planned. Think Fourier.

Complete interference is shown in Figure 8. Here two waves 1 and 2 of the same intensity but different frequency interact. Starting half a wavelength out of phase, where their resultant strength is zero (wave A), the waves move on, alternately reinforcing and canceling each other. The ear hears this interference as a faint throbbing as, for example, when two adjoining notes on the piano are played together.

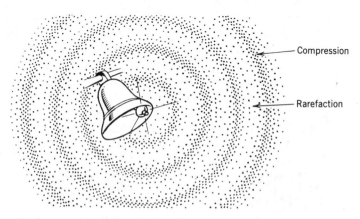

Figure 5. Compression and expansion of air particles upon striking a small bell.

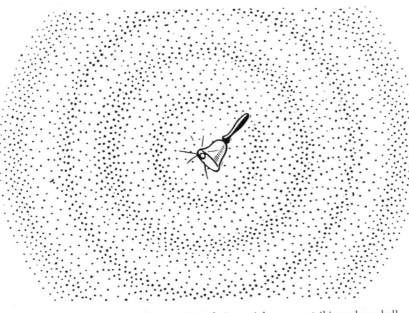

Figure 6. Compression and expansion of air particles upon striking a large bell.

Sound waves vibrate at various frequencies. The number of vibrations per second (when the sound waves emitted from a source exert a pulsating pressure on the ear) is the frequency of the sound, usually described as cycles per second (cps), hertz (Hz), or vibrations per section (vps).

An essential characteristic of sound as it affects the perception of noise is loudness or intensity. Loudness is a highly subjective perception interpreted by the brain from the nerve stimulus it receives. Each person experiences his or her own degree of loudness because the stimulus received depends on the intensity of the sound energy reaching the nerve

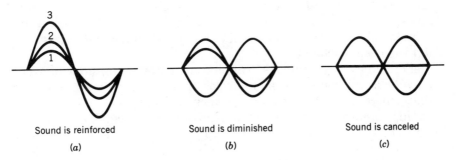

Sound is reinforced   Sound is diminished   Sound is canceled

(a)   (b)   (c)

Figure 7. Sequence of wave interactions.

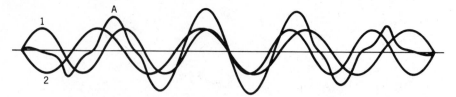

Figure 8. Complete interference of sound waves of different intensity.

and the sensitivity of the nerve. Here again, wide ranges of individual differences are found. The human ear can translate sound between approximately 20 and 2000 cps into nerve impulses. This means that the ears of healthy young adults are sensitive to a wide range of frequencies. For the most part, our ears are most sensitive to the range of 500–4000 cps, which includes conversational speech.

The range of a piano may help give an idea of the frequencies of the various sounds (see Figure 9). The keys range from the lowest note, A_0, which emits pulses that vibrate at the rate of 27.5 cps, through middle C, with a frequency of 261.63 cps, to the highest note, C_8, which generates 4186 cps. Men's voices range from the lowest basso, at 80 cps, to tenor, producing sounds at 300 cps—just above middle C. It is not uncommon for sopranos to reach 850 cps. Some exceptionally trained voices have been known to exceed even this.

Technically, noise can be defined as "a class of sounds that do not exhibit clearly defined frequency components." No matter the definition, it is widely held that a sound wave means nothing until it strikes the ear. Could a philosopher, however, properly wonder whether there would be a sound if a tree fell in a forest, and no one was there?

II.   COMMUNITY CONCERN

The question of how much noise communities will tolerate or at what point the citizenry will have reached its threshold has never had greater currency. The number of towns enacting strict and enforceable ordinances to reduce and control noise levels, both day and night, suggests the point has been reached.

Barking dogs, lawn mowers, leaf blowers, power saws, snow blowers, church bells, jackhammers, motorcycles, mopeds, airplanes, radios, and traffic generally have combined to such a degree that noise-induced irritation, annoyance, discomfort, and hearing impairment have become a significant public health issue—certainly enough of one to motivate a political response.

In Brookhaven, New York, an ordinance that recently became effective stipulated that a disturbance created by dogs barking for more than 15 min

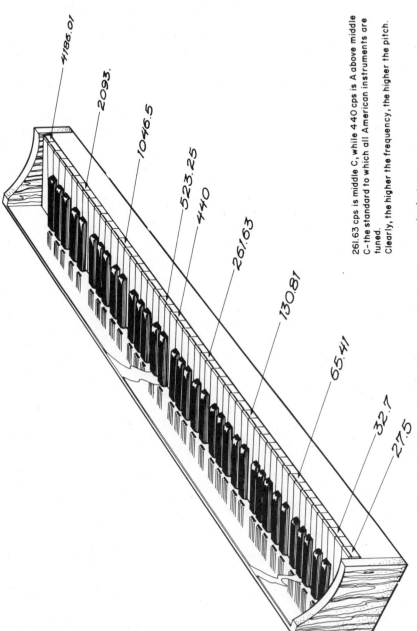

4186.01

2093.

1046.5

523.25

440

261.63

130.81

65.41

32.7

27.5

261.63 cps is middle C, while 440 cps is A above middle C – the standard to which all American instruments are tuned.
Clearly, the higher the frequency, the higher the pitch.

Figure 9. Spectrum of sound frequencies defined by a standard piano.

is illegal and punishable by a fine of $250 for each offense. This ordinance also sets acceptable noise levels for residential areas for 55 dB during the day and 50 dB during the evening, measured at the receiving property line. By setting decibel (dB) levels, the statute seeks to protect health as well as prevent interference with sleep and speech, the precipitating problems.

In the adjoining town of Babylon, the ordinance being drafted will include a clause that states noise from a disco or bar cannot exceed 95 dB for more than 30 continuous seconds measured from any point where a customer may be inside the building, unless a sign, posted in an obvious place, stipulates that "sound levels within may cause permanent hearing loss." Clearly, the new ordinances are being drafted with a new wrinkle, the capability of enforcement.

Tables 1 and 2 indicate the magnitude of the U.S. population exposed to noise, and the percentage expressing annoyance with specific sources of noise. Considering that 60 dB is akin to the sound of an air conditioner at a distance of 20 ft, it is evident that with a population in excess of 246,000,000, approximately 7%, or 17+ million people are exposed to noise levels, from traffic alone, of from 70 to over 80 dB.

It is also evident that no matter the density of the population, motorcycle noise is considered the outstanding annoyance. Clearly, the gasoline-driven engine whether in a truck, car, bus, or other vehicle is a major source of annoyance in small and large communities. There must be a lesson there for those concerned with noise prevention and control.

Table 1
U.S. Population Exposed to Noise, by Level and Source, 1980

Decibels[b]	Million People Exposed[a]				
	Traffic	Aircraft	Construction[c]	Rail	Industrial
More than 80	0.1	0.1	—	—	—
More than 75	1.1	0.3	0.1	—	—
More than 70	5.7	1.3	0.6	0.8	—
More than 65	19.3	4.7	2.1	2.5	0.3
More than 60	46.6	11.5	7.7	3.5	1.9
More than 55	96.8	24.3	27.5	6.0	6.9

[a] There is some overlap among populations exposed to different noise sources. For example, some of the 96.8 million people exposed to traffic noise levels of 55 dB and above are also exposed to aircraft noise.
[b] An increase of 10 dB means that loudness has apparently doubled; an increase of 20 dB would quadruple apparent loudness. Levels here refer to average annual outdoor exposure levels.
[c] Includes both residential and nonresidential exposure.
Source: U.S. Environmental Protection Agency, Noise in America: The Extent of the Noise Problem.

Table 2
Population Reporting "Highly Annoying" Noise Sources[a]

| Source | Population per square mile | | | | | |
| | More than 20,000[b] | | 3,000–20,000[c] | | Less than 3,000[d] | |
	Rank[e]	Percentage of Respondents Highly Annoyed by Source	Rank[e]	Percentage of Respondents Highly Annoyed by Source	Rank[e]	Percentage of Respondents Highly Annoyed by Source
Motorcycles	1	12.7	1	13.2	1	9.4
Automobiles	2	9.4	3	7.4	3	4.2
Large trucks	3	7.3	2	10.0	7	2.6
Construction	4	6.5	4	7.2	4	3.7
Sport cars	5	5.9	5	7.0	6	3.1
Constant traffic	6	4.7	6	5.5	10	1.5
Buses	7	4.7	8	3.5	11	1.1
Small trucks	8	4.1	7	4.1	9	1.5
Helicopters	9	3.9	10	3.1	2	5.3
Airplanes	10	3.6	9	3.4	5	3.2
Power garden tools	11	1.2	11	2.1	8	1.8
		(66.0)		(62.2)		(55.9)

[a] Areas surveyed were away from the direct impact of freeway or aircraft noise.
[b] Represents 12% of the population sample surveyed.
[c] Represents 51% of the population sample surveyed.
[d] Represents 37% of the population sample surveyed.
[e] Rank=Rank order of noise sources that highly annoy respondents.
Source: Council on Environmental Quality, Environmental Quality—1979, U.S. Government Printing Office, Washington, DC, 1980.

Vexation and anger with increasing noisy communities is not a uniquely American problem. Between May and July 1984, over 70% of the adult population of Yamoto, Japan, signed petitions demanding cessation of the U.S. Navy's flight activities over their city. Yamoto, with a population of 175,000, is a bedroom community for Tokyo, some 20 mi to the northeast. U.S. Navy fighter aircraft, F-4, A-6, and E-2 jets, 100 of them on station aboard the aircraft carrier *U.S.S. Midway,* fly over Yamoto as they come in for landings at nearby Atsugi Air Base. As the jets sweep low they rattle houses and occupants. The evenings are the worst with more takeoffs and landings at full power and thus maximum noise, at a time, of course when the people are trying to sleep.

Kenichi Ohsaka, a Yamoto City official who keeps track of noise levels, has been reported as saying "it's just like living inside a subway car." According to spokespeople for the U.S. Navy, the *Midway,* stationed off Yamoto, holds regular weekly takeoff and landing exercises to keep its pilots' skills honed, and night sessions are particularly important. Be that as it may, the residents are unimpressed, cannot sleep, and prefer the training sessions to be moved elsewhere. But where else? No one wants them and their wretched noise.

III.   STUDIES OF THE EFFECTS OF NOISE

Aircraft noise is not simply a problem for those trying to sleep. Well-designed, well-controlled studies have demonstrated that exposure to high levels of aircraft and environmental noise can adversely affect reading ability in school-age children. For example, Maser and co-workers (1) reported that children who attended school beneath the Seattle–Tacoma airport flight paths showed a deficit on standardized tests of scholastic achievement compared to students in quiet schools.

According to the investigations of Cohen and colleagues (2), reading and math scores of third-grade students in noise-abated classrooms were higher than those in classrooms without that quality.

Most recently, Green and co-workers (3) of New York University's Institute of Environmental Medicine found that for all elementary schools in the boroughs of Brooklyn and Queens "an additional 3.6% of the students in the noisest schools read at least one year below grade level." They went on to remark that "the dose–response relationship indicated that the percent reading below grade level increased as noise level increased." Figure 10 suggests the relationship between aircraft noise and percentage of children below grade level. In the figure, point estimates and 95% confidence limits for the effect at each level are plotted along with a dotted line representing the linear dose–response relationship assumed by the use of the noise scale values.

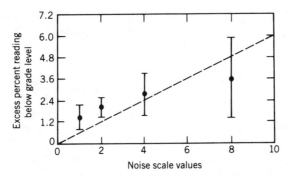

Figure 10. Relationship between aircraft noise and reading level achievement.

Others have found similar relationships. Cohen and colleagues (4) determined that elementary school students living for at least 4 years in the lower floors of an apartment complex near heavy traffic showed greater impairment of reading ability than children living on higher floors away from the traffic. In their studies, indoor sound levels varied from 66 dB on the lower floors to 55 dB on the higher ones. Interestingly enough, in a recent U.S. EPA classification, "noisy residential areas" averaged 58 dB and were rated low socioeconomic, while "quiet residential" averaged 38 dB and were rated affluent neighborhoods. These, of course, were outdoor sound levels. With indoor levels of 55–66 dB, concentration, the ability to pay attention, may well be difficult to nonexistent. If that is true, it may be pertinent to ask why far more children are not reading impaired. This is not a frivolous question.

Studies of both traffic noise and noise in communities hard by major airports have concluded that elevated blood pressure, heart disease, and psychological trauma are direct consequences of noise exposure. Although these associations have been reported, others contradict or do not bear them out. Hypertension, heart disease, and psychological trauma, as well as irritation and annoyance, can be engendered by a variety of risk factors or by several operating simultaneously. Accordingly, establishing direct causal relationships can be exceedingly difficult. If research studies are methodologically deficient, complications arise to further confound relationships between the independent variable noise, and such dependent variables as elevated blood pressure, reading impairment, annoyance, anxiety, accidents, and heart disease. Obviously, with the number and variety of factors known to contribute to these events, there is good reason for contradictory results.

As an example, Jenkins and his group at the London Institute of Psychiatry (5) reviewed the findings of two studies conducted in the area of London's Heathrow Airport. These studies had compared rates of admission to Springfield Psychiatric Hospital (in West London) among residents living near Heathrow. Findings suggested that areas closest to the airport, with presumably higher levels of noise, also had the highest rates of hospital admissions.

Using similar residential areas and admissions to the same hospital, the Jenkens group obtained quite different results. Their study included a wider area, studied the relationship over a longer period (4 years), and used four levels of noise exposure. Among their conclusions they reported that previous studies had used poor admissions data on their patient subjects, studies were not continued long enough, and exposure to noise was poorly documented. Further analyses suggested that while both mentally disturbed and normal individuals are annoyed by aircraft noise, the mentally disturbed are more likely to be considerably more annoyed and even show annoyance in low-noise areas. Mental disturbance may produce both sensitivity to noise as well as fervent expressions of annoyance.

In studies of this type, it is not only necessary to utilize control groups, but to be sure that the number of mentally agitated are not overrepresented in the sample. Finally, it is essential that decisions to modify existing conditions, whatever they may be, are based on appropriate data. Thus far they do not appear to be.

Aircraft, traffic, and industrial noises have been much discussed and implicated as a potential risk factor for elevation of blood pressure (essential or primary hypertension) and cardiovascular disease generally. However, here again equivocal results of studies, many with methodological deficiencies, have left these associations hotly disputed.

Recently, Peterson and co-workers (6) of the Department of Otolaryngology, University of Miami School of Medicine, appeared to demonstrate in rhesus monkeys that moderate levels of realistic noise ("meant to resemble the daily pattern of noise to which a worker in a noisy industry might be subjected") can produce sustained elevations in blood pressure *without* significant alterations in the auditory mechanism. The unique aspect of this investigation was the finding that changes in auditory sensitivity did not necessarily follow changes in such physiological parameters as blood pressure. However, translation of animal-derived data to the human experience must be done cautiously.

This is highlighted by recent studies in Switzerland and West Germany. From their investigation of physiological effects, including blood pressure and cardiac output (ECG), Bättig and colleagues (7) of the Institute of Behavioral Sciences of the Swiss Federal Institute of Technology were unable to verify the relationship of aircraft noise to elevated vital signs. Their subjects, living close to the Zurich Airport and tested in their own homes, were required to perform a number of mentally related activities at the height of airport activity. They concluded that other annoyances and uncontrolled variables could explain the increased physiological effects.

Neus and co-workers of the Medical University of Bonn, West Germany (8), were of a similar mind with respect to traffic noise and hypertension. They noted that a family history of hypertension had to be controlled before conclusions could be made. When that was done, elevated blood pressure seemed to follow increased sensitivity to noise. Again, studies of this type must deal with people's adaptability to noise.

Sleep disburbances are probably the most widespread source of annoyance caused by noise, if anecdotal responses are any criteria. Recently, French investigators (9) studied the problem under real-life conditions in bedrooms of people living close to freeways and airports. Using miniaturized electronic units, they recorded EEG, eye movements, muscular activity, and heart rhythm with remote-reading equipment. Noise inside the rooms was recorded continuously. With the noise from the highways, subjects took longer to fall asleep and had less deep sleep so that the young to middle-aged group became more like the 50–60-year-old group in their depth of sleep. Rapid eye movement (REM) sleep was also reduced. If both deep and REM sleep are physiologically and psychologically important, this type of alteration may well be damaging. But this remains to be substantiated by further study. Bear in mind, surveys suggest that as much as 25% of sleep disturbance may be due to causes other than noise.

The problem is indeed complex, but progress is being made. Joseph Sataloff and co-investigators at the Jefferson Medical College in Philadelphia studied the effect of prolonged but intermittent exposure to loud noise on the hearing acuity of industrial workers (10). Theirs was a particularly well-designed investigation with strict criteria: so strict in fact that they could enlist only 295 subjects, of some 12,000 screened, who qualified as being exposed to intermittent noise. Their results showed that "years of exposure to intense but intermittent noise up to 188 dB damages the high frequencies of hearing but has surprisingly little effect on the lower frequencies (<3,000 Hz)." They suggested that frequent rest periods protected the auditory mechanism from damage in the speech frequencies. This type of finding opens up a variety of avenues for both prevention and control. It may also suggest an approach for discos, if anyone is listening.

Given the concern over noise, one wonders just how desirable a quiet town would be. Darlington, near Newcastle, England, was almost such a place. Between 1976 and 1978, Darlington was designated a "quiet town experiment" (11). Noise abatement zones and better traffic management were instituted, as were vehicle noise testing and stricter enforcement of noise regulations.

To make people more aware of noise, exhibitions, lectures, and school projects were set up. Leaflets were circulated, press and radio coverage was extensive, and free advice on noise control was made available to both industry and the public. Unfortunately, the main objective turned out to be more educational than quiet; lack of funds curtailed the effort prematurely. It remains an experiment waiting to be done.

A. Fourier Revisited

Mathematics and the computer may just resolve these many questions and uncertainties. Diagrammed in Figure 8 are interactions of waves. Think of a tuning fork. If you have ever struck one, you have heard a pure tone—the rare sound of a single frequency. This pure tone, or single-frequency wave, is referred to as a sine wave by physicists and mathematicians.

Sine waves are the building blocks of sound. The many different tones of a rock band or a symphony orchestra or a Manhattan subway station are an aggregation of sine waves. Scientists have learned to single out and neutralize unwanted sounds fighting noise with "antinoise"—sound whose wavelengths are exactly opposite the unwanted noise. This antinoise can cancel out the roar of engines, the whine of industrial fans, power transformers, and the throb of heavy machinery, leaving desired sounds sharp and clear.

Using powerful computers and delicate sensors, scientists can, for example, pick out the distinctive auditory signatures of faulty parts deep inside nuclear reactors. The technique breaks apart a complex auditory signal—an aggregation of sine waves—and plots a curve or graph. Adding antinoise—a noise itself—to noise, doubling the amount and obtaining silence, must be difficult to conceive. The technique works on predictable noises such as the repetitive roar of engines, rotating machinery, and fans. Microphones sample the rhythm of the periodic noise, microchips design its mirror image, and the resulting antinoise is broadcast by loudspeakers or headphones. Quite results (12).

At the heart of this technological advance is a 2000-year-old relationship described by Pythagoras and rediscovered in the 19th century. In 1801, Jean Baptiste Fourier described a set of trigonometric functions that reduced any series of complex waves to a series of simple sine waves. He related the complexity of wave motion to an idea Pythagoras had discovered in the sounds of music and developed a theory of vibratory motion—the Fourier theorem.

With the advent of fast computers and appropriate software, scientists were able to efficiently perform the fast Fourier transform. Powerful computers can now split a complex noise into over 1000 separate frequencies in a fraction of a second.

The essence of the antinoise technique is the measurement of the frequency of an unwanted sound using Fourier analysis and then broadcasting its mirror image. The peaks of one wave coincide with the troughs of the other, canceling each other out. This analysis works in noise cancellation because it isolates the periodic element of a target noise,

(a) The whine of an engine, picked up by a microphone and displayed on a graph, is a confused jumble that is nearly impervious to analysis and manipulation by computer.

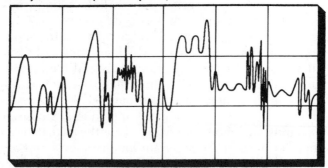

(b) The technique known as Fourier analysis allows engine noise to be broken into its underlying harmonics, which can be analyzed, amplified or turned upside down as "antinoise."

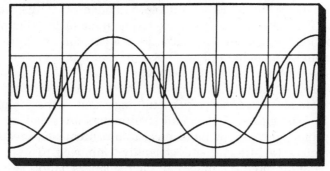

Figure 11. Breaking noise into its component parts. Reproduced by permission of *The New York Times*.

ignoring everything else. A fraction of the wanted sound may also be canceled, but the effect is usually undetectable. Figure 11a is a plot of the whine of an airplane engine. Figure 11b plots the results of a Fourier transformation of that noise.

A company in Wisconsin has developed an instrument that can place antinoise into a fan duct, canceling unwanted sounds that would otherwise spread through a building. And a company in Massachusetts is experimenting with a unit that will cancel unwanted sound in aircraft and industrial settings while allowing normal speech to continue.

An antinoise unit in every house and office could have a beneficial impact on productivity as well as our personal relationships. We owe a debt of gratitude to whoever it was that saved Fourier from the guillotine during the French Revolution.

IV. MEASUREMENT OF SOUND AND NOISE

At this juncture, it is appropriate to ask how sound and noise are measured. Few, if any, people can hear sound above 20,000 cps. This is beyond the audible, the region of "silent" sounds or ultrasound. Here, it is the auditory mechanisms of animals which are unusually keen. Table 3 compares the range of frequencies animals are capable of hearing.

"What noise annoys an oyster?" Although no adequate way to measure the annoyance level of noise has been devised, we have seen that the annoyance level is often directly related to loudness. Studies have found that loud sounds are more annoying than those that are of similar character but not as loud. To ascertain loudness level and to determine whether a noise is a potential hazard to the hearing mechanism, an accurate sound-measuring system must be employed. Several general types are available.

For preliminary screening of a site for possible hazards, the sound level meter shown in Figure 12 is suitable, although other instruments are available. This meter measures noise levels in manufacturing plants as a means of locating potentially hazardous areas and is appropriate for checking environmental (outdoor) noise. It records noise levels of from 40 to 140 dB with three weighting networks.

Often used by architects, planners, and engineers to study possible sites for schools, hospitals, office buildings, and express highways in relation to existing housing, sound-level meters are small, simple, and relatively inexpensive. For more precise measurements, precision integrating meters (Figure 13) are generally employed. These are more sensitive and far more accurate than their smaller counterparts.

Recall that sound is a transmission of energy in the form of vibrations which create variations in the pressure of air, liquid, or solid media. The amount of energy being transmitted and the relative intensity of different sounds can be measured and expressed as decibels—"deci" for one-tenth and "bel" for Alexander Graham Bell. A sound ten times as powerful as

Table 3
Range of Hearing Frequencies of a
Sampling of Animal Species

Grasshoppers	100– 15,000
Dogs	15– 50,000
Cats	60– 65,000
Moth	3,000–150,000
Bat	1,000–120,000
Porpoise	150–150,000
Emission of "silent" dog whistle	12,000– 20,000

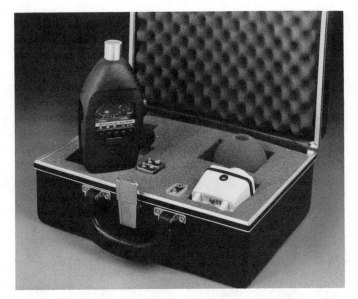

Figure 12. A GenRad sound-level meter model 1565-B.

another is said to be 10 decibels (dB) more intense, and each ten-fold increase adds another 10 dB to the level of the sound.

The magnitude of these pressures is relatively small and is measured in dynes per square centimeter. The standard reference level used in measuring sound levels in decibels is equivalent to a sound pressure of 0.0002 dynes/cm^2.* This is the faintest sound that can be heard by a healthy young adult in a quiet location. At the high end of the sound spectrum, the ear can respond to pressures as high as 2000 dynes/cm^2, considered intolerable. The ratio between high and low is 10,000,000:1. To render it manageable, a logarithmic scale is used. Therefore, if one sound pressure (SP) is ten times another SP$_2$, the SP level is 20 dB higher. Consequently, an increase of SPL of ten times means that loudness has doubled, while an increase of 20 dB quadruples the apparent loudness.

For translating sound pressure levels (SPLs) into decibels the equation is

$$SPL = 20 \log_{10} P/P_0 \quad dB$$

where P = pressure measured by a sound level meter.
P_0 = a reference pressure generally understood to be the lowest audible pressure or 0.0002 μbars.

The 0.002 μbars (dynes/cm^2) noted above, the faintest perceptible

* Often the term *microbar* (μbar) is used as the reference; 1 μbar = 1 dyne/cm^2.

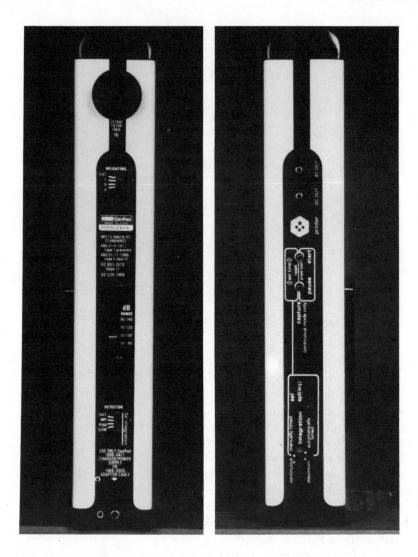

Figure 13.

sound, is equal to 0 dB. The decibel is a dimensionless unit, expressing a ratio between two sound intensities: the reference pressure and the source being measured, whether it is a pneumatic hammer, a speech, a food blender, or a subway train.

Between the threshold of hearing and the greatest pressure intensity the ear can interpret as sound but not as pain, the average individual can distinguish about 130 steps. In other words, the range of hearing covers approximately 130 dB (see Figure 14 and Table 4, which lists some commonly encountered sound-pressure levels).

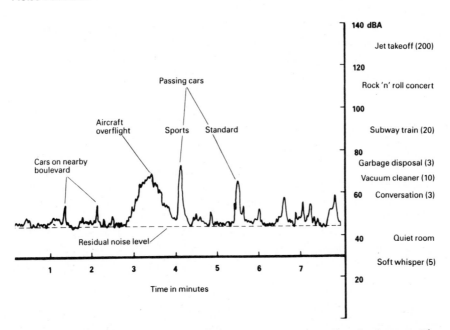

Figure 14. Graph showing 8 min of noise sampled on a city street in the early afternoon. The volume ranges from a base of about 45 dBA to a peak of more than 70 when a sports car drives by. The volume of some familiar sounds, as measured by the decibel "A" scale, is indicated on the graph at the right. Distance (in feet) from the source is given in parentheses. An increase of 10 dBA indicates an approximate doubling in perceived loudness. Courtesy of EPA, *Community Noise*, 1971.

Table 4
Sound Pressure Level and Related Sound Pressure

Pressure (μbars)	SPL (dB)
2000.0	140
200.0	120
63.0	110
20.0	100
6.3	90
2.0	80
1.0	74
0.63	70
0.20	60
0.063	50
0.02	40
0.0063	30
0.002	20
0.00063	10
0.0002	0

Although the decibel is used to express the sound level associated with noise measurements, sound-pressure level cannot be taken at face value as an indication of loudness, because frequency has much to do with apparent loudness. It is therefore important to know the frequency of the noise being measured. This requires consideration of weighting networks.

A. Measurement Scales

Weighting networks enable a sound-level meter (nothing more than "electronic ears" that change noise into electrical signals which can be displayed on an indicator) to respond to some frequencies more than others with a bias akin to that of the human ear. Three weightings, standardized as A, B, and C, are used in the environmental and occupational settings. The primary difference between them is that very low frequencies are severely discriminated against by the A network, moderately by the B, and very little by the C. Therefore, if the measured sound level of a noise is much higher in C weighting than an A weighting, the noise is probably of low frequency.

Currently, federal regulations require use of the A network. Consequently dBA is the preferred notation and calculation rather than dB. In state-of-the-art sound-level meters, A weightings are built in, eliminating the need for recalculations. The dBA is usually considered to be equal to 45 dB.

It is critical for noise measurements to recall that the dB scale is logarithmic. Consequently, the values obtained for two or more sound levels are not a function of addition. Accordingly, an SPL of 70 dB from one vibrating source and an SPL of 74 from another source do not combine to yield 144 dB.

To determine the level impacting on the ear, it is necessary to convert decibel readings into sound pressures, add them, and reconvert to decibel levels. Knowing the difference between two levels, the figures in Table 5 can indicate the number to be added to the highest of the two sounds. Thus, a difference of 4 indicates that 1.5 is to be added to 74, for a total or final dB level of 75.5.

The number of decibels, while indicative of the sound level, tells nothing about the distribution of the component frequencies. Deafness, whether temporary or permanent, is better correlated with frequencies than with overall intensities. While overall noise levels can indicate the need for noise reduction or control, they provide no information about the particular frequencies that are causing the noise problem. Most noises in our urban communities are complex, since the noise emanating from a single source is usually composed of sounds of many frequencies, varying in intensity.

The response of the human ear to a certain sound pressure depends on

Table 5
Conversion Factors for Obtaining Total Decibel Level
When Differences Between Sound Levels Are Known

Difference Between Levels	Number of dB to be Added to Highest Level
0	3.0
1	2.6
2	2.1
3	1.8
4	1.5
5	1.2
6	1.0
7	0.8
8	0.6
10	0.4
12	0.3
14	0.2
16	0.1

the frequency of the sound. While sensitivity is greatest between 500 and 4000 Hz (cps), it falls off for both lower and higher frequencies.

When adolescents with good hearing are tested, a characteristic profile of the efficiency of hearing at several frequencies is obtained. This is portrayed in Figure 15. The curve shows that at low frequencies the sound-pressure level must be relatively high before the tone can be heard. By contrast, tones in the range 200–10,000 cps can be heard even though the levels are very low. This variation in hearing acuity depending on frequency is one of the reasons it is essential to know the frequency rather than the overall noise level if a noise problem is to be dealt with intelligently. For example, a value of 90 dB was obtained near a steel-rolling machine. Analyses of the frequencies involved showed that the greatest intensity was in the low range—where the ear is least sensitive. The intensity was 83 dB, almost ten times less, in the higher ranges, where the ear is more sensitive.

To resolve the spectrum of frequencies of a sound, a frequency analyzer must be used. For the most part, an octave-band analyzer is employed. Octave bands are arbitrary spreads of frequencies in which the upper limit is twice the lower. Generally, the bands chosen are 20–75, 75–150, 150–300, 300–600, 600–1200, 1200–2400, 2400–4800, and beyond 4800. With both the sound-level meter and the octave-band analyzer, readings should be taken at several points in and around the area being studied and, if for an industrial machine, as near to the ear of the operator as practically feasible.

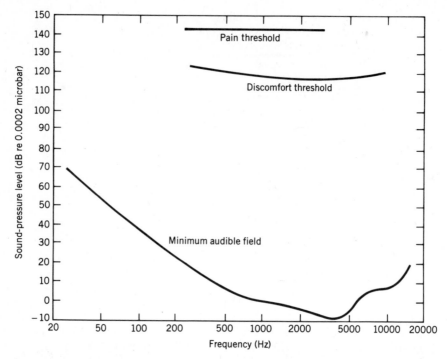

Figure 15. Thresholds of hearing and tolerance. MAF is the minimum audible field. Reproduced from *Handbook of Noise Measurement*, General Radio Company, West Concord, MA.

V.   HEARING LOSS

Hearing loss can be insidious in onset. Years of traumatic exposure to high levels can occur before symptoms become manifest. The vogue for portable sound equipment—such as Walkman-type radios and tape players—has already produced a sharp increase in clinically verified hearing loss, especially among rock music addicts who prefer their music loud—very loud.

As predictable, the audio industry believes it is not their products that are the problem; rather it is improper use. If, they say, the volume is kept down, there would be no problem, which is tantamount to saying that if we all drove cautiously there would be no accidents.

Considering that earphone listening has been around for some 20 years, why has the problem only recently surfaced. Apparently the pattern of listening has changed. Currently, earphones are used while walking or running on noisy city streets rather than in the privacy of the home or other relatively quiet area where the listener did not wish to disturb others. Now the volume must be turned up to overcome the noise of city traffic. The listener wants the Walkman to blot out the "noises of the city." Doing that courts aural disaster. A similar result occurs to users in noisy factory or industrial environments.

Since to these people louder is better, the makings for an epidemic of hearing loss are at hand.

John Koss of the Koss Corporation of Milwaukee, a leading manufacturer of the Koss Music Box, has designed into his units a warning light that flashes whenever the threshold of danger to the ear is reached. A similar light is now avaliable on the Koss pocket cassette player.

The Occupational Safety and Health Agency (OSHA) has set the danger level at 95 dB and above for 4 or more hours per day as likely to induce permanent hearing impairment.

Related to this is a recent study conducted in New York City by Jane Mandell. Her data showed that much of the ambulatory music is played at levels well beyond the 95-dB—upward of 100 and 125—close to the sound level of unmuffled jackhammers. It may well be that OSHA's 95-dB level is far too lenient and that damage occurs below 90.

A study at University Hospital in Iowa City supports the headphone–hearing loss association.

At a recent meeting of the American Academy of Otolaryngology, Phillip Lee reported that 16 teenagers in his test group had normal hearing and that they listened to music on their headsets for an average of 2–3 h/day. Audiograms revealed that seven of the 16 had incurred temporary threshold shifts (reversible damage to the cochlea) of 10–30 dB after 3 h of listening to rock or strong rhythmic "fusion" music at volumes of 98–104 dB.

Tinnitus (the perception of sound in the absence of an acoustic stimulus—hearing sound when there isn't any) was reported by 11 of the 16. Lee recommended that the standards set by OSHA, limiting noise exposure of 100 dB to 2 h/day for industrial workers, be used for recreational exposure. Unfortunately, he did not suggest a method of enforcement.

David M. Lipscomb of the University of Tennessee's Noise Study Laboratory reported on the effects of rock music on the ears of guinea pigs. Lipscomb and his colleagues recorded music in a Knoxville discotheque and then played it back to an audience of guinea pigs (13). After some 90 h of intermittent exposure to the music, the cells of the cochlea were photographed. They "had collapsed and shriveled up like peas."

Lipscomb unfortunately exposed his animals to continuous sound of approximately 119 dB. This did not simulate the brief, but intermittent rest periods between records, as well as time needed for changing groups, which does occur in discos. In addition, his guinea pigs were subjected to taped music rather than the full impact of amplified sound.

To redress these methodological deficiencies, Bohne and co-workers (14) took six chinchillas to a discotheque where they were exposed (subjected?) to live rock music for 2.5 h. They found considerable variability in the degree of tissue damage among the rodents. This was comparable to the differences in the losses suffered in people exposed to similar levels and periods of loud music—which supports the concept of a range of susceptibility or sensitivity in the general population. Interestingly enough, this follows the pattern of both infectious and noninfectious disease. Nevertheless, based on the degree of permanent hearing losses induced in the chinchillas, these researchers predicted that damage to people would be widespread, and they recommended that the same damage-risk criteria used for industrial settings be extended to music exposure.

Brass Bancroft, Secret Service Agent and movie actor, was playing a scene when a member of the sound effects crew fired a .38 calibre blank-loaded pistol too close to his right ear. Over the ensuring 45 years, his hearing deteriorated. Early in September 1983, President Ronald Reagan, Brass Bancroft's alter ego, appeared in public for the first time wearing a state-of-the-art hearing aid.

Adding to President Reagan's noise-induced hearing loss is his 74 years. Simply put, as people age, their auditory acuity diminishes. With steadily increasing numbers of our population attaining 65 and over, hearing loss and deafness are also expected to increase. That is a given, without the excess ascribed to environmental noise, which is also increasing.

Without hearing deficits—normal acuity—a person should hear sound at the level of 20–25 dB. Subdued or sedate conversation proceeds at about 35 dB. Difficulty hearing at that level is considered impaired and anyone unable to hear at 90 dB or above is, for all intents and purposes, deaf.

Approximately 8%, or 20 million Americans, suffer degrees of hearing impairment, and 1% are completely deaf. This does not include those who were born deaf. Serious hearing disability affects more people than most neurological problems combined.

Again, help is on the way. Improvements in hearing aids—from digital sound systems to electromagnetic implants—are helping some of the most serious cases. The main advance enabling analog devices to distinguish unwanted noise from speech has been a miniature computer. The Zeta noise blocker, developed by Daniel Graupe, professor of electrical engineering, University of Illinois, Chicago, can recognize speech amid other sounds and uses a series of filters to reduce background noise. When the device detects noise—usually an uninterrupted band of sound as opposed to the rapidly changing frequency of speech—it analyzes the level and activates specific filters to remove the noise. Figure 16 shows oscilloscopic traces of noise, speech, a mixture of the two, and the mixture filtered by the Zeta blocker (15).

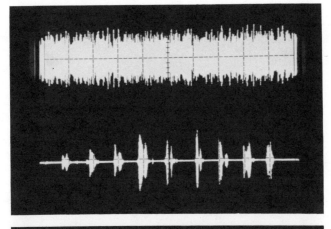

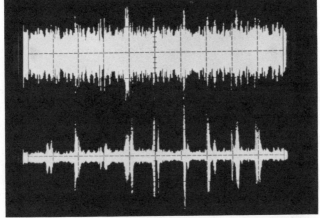

Figure 16. The performance of the Zeta Noise Blocker™ is shown in these four traces of sound over time as shown on an oscilloscope. The top trace illustrates low-frequency noise, 400–600 Hz; second, a person counting from two to ten; third, a mixture of the first two traces, noise with speech; and last, the same mixture as filtered through the Zeta Noise Blocker, which reduces the noise and allows speech to be heard more clearly.

A. Mechanism of Hearing

When sound waves reach the outer ear, they initiate vibrations in the eardrum—the tympanic membrane, which can be seen in Figure 17. These vibrations are transmitted to three ossicles, the bones of the middle ear: the hammer (malleus), anvil (incus), and stirrup (stapes). Figures 18 and 19 show the relation of these three to the tympanic membrane. From the position and shape of the bones, it is not difficult to understand how they transmit vibrations to the inner ear.

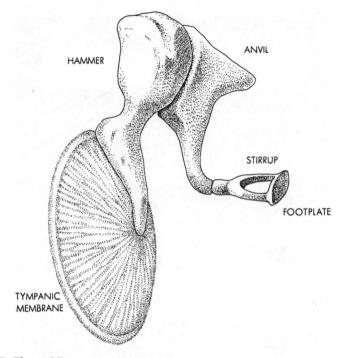

Figure 17. The middle ear. Reproduced from Georg von Békésy, The Ear, *Scientific American* **197:** 61–67 1957.

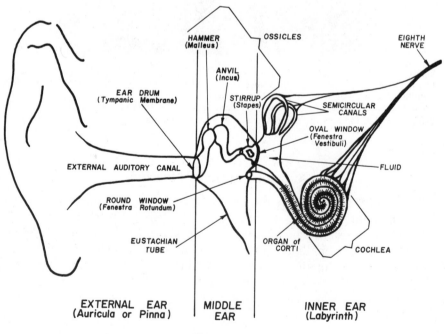

Figure 18.

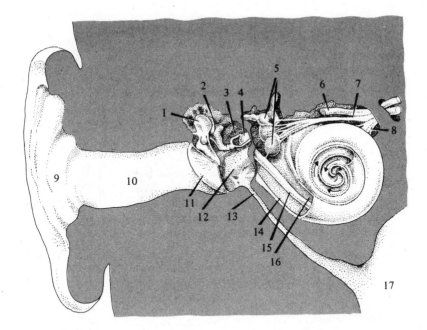

Figure 18. Pathways of sound reception.

1	Malleus	10	External auditory meatus (ear canal)
2	Incus	11	Ear drum (tympanic membrane)
3	Crura of stapes	12	Promontory
4	Footplate of stapes in oval window	13	Eustachian tube
5	Semicircular canals, utricle, and saccule		Cochlea
		14	Scala tympani
6	Cochlear nerve	15	Cochlear duct containing organ of Corti
7	Vestibular nerve	16	Scala vestibuli
8	Facial nerve	17	Pharynx
9	Pinna		

The stirrup, a tiny bone weighing approximately 1/30,000 of an ounce, drives the perilymph (fluid) of the cochlea back and forth, piston fashion, according to the rhythm of the vibrating sound pressure. These movements of the stirrup, and particularly of its footplate, magnify the original vibrations some 20 times and initiate vibrations in the basilar membrane,* which in turn transmits the vibrations to the haircells of the organ of Corti, within the cochlea; the endings of both branches of the auditory

---

* Because of its filtering and analytical functions, the basilar membrane is of interest in hearing research. Since it is embedded in the skull, however, direct study is extremely difficult.

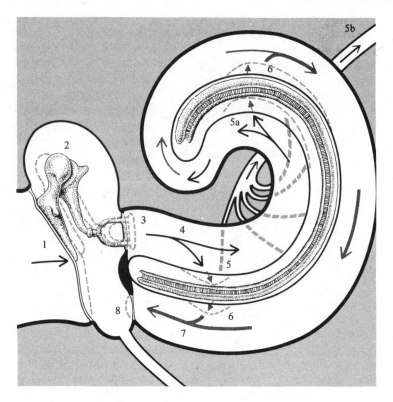

1 Sound waves impinge on ear drum, causing it to vibrate. 2 Ossicles vibrate as a unit. 3 Stapes moves in and out of oval window. 4 Sound waves transmitted up scala vestibuli in medium of its contained perilymph. 5 Short waves (high frequency, high pitch) act at base of cochlea. 5a Long waves (low frequency, low pitch) act at apex of cochlea. All waves act by distorting Reissner's membrane and basilar membrane of cochlear duct and its contained organ of Corti, thus stimulating hair cells which are in contact with the tectorial membrane. 5b Impulses then pass up cochlear nerve. 6 Wave transmitted across cochlear duct in medium of endolymph, from scala vestibuli to scala tympani. (Note: waves may also travel around helicotrema at apex of cochlea.) 7 Waves descend scala tympani in medium of its contained perilymph. 8 Impact of wave on membrane of round window causes it to move in and out at round window in opposite phase to oval window.

Figure 19. Transmission of vibrations from drum through the cochlea.

nerve are also contained within the spiral cochlea (see Figure 19). Here, mechanical motion is converted to nerve impulses. These impulses, carried by the auditory nerve to the brain, are then perceived as sound. It is in the inner ear that the vestibular and cochlear branches of the acoustic nerve receive their signals, and it is within the cochlea that translations of sounds are made. If the translations are muddled or weak, the acoustic nerve obtains and transmits faulty impulses to the brain. This is manifested in some degree of hearing loss.

It may therefore be concluded that noise-induced hearing loss is due not only to ill effects on the eardrum, as so many erroneously believe, but primarily to damage to the internal ear—to the cochlea, the organ of Corti, and the acoustic nerves.

The eardrum is not the only pathway by which sound moves. Sound energy can be carried to the inner ear by way of the bones of the skull. The sounds we hear when we click our teeth or move the tongue around the mouth come through the skull. Much of the sound we receive when the ears are plugged or covered also comes in this way. This distinction is an aid in diagnosing deafness. Von Békésy noted that "if a person can hear bone-conducted sounds but is comparatively deaf to airborne sounds, we know that the trouble lies in the middle ear. But if he hears no sound by bone conduction, then his auditory nerves are gone and there is no cure for his deafness. This is an old test long used by deaf musicians. If a violin player cannot hear his violin even when he touches his teeth to the vibrating instrument, then he knows he suffers from nerve deafness, and there is no cure."

Deafness is far from being solely of occupational origin. Any diagnosis of occupational deafness must first rule out the possibility of loss of hearing accompanying advancing age. Sensory presbycusis, one form of the natural aging process, is characterized by hearing losses in the high-frequency ranges. Other forms produce losses of hearing in all frequency ranges, as evidenced by flat audiometric curves. (This point is discussed further on.) Thus, it is difficult to prove that a case of deafness is solely of occupational origin. In addition, tumors, infections, or blows on the head can also produce loss of hearing.

B. Tests of Hearing Acuity

In the Weber test (see upper portion of Figure 20), the stem of a vibrating tuning fork is placed on the midline of the head. The individual being tested is asked to indicate in which ear the tone is heard. With conductive hearing loss, the tone is loudest in the affected ear. With sensorineural loss, the tone is loudest in the unaffected ear, because the fork stimulates both inner ears equally, and the person perceives the stimulus with the more sensitive, unaffected end organ and nerve.

In the Rinné test (see bottom half of Figure 20), hearing ability by air conduction is compared with bone conduction. The tines of a vibrating fork are held near the pinna (air conduction); then the stem of the fork is placed in contact with the mastoid bone (bone conduction) and the individual is asked which of the two stimuli is loudest. To the normal ear, the stimulus is both longer and louder by air conduction; 40 s by air, 20 s by bone. With conductive loss, this ratio is reversed. With sensorineural

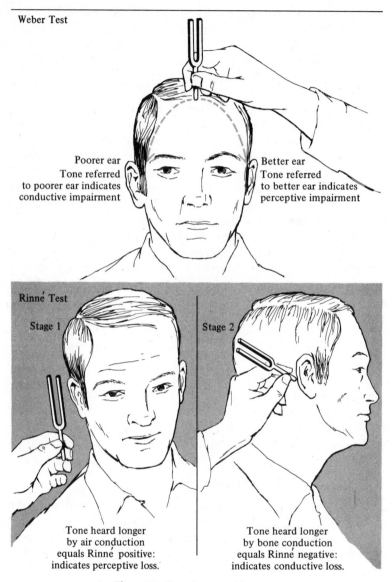

Weber Test

Poorer ear
Tone referred
to poorer ear indicates
conductive impairment

Better ear
Tone referred
to better ear indicates
perceptive impairment

Rinné Test

Stage 1

Stage 2

Tone heard longer
by air conduction
equals Rinné positive:
indicates perceptive loss.

Tone heard longer
by bone conduction
equals Rinné negative:
indicates conductive loss.

Figure 20. Simple tests of hearing.

impairment, both air and bone conduction are reduced, but the ratio remains the same.

Progressive noise-induced deafness is known to occur through continuous exposure to sound levels above 80 dB over an 8-h day. Exposure to excessive noise is initially seen as a temporary threshold shift (TTS), which is the difference between the postexposure threshold and the preexposure threshold.

Accumulating evidence indicates that levels of noise below 75 dB are not dangerous. Levels about 80–85 dB in the frequency range of 1200–4800 cps, however, appear unsafe. In fact, as noise-induced deafness progresses, the ability to hear the high-pitched sounds of speech is lost first. The most common complaint of people with noise-induced deafness is "I can hear but I don't understand." The first sounds to be lost by the nerve deaf are the fricative consonants *f*, *s*, *th*, *ch*, and *sh*. It becomes increasingly difficult to discriminate between such words as *sick*, *thick*, *flick*, and *chick*. With greater hearing loss the explosive consonants *b*, *t*, *p*, *k*, *g*, and *d* become difficult to distinguish.

Recovery from a temporary shift of from 30 to 40 dB may take several hours. For a shift of 50–60 dB, even several days may be insufficient. Despite these figures, it should not be assumed that nerve deafness depends solely on the noise level. A good deal depends on the total noise exposure. This includes the overall noise level, the time distribution (whether the noise is continuous or intermittent), and the total duration of exposure during a lifetime. It is this combination of factors working in concert that constitutes the hazard of noise.

A step in the direction of controlling or preventing nerve deafness has been the development of damage risk criteria. Figure 21 indicates the

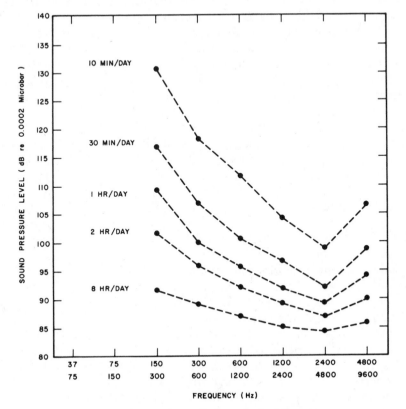

Figure 21. Damage-risk criteria for a single daily exposure.

combination of sound-pressure level and frequency that may be tolerated without danger. From the graph, it is clear that the tolerable exposure periods decrease with an increase in decibels for the same octave band. For example, at 150–300 Hz, sound intensity is 10,000 times greater at 130 dB than at 90 dB. Thus, the difference between 10 min and 8 h of tolerable exposure becomes understandable. Briefly then, a damage-risk criterion specifies the maximum sound-pressure level of a noise to which a person may be exposed without risk of hearing loss.

C. Types of Hearing Loss

Two major types of hearing loss are usually recognized: conductive and sensorineural. In conductive deafness sound-pressure waves never reach the cochlea—most often as a consequence of a ruptured eardrum or a defect in the ossicles of the middle ear.

The three bones form a system of levers linked together, hammer pushing anvil, anvil pushing stirrup. Working together, the bones amplify the force of sound vibrations. Taken together, the bones double, often treble the force of the vibrations reaching the eardrum.

Mitigation of potentially harmful amplification occurs via muscles of the middle ear. These muscles act as safety devices protecting the ear against excessive vibrations from very loud noises—very much like an automatic damper or volume control.

When jarring sounds with their rapid vibrations strike the eardrum, the muscles twist the bones slightly, allowing the stirrup to rotate in a different direction. With this directional shift, less force is transmitted to the inner ear: less, not all.

Remember, we are dealing with delicate and fragile anatomical structures on the one hand and fairly powerful physical forces on the other. These muscles act quickly (especially in the young) but not always quickly enough. A sudden blast—a .38 caliber pistol shot close to the ear as in "Brass Bancroft's" case—catches the muscle relaxed and unprepared. When the shot rang out, they had been dealing with sounds at the level of normal speech. But it need not be a gun or a cannon. Thomas Alva Edison had a similar loss as a consequence of his being boxed on the ear "by an irate teacher."

Conductive hearing loss can be minimized, even overcome, by use of the familiar hearing aids. The most common is worn over the mastoid bond behind the pinna. It picks up sound waves and transmits them through the skull to the cochlea.

Sensorineural hearing loss, the most common form in the United States, occurs as a result of advancing age as well as exposure to loud noises. In both instances there is a disruption of the organ of Corti. The organ serves two functions: converting mechanical energy to electrical and dispatching to the brain a coded version of the original sound with information about frequency, intensity, and timbre. The hair cells of the organ of Corti send their electrochemical signals into the central nervous system, where the signals are picked up by thousands of auditory nerve fibers and transmitted to the brain.

It is the decoding of all the information that enables a person to distinguish the unique and separate sounds of a violin, trumpet, and clarinet, even though all three are playing the same note.

The organ of Corti, a gelatinous mass, is one of the best protected parts of the body, encased as it is within the cochlea which in turn is deeply embedded in the temporal bone—perhaps the hardest of the 206. Nevertheless, harsh noise can damage the hair cells and the auditory nerve, producing at times, depending on the type of noise, sudden and often total deafness.

Sustained noise over a period of time can also engender sensorineural deafness in the form of gradual losses in hearing—presbycusis. This is President Reagan's problem. It is also Pete Townshend's, Brian Wilson's and Frankie Vale's problem. And it will be the problem of thousands of teenagers if they continue with the heavy metal music of Twisted Sister, probably the loudest music in America today—140 dB—similar in intensity to a gun shot close to the ear.

Until recently, sensorineural deafness could not be helped by hearing aids. However, with advances in electronic wizardry and miniaturization, devices for insertion into the auditory canal are available. President Reagan uses an "intracanal" device manufactured by Starkey Laboratories of Minneapolis. This one consists of a microphone, amplifier, and speaker engineered into a 1-in-long plastic case that fits easily into the eustachian tube.

The microphone converts sound into electrical impulses that are augmented by the amplifier, then reconverted to sound by "the speaker." In sensorineural type of hearing loss, unlike conductive, amplification is not a solution. No matter how loud the sound, the wearer will probably not understand speech because hair cells are lacking.

A healthy infant comes into the world with approximately 20,000 hair cells lining the cochlea. Each hair cell responds best to sound of a given frequency. At the base of the cochlea the cells appear to respond to higher frequencies, while cells in the topmost coils respond to lower frequencies.

Figure 22 is an electron photomicrograph of clusters of hairs standing erect as a fringe on top of auditory cells in the inner ear. The angle at which this photomicrograph was taken shows their fiberlike architecture. Figure 23 is a view of hair cells magnified 17,000 times. Obtained by scanning electron microscope, it clearly shows each fiber and the fibrous nature of the "hairs." The cottonlike balls seemingly scattered on the surface are in fact microvilli projecting from a hair cell beneath the surface. The puff-ball near the apex of the bundle is a kinocilium in contrast to the other fibers known as stereocilium.

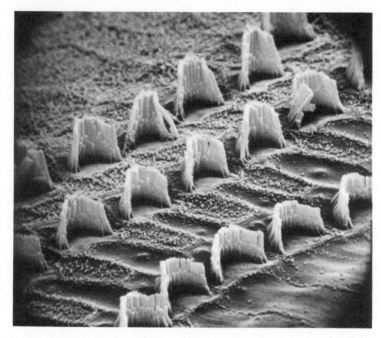

Figure 22. Electron photomicrograph showing clusters of hair cells standing erect. Courtesy of Dr. Patricia A. Leake, Coleman & Epstein Laboratories, School of Medicine, University of California, San Francisco. (Magnification is approximately 5500×.)

Normal and damaged cells are compared in Figure 24. Although the hairs are seen from above, the erect condition of the normal, healthy hairs is obvious. On the right, hairs damaged by excessive exposure to noise are lying flat (upper layer) while the lower layer has been totally destroyed. Apparently, the flaccidity occurs in response to denaturation of the protein of which the fiber is composed.

VI. CONTEMPORARY PROBLEMS

A. Noise and Accidents

Can noise cause accidents? Anecdotally, it may appear that way. Not so simply answered, however, is a somewhat different question: To what extent does noise contribute to occupational accidents? Before answering, it would be well to recall the concept of "accident syndrome" noted in Chapter 6. Another look at Figures 2 and 3 may be helpful. Also worth recalling is the fact that accidents are the leading cause of death among the 18–35-year age group, who by the way, constitute a major component of the work force. Thus, the problem becomes one of ascertaining if noise alone, or in combination with other factors, can cause accidents. It may

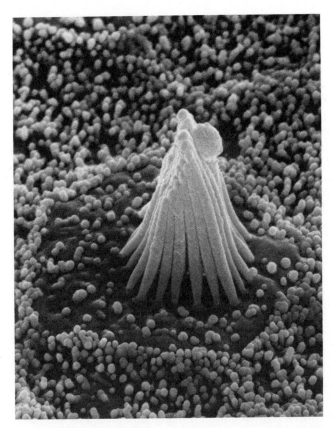

Figure 23. A bundle of fibers at the tip of a hair cell in the inner ear. The rice crispie-like particles scattered about are microvilli projecting from underlying supporting cells. This is an electron scanning photomicrograph enlarged 16,800× and is from the inner ear of a bullfrog. Courtesy of Drs. A. J. Hudspeth and R. Jacobs, University of San Francisco Medical School.

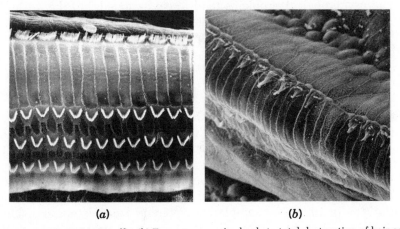

(a) (b)

Figure 24. (a) Normal hair cells. (b) Exposure to noise leads to total destruction of hair cells (lower layers) or damage (upper layer).

also be that noise is not a contributing factor. The issue is neither straightforward nor simple.

It has been suggested that reducing excessive noise may reduce the accident rates in the occupational setting. Cohen (16), for example, studying men working in a plant manufacturing boilers, reported that those working in high-noise areas were involved in more accidents than those working in low-noise areas. Thirty-five percent of the high-noise group (exposed to noise levels of 95 dBA and above) had 15 or more injuries over a 5-year period as against 5% of workers in the low-noise group (exposed to less than 80 dBA) with a similar injury rate.

The questions, of course, come quickly. Were the two environments similar? Were the jobs performed by each group the same with respect to injury potential? Unfortunately, positive statistical correlations are by themselves not proof of cause and effect.

At a meeting of the Acoustical Society of America, Schmidt and co-workers reported significant reductions in injury rates in a cotton yarn plant in periods before and after the introduction of a hearing conservation program (17).

Curiously enough, women workers had greater reductions in injury than their male counterparts. To this study's detriment, it did not include a control group, and it seems that the women may have been more judicious in the use of ear defenders than the men.

The medical and scientific literature is replete with reports similar to these. While not adequate for establishing a causal relationship between noise and accidents, they do suggest that noise may be contributory. But this too requires additional well-designed studies to ascertain.

Donald Broadbent of the Medical Research Council's Department of Experimental Psychology (Oxford) recently reviewed 50 years of reports dealing with annoyance and noise. In his words, "the causal relations seem to run this way. People in noise tend to get annoyed; people who are having psychological problems tend to get annoyed, but the noise does not increase the frequency of people with problems" (18). The relationship, he suggests, between noise, annoyance, and mental health will be fully clarified in another 50 years.

B. Concern for Hearing Loss in the Armed Forces

Earlier, I noted the relationship between hearing loss and national security. Sweden, faced with an imminent emergency, learned of a hearing deficit among its sailors of which it had not been aware. The U.S. military establishment is concerned about hearing disability for a number of reasons. Hearing losses can, of course, adversely affect mission accomplishment. It is, however, the most costly of all occupationally related disabilities. In 1982, for example, the Veterans Administration disbursed some $145 million to 62,000 retired veterans with hearing losses as their

Table 6
Percentage of Clinically Significant Hearing Losses
Within Selected U.S. Army Branches

Time-in-Service (years)	Branch		
	Infantry	Armor	Artillery
1.5–2.4	11.5	9.0	13.0
2.5–7.4	20.0	21.0	22.5
7.5–12.4	41.0	35.0	43.5
12.5–17.4	46.0	48.0	53.5
17.5–22.4	50.5	56.0	64.0

Source: Adapted from tables contained in Research Report: The Prevalence of Hearing Loss Within Selected U.S. Army Branches, supported by U.S. Army Medical Research and Development Command, Interagency No. 1A0 4745, 31 August 1975.

primary disability. Most of these did not occur in battle. The vast majority are the consequence of routine training.

During the past 10 years, tests conducted at bases around the country have indicated that as many as 30% of personnel with two or more years of service in any of the combat branches had significant hearing losses. Table 6 shows the direct relationship between length of service and increasing severity from infantry to armor to artillery. Motivation to reduce this loss is great.

The damaging effects of impulse noise from a variety of weapons is an area of major concern. Current damage-risk criteria (DRC), established to set limits on the amount of noise a worker or soldier can be exposed to during a determined period, require reevaluation, especially as rifle fire has been the primary source of data for the DRC and hence protective measures.

Recently , Price of the U.S. Army Human Engineering Laboratory, Aberdeen, Maryland (19), developed experimental data that should assist in establishing more accurate DRC* for intense impulsive sounds, as well as generating a new view of the relative hazardousness of weapons.

In an attempt to establish relative hazards from rifle and cannon impulses, he exposed cats to 60 rounds of fire from either a 105-mm Howitzer or a 7.62-mm rifle, firing at 2–5-s intervals. Alterations in hearing acuity were monitored for up to 2 months following this exposure.

Threshold shifts from the day of exposure to the rifle impulses are shown in Figures 25 and 26. Parallel data for exposure to the Howitzer are shown in Figures 27 and 28. The trend lines in the figures are the least-squares regression lines fitting the data. The important fact to note is

---

* Similar DRC are used by West Germany, the United Kingdom, and the United States. Military personnel of all three countries depend on these criteria for their protection.

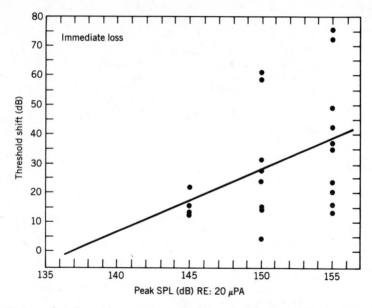

Figure 25. Immediate losses for 22 animals (both ears, four frequencies) exposed to 60 rifle impulses. Reproduced with permission of *Scandinavian Audiology*.

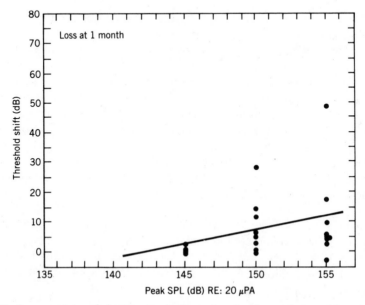

Figure 26. Losses at 1 month for 20 animals. Reproduced with permission of *Scandinavian Audiology*.

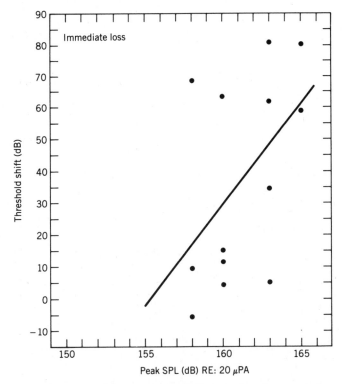

Figure 27. Immediate losses for 13 animals (both ears, four frequencies) exposed to 60 impulses from a 105-mm Howitzer. Reproduced with permission of *Scandinavian Audiology.*

that the permanent losses begin to grow for rifle fire at about 140+ dB, and for the cannon at 150+ dB. Consequently, it would appear that the rifle is more hazardous than the cannon.

These data, of course, are for cats' ears, which are known to be more susceptible to intense noise than the human ear. Price remarked that current DRC are much too conservative for large caliber weapons because they are based on rifle-fire data. He also believes that because the energy of greatest interest is that in the mid range, ear protection for all classes of weapons should be possible "with a single hearing protector of good quality, properly fitted and worn." In addition, he found an almost total loss of hair cells in the middle of the cochlea for the cats suffering the greatest hearing loss–additional proof of the mechanism of sensorineural impairment.

C. Music to Their Ears

Senza Sordino, the official publication of the International Conference of Symphony and Opera Musicians, is not anyone's garden variety of journal devoted to the health effects of noise. Nevertheless, the editor was

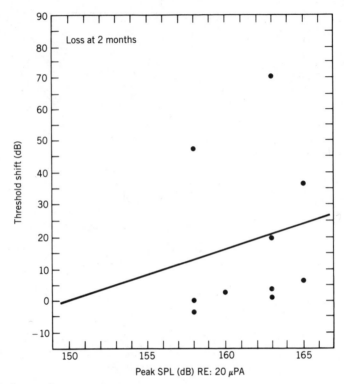

Figure 28. Losses after 2 months for 10 animals. Reproduced with permission of *Scandinavian Audiology*.

sufficiently impressed with Karla Holland-Moritz's data to open the pages of the June 1985 issue for it in order that orchestras everywhere benefit from the findings. Holland-Moritz, herself a musician with the San Diego Symphony Orchestra, had sent a survey questionnaire to members of 23 orchestras. The results revealed that of the 900 musicians responding, more than half suffered nervousness, tension, anger, and irritability. Seventy-five percent believed the sound level (noise?) had an adverse effect on their professional ability.

Fear of deafness was a major concern of orchestral musicians. Earplugs, of the types shown in Figure 29, are currently worn by those sitting directly in front of brass or percussion sections. Although they do cut down noise, they also prevent the musicians from hearing the music. Some orchestras have experimented with plexiglass shields attached to the backs of chairs. Others use acoustic paneling around the percussion section when the orchestra is in the pit. Some have placed the brass on risers to allow sound waves to travel over the heads of those in front.

Is symphonic music appreciably louder today than, for example, during the Stravinsky, Wagner, Strauss, and Bartok era of the 1930s? It is a

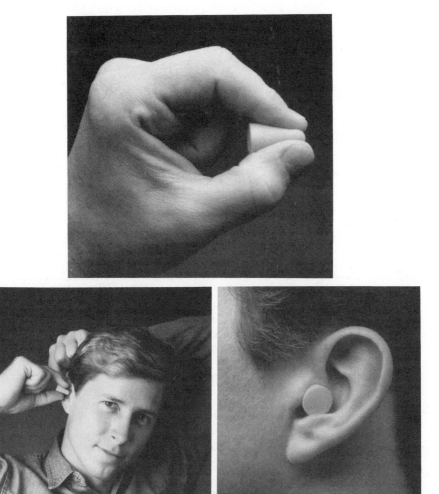

Figure 29. Sample of earplugs.

distinct possibility—a responsibility to a changing environment; though few would recognize or admit it. Audiences today come to music halls from terribly noisy environments. They live with traffic noise, air hammers, subways, electronic music, and frequent bouts of sound amplification. To make any emotional impact on such audiences orchestras do increase volume.

Do bear in mind, however, that this survey was an uncontrolled study and that the behavioral responses of tension, nervousness, and anger, especially among musicians, can come from a number of other sources.

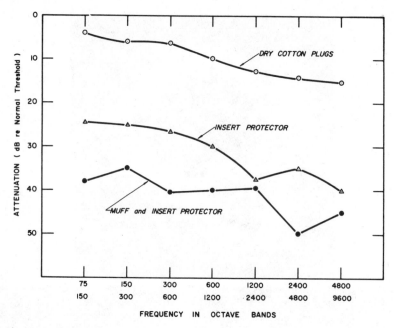

Figure 30. Degree of noise attenuation achieved with three types of ear protector.

VII.   SOUND BARRIERS

When it has been established that a loss of hearing has resulted from excessive noise, it should be possible to reduce the level of noise at its source. In addition, acoustical barriers can be imposed between the noise source and the ear. Individual acoustical barriers run the gamut from a wad of cotton placed at the entrance to the auditory canal to muff-type ear defenders, used with increasing frequency by airport mechanics.

Although dry cotton plugs provide little, if any, attenuation, many people continue to use them. Insert-type plugs made of rubber, plastic, or wax are far superior to cotton; muff protectors are better yet. Figure 30 shows the degree of attenuation achieved with a muff, with an insert-type plug other than cotton, and with cotton. Unfortunately, many workers in high-noise areas discard or fail to wear plugs or muffs. This should not be permitted.

In addition to employing barriers at the entrance to the auditory canal, environmental control can be instituted to protect people from noise. This includes redesigning equipment so as to reduce its noise level, muffling apparatus with sound-absorbing materials, or isolating noise processes or machines. Any inelastic material of low density, such as ground cork, hairfelt, fiberglass, or rubber compositions, are good insulators.

Sound waves travel with greater velocity through many solid materials than they do through air. For example, they have a velocity of 11,000–16,000 f/s in wood—14,000 in oak and approximately 16,000 in Norwegian spruce—16,000 in glass and steel, and 17,000 in aluminum. Through air, sound travels at a velocity of 1130 f/s. Small wonder then that street traffic noise can be heard so easily in homes and offices.

According to the mass law, the heavier the wall, the more effectively it will damp noise. But heavy walls are costly; hence the widespread use of lightweight materials, which permit noise to pass through. Less costly and reasonably effective in dampening passage of sound are double walls with an air barrier between—sandwich fashion.

Be that as it may, reducing extraneous, annoying, even harmful noise is expensive, a major reason for foot-dragging. Another is apathy: apathy on the part of public officials, government agencies, and the public generally, who must all make choices and set priorities, given the competition for scarce tax dollars. Impetus for noise reduction will come when substantial evidence links hearing and other physical impairments with excessive noise. Currently, such evidence is weakly circumstantial.

The quietest city in the United States, if annual wards are any judge, is Memphis, Tennessee. Memphis has won more than a dozen annual awards for its noise abatement program. City officials add with some pride that the impetus for a quieter city (Memphis long had a reputation as a brawling river town) came from "Boss" Crump and his powerful political machine. E. H. "Boss" Crump, while a personification of the political boss, decided in the late 1930s that he wanted to live in a quiet city. He had several strict noise-control ordinances passed that are still in force. Consider a few of Crump's ideas. In part, the ordinances state the unnecessary horn blowing is illegal and that vehicles making "loud and unnecessary grating, grinding, rattling, and other noises are outlawed." It's as simple as that. But also prohibited are yelling, shouting, hooting, whistling, and singing in the streets between 11:00 p.m. and 7:00 a.m., playing radios or television sets loudly enough to be heard outdoors between the same hours, and the use of noisy construction equipment between 7:00 p.m. and 6:00 a.m. Is it really necessary to bring back the political boss and his controlled votes in order to obtain quieter cities? It's something to dwell upon.

Several years ago, the slogan "Keep New York Plastered" appeared in newspaper ads, as well as subways and buses. The reference was not to the salubrious effects of alcohol but to demands for greater use of plaster as an effective coating for walls and ceilings to reduce noise.

In some quarters, noise is perceived as an intolerable nuisance, and people are organizing to reduce the clatter in their communities. But it remains to be established that adverse health effects—other than loss of hearing acuity—are by-products of elevated noise levels.

If the contemporary music scene is any bellwether, huge numbers of people not only readily accept but eagerly subject themselves to sound levels bordering on physical pain. This indifference or denial may be the reason for the meager progress of noise prevention. Adverse health effects could make the difference. It may be a noisy time before such data are forthcoming. Then again, it may be Fourier to the rescue.

REFERENCES

1. A. Maser, P. Sorensen, and K. Krypter, Effects of Intrusive Sound on Classroom Behavior: Data from a Successful Law-Suit. Presented at the Annual Meeting of the Western Psychological Association, San Francisco, 1978.

2. S. Cohen, G. Evans, D. Krantz, and S. Kelly, Aircraft Noise and Children; Longitudinal and Cross-Sectional Evidence on Adaptation to Noise and the Effectiveness of Noise Abatement. *J. Pers. Soc. Psychol.* **40**: 331–345, 1981.

3. K. B. Green, B. S. Pasternack, and R. E. Shore, Effects of Aircraft Noise on Reading Ability of School-Age Children. *Arch. Environ. Health* **37** (1): 24–31, 1982.

3. S. Cohen, D. C. Glass, and J. E. Singer, Apartment Noise, Auditory Discrimination and Reading Ability in Children. *J. Exp. Soc. Psychol.* **9**: 407–422, 1973.

5. L. M. Jenkins, A. Tarnapolsky, D. J. Hand, and S. M. Barker, Comparison of Three Studies of Aircraft Noise and Psychiatric Hospital Admissions Conducted in the Same Area. *Psychol. Med.* **9**: 681–693, 1979.

6. E. A. Peterson, J. S. Augenstein, D. C. Tanis, and D. G. Augenstein, Noise Raised Blood Pressure Without Impairing Auditory Sensitivity. *Science* **211**: 1450–1452, 1981.

7. K. Batig, H. Zeier, R. Muller, and R. Buzzi, A Field Study on Vegetative Effects of Aircraft Noise. *Arch. Environ. Health* **35** (4): 228–235, 1980.

8. H. Neus, H. Ruddel, and W. Schulte, Traffic Noise and Hypertension: An Epidemiological Study on the Role of Subjective Reactions. *Intl. Archives Occup. Environ. Health* **51**: 223–229, 1983.

9. M. Vallet, Psychophysiological Effects of Exposure to Aircraft or to Traffic Noise. *Proc. Inst. Acoustics* **3**: 1–4, 1979.

10. J. Sataloff, R. J. Sataloff, H. Menduke, R. Yerg, and R. P. Gore, Hearing Loss and Intermittent Noise Exposure. *J. Occup. Med.* **26** (9): 649–656, 1984.

11. D. Gloag, Noise and Health: Public and Private Responsibility. *Br. Med. J.* **281**: 1404–1406, 1980.

12. W. J. Broad, New Technology Detects Unwanted Noise. *The New York Times*, June 30, 1987, pp. C1 and C9.

13. David M. Lipscomb. Non-Occupational Noise and the Effect upon Hearing of Young Persons: A Report Presented to the House of Representatives, Committee on Health and the Environment, Washington, D.C., June 22, 1971 [University of Tennessee].

14. B. A. Bohne, P. H. Ward, and C. Fernandez, Irreversible Inner Ear Damage from Rock Music. *Trans. Am. Acad. Ophthalmol. Otolaryngol* **82** (Jan./Feb.): 50–59, 1976.

15. J. K. Grosspietsch, The Zeta Noise Blocker: A Basic Reintroduction. *The Hearing J.*, 19–21, May 1987.

16. A. Cohen. *Industrial Noise and Medical Absense and Accident Record Data on Exposed Workers.* Proceed. Int'l. Congress on Noise as a Public Health Program. U.S. Environ. Protection Agency, Rept. No. 550/9-73-008PGS, 441–453, Washington, DC, 1975.

17. J. A. W. Schmidt, L. H. Royster, and R. G. Pearson Impact of an Industrial Hearing Conservation Program on Occupational Injuries for Males and Females. Paper presented at 99th Meeting of the Acoustical Society of America, Atlanta, GA. Abstract in *J. Acoust, Soc. Am.* **68** 17:559, 1980.

18. D. Broadbent, Noise in Relation to Annoyance, Performance and Mental Health. *J. Acoust. Soc. Am.* **68** (1): 15–17, 1980.

19. R. G. Price, Rating the Hazard from Intense Sounds: Putting Theory into Practice. *Scand. Audiol. (Suppl.)* **16:** 112–122, 1982.

SUGGESTED READINGS

Fields, J. M. A Catalog of Social Surveys of Resident Reactions to Environmental Noise (1943–1980). NASA, Washington, DC, August 1981 (Tech. Memo 83187).

Helper, E. L., Moul, M. J., and Gerhardt, K. J. Susceptibility to Noise—Induced Hearing Loss: Review and Future Directions. *Milit. Med.* **149:** 154–159, 1984.

Hudspeth, A. J. The Cellular Basis of Hearing: The Biophysics of Hair Cells. *Science* **230:** 745–752, 1985.

Lang, J. *Assessment of Noise Impact on the Urban Environment*, Environ. Health Series No. 9. WHO, Copenhagen, 1986.

Ward, W. D. General Auditory Effects of Noise. Symposium on Noise—Its Effects and Control. *Otolaryngol. Clin. N. Am.* **12** (3): 473–729, 1979.

Wilkins, P. A., and Acton, W. I. Noise and Accidents—A Review. *Ann. Occup. Hyg.* **25** (3): 249–260, 1982.

8

Occupational Health

The protection of health of the individual workers is not an individual question. It is a social question demanding social regulation. The prevention of occupational diseases is essentially a function of government.

—John B. Andrews

Gauley Bridge is but a dim memory. Perhaps it should remain so. Or perhaps it should live in infamy. The human cost of constructing the water diversion tunnel at Gauley Bridge, West Virginia, in 1930–1931, was another great American occupational tragedy.

Interest in occupational safety declined precipitously with the Great Depression of 1929. As unemployment rose, conditions became so bad that workers did almost anything to get or keep jobs. And employers put them to the test.

In 1929, the Rhinehart–Dennis Company won a contract to drive a tunnel through the mountains near Gauley Bridge. This would divert water from both the New and Kanawha Rivers. The waters so diverted would be made to flow to a hydroelectric power plant* that was to supply badly needed electricity to the surrounding county.

Workers, mostly young black men, were recruited from nearby states and paid 50¢ per hour (which dropped to 35¢ as the depression deepened) from which they were charged board and medical insurance.

As the record of the Senate Appropriations Committee spells out, conditions bordered on the obscene (1). The silica dust in the tunnel was so thick the workers could barely see their feet in front of them, even

---

* The record indicates that the diverted water would be used primarily to power the Electro-Metallurgical Co., a Union Carbide subsidiary that was in the business of processing steel electrolytically.

under the headlight of the work train, which itself was spewing carbon monoxide from its gasoline-driven engines. Though West Virginia law required a 30-min waiting period after blasting to allow dust to settle, the men were immediately herded back into the tunnel. Here too the House subcommittee record is clear about the fact that the foreman beat them with axe handles to get them into the dusty tunnels. Neither respirators nor masks of any kind were used. Workers were dying from respiratory failure. Although silicosis was the cause of their impairment, "official" death certificates reported bacterial agents—tuberculosis and pneumonia—as causes of death.

In 1930, it was well known that prolonged exposure to dust with a silica content of 25% would produce silicosis. At Gauley Bridge, the silica content reached 95%. Ventilation was almost nonexistent. Men died within 9 months of exposure. Others lingered for 18 months. By 1931, 50% of the workers died or had become silicotic.

Not only did workmen's compensation not cover silicosis, but for those who sought relief through the courts, both government inspectors and company officials testified that dust did not exist in the tunnels at Gauley Bridge. A cadaverous Rhinehart–Dennis foreman struggling for breath so testified. He died of silicosis within a year. One of the jurors was held in contempt of court for accepting daily rides home in a company car.

In January 1936, the House Committee on Labor held hearings on the Gauley Bridge disaster. The subcommittee concluded that "the driving of the tunnel was begun, continued, and completed with grave and inhuman disregard of all considerations of health, lives, and future of employees." Senator Rush Holt of West Virginia called the Gauley Bridge affair "the most barbaric example of industrial construction that ever happened in this world. That company well knew what it was going to do to those men. The company openly said that if they killed off those men, there were plenty of other men to be had." No punitive action was ever taken against Rhinehart–Dennis, New Kanawha Power, or the parent company, Union Carbide.

Respiratory safety, however, has taken major strides since that infamous affair.

I. THE WORKPLACE AND HEALTH: EARLY AND CONTEMPORARY ASSOCIATIONS

"Of what trade are you?" "Yes, what occupation do you follow?" Because we spend the greatest proportion of our lives at work, physicians have long asked their patients, "Of what trade are you?" "Where do you work?" "What do you do?" Although the association between occupation and ill health was first recorded by Pliny the Elder (A.D. 23–79) who advocated the use of protective masks for workers in mining and grinding operations in the first century A.D., it was Bernardino Ramazzini (professor and physician at the University of Padua, Italy) who is considered the father of occupational health and industrial hygiene. In his book, *DeMorbis Artificum Diatriba (The Diseases of Workmen)*, published in 1700, he described the many occupational diseases of his day and suggested preventive measures.

In the 1690s he devoted his spare time to visiting the workshops of Modena (40 mi northwest of Bologna), where he lived and practiced medicine, to gather information on the hazards of various occupations. In the first (and Latin) volume published in 1700, there were 41 chapters. In the second, enlarged edition, published in 1713, which is the one available in English, he added another 12 chapters, discussing such workers as bakers, laundresses, weavers, chemists, painters, printers, glassmakers, blacksmiths, brickmakers, and runners, the forerunners of our letter carriers, whose job it was to carry letters around town and return quickly with an answer.

A. The 18th Century

Industrial or occupational health in the United States may just have begun with the observations of Dr. Benjamin Franklin.* In a letter to Benjamin Vaughn† in 1786, he wrote:

> The first Thing I remember of this kind was a general Discourse in Boston, when I was a Boy of A complaint from North Carolina against New England Rum, that it poisn'd their People, giving them the Dry Bellyach, with Loss of the Use of their Limbs. The Distilleries being examin'd on the Occasion, it was found that several of them used leaden Still-heads and Worms,* and the Physicians were of Opinion, that the Mischief was occasioned by the Use of Lead. The Legislature of the Massachusetts thereupon pass'd an Act, prohibiting under severe Penalties the Use of such Still-heads and Worms thereafter. Inclos'd I send you a Copy of the Acct. then from my printed Law-book.

> In 1724, being in London, I went to work in the Printing-House of Mr. Palmer, Bartholomew Close, as a Compositor. I there found a Practice, I had never seen before, of drying a Case of Types (which are wet in Distribution) by placing it sloping before the fire. I found this had the Additional Advantage when the Types were not only dry'd but heated, of being comfortable to the Hands working over them in cold weather. I therefore

* Franklin is often referred to as Doctor. Although largely self-taught, St. Andrews University conferred the honorary Doctor of Laws degree on him in 1759 for his work on electricity.
† Benjamin Vaughn, a British M.P., was a strenuous supporter of the French revolution. Forced to leave England, he settled in Maine and worked closely with Franklin, publishing a number of his political essays.
* "Worms" in Franklin's time were the spiral tubing used in distillation equipment to condense the vapor.

sometimes heated my case when the Types did not want drying. But an old Workman, observing it, advis'd me not to do so, telling me I might lose the use of my Hands by it, as two of our Companions had nearly done, one of whom that us'd to earn his Guinea a Week, could not then make more than ten shillings, and the other, who had the Dangles but seven and six-pence. This, with a kind of obscure Pain, that I had sometimes felt, as it were in the Bones of my Hand when working over the Types made very hot, induced me to omit the practice. But talking afterwards with Mr. James, a Letter-founder in the same Close, and asking him if his People, who work'd over the little Furnaces of melted Metal, were not subject to that Disorder; he made light of any danger from the effluvia, but ascribed it to the particles of the Metal swallow'd with their Food by slovenly Workmen, who went to their Meals after handling the Metal, without well washing their Fingers, so that some of the Metalline Particles were taken off by their Bread and eaten with it. This appeared to have some Reason in it. But the Pain I had experienc'd made me still afraid of those Effluvia.

When I was in Paris with Sir John Pringle in 1767, he visited *La Charite*, a Hospital particularly famous for the Cure of that Malady, and brought from thence a Pamphlet containing a List of the Names of Persons, specifying their Professions or Trades, who had been cured there. I had the Curiosity to examine that List, and found that all the Patients were of Trades, that, some way or other, use or work in Lead, such as Plumbers, Glaziers, Painters, etc., excepting only two kinds, Stonecutters and Soldiers. These I could not reconcile to my Notion, that Lead was the cause of that Disorder. But on my mentioning this Difficulty to a Physician of that Hospital, he inform'd me that the Stonecutters are continually using melted Lead to fix the Ends of Iron Balustrades in Stone; and that the Soldiers had been employ'd by Painters, as Labourers, in Grinding of Colours.

You will see, that the Opinion of this mischievous Effect from Lead is at least above Sixty Years old; and you will observe with Concern how long a useful Truth may be known and exist, before it is generally receiv'd and practic'd on. (2)

These were some of the problems extant 200–300 years ago during "the good old days." Future historians will record that on July 21, 1984, the first American worker was killed by a robot.

B. Death by Robot

A 34-year-old man in Michigan was crushed to death by an out-of-control robot. This young man was operating an automated die-casting system, which included an industrial robot. The robot stalled, pinning him against a steel pole while continuing to apply pressure to his chest. Cardiopulmonary arrest occurred. He died 5 days later.

The operator, who had 15 years experience in die-casting operations, had completed a 1-week robotics training course a month earlier. On the day of the accident, the robot was programmed to extract a casting from a die-cast machine, dip it into a quench tank, and insert it into a trim press. This cycle, involving 27 programmed steps, required 60 s to complete. Two sides of the robot's "work envelope" were surrounded by safety rails with an electrical interlock gate. Apparently, the young man had entered the robot's work envelope by climbing over or under or around the safety rail.

Although he had received instructions during the training course and warnings on the job about such ill-advised entry, fellow workers had seen him inside the work envelope from time to time and had cautioned him against this. Can the robot be blamed? Was this an accident? Is it an example of human error? This was the first robot-related fatality in the United States. Japan has had two. We can anticipate others in every country that employs robots. High-technology occupations are only part of the problem. As we approach the millenium more mundane occupations may be our biggest problem.

C. Agriculture: The Crops May Change But Not the Conditions

Statistics suggest that four of every five American farm employers show little concern for the welfare of their workers. Most farms provide no toilet facilities. But are toilet facilities an occupational concern?

When the Occupational Safety and Health Administration (OSHA) denied a proposal in the spring of 1985 requiring farmers to provide such facilities, the agency left the job of improving conditions for the nation's 5 million migrant agricultural workers to the states. Labor unions appear to doubt that state legislatures, often dominated by agricultural interest, will adopt field standards without federal prodding.

Across the country, in our high-tech society, thousands of field hands work long days in the sun, often with no access to toilets and only a bucket of water for drinking and hand washing. Data have been compiled showing that farm workers suffer higher than normal rates of infectious diseases. These, it is believed, would decline dramatically if migrant workers had access to toilets and fresh water. It is also maintained that diseases of poor sanitation among migrants can be transmitted to the communities through which the workers travel, as well as to the consumers who purchase the crops. Eugene J. Gangerosa of Emory University found that migrants have 85 times more diarrhea and five times more skin rashes than workers in other high-risk industries (3).

This waterless condition, for drinking, toilets, and washing, has gone on for the past 13 years. Simply stated, it means that for 13 years these field workers have lacked basic human dignities.

As of July 1987, farm workers came under a new OSHA regulation, which entitled them to wash when they needed to, drink when they needed to, and relieve themselves in toilets at their places of work. The new rule covers farms with 11 or more workers. Unfortunately, it is all too evident that with tens of thousands of small farms and few OSHA inspectors, the new rule will be regularly violated.

D. Heat Stroke

Available cool drinking water is an established preventive measure for heat stroke—a common problem among field workers. Performing heavy work in a hot environment adds measurably to a workers heat load by generating more metabolic heat. When a worker's physiologic capacity to compensate for added heat is exceeded, a spectrum of untoward ill effects can occur. Heat stroke is one of them.

The problem is failure of the central nervous system's sweat control mechanism, leading to a loss of evaporative cooling. With the resulting rapid rise in body temperature, headache, dizziness, nausea, and hot-dry skin follow rapidly. This progresses to confusion, collapse coma, and at times death, if not treated. An available water supply is both the optimum preventive as well as therapeutic measure.

Indeed, the question "What occupation do you follow?," remains pertinent today because new industries, new chemicals, new processes, and new stresses confront each generation of workers, while traditional diseases continue in old industries and reappear in new ones.

II. OCCUPATIONALLY INDUCED DISORDERS

A. Pulmonary Disorders

The lungs and other parts of the respiratory tract come in contact with gases, vapors, fumes, fibers, and particles in the workplace. This is not a new idea. Georgius Agricola (1494–1556), physician and mineralogist, is remembered for his *Treatise on Mining,** which described all aspects of mining in the mountains of Silesia and Bohemia (Germany). Agricola's observations that men became sick and died from breathing dust led to the recognition of dust and airborne particles as occupational hazards. In the centuries that followed, observers recorded the relationship between occupations such as mining and quarrying to cough, dyspnea (shortness of breath), and wasting diseases.

Most causes of occupational pulmonary disease are the result of inhalation of particulate matter by individuals exposed to respirable material in milling, mining, cleaning, mixing, grinding, and shaping of industrial materials.

---

* Agricola's great work was published in 1556, the year he died.

Table 1
Respiratory Conditions Caused by Toxic Agents, 1982

Area	×1000
Manufacturing	4.7
Services	1.6
Transportation and public utilities	0.7
Wholesale and retail	0.7
Construction	0.6
Finance and insurance	0.5
Agriculture	0.1
Mining	0.1

Of all health effects, occupationally related cancers of the respiratory tract receive the most attention, but they are not the only serious respiratory conditions associated with the workplace. Other responses may be acute irritation, immunologic or allergic reactions, or chronic changes in tissue that line the tract.

Some conditions that begin as acute problems progress to chronic states: "brown lung" (byssinosis†), a widespread lung ailment of textile workers, is an excellent example.

As shown in Table 1, it is manufacturing not mining that currently produces the greatest number of respiratory illnesses.

1. Acute Reactions

The upper respiratory tract (nose, throat, and larynx) is the most frequent site of irritation. It is particularly susceptible to the effects of soluble gases and to particles larger than 10 μm in diameter. *Soluble gases* such as chlorine, sulfur dioxide, and ammonia quickly cause mucosal irritation in the conjunctivae and upper respiratory tract that is sufficiently uncomfortable to alert exposed workers to danger. If exposure to soluble gases is brief, only simple irritation of the eyes and of the nose and throat may result, leading to cough and expectoration. Severe and prolonged exposure to substances such as ammonia and sulfur dioxide may cause keratoconjunctivitis and bronchitis and can lead to chronic suppurative bronchopneumonia, bronchiectasis, and progressive respiratory insufficiency.

If the endangered worker is unable to escape, excessive concentrations of gas are inhaled into the lower airways and alveoli, damaging bronchial mucosa as well as the alveolocapillary membranes. Symptoms of injury to the lower respiratory tissues include severe dyspnea, wheezing, and orthopnea, usually followed by fever and purulent expectoration, indicating secondary infection or acute pneumonia. Overwhelming exposure to toxic gas may lead to early death as a result of pulmonary congestion,

† From the Latin *byssus,* meaning flax, and *osis,* meaning condition.

edema, and pneumonia, associated with diffuse damage to the mucous membranes.

Equipment such as that shown in Figure 1 provides protection against potentially harmful effects of a variety of gases. Figure 1 displays components of the Breathe-Easy System, which provides variable-flow protection from gases, vapors, and particulate matter. It also provides positive pressure personal protection with cartridge filtration. Figure 1a shows the variable-flow face mask and the light weight turbo unit strapped to the technician's waist. Figure 1b shows the turbo unit with filters and battery pack. Figure 1c shows additional types of personal protective devices available in the Breathe-Easy System.

These easy-to-wear, lightweight units are more acceptable to workers who not only require protection, but who must also function appro-

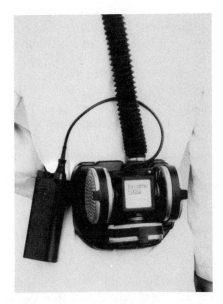

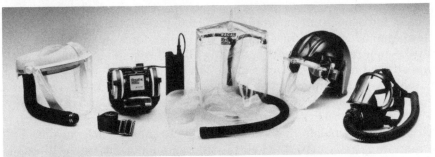

Figure 1. The components of the Breathe-Easy Respirator System. Courtesy of Racal Airstream, Inc., Frederick, MD.

priately. In the past, protective gear was unwieldy and cumbersome so that workers eschewed its use.

Acute reactions in the airways can also be a consequence of *hypersensitivity*. Hypersensitivity to specific industrial agents is an acquired condition. It is most likely to occur in allergic individuals but does occur in individuals with no history of familial hypersensitivity. The symptoms may be *immediate* or *delayed* asthmatic reactions, including nocturnal wheezing, dyspnea, and fever. Delayed reactions may occur hours after exposure or after the worker has left the workplace. Thus, the time lapse may obscure the relationship between the hypersensitivity reaction and the occupational exposure. Substances known to cause hypersensitivity reactions include toluene diisocyanate, western red cedar dust, and textile dusts.

Toluene diisocyanate (TDI), used in the production of polyurethane foams, is widely used as a coating agent, insulation and in upholstery. The threshold limit value (TLV) of TDI has been set at 0.02 ppm (on a time-weighted average) for the average worker. However, some workers have become sensitized to concentrations of TDI well below this value. Asthmatic reactions to TDI in sensitized individuals are severe and may be prolonged.

Cotton dust or linters may provoke acute bronchospasm in textile workers (especially those working in carding rooms) and cottonseed oil workers. The disease known as byssinosis or brown lung generally appears in workers who have been exposed to the dust for several years. The presenting symptoms are chest tightness and a low-grade fever, which appear on the first day of work following a weekend or short layoff and, as a result, are known as "Monday fever." Later in the week, the symptoms disappear. As the disease progresses, the symptoms extend beyond Monday to the rest of the week, and bronchitis with productive cough and dyspnea develops. If contact with the dust is not broken, byssinosis may lead to chronic obstructive pulmonary disease, even in workers who are not cigarette smokers.

It has been postulated that the cycle of symptoms in byssinosis is a pharmacologically mediated phenomenon. Cotton dust stimulation causes a discharge of naturally produced histamine, leading to bronchospasm, which continues until the histamine stores are exhausted. This results in loss of reactivity, allowing the affected individual to work undisturbed for the remainder of the week.

Inhaled metal fumes, notably those of copper and zinc, may cause a disease known as "metal fume fever." This reaction particularly occurs in welders employed in closed spaces, who refer to the disease as "galvanization." Other metals that cause this illness include magnesium, cadmium, iron, manganese, nickel, tin, and antimony. The acute illness is characterized by fever, malaise, nausea, and aching muscles.

2. Chronic Pulmonary Disease

Chronic occupational pulmonary disease is usually the result of long-term exposure to toxic dusts. The histories of workers who develop the chronic pulmonary disorders commonly reveal exposure to dusts for 20 or more years before the disease becomes symptomatic. Because of the insidious nature of these chronic diseases, measures must be taken to protect workers before the diseases develop. Such measures should include reduction of exposure to dust and periodic examination of the exposed workers.

About 10% of coal workers develop simple pneumoconiosis without symptoms, while a smaller number develop the complicated form, a progressive massive pulmonary fibrosis. The disease is progressive, and length of life is definitely shortened.

The damage in simple coal worker's pneumoconiosis develops as a result of the inhalation of respirable coal dust that first settles within the alveoli and later accumulates near the respiratory bronchioles. There, coal macules form with limited scarring, leading to disease of the respiratory bronchioles and emphysema. In nonsmokers, simple pneumoconiosis may be asymptomatic. Cigarette smoking is the usual cause of increased cough and expectoration in individuals with simple coal worker's pneumoconiosis; but in coal miners with many years of exposure, specific industrial bronchitis related to dust inhalation can contribute to the bronchial symptoms. The complicated forms of coal worker's pneumoconiosis are disabling, and a decision must be made to remove workers from the mining environment when x-rays show excessive reaction to dust.

B. Carcinoma

Cancers caused by occupational exposure are significant because, like many work-related diseases, many of them can be prevented. The forms of lung cancer linked to occupational exposure are bronchogenic tumors and mesothelioma. As in other occupational pulmonary diseases, the important etiologic factors are the nature of the inhaled substance and the duration or intensity of exposure. Researchers have identified many industrial agents as carcinogenic, and undoubtedly more will be identified. Curiously enough, coal is one recognized industrial hazard that is not associated with an increased incidence of lung cancers.

Table 2 lists a range of substances known or suspected of being human carcinogens for the lungs as well as other anatomic sites. The shortness of this list may be surprising. Actually, less than 30 chemicals are currently known to induce cancer in the human population.

Table 2

Occupational Cancer Hazards

Agent	Cancer Site or Type	Type of Workers Exposed
Acrylonitrile	Lung, colon	Manufacturers of apparel, carpeting, blankets, draperies, synthetic furs, and wigs
4-Aminobiphenyl	Bladder	Chemical workers
Arsenic and certain arsenic compounds	Lung, skin, scrotum, lymphatic system, hemangiosarcoma of the liver	Workers in the metallurgical industries, sheep-dip workers, pesticide production workers, copper smelter workers, vineyard workers, insecticide makers and sprayers, tanners, miners (gold miners)
Asbestos	Lung, larnyx, GI tract, pleural and peritoneal mesothelioma	Asbestos factory workers, textile workers, rubber-tire manufacturing industry workers, miners, insulation workers, shipyard workers
Auramine and the manufacture of auramine	Bladder	Dyestuffs manufacturers, rubber workers, textile dyers, paint manufacturers
Benzene	Leukemia	Rubber-tire manufacturing industry workers, painters, shoe manufacturing workers, rubber cement workers, glue and varnish workers, distillers, shoemakers, plastic workers, chemical workers
Benzidine	Bladder, pancreas	Dyeworkers, chemical workers
Beryllium and certain beryllium compounds	Lung	Beryllium workers, electronics workers, missile parts producers
Bis(chloromethyl)ether (BCME)	Lung	Workers in plants producing anion-exchange resins (chemical workers
Cadmium and certain cadmium compounds	Lung, prostate	Cadmium production workers, metallurgical workers, electroplating industry workers, chemical workers, jewelry workers, nuclear workers, pigment workers, battery workers
Carbon tetrachloride	Liver	Plastic workers, dry cleaners
Chloromethyl methyl ether (CMME)	Lung	Chemical workers, workers in plants producing ion-exchange resin

278

Agent	Cancer site	Workers
Chromium and certain chromium compounds	Lung, nasal sinuses	Chromate-producing industry workers, acetylene and aniline workers, bleachers, glass, pottery, pigment and linoleum workers
Coal tar pitch volatiles	Lung, scrotum	Steel industry workers, aluminum potroom workers, foundry workers
Coke oven emissions	Lung, kidney, prostate	Steel industry workers, coke plant workers
Dimethyl sulfate	Lung	Chemical workers, drug makers, dyemakers
Epichlorohydrin	Lung, leukemia	Chemical workers
Ethylene oxide	Leukemia, stomach	Hospital workers, research lab workers, beekeepers, fumigators
Hematite and underground hematite mining	Lung	Miners
Isopropyl oils and the manufacture of isopropyl oils	Paranasal sinuses	Isopropyl oil workers
Mustard gas	Respiratory tract	Production workers
2-Naphthylamine	Bladder, pancreas	Dyeworkers, rubber-tire manufacturing industry workers, chemical workers, manufacturers of coal gas, nickel, refiners, copper smelters, electrolysis workers
Nickel (certain compounds) and nickel refining	Nasal cavity, lung, larynx	Nickel refiners
Polychlorinated biphenyls (PCBs)	Melanoma	PCBs workers
Radiation, ionizing	Skin, pancreas, brain, stomach, breast, salivary glands, thyroid, GI tract, bronchus, lymphoid tissue, leukemia, multiple myeloma	Uranium miners, radiologists, radiographers, luminous dial painters
Radiation, ultraviolet	Skin	Farmers, sailors, arc welders
Soots, tars, mineral oils	Skin, lung, bladder, GI tract	Construction workers, roofers, chimney sweeps, machinists
Thorium dioxide	Liver, kidney, larynx, leukemia	Chemical workers, steelworkers, ceramic makers, incandescent lamp makers, nuclear reactor workers, gas mantle makers, metal refiners, vacuum tube makers

Table 2 (*Continued*)

Agent	Cancer Site or Type	Type of Workers Exposed
Vinyl chloride	Liver, brain, lung, hematolymphopoietic system, breast	Plastics factory workers, vinyl chloride polymerization plant workers
Agent(s) not identified	Pancreas	Chemists
	Stomach	Coal miners
	Brain, stomach	Petrochemical industry
	Hematolymphopoietic system	Rubber industry workers
	Bladder	Printing pressworkers
	Eye, kidney, lung	Chemical workers
	Leukemia, brain	Farmers
	Colon, brain	Pattern and model makers
	Esophagus, stomach, lung	Oil refinery workers

Source: Office of Technology Assessment, Congress of the United States, 1985.

Although bronchogenic carcinoma (lung cancer) and mesothelioma and asbestos are currently of greatest concern in the occupational setting, it was the observation of cancer of the scrotum in chimney sweeps that led to the identification of the first carcinogen known to humans.

1. Scrotal Cancer

When Percival Pott was 62 years old (1775), he sent his publisher a manuscript on cancer of the scrotum. He had found that it occurred almost exclusively in chimney sweeps and ascribed its origin to contact with soot. "The disease in these people," he wrote, "seems to derive its origin from the logement of soot in the rugae of the scrotum." Not only did he identify the first carcinogen some 210 years ago, but he set off a reaction that led ultimately to the identification of the carcinogenic hydrocarbon benso[a]pyrene by Cook and his associates in 1933 (4).

The annoying aspect of his work was the part left unspoken—prevention of the problem once he knew the cause. Sweeps in Europe, France, Germany, and the low countries, Belgium and Holland, contacted soot yet they had little or no scrotal cancer.

The mystery evaporated when it was learned that sweeps in those countries were required by their guilds to shower each day. The simple expedient of washing removed the soot and prevented the cancer. In England, bathing was not a particularly common practice, and Pott did not make the connection.

Chimneys are still being cleaned by small, young boys, most of whom wash properly. Nevertheless, occasional cases of scrotal cancer are seen in those who have not taken that simple precaution.

C. Asbestos

Public concern for the potentially severe health effects of asbestos escalated sharply in the 1970s with the declarations by the U.S. EPA that asbestos would be banned as a building material.

Asbestos is a generic term for a group of naturally occurring hydrated silicates existing in fibrous form. Figure 2 shows the curly fibers of chrysotile and the straight fibers of amosite ores. These ores have been classified in two minerological groups: the serpentines and the amphiboles.* Over 90% of the asbestos used in the United States was chrysotile, a serpentine. The remainder consisted of small amounts of crocidolite, amosite, tremolite, and anthophyllite, all amphiboles. These are strong, durable, inert materials, which are resistant to heat and acid. Nothing can match their ability to withstand abrupt increases in temperature.

---

* Chrysotile [$Mg_3Si_2O_5(OH)_4$] is the only member of the Serpentine class. Amosite [(Fe-$^{+2}$Mg)$_7Si_8O_{22}(OH)_2$], brown asbestos, and crocidolite blue are amphiboles.

Figure 2. The distinct fiber types of chrysotite and amosite.

It was not until the Industrial Revolution, especially with the widespread use of steam which required heat-resistant materials for pipe wrappings, joints, and seals, that asbestos use burgeoned. In 1896, the first asbestos textile mill opened in the United States and just prior to the United States entering World War II, an amosite factory was established in Patterson, New Jersey, to supply ships of the U.S. Navy with asbestos insulation for pipes, boilers, and turbines.

Asbestos' heat- and corrosion-resistant qualities have been so beneficial and so desirable that between 1900 and 1980, some 36 million metric tons were used in over 3000 products (see Figure 3). Its outstanding fire-resistant protection has benefitted literally millions of homes, schools, state and federal office buildings, and commercial and industrial buildings.

Why then the sudden flurry of interest and concern for asbestos? Concern focuses on its fibrous nature. These are mineral fibers, inorganic fibers, not the ordinary, carbon-containing garden variety of plant or animal fiber such as cotton, wool, linen, or flax, that the body can readily metabolize. The qualities that make asbestos fibers so valuable as building materials—its heat and acid resistance and its great tensile strength—are qualities that also make them resistant to metabolism.

The ancient Greeks, who had a word for everything, referred to the wicks used in their oil lamps as *sasbestos*, meaning inextinguishable or unquenchable, because the flames did not consume them. And Char-

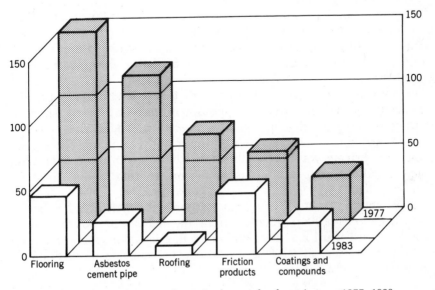

Figure 3. Major asbestos end uses in thousands of metric tons, 1977–1983.

lemagne, Charles the Great, the 9th century Emperor of Gaul, was known to clean his asbestos table cloth by throwing it into a fire.

For the most part, asbestos-containing materials (ACMs) do not become a hazard until sufficient drying or mechanical damage occurs, making their matrix materials—vinyl, plaster, paper, pitch, cellulose, or even cement—friable (crumbly on hand pressure), thus permitting escape of fibers.

Asbestos is not one of those chemicals whose carcinogenicity and other ill effects are extrapolated from animal studies to humans. Unfortunately, our knowledge of asbestos' ill effects comes directly from exposed workers, especially those who applied asbestos to ships during World War II. During those war years, 1942–1945, workers in holds of ships were exposed to unusually heavy doses of asbestos fibers, many without benefit of protective clothing and respirators. It was from among these workers that greater than expected numbers of cases of asbestosis, lung cancer, and mesothelioma began to appear in the 1960s and 1970s—20–30 years following their heavy exposures.

1. Asbestos and Smoking

There is, however, another complicating factor uniquely related to asbestosis and lung cancer. As noted in Table 3, those workers who smoked a pack of cigarettes or more per day had an even greater risk. Their risk of lung cancer was multiplicative, not additive. The combination of smoke and asbestos exposure multiplied the risk tenfold. That is a point that must be borne in mind.

In their seminal paper, Asbestos Exposure and Neoplasia, Selikoff and co-workers clearly revealed the asbestos–lung cancer risk* (5). In subsequent studies, they and their co-workers also estimated that some 4.5 million workers could have been exposed to asbestos in shipyards during World War II (6).

In addition, it is now recognized that wives, children, and even household pets were indirectly exposed, with resulting illness. However, the current concern that produced federal involvement and action for asbestos abatement activities in all primary and secondary schools, both public and private, was the possible health hazard to millions of children, teachers, administrators, and custodial workers from friable asbestos in their schools.

The Asbestos Hazard Emergency Response Act (AHERA, Public Law 99-519, October 1986) proceeds from the belief that any amount of asbestos is a hazard and thus mandates the detection, assessment, and control of asbestos in all schools. Is that a reasonable proposition and undertaking? Let us look at the evidence.

* The association between asbestosis and lung cancer and their association with asbestos exposure were first documented by K. M. Lynch and W. A. Smith in 1935. (7).

Table 3
Number of Times Lung Cancer Risk Is Increased

Nonsmoker asbestos negative (−)	1
Nonsmoker asbestos positive (+)	5
Smoker, asbestos negative (−)	10
Smoker, asbestos positive (+)	50–90

2. Portals of Entry

Asbestos can enter the body via the skin and the digestive and respiratory tracts. The skin can be pierced by spicules, splinters of asbestos fibers from the raw ore. If not properly tended, these can give rise to asbestos corns, which if present in sufficient numbers can induce clubbing of the fingers. This is a rare condition generally limited to miners and those handling ores.

Because of their long history of consumption of drinking water containing taconite, an asbestos-containing ore, the people of Duluth, Minnesota, represent a type of natural experiment. In 1983, Eunice Sigurdson of the State of Minnesota Department of Health, reported on the health effects of ingestion of asbestos among the residents of Duluth. She tells that in 1973 amphibole asbestos was discovered in the Duluth municipal water supply. Evidently the entire city, some 100,000 people had been exposed for twenty years—from 1956 to 1976, at levels of 1-65 million fibers per liter of water. Consider for a moment the number of liters and fibers consumed. For an individual who drank no more than one liter of water per day over one year, 7,280 liters would have been consumed. Those who drank 2 and 3 liters per day, would have consumed approximately 15 and 22,000 liters respectively. Multiply those numbers by 1 to 65 million fibers per liter and you have an idea of the astronomic numbers of fibers ingested. Thus the outcome of her study is all the more remarkable. Talking about cancer incidence she tells us, and these are her words, "statistically significant excesses are observed in several primary sites in Duluth residents. However, lung cancer in Duluth females is the only primary site considered also of biological significance, and this is primarily due to their increased smoking of cigarettes. The mesothelioma incidence rate (of the entire city) is no more than expected." (8).

An additional observation is also worth considering. If the facts of Figures 11 and 12, Chapter 14, are correct, stomach cancer has been declining in this country since the 1930's—when the use of asbestos began its steep climb into the 1970's. One can not be said to have caused the other, but it does suggest that taken orally, asbestos does not pose a significant health hazard. The hazard comes by way of the respiratory tract.

3. The Respiratory Pathway

Asbestos-related illness begins with the inspiration of air containing asbestos fibers. There is no way of knowing of their presence. They are odorless, tasteless, and nonirritating. However, the body is not a passive recipient. Its defenses begin with the hairs in the nose which filter out the larger particles.

As the airstream moves toward the throat, the sudden changes in direction of the high-velocity stream promote impaction on the surface (mucosa). This causes additional deposition of particles and further cleanses the airstream. In addition, because of the size of the medium-size particles (2–5 μm), many settle on the mucous surfaces. The very small particles (1.0–0.2 μm) are often exhaled with the next expiration. Mucus produced by the goblet cells embedded among the cilia—the hairlike fibers lining the tracheobronchial tree—are a second level of defense (see Figure 4). Figure 4a shows goblet cells (GC) among a carpet of cilia. In Figure 4b, goblet cells are shown in cross section. The cilia have a characteristic biphasic beat: a fast-forward flick and a slow recovery phase. This beating produces a wavelike motion—the mucociliary elevator—which carries entrained fibers up to the throat to be swallowed or spit out. This defense works well in those who do not smoke. Cigarette smoke paralyzes the cilia and reduces output of mucus. Loss of the mucociliary elevator can permit asbestos fibers to slip deeper into the respiratory tract.

By the time the air reaches the major bronchi, it has been substantially cleaned. Fibers such as chrysotile, the unusually silky and slippery type, may pass through, especially if they are smaller than 1 μm: smaller than many bacteria and of course invisible without benefit of a microscope.

Particles in the airstream can continue into the bronchioles and finally into the alveoli—the air sacs—where transpiration of gases occurs. The walls of the 300–400 million alveoli, no more than 0.5 μm thick, facilitate the diffusion of carbon dioxide and oxygen from and to the plexus of capillaries closely approximating the air sacs. This relationship is shown in Figure 5. It is here in the alveoli that trapped fibers can initiate ill effects.

4. Asbestosis

Foreign particles elicit yet another defense mechanism—white cells, macrophages, which move to the alveoli and bronchioles attempting to engulf and destroy the fibers. Figure 6 shows a macrophage moving to

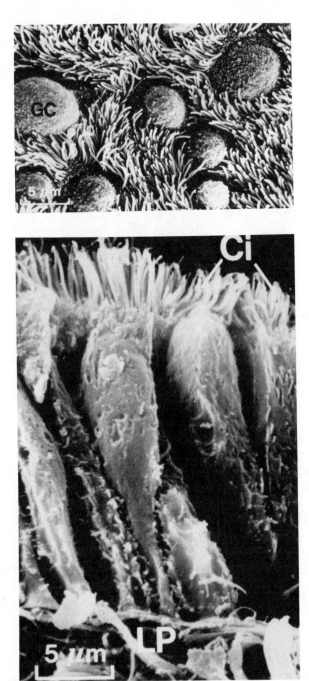

Figure 4. The shag-rug-like appearance of the cilia lining the trachea. GC refers to the mucus-secreting goblet cells.

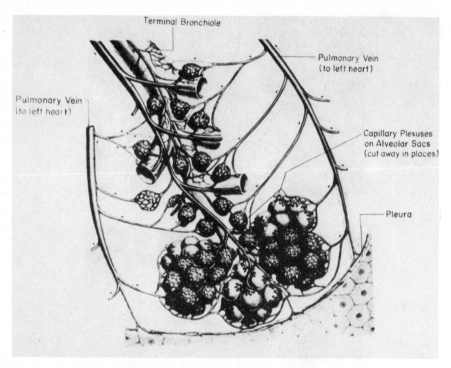

Figure 5. The alveoli and oxygen–carbon dioxide transport system showing the relationship between the alveoli and the capillary plexus. Reproduced with permission of Ciba-Geigy.

engul a foreign particle. Figure 7 shows one such macrophage pierced by asbestos fibers. But as noted earlier, these are mineral fibers which can resist breakdown. A third line of defense is called upon. The fibers are coated with hemosiderin, an iron–protein compound. Figure 8 shows one of the iron–protein-coated ferruginous bodies in a section of human lung tissue. This coating produces the beadlike appearance. On further magnification (Figure 9) the beads resemble "brushes." The bulblike or dumbbell-shaped ends are the result of the staining process. These ferruginous bodies are pathognomic of the presence of asbestos.

This process of walling off fibers produces scar tissue—deposits of collagen. As seen in Figure 10, sufficient scarring thickens the alveolar walls. When this thickening or fibrosis is caused by asbestos, it is called asbestosis.

If enough of the hundreds of millions of air sacs are affected, gas exchange is severely reduced, and rather than breathing normally, affected individuals must strain to get a breath of air. Shortness of breath occurs on walking, climbing steps, and later on lying down. Progressive loss of alveoli to fibrosis produces a stiffening of lung tissue, which can

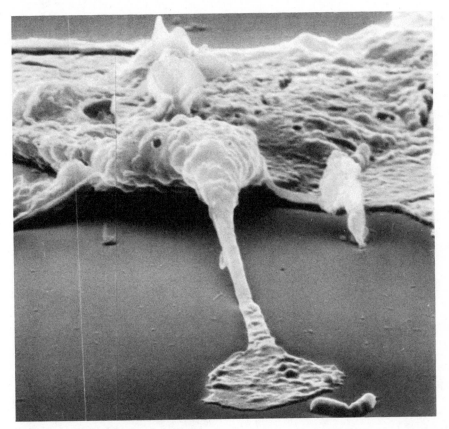

Figure 6. A macrophage has extended a pseudopod as it moves to contact and engulf a foreign particle.

leave a person gasping for air. It is nothing less than a form of suffocation. And this straining to breathe can produce an enlarged heart at the same time that reduced pulmonary function is occurring.

For the most part, a person with asbestosis has had a long history of exposure to high concentrations of asbestos. The belief that asbestosis progresses to lung cancer is one of a number of issues which currently enmesh asbestos-related health effects in controversy.

5. Bronchogenic Carcinoma

The bronchogenic carcinoma, lung cancer, found in asbestos workers is indistinguishable from lung cancer seen in workers without asbestos exposure. In either instance, the cancerous condition is characterized by the uncontrolled growth of cells in lung tissue. Figure 11 shows a typical

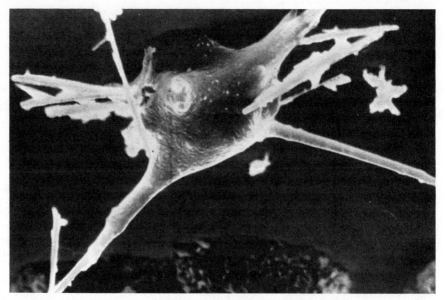

Figure 7. A macrophage pierced by an asbestos fibers.

mass localized in the area of the major bronchus. The space normally occupied by noncancerous tissue has been taken over by a massively growing tumor.

For the most part, workers with a long history of asbestos exposure who have lung cancer have also been inveterate smokers. As noted earlier, the risk of lung cancer among asbestos workers who smoke is increased at least tenfold. It is a pernicious combination (recall Table 3). For the physician, the determination of whether smoking or asbestos was the inciting cause of the tumor is all but impossible. The decision is heavily influenced by the information given the physician by the worker or his family.

Currently, the underlying mechanism by which long-term asbestos inhalation initiates the malignant transformation remains unknown. What is known, is that asbestos is a low-grade carcinogen, certainly not as potent as cigarette smoke or radon, for example, and therefore the risk of asbestos-related lung cancer is low. Given the widespread use of asbestos over the past 50 years, it is obvious that if asbestos were a more potent toxin or carcinogen, we should have seen far more asbestos-related illnesses than we have.

A major question at issue today, and one that is crucial to regulatory decisionmaking such as the AHERA, is whether there is a dose or an exposure level beneath which asbestos is noncarcinogenic. Of course, arguments for and against a threshold can be mustered. But as Mark R.

Cullen of Yale University, School of Medicine, has shown, the arguments for and against all "suffer from a lack of substantial data" (9). Consequently, the decisions that are being made concerning probable effects of low-level exposure have the unmistakable failing of being made using unsatisfactory data. But as is often the case, decisions are made no matter the quality of the data.

6. Mesothelioma

Both the lungs and the organs of the abdomen sit in hollow cavities. For the lungs, it is the pleural cavity and for the abdominal organs it is the peritoneum. Both cavities are lined and covered by a clear, saranlike

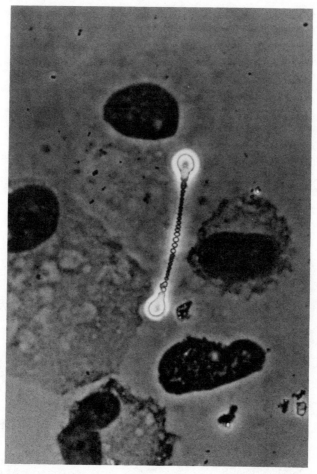

Figure 8. An iron-coated ferruginous body in a section of lung tissue.

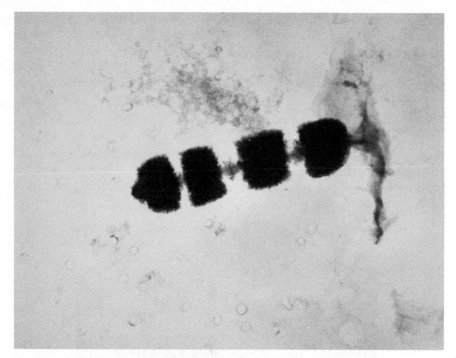

Figure 9. Transmission electron photomicrograph of an asbestos body at a magnification of 40,000×. Courtesy of Dr. David H. Groth, National Institute for Occupational Safety and Health.

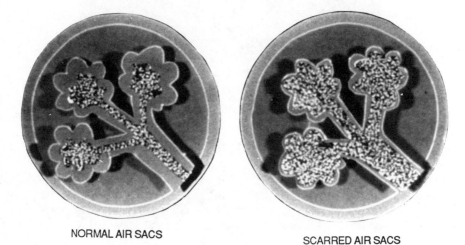

NORMAL AIR SACS   SCARRED AIR SACS

Figure 10.

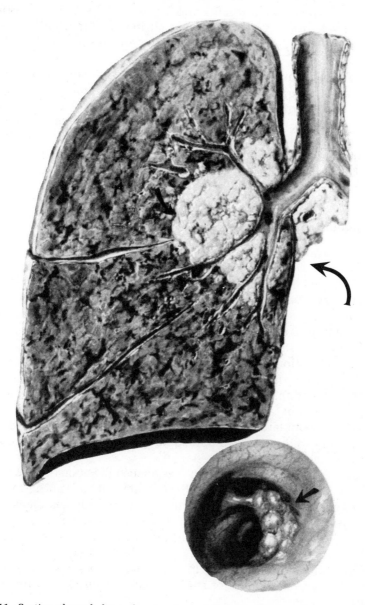

Figure 11. Section through lung showing (arrow) the massive tumor in the area of the bronchus. Lower photo shows tumor growing along wall of the bronchus. Reproduced with permission of Ciba-Geigy.

tissue. It is in this mesothelial tissue that the rare malignancy, mesothelioma, arises. Mesothelioma was first described and accepted as a nosologic entity as late as 1960. This early report attributed the mesotheliomas to heavy exposures of crocidolite received by the workers in South African mines (10).

In the United States, mesothelioma occurs in some two to five people for every million population. And 10–15% of these tumors do not appear to have a history of asbestos exposure. Unlike bronchogenic cancer, it does not appear to be smoking related, nor does it appear to be dose related. But like lung cancer it takes 20–40 years (of latency) before expressing itself. Pleural mesothelioma is a rapidly progressing malignant tumor. The resulting disability is total, and it is usually fatal in 12–24 months after diagnosis.

7. Extrapolation to the Urban Environment

Most asbestos-related health problems have occurred in occupational settings where, over long periods, exposure has been heavy. What then is the justification for attempting to extrapolate this type of exposure to such nonoccupational settings as schools, in which asbestos can be both friable and nonfriable, and in which asbestos levels in air may approach 0.001 fiber/cc—not unlike the urban environment.

In a recent report, J. Corbett McDonald of McGill University stated that "linear extrapolation to very low fiber concentrations almost certainly overstates the true risk" (11). And Brooke Mossman, a pathologist at the University of Vermont College of Medicine, speaking at a congressional hearing declared that "if one fiber of asbestos could kill, we'd all be dead, as the general population all contain asbestos fibers in their lungs" It is, however, from well-controlled epidemiological studies that an answer to the original question may arise.

As noted earlier, a textile mill Unarco, opened in Patterson, New Jersey, just before World War II. This mill is of more than passing historical importance, especially for the message it may hold for us today. In the 1970s, medical scientists, in this instance, epidemiologists, became concerned about possible untoward health effects in families living in the vicinity of the plant. The results of a study they conduced were reported at a major scientific conference in 1979, but it does not seem to have impressed or guided federal or state officials, or our elected representatives (12).

Because the mill was located in a residential area and because the heavily polluted factory air, described as resembling falling snow, was exhausted outdoors, where the prevailing winds carried the fiber-filled plumes directly past nearby homes, it was reasonable for them to be concerned about airborne exposure, albeit at lower levels, to residents who did not work in the mill.

On inspecting homes downwind of the plant, they found ample evidence of amosite asbestos in attics around Patterson. With this finding, the study moved forward. They then compared death certificates and causes of death among Patterson residents with residents of Totowa, New Jersey, a town miles away but of comparable size, and with normal ambient levels of airborne asbestos. Table 4 speaks for itself. In almost every disease category, Totowa had higher rates. Even if the differences were not statistically significant, they surely were not lower. Indirect or inadvertent contact with asbestos did not appear to impose additional risk on the residents of Patterson.

Shortly thereafter, Wagner, and co-workers of the Pneumoconiosis Unit, University of Sydney, using transmission electron microscopy (TEM), examined lung tissue of former employees of an asbestos textile factory in northern England (13). Those who had died of mesothelioma were compared with a matched sample of individuals who had died of nonasbestos-related causes. Their findings are shown in Table 5. Column 2, total fiber, of both the patient and control (non-patient) groups is especially trenchant. It is evident that both groups had substantial contact with asbestos fibers. Curiously enough, the range of total fibers is greater in the controls (cases 13–24), 20–762 million versus 17–480 million in the patients with mesothelioma (cases 1–12). Since the controls did not work with asbestos, environmental exposure, that is, air, and water, must be considered. Nevertheless, exposure to what must be considered as substantial numbers of fibers did not engender asbestos-related illness in all those actually exposed (13).

In 1986, Andrew Churg, professor of pathology, University of British Columbia Health Sciences Center Hospital, reported the results of his

Table 4
Number and Percentage of Deaths for Cancer Sites: Riverside and Totowa Neighborhood Study

	Riverside		Totowa	
	Number	Percentage	Number	Percentage
Lung	41	2.31	98	2.59
Colon–rectum	24	1.35	74	1.96
Prostate	17	0.96	37	0.98
Leukemia	11	0.62	15	0.39
Stomach	9	0.51	22	0.58
Pancreas	9	0.51	13	0.34
Kidney	5	0.28	7	0.19
Bladder	5	0.28	14	0.37
Lymphoma	5	0.28	9	0.24
Esophagus	4	0.22	12	0.32
Other specified	23	1.29	41	1.09
Unspecified	10	0.56	11	0.29

Table 5
Lung Contents of Those Dying of Mesothelioma and Controls[a]

Patients with Mesothelioma				Controls			
Case Number	Total Fiber	Chrysotile	Crocidolite	Case Number	Total Fiber	Chrysotile	Crocidolite
1	384	77.7	277.2	13	216	62.9	85.5
	480	48.0	393.6				
2	59	6.7	46.1	14	125	89.3	22.8
3	17	12.5	1.2	15	762	228.6	255.3
	59	19.5	1.8				
4	50	21.4	19.5				
	121	43.0	57.2	16	496	181.0	259.4
5	124	90.0	25.2	17	160	118.4	16.8
	49	37.0	8.6				
6	25	17.2	4.5	18	20	16.8	0
7	161	16.7	139.8	19	116	24.0	12.4
8	466	218.6	86.7	20	56	42.8	6.1
9	284	22.4	190.2	21	252	46.3	109.9
10	38	25.9	0.7	22	22	18.5	0.7
11	40	24.1	7.0	23	74	39.4	24.5
12	45	27.9	0.6	24	178	106.8	66.4

[a] Fiber counts expressed in millions/gram (and percentage of total fibers).
Source: Reproduced with permission of the British Medical Journal.

study of the lung asbestos content of long-term residents of a chrysotile mining town (14). Perhaps of greatest significance for regulatory decisionmakers, as well as the public generally, were the data comparing the findings of the residents of Thetford Mines, Quebec, with that of long-term Thetford asbestos miners, as well as residents of Vancouver—over 1000 mi to the west.

Churg found that the median concentrations of asbestos fibers in the lungs of Thetford residents were about 1/50th of those of the miners, but about ten times greater than that of the population of Vancouver. Tables 6 and 7 require attention. Table 6 shows the diversity of occupations of the resident series, the content of asbestos particles in their lung tissue, along with their smoking history. That in itself should have substantially increased their cancer risk. Table 7 shows the asbestos fiber levels in the lung tissue of each of the three groups studied. Of course, no group is without fibers. And as Mossman noted, if one fiber were sufficient, we all would indeed be ill or dead. Churg's conclusions are noteworthy. "The Thetford residents," he tells us, "are exposed to and retain not only more fibers but potentially more dangerous fibers than the residents of Vancouver and (assuming that Vancouver is reasonably representative) the residents of other North American cities. Nonetheless, none of the

Table 6

Demographic and Fiber Data on Thetford Resident Cases

Case Number	Age (years)	Sex	Smoking (pack-years)	Occupation/Years in Thetford	Cause of Death	Chrysotile/Tremolite ($\times 10^6$/g dry lung)
1	70	M	15	Farming/construction/lifetime	Myocardial infarct	1.2/12
2	50	M	40	Service station attendant/1958–1983	Myocardial infarct	0.7/1.2
3	59	M	45	City laborer/1957–1983	Sepsis	1.2/0.3
4	69	M	Pipe	Merchant/1953–1983	Cirrhosis	2.7/0.4
5	59	F	40	Housewife/lifetime	Carcinoma, most likely colonic	3.8/20
6	52	M	60	Telegraph operator/1954–1984	Myocardial Infarct	0.3/0.2
7	55	F	NS[a]	Housewife/lifetime	Myocardial infarct	2.2/2.8

[a] Nonsmoker.

Source: Reproduced with permission of the *American Review of Respiratory Diseases.*

Table 7
Fiber Concentrations ($\times$ 10^6/g Dry Lung)

Group	Mean	Median	Range
Chrysotile			
Vancouver residents	0.3	0.2	0–1.3
Thetford residents	1.7	1.2	0.3–2.7
Workers	65	46	3.3–470
Tremolite			
Vancouver residents	0.4	0.2	0–1.2
Thetford residents	5.3	1.2	0.2–20
Workers	218	85	4–2300

Source: Reproduced with permission of the American Review of Respiratory Disease.

Thetford residents in this study had evidence of asbestos-induced disease." He goes on to say that "the implications of these observations, therefore, is that substantially higher burdens of chrysotile and tremolite than those to which the general population are exposed can be tolerated for longer periods (even lifetimes) of continuous exposure with no obvious harm. These observations should provide reassurance that exposure to chrysotile asbestos from urban air or public buildings will not produce detectable disease" (14). This type of human-derived data should also be helpful in establishing the fact that a threshhold for asbestos, as well as dispelling the notion that one, two, four, or eight fibers are a real risk of asbestos-related disease.

It is becoming evident that insult from asbestos fibers is not unlike cigarette-smoke-induced lung cancer. Those who smoke a pack a day or more are at significantly increased risk of developing lung cancer, but all those who smoke this amount will not get lung cancer or cancer of other anatomical sites. *Exposure is not synonymous with disease.* Similarly with asbestos: exposure to heavy or light doses (levels) of fibers does not appear to induce illness in most people. Innate susceptibility or resistance must play a major role in the process. If that were not true, far more asbestos-related disease would be readily evident.

Janet Hughes and Hans Weil of Tulane University School of Medicine, responding to public concern for asbestos exposure, estimated the lifetime risk for school children exposed for 5 years, beginning at age 10, to an asbestos concentration of 0.001 fiber/mL of air. Their calculations predicted 0.02–0.37 deaths per million children annually. If this is reality based, the risk to children should be minimal (15). With levels of fibers in schools of 0.001–0.004 fiber/cc, and with many of these other than asbestos,* the crisis effort currently under way to clear asbestos from all schools, K–12, seems little more than a bad case of hysteria tinged with national hypochondria.

Discussing the risk of asbestos-related illness from the viewpoint of appropriate public policy decisionmaking, Weil and Hughes raise a salient consideration. They say that "attempting to revise rather than to use [the] science in order to meet predetermined social or economic objectives is inefficient and unreasonable. It does not lead," they go on to say, "to the most rational allocation of our national resources and often places disease or economic burdens on those least able to deal with either" (16).

Although former Secretary of Health, Education and Welfare (now Health and Human Services) Joseph Califano created the erroneous impression in 1978 that cancer related to occupation accounted for some 20% of all cancers, that frightening figure has now been shown to be far less. Estimates generated since then indicate that occupationally related cancers account for some 1–3% of the total (17). Figure 12 gives an average of 4% with a range of from 2 to 8%. Although these newer estimates are less depressing, it is still a large number of deaths. But most

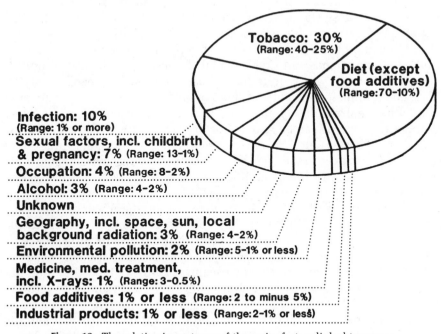

Figure 12. The relative importance of the major factors linked to cancer.

* Recall that OSHA does not define the nature of a fiber. Fibers counted with a light microscope may be cotton, wool, paper, cellulose, or glass, along with asbestos, if it is present. The number of fibers reported is not specified as asbestos.

important is the strong belief by all parties concerned, industry, government, and academia, that most, if not all, are preventable.

III. ESTIMATION OF TOXICITY

Chemicals vary markedly in the degree of their inherent toxicity. Consequently, responses to varying exposure will also differ. This variation, the spectrum of toxicity, is described in Table 8.

Thus, it can be seen that quite a wide range (an increase of more than 15,000 times between <1 mg and >15 g) exists in degree of toxicity. To be harmful to health, some chemicals must be consumed in large doses, while others need be taken only in small amounts. This is why there is such difficulty in determining and establishing levels of chemicals that workers can be exposed to without harm. But bear in mind that these levels are based on data obtained from animal studies, and as we have said earlier, it is difficult to draw human conclusions from animal studies.

The American Conference of Governmental Industrial Hygienists (ACGIH)* each year publishes lists of standards it has set for allowable occupational exposure to chemicals based on an 8-h working day and a 5-day deek. These standards are called Threshold Limit Values (TLVs). Although TLVs exist for hundreds of chemicals, they do not exist for hundreds of others. These values are based on information gathered from industrial experience and experimental studies with animals and human subjects. Thus, it will require many years and much research to gather data on all industrial chemicals. It is important to note that TLVs are only guidelines, not absolute values. As new data become available, a value published one year can be reduced or its threshold increased the following year. These standards are pegged to allow the concentration of airborne substances to fluctuate a reasonable amount above and below the listed TLV, with the understanding that the average value for the 8-h day does not exceed the standard.

In remarking on the difficulty of determining these values, one must consider the age differences among workers in a factory, as well as the differences in body type and genetic constitutions. These will all have an effect on the individual's reaction to a specific concentration of chemical. The TLVs are based on the reasonably healthy adult; they are not applicable to continuous exposure to the very young and very old, the indisposed, or the diseased.

The American Standards Association (ASA) has also studied the problem of occupational exposures and has set standards called Maxi-

---

* The ACGIH, organized in 1938, is a private, non-profit organization concerned with administrative and technical aspects of worker health protection. Membership is limited to professionals in governmental agencies or educational institutions.

Table 8
Spectrum of Toxicity[a]

	LD_{50}[b]	Lethal Dose
Extremely toxic	<1 mg[c]	Taste
Highly toxic	<50 mg	Teaspoon
Toxic	50–500 mg	An ounce
Moderately toxic	0.5–5.0 g	A pint
Slightly toxic	5–15 g	A quart
Nontoxic	>15 g	More than a quart

[a] Bear in mind that 1000 milligrams (mg) equal 1 gram (g) and that 30 g are just about an ounce.
[b] LD_{50} is the notation used by biological scientists to express that dose in milligrams per kilogram of body weight that kills 50% of the test animals.
[c] The notation $<$ means "less than," and $>$ means "more than."

mum Allowable Concentrations (MACs). These are issued with the proviso that they are only guides in the control of health hazards and are not to be regarded as fine lines between safe and dangerous levels. Rather, they represent a ceiling above which concentration should not be allowed to rise at any time during the working day. Below these MACs only the most susceptible or sensitive individual should show ill effects.

It is the rare industrial environment that has only a single chemical circulating. It is much more likely that several chemicals are present at the same time. How then is an estimation of a safe or hazardous condition to be made?

Experience and experiment have shown that certain combinations of chemicals have an additive effect. That is, their combined effect, rather than the effect of either individually, must be evaluated. To express this, an equation relating observed atmospheric concentration (C) to the TLV of that specific chemical (T) can be set up. For example,

$$\frac{C_1}{T_1} + \frac{C_2}{T_2} + \frac{C_3}{T_3} + \cdots \frac{C_n}{T_n} = 1$$

This expression says that when the ratios of each of the chemicals are added together, their value can be equal to unity. If their value is less than unity, the TLV is not exceeded; if it is unity, the condition is borderline; and if it is more than unity, the TLV is exceeded and the chemicals can be assumed to be potentially hazardous.

If experience indicates that the effects of each chemical are in fact independent of those of the other, the ratio $C/T = 1$ must be ascertained for each substance independently, and the TLV is exceeded only when at least one constituent has a value higher than 1.

Example. Air contains 400 ppm of acetone (TLV = 1000 ppm), 150 ppm of sec-butyl acetate (TLV = 200 ppm), and 100 ppm of 2-butanone (TLV = 200 ppm). The atmospheric concentration of the mixture is 400 + 150 + 100 = 650 ppm of mixture.

$$\frac{400}{1000} + \frac{150}{200} + \frac{100}{200} = 0.4 + 0.75 + 0.5 = 1.65 \quad \text{TLV is exceeded}$$

TLVs represent conditions under which it is believed most workers may be repeatedly exposed day after day without adverse effect. As issued by ACGIH, they are recommendations and are no more than guidelines of good practice. They are neither official nor do they have legal status. However, OSHA has adopted many of these standards. When they are included in federal and state statutes, the TLVs do have the force of law.

Three categories of TLVs are currently specified:

1. TLV-TWA. This is a time-weighted average. It is the concentration for a normal 8-h day to which all workers may be repeatedly exposed day after day without adverse effect.

2. TLV-STEL. This is a short-term exposure limit. It is the maximum concentration to which a worker can be exposed for 15 min without suffering irritation, tissue changes, or drowsiness to a degree that would impair self-rescue, and to which a worker will be exposed no more than four times per day, without at least 1 h between exposures

3. TLV-C. This is a ceiling limit. It is the concentration that should not be exceeded for an instant.

Each year, the ACGIH publishes a list of TLVs. These are recommended to the ACGIH membership by the TLV Airborne Contaminant Committee. Labor unions have not participated on this committee because they believe standard setting should be a government activity. Changes in ACGIH TLVs are not automatically incorporated by OSHA.

A. Integument

The skin, the largest organ of the body, also provides the body with its first line of defense against environmental risks. Because it is readily observable, problems are quickly recognized. Given these two factors, it is understandable why skin disorders account for nearly half of all reported job-related illnesses. As shown in Figure 13, skin diseases surpass all other health-related industrial problems. The relative importance of skin disorders in the eight major industrial categories is shown in descending

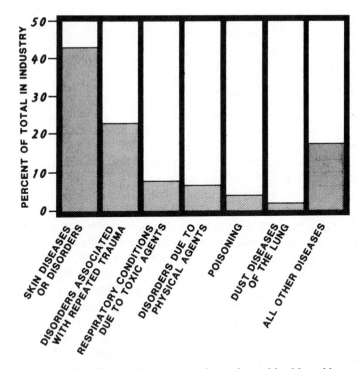

Figure 13. Skin diseases that surpass other industrial health problems.

order in Table 9. Within the category manufacturing, the greatest number of skin disorders among their workers comes from leather and tanning, poultry dressing, and boat building. This is shown in Table 10.

Skin disease is also responsible for some 200,000 lost workdays each year. That is approximately 25% of the total time lost by all occupationally related illnesses. Obviously, the question that begs response is why skin diseases are the most prevalent of all illnesses. Part of the answer is that it is the most exposed. But that needs further elucidation. Any

Table 9
Skin Disorder by Industry: Number of Illnesses
(×1000), 1982

Manufacturing (all types)	22.5
Services	7.6
Wholesale and retail trades	4.0
Agriculture and fishing	2.5
Construction	2.2
Financial services	0.5
Mining	0.4

Table 10
Skin Disorders in Manufacturing, 1982

Leather and tanning
Poultry dressing
Boat building
Ophthalmic goods
Plating and polishing

abnormality of the skin induced or aggravated by the work environment is referred to as an occupational dermatosis.

Meat or fish cutters may have a cut on their hands into which pathogenic microorganisms enter and set up an infectious skin disease; a machine operator may have acne or contact dermatitis as a result of exposure to cutting oils used in metal cutting and machine processes to dissipate heat while lubricating metal surfaces; tannery workers may get chemical burns from contact with strong alkaline or acidic solutions or may develop "chrome ulcers" from exposure to chromates in the tanning solutions. Rubber workers may develop skin depigmentation from exposure to a variety of antioxidants and other esoteric chemicals used to prevent rubber from decomposing or may develop chronic eczema from handling rubber compounds. Exposure to tar fumes or its dusts may induce melanosis or a photosensitivity that can give rise to a sunburnlike condition.

Dermatosis thus represents a broad spectrum of skin conditions. Dermititis, on the other hand, represents a superficial inflammation of the skin, characterized by redness, swelling, scaling, crusting, and often itching. Eczema and dermatitis are synonymous and can be used interchangeably. Table 11 lists the percentage contribution of a variety of agents involved in almost 1400 cases of skin disorders. Chemicals along

Table 11
Percent Distribution of Skin Disease in Manufacturing

Chemicals	42
Coal and petroleum products	10
Animal products (not food)	4
Soaps, detergents, and cleaning compounds	4
Textile items	4
Food products (including animal food)	3
Glass	2
Plants, trees, and vegetation	2
Plastic items	1
Wood items	1
Others	24

with coal and petroleum products account for more than 50% of the inducing agents. Other sources of skin disorders, such as detergents, textiles, and plastics, are probably no less hazardous because of their chemical constituents.

Inflammatory disorders of the skin often occur in individuals who have hypersensitivity reactions to the many chemical substances they contact daily in the environment. Infectious agents ranging from viruses to insects often infect skin. The combination of chemicals and microbes is often more than enough to breech defensive barriers. Where an infectious agent can gain entrance to a cut on an otherwise healthy individual, little may occur. The presence of an organism is insufficient by itself to produce an infection. But add the additional insult of hypersensitizing chemicals and the organisms may just gain enough of an edge to overcome otherwise sound defenses.

B. The Hand Hammer Syndrome

Although not specifically a skin disorder, the hand-hammer syndrome does inflict trauma on the skin. But it goes beyond. The use of the hand as a substitute for a hammer or mallet when they are not immediately handy is an established practice. Even when they are handy, carpenters and plumbers often prefer whacking tools with their open palms. As a matter of course, painters use their palms to unstick newly painted windows. Physicians call this use of the hand as a hammer the hypothenar syndrome, referring to the trauma to the fleshy mass of the palm in the ridge between the thumb and pinky.

In most instances, it is the dominant hand that suffers most. However, both hands can be involved. Primarily involved and damaged is the superficial branch of the ulner artery in the palm as it winds around the hamate bone (os hamatum). Symptoms produced are a reduction in the blood supply to the hand, which will depend on the degree to which the artery has been hammered shut (occluded). With an inadequate blood supply, one or all (or several) of the fingers may be affected. The skin can be pale or blue and sensitive to cold. Numbness and aching, tenderness, tingling, stiffness, and ulceration have been reported. Some hands have become gangrenous, requiring major surgery.

Of course, prevention can take two forms. Education away from such practices is first needed, followed by the redesigning of tools to deal with the specific needs so that hands are not used as tools.

C. Carpal Tunnel Syndrome

The meat-packing industry, the most hazardous in the country, has topped the list of most dangerous trades for the past 10 years. Meat cutters and packers work in extreme heat and/or refrigerated temperatures. They often work with unprotected knives and powersaws, and grease and blood make floors and walls slippery.

The speed with which repetitive operations must be done can cause their problems. Not long ago, hog carcasses were carried past work stations at the rate of 365 per hour—killing hogs at the rate of 6 per minute—all day. The rate was increased to 640 per hour. Recently, according to union figures, some slaughtering plants increased the rate at which the hog chain moves past workers to 1065 per hour—16 per minute, up from 6. Not quite a 300% increase.

Under these conditions, it is little wonder that these difficult and repetitive movements, done at the pace the machines fly by, cause the tendons within workers wrists to swell and pinch the adjacent nerve, causing pain and weakness. Workers in this industry often force their fingers open in the morning before work, then force them back to grip the knife handle or saw.

Carpal tunnel syndrome (CTS) is one of a group of muscular–neurologic conditions resulting from compression of the median nerve between the longitudinal tendons that serve the wrist and hand muscles and the transverse superficial carpal ligament.

The carpal tunnel, shown in Figure 14, is a channel through which pass both the median nerve and the flexor tendons. Constant flexion and extension of the wrist move the tendons back and forth against the canal, producing tendonal irritation and swelling. This swelling places undo pressure on the median nerve.

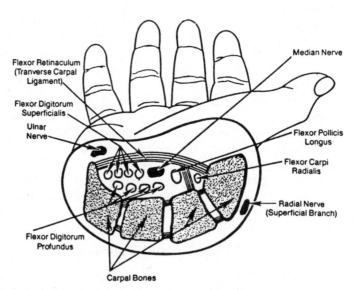

Figure 14. Carpal tunnel syndrome. Reproduced with permission of the American Industrial Hygiene Association.

The first sign of nerve entrapment is numbness and tingling at the tips of the fingers, often at night, hours after work, making it difficult to relate work practices with numbness. Progressive involvement leads to loss of grip strength and constant wrist pain. CTS is widespread among meat cutters, most of whom are young men who will be faced with this crippling disability for the rest of their lives.

D. Cardiovascular Illness

Cardiovascular diseases (ICD* codes 401–405), ischemic heart disease (ICD codes 410–414), and cerebrovascular disease (ICD codes 430–438) are responsible for more deaths each year than any other types of illness.

In 1986, cardiovascular disease took 967,930 lives, with ischemic heart disease responsible for 524,080. Although the rates of death from cardiovascular disease have declined markedly since 1968 (cf. Chapter 14), coronary atherosclerosis and acute myocardial infarction remain the leading causes of death in the United States.

The role of occupation as a factor is far from clear. However, the categories of finance, services, and manufacturing are high in their apparent contribution to heart attacks. Epidemiologists are convinced that personal risk factors such as cigarette smoking, blood pressure, diet, personality, and heredity are more important than environmental factors. However, it is also not clear whether smoking, high blood pressure, and poor diet, for example, are the consequence of stressful occupations. Nevertheless, a number of factors have been directly related to heart disease.

In 1981, the American Heart Association issued a report on the impact of the environment on cardiovascular disease (18). In it, six factors with a potential impact on CVD were identified:

1. *Carbon Monoxide.* Decreases the oxygen-carrying capacity of hemoglobin and thus reduces the oxygen supply available to the heart muscle. Carbon monoxide can induce changes in heart rhythm.

2. *Carbon Disulfide.* A widely used solvent, carbon disulfide has been shown to increase the risk of cardiovascular disorder.

3. *Halogenated Hydrocarbons.* Acute exposures to such solvents as trichloroethylene, chloroform, and fluorocarbon aerosol propellants have precipitate sudden death in workers exposed to high levels. These may also affect heart rhythm.

---

* ICD codes are those developed as the International Classification of Disease in which every disease or condition has been assigned a number. These range from cholera 001 to 999.9—other unspecified complications of medical care. The idea is for all nations of the world to adopt this classification so that there will be worldwide uniformity of reporting and thus no confusion about what condition is actually being described.

4. *Nitroglycerin and Nitrates.* Workers exposed to these chemicals during the manufacture of explosives have experienced chest pain, myocardial infarction, and sudden death.

5. *Noise.* Millions of workers are regularly exposed to high levels of sustained and/or intermittent noise. These may produce transient increases in blood pressure. Several studies, still to be more fully evaluated, have demonstrated increasing cholesterol with increasing noise.

6. *Psychological Stress.* Although stress has long been thought to adversely affect CVD, the association has remained elusive. It has been suggested that "work overload," role conflicts, and thwarted career goals were related to heart disease. In a recent study, air traffic controllers showed an increase in hypertension attributed to difficulties in coping with working conditions. In addition, men whose wives worked in "white collar" jobs appeared to experience heart disease at a rate three times greater than men whose wives worked in clerical or "blue collar" positions or were housewives. Indications are that inability to adjust to the conflicting demands of job and family may be related to the risk of CVD (19).

E. Auditory Problems

Exposure to loud and continuous noise is known to produce progressive loss of hearing. The impact of occupational noise is often difficult to isolate and separate from progressive deterioration that can accompany aging. Nonetheless, exposure to noise levels above 90 dB (cf. Chapter 7), similar to the noise of a diesel truck moving at 40 mi/h 50 ft from the ear, may be the level associated with hearing impairment. Continuous sound (noise) above 90 dB is clearly associated with hearing loss. The Occupational Safety and Health Administration requires employers to reduce workplace noise levels to 90 dB, similar to the sound level encountered in a newspaper room.

In the discussion of the effects of noise in Chapter 7, it was noted that the primary site of auditory injury is the receptor organ of the inner ear, the organ of Corti, in the cochlea. Several of the accompanying photomicrographs showed the damage to the sensory hair cells. The greater the destruction of these exquisitely delicate cells, the greater the degree of hearing loss to be expected.

According to OSHA regulations, occupational exposure limits are governed by the factors of time and intensity. The current limit has been set at 90 dBA for an 8-h day. But the regulations do permit sounds about 90 dB for shorter periods. However, at no time is exposure to continuous levels of 115 dBA permissible. Given the known relationship of hearing loss at the 90-dB level, one must wonder why OSHA set 90 dB as its limit rather than 85. In fact, OSHA did try for legislation that would have

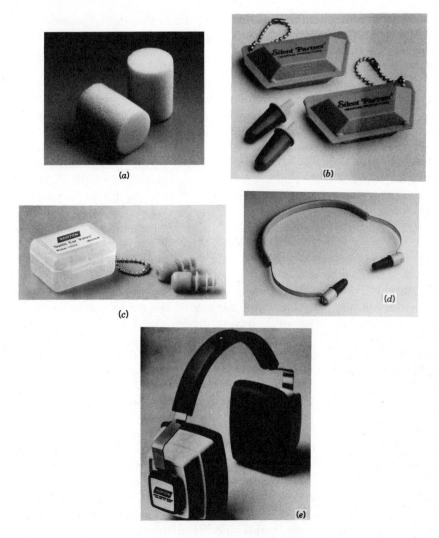

Figure 15. Types of ear protectors: (a) Deci Damp earplugs; (b) Silent Partner; (c) Sonic Ear Valvs; (d) Silent Bandit; and (e) 4540 hearing protector. Courtesy of Norton Corporation.

accomplished that, but industry objected on the grounds of economic hardship. To reduce noise levels 5 dB would, they said, cost billions of dollars. Remember that much of industrial noise is generated by excessive vibration of machinery, steam leaks, worn bearings, slapping belts, and unbalanced rotating parts. To damp these down would entail large financial outlays. I suspect that as long as compensation damage claims are also kept low, there will be little incentive to alleviate the noise problem. Less expensive are ear defenders, in the form of devices shown in Figure 15.

Interestingly enough, the earplugs have a greater noise reduction rating (NRR), at 29, than any of the other devices pictured. NRR is a single number which is an A-weighted approximation of the protection that a hearing protector will offer. To ascertain the effectiveness of a protector, the NRR rating of the protector is subtracted from the measured noise level in the particular environment. The resulting figure is the approximate dBA noise level to which a person is then exposed while wearing the unit. This final number should not exceed OSHA's exposure limits.

The Silent Partner, also shown in the Figure 14, has a NRR of 25. The Cona-Fit plugs made of silicone rubber with a triple flange, designed to provide sealing over a wider range of ear canal sizes, has a NRR of 26. The Sonic Ear Valvs with a NRR of 6 are designed to attenuate harmful, high-level impact noises while permitting reception of normal background sound. It is specifically helpful against the effects of impulse or repetitive impulse noises such as drop forges, riveting operations, punch presses, stamping and chipping, and piston-type engine noises. The Silent Bandit, with a NRR of 25, is useful in areas where noise is intermittent. Food processing areas and foundries are good examples. The Industrial Earmuff with a NRR of 23, behind the ear, is useful in shops, mills, construction sites, and airports, where high-frequency noises can be hazardous. Unfortunately, accuracy in determining NRRs may be less than optimum. If so, the current values for each of these devices may be higher than deserved. This too remains to be evaluated.

F. Heat Effects

The human body regulates its internal temperature within narrow limits; for the most part by sweat production and evaporation. Occupations requiring heavy work in hot atmospheres place workers at greatest risk of heat-induced cramps, rashes, and stroke because they add to the individual workload by generating more metabolic heat. The most troublesome conditions usually occur in warm, moist environments, such as canneries, laundries, deep mines, and textile manufacturing.

Intense exposure to heat in the workplace may be the most important environmental cause of male infertility. According to Richard J. Levine of the Chemical Industry Institute of Technology, workers at risk of heat-provoked infertility include men involved in smelting of metals and fabrication of glass, as well as those sweltering in the engine room of ships. Others at risk are bakers, farm laborers, and long-haul truck drivers who are continually exposed to the heat generated by truck engines.

Under normal conditions, testicular temperature is between 34 and 35°C. When testicular temperature rises to 37°C (98.6°F), normal body heat or higher, sperm production can be depressed. Although this condition is usually reversible, the effects can persist and there is usually a lag time before sperm counts rise again. If a worker is exposed to heat daily for long periods, impairment can be sustained.

The largest occupational study of fertility was conducted by researchers at the University of Odense in Denmark. They tested 10,748 people, 1069 infertile couples and 4305 fertile control couples. They were looking for possible associations with insecticides, paint, glue, organic solvents, dry cleaning chemicals, and asbestos. Their data strongly supported a link between infertility and heat exposure, providing "the most convincing evidence" of an association in men. Female infertility does not appear to be heat related. Rather, it appears to be associated with exposure to the heavy metals lead, mercury, and cadmium.

Additional research indicates that testicular temperature and depressed spermatogenesis is associated with sitting as opposed to standing. Testicular temperature climbs when a man sits. Thus, other workers at increased risk may be long-distance drivers of buses and trucks. These studies tend to support nonoccupational findings indicating that increased testicular temperature, as with hot baths, can be a suitable method of birth control. Here again, the problem appears amenable to preventive measures. The obvious approach in this instance is to "cool it."

G. Construction Sites

Although statistical data inform us that industries vary in their safety practices, the data do not mention that safety of workers appears to have different meanings in different areas of the country.

For example, the data presented in Table 12 are revealing as to the most and least dangerous areas with respect to building construction—always near the top of the riskiest industries. From this classification we learn that New York City may have the least desirable working conditions.

Over the period 1979–1986, buildings in New York were issued 1.1 times more permits than San José, California. Yet the number of fatalities in New York related to unsafe conditions was greater than San José's by a factor of 83 : 1. And deaths per billion dollars of construction was 38-fold greater. Whatever it is they are doing in San José to keep their job sites safe has not yet reached the East Coast. But New York is not alone in its poor safety record. Philadelphia, Denver, and Chicago could well learn from Colombus, Phoenix, and Long Beach—all with good records.

Obviously, the number of building permits (job sites) is in itself not necessarily a reservoir of accidents, although one would suspect that the more sites, the greater the opportunity for injury and death. Nevertheless, Los Angeles with almost 300,000 permits has one of the country's lowest rates, whereas Houston with over 200,000 has one of the highest rates.

Although job sites hold the clue, the number of sites is not the answer. Onsite work practices are the more likely source of differences between

Table 12
A Dangerous Place to Work[a]

City	Population Ranking	Building Permits	Cost (in millions)	Deaths Related to Unsafe Conditions	Deaths per Billion Dollars Worth of Construction
New York City	(1)	53,499	$10,912	83	7.61
Philadelphia	(5)	42,384	3,709	23	6.20
Denver	(24)	60,633	4,450	26	5.84
Chicago	(3)	34,118	8,927	52	5.83
Jacksonville	(19)	70,259	4,115	24	5.83
St. Louis	(30)	18,424	2,657	14	5.27
Houston	(4)	203,445	23,751	125	5.26
Atlanta	(31)	54,183	5,394	28	5.19
Memphis[b]	(16)	46,119	3,667	19	5.18
Seattle[b]	(25)	34,054	5,033	28	4.97
Kansas City	(29)	30,413	2,853	14	4.91
Boston	(20)	21,393	3,264	15	4.60
Baltimore[b]	(12)	133,144	4,186	19	4.54
Pittsburgh	(33)	31,182	2,924	13	4.45
Cleveland	(23)	48,501	1,962	8	4.08
Washington	(17)	28,296	3,710	15	4.04
Honolulu[b,c]	(11)	47,155	1,854	7	3.78
Dallas	(7)	120,629	19,512	66	3.38

City	Rank		Deaths	Ratio
Milwaukee	(18)	53,359	6	3.29
El Paso	(26)	63,110	11	3.02
Detroit[b]	(6)	47,209	6	2.81
Oklahoma City	(28)	51,933	16	2.56
New Orleans	(22)	38,722	12	2.51
San Francisco[b]	(13)	82,982	18	2.51
Austin	(34)	72,810	22	2.27
Fort Worth	(32)	47,805	10	2.22
Indianapolis[b,c]	(14)	45,467	7	1.96
San Antonio	(10)	104,930	14	1.82
Long Beach	(35)	45,790	3	1.50
Nashville[b]	(27)	50,181	11	1.48
San Diego[b]	(8)	93,732	13	1.19
Phoenix[b,c]	(10)	40,747	5	1.03
Los Angeles[b]	(2)	298,318	21	0.88
Columbus	(21)	52,017	4	0.73
San Jose[b]	(15)	46,411	1	0.20

[a] Deaths at construction sites in the 35 largest U.S. cities from 1979 through 1986. Cost is adjusted to equal 1987 price for New York City. Deaths related to unsafe conditions are those investigated by Federal and state authorities.

[b] Programs run by states.

[c] Incomplete figures: Indianapolis, 1982–1986; Phoenix, 1984–1986; Honolulu, 1983–1986; Seattle, 1981–1986. Phoenix total construction for 1979–1986 was $10.379 billion.

Source: U.S. Occupational Safety and Health Administration; state agencies in Arizona, California, Hawaii, Indiana, Maryland, Michigan, Tennessee, Washington; Commerce Department, Bureau of Labor Statistics; Census Bureau; Dodge Building Cost Index of McGraw-Hill.

cities. Apparently, there needs to be greater dissemination of information between cities. Political considerations might be another.

As seen in Table 12, the construction industry is one of the deadliest. Curiously enough, its injury rate (see Table 13) is among the lowest. Does this suggest that workers know how to use their equipment, but that they are unprotected with respect to falls, for example, from great heights? The table is fraught with such questions as the inherent concern of employers for their employees, cost versus profits, and the rapidity of project completion.

Compare the overall industry rate with the rates of the riskiest industries. None of those listed in the table produces less than 20 injuries and illnesses per 100 full-time workers. By not separating out rate

Table 13
The Riskiest Industries

Injuries and Illnesses per 100 Full-Time Workers, 1986[a]	
Meatpacking	33.4
Mobile home manufacturing	29.8
Vending machine manufacturing	28.1
Structural wood manufacturing	27.1
Raw cane sugar processing	26.2
Prefabricated wood building	26.0
Rubber recycling	25.7
Sawmills making wood shingles and other special products	25.4
Boat building and repairing	24.2
Vitreous plumbing manufacturing	23.5

Annual Fatalities per 100,000 Workers, 1980–1984[b]	
Mining	30.1
Construction	23.1
Agriculture, forestry, and fishing	20.3
Transportation, communication, and public utilities	19.5

[a] Data from Bureau of Labor Statistics
[b] Data from National Institute for Occupational Safety and Health.

constituent parts, we often come away with misleading interpretations as well as lack of real understanding.

H. Motivating Worker Safety: New Directions

Too often incentives for protecting workers fail to motivate negligent or otherwise disinterested employers. Recently, however, district attorneys in cities around the country have taken the view that when employers endanger or cause injury to their employees by making economic decisions not to use available safety measures, charges of criminal negligence may be an appropriate procedure. Filing criminal rather than civil charges not only sets a higher standard of responsibility for worker protection, but the idea or fact of going to jail is a remarkably motivating incentive for employers.

In July 1986, Gilberto Torres was buried alive when a trench in which he was working collapsed. The district attorney of Los Angeles County charged Torres' supervisor with involuntary manslaughter, alleging that the supervisor kept Torres working in the trench knowing it was improperly supported. If convicted, the supervisor faces a 4-year prison term.

In another case, the district attorney's office charged the Southern California Gas Company with violating the state labor code by failing to ascertain the location of underground utility lines prior to undertaking excavations. Had they done so, 23-year-old Richard Bergman may not have been electrocuted when his shovel struck a buried electrical line. Southern Gas was found guilty, ordered to pay a fine, and was placed on summary probation for 2 years in addition to being ordered to institute a full-scale accident prevention plan (20).

Two additional cases are of special significance because the decisions rendered will have far-reaching implications for the labor force throughout the country. Both cases occurred in Chicago and involved death or injury from exposure to potentially toxic chemicals.

In what was the first decision of its kind, an Illinois judge found three corporate officials guilty of murder in the death of an employee who regularly and continually inhaled gas while working.

Cook County (Chicago) Circuit Judge Ronald J. P. Banks, who heard the case, said the death of Stefan Golab in February 1983 "was not accidental but in fact murder." He also noted that the conditions faced by other employees of Film Recovery Systems, Inc., which operated a silver-reclamation plant in the Chicago suburb of Elk Grove Village, was "totally unsafe." Figure 16, offers a glimpse of the plant with its many open drums of cyanide. The plant was closed after the death of Mr. Golab, a Pole who was in the United States illegally. Prosecutors said Mr. Golab was poisoned by cyanide used to recover silver from used x-ray film and died after his complaints of nausea and dizziness were ignored. The state also maintained that the defendants had been aware that conditions in their

Figure 16. Open drums of cyanide solution and the lack of appropriate ventilation in the Film Recovery Systems plant. Photo courtesy of The Office of the District Attorney, City of Chicago.

plant posed a risk of death or great bodily harm and were thus playing "Russian roulette" with the lives of workers. Many of them were illegal aliens who spoke little or no English and thus could not readily complain about conditions in the plant or understand the hazardous nature of their work.

Having been found guilty of 14 counts of reckless conduct, each of the three defendants faces prison sentences ranging from 20 to 40 years. Such a verdict would have been unheard of 10 years ago. However, the case has been on appeal for 2 years. A final decision must await the outcome of *People v. Chicago Magnet Wire Corporation.* In this case, the Cook County states attorney charged that the company intentionally injured employees by exposing them to toxic chemicals. The defendants are charged with aggravated battery, reckless conduct, and conspiracy to commit aggravated battery.

Both of these cases are being held up by a technicality raised by the defense counsel, that OSHA has no jurisdiction in these cases. An Illinois court of appeals found that OSHA (1970) preempts state efforts to prosecute employers for workplace accidents. The court ruled that unless a state receives prior approval from OSHA to administer its own workplace health and safety program, the state cannot prosecute employers for workplace injuries.

As of January 1988, both cases await the outcome of the "preemption" consideration. The outcome of the Chicago Wire Magnet case could seriously affect the Film Recovery Systems case.

The preemption problem has become a national concern. If upheld, it could signal employers that they are literally free of any prosecution under OSHA. It could become the "defense of the year." Do as you please you needn't worry (21). Whichever way the issue is decided, worker protection will be significantly affected for years to come.

IV. OSHA: THE LAW

Members of both the House of Representatives and the Senate tried unsuccessfully for years to move the federal government into the arena of occupational safety. By 1969, however, draft legislation was developed. Congress had finally decided to involve the government in occupational research and regulatory activity. On December 29, 1970, President Nixon signed into law the Occupational Safety and Health Act. Congress' declared purpose was "to insure so far as was possible every working man and woman in the nation safe and healthful working conditions and to preserve our human resources." The act also created the Occupational Safety and Health Administration, within the Department of Labor, as the enforcement agency to carry out the mandate of the act.

A subsection of the act defines health and safety standards. Section 3 (8) of the act states: "The term 'occupational safety and health standard' means a standard which requires conditions, or the adoption or use of one or more practices, means, methods, operations, or processes, reasonably necessary or appropriate to provide safe or healthful employment and places of employment."

In addition, the act created the National Institute for Occupational Safety and Health (NIOSH) as the research arm of OSHA. NIOSH was placed within the Department of Health and Human Services to be administered by the Centers for Disease Control.

OSHA's activities can be seen as standard setting, enforcement, and public education and service. To carry out its mandate it has set up ten regional offices as shown on Figure 17.

In its 18-year history, has OSHA been effective? That is, has it performed as Congress hoped it would? From its inception it has never had adequate funds with which to set standards. Less than half a dozen emerge each year. Many believe it was concerned with the minutiae of health and safety rather than major workplace hazards. And it has been criticized for the slow pace of its inspection activities, the small number

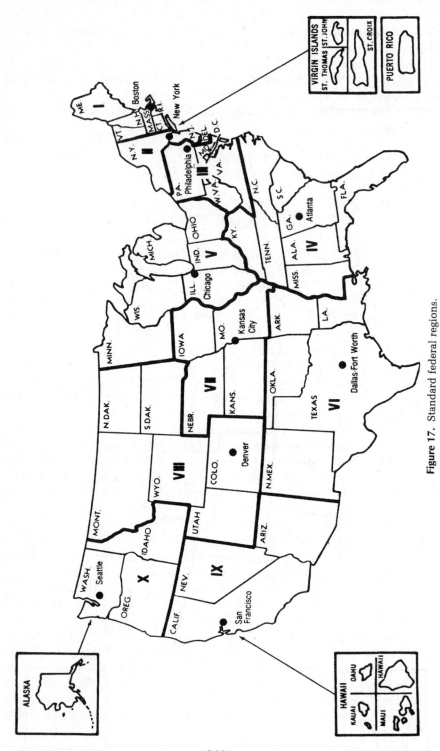

Figure 17. Standard federal regions.

of onsite inspections, and the low penalties levied. On the other hand, many credit it with increasing awareness of occupational safety and health on the part of industry managers.

OSHA's greatest needs are better data on the causes of injury. Currently, the difficulties in developing accurate statistics on workplace accidents and illness and in assessing OSHA's impact on workplace safety leave a good bit to be desired. Neither those who feel OSHA has done some good in protecting workers nor those who do not can make their case. The data are simply not available.

There is also widespread belief that the Reagan administration "cultivated a sense" in the country that employers did not have to worry about OSHA. Such an attitude could easily undermine compliance. And voluntary compliance on the part of employers, large and small, is probably the only way 70 million workers can be protected.

The following account suggests that worker protection is not a priority item on the part of the President or Congress.

A. The Pymm Thermometer Company

In October 1986, New York Attorney General Robert Abrams and Brooklyn District Attorney Elizabeth Holtzman filed criminal charges against William Pymm, Edward Pymm, Jr., and the Pymm Thermometer Company, Pak Glass Machinery Corporation, a related firm, and Thomas Daniels. The charges include assault, reckless endangerment, falsifying business records, and conspiracy.* If found guilty, each of the defendants faces up to 15 years in jail and $5000 in fines. And using a 1984 forfeiture law, prosecutors, in February, froze $167,951 in assets of Pymm Thermometer profit that the Pymms allegedly gained by operating an illegal business. If Pymm and its officials are convicted, that money could be used to compensate the employees who suffered mercury poisoning.

According to the indictment, Pymm Thermometer had conducted a secret mercury recovery operation in a cellar which had no windows and was poorly ventilated. Vidal Rodriquez, a Pymm employee, worked for 11 months in the cellar, where broken thermometers were crushed to recover mercury. Prosecutors allege that Rodriquez suffered permanent brain damage from his exposure to mercury. They also claim that the health of other workers was imperiled by conditions in the factory.

Moreover, prosecutors accuse Pymm Thermometer officials of hiding the existence of the cellar operation from OSHA inspectors. During the 1984 OSHA inspection, Daniels, the company foreman, allegedly removed workers from the cellar area in order to conceal the mercury recovery operation. During a 1985 inspection, William Pymm allegedly

---

* This account was taken from the journal *Occupational Hazards*—specifically, the September 1987 issue and is excerpted here by permission of the editor.

told OSHA inspectors that there was no work area in the factory below street level. The next day, when asked directly about the mercury reclamation machine, he allegedly first denied its existence and then delayed the inspector to prevent him from seeing the operation while Edward Pymm, Jr. and Daniels attempted to clean up the area. The Pymms and Daniels also allegedly lied to inspectors about when the recovery machine was first used in the cellar.

Accusing the defendants of "sacrificing worker safety for greed," Brooklyn District Attorney Holtzman said: "A chemical assault on a worker is just as serious as a physical assault on the street. The weapon here was a dangerous chemical, and it inflicted just as much damage as a gun or knife."

1. Troubled History

While prosecutors brand the Pymms' and Daniel's conduct as criminal, OSHA is not spared their invective. Holtzman said the year-long investigation by her office and Abram's had "uncovered an appalling pattern of conduct by OSHA that was both cavalier toward enforcement of Federal laws and callous to worker safety."

According to prosecutors and safety officials, the Pymm Thermometer story surfaced in the early 1970s, when state safety and health inspectors found mercury exposures in violation of state standards. Federal OSHA assumed enforcement responsibilities in New York in 1975, but no inspections took place at the Brooklyn factory until 1981. That January, an OSHA compliance officer tried to inspect the plant but was denied entry. OSHA then obtained a warrant and inspected the plant in March 1981. OSHA found a number of alleged serious violations, including lack of personal protective equipment, mercury contamination of work surfaces, and workers eating in areas exposed to mercury. The agency proposed fines of $1400 and set an abatement deadline of October 1981.

In November, OSHA settled the case with Pymm Thermometer, reduced the fine to $350, and extended the abatement deadline to October 1982.

In August 1983, responding to an employee complaint, OSHA inspected Pymm Thermometer and cited the company for a repeat violation for continuing to allow workers to eat and drink in areas contaminated by mercury. The company was fined $100.

In September 1983, a Pymm Thermometer official asked that the company be allowed more time to comply with the citations it received in June 1981. OSHA granted a 1-year extension to October 1984. In September 1984, a similar request by Pymm Thermometer resulted in the fourth extension granted to the company—this time to October 1985.

2. Health Department Involvement

In September 1984, a physician contacted the New York City Department of Health and reported that one of his patients might be exhibiting signs of mercury exposure. The department analyzed blood and urine samples from the employee, Vidal Rodriquez, and found that his urine mercury levels were 30 times higher than the normal level of 20 μg/L.

The health department contacted Pymm Thermometer, told them about the case of mercury poisoning, and recommended that medical screening be conducted for the employees. Both the health department and Pymm conducted screenings. According to Dr. Andrew Goodman, the health department's director of environmental epidemiology, Pymm's testing of more than 90 employees showed that half had urine mercury levels over 200 μg/L. Such levels, said Goodman, showed employees were "at fairly high risk of developing symptoms of mercury poisoning." And when physicians subsequently examined 40 of those employees, half had tremors, a sign of nervous system damage associated with mercury toxicity.

In November 1984, the health department inspected the factory to assess working conditions and take air samples. Said Goodman: "At many of the work stations where breakage was evident there was no local exhaust ventilation. None of the employees were using personal protective equipment. Mercury-contaminated glass and trash were stored without any controls for mercury vapors. In addition, the mercury vacuum cleaning device exhausted high concentrations of mercury into the workroom." And of 45 air samples taken in the factory, 94% exceeded the OSHA mercury standards of 0.10 mg/m^3.

Based on the inspection results and the medical tests, the health department drew up a set of industrial hygiene recommendations for Pymm to follow. Goodman said Pymm officials agreed to implement them, but a follow-up inspection in late December showed that the company was not in compliance. The health department then issued a city enforcement order requiring an immediate clean-up of the factory and proper personal protective equipment for employees. That order was quashed in federal court by Pymm's attorneys, who argued that the city was preempted from acting because OSHA had jurisdiction over workplace safety.

3. OSHA Returns

According to Holtzman, media coverage of the health department's investigation prompted OSHA to conduct another inspection of Pymm Thermometer in December 1984. OSHA found mercury air samples continued to exceed the 0.10-mg/m^3 levels, work surfaces were contaminated by mercury, adequate engineering controls had not been installed, workers were eating and drinking in mercury-contaminated areas, and operators were not wearing respiratory protection. OSHA proposed penalties of $2880, Pymm contested the citations and penalties, and the case was eventually settled, with penalties reduced to $960.

In October 1985, OSHA again inspected Pymm Thermometer and this time found the cellar operation that eventually led to the criminal indictment by local authorities. Among other conditions cited, Pymm was accused of willfully violating standards for respiratory protection, personal protective equipment, and sanitary conditions.

An OSHA inspector found that workers in the cellar were overexposed to mercury vapor. Workers were wearing dust/mist respirators, but they were models that were not certified for use against mercury vapor. Workers had not had their respirators fit-tested, had not received training in the use of respirators, and at least one worker was wearing his respirator over a full beard which nullifies its protection. OSHA also found workers with elevated urine mercury levels which were omitted from the log, certain air monitoring records had not been retained, and employees had not been told of the availability of either air monitoring or medical records.

OSHA issued three citations and proposed fines of $30,100. Pymm contested the citations, and later OSHA settled the case, reducing the fine to $21,000.

OSHA's most recent inspection of Pymm, a November 1986 follow-up inspection, resulted in alleged willful violations against the company for allowing workers to store their personal belongings in mercury-contaminated areas. The thermometer manufacturer also was cited for failing to implement necessary engineering or administrative controls of mercury, and for failing to provide, and ensure the use of, necessary respiratory protection.

"This employer has been cited in the past for similar violations and should be well aware of the OSHA standard on mercury exposure," said James Stanley, acting OSHA regional administrator. Stanley said work practices at the plant not only contributed to employees' exposure, but also could lead to health disorders among their families and members of the community.

Additional citations, unrelated to mercury exposures, were issued. OSHA proposed penalties totaling $75,900. Pymm has contested the citations and penalties.

Because workers might be carrying home mercury on their personal belongings, the health department has contacted workers and offered to test their children for mercury poisoning. So far, said Goodman, none of the workers has responded to the offer. Goodman said his office has also contacted both the State Department of Health and OSHA to try to get a roster of workers at the plant, but that so far that information has not been forthcoming.

According to DA Holtzman, the Pymm incident shows that the federal government "utterly failed to protect the workers in this Brooklyn factory." She added that OSHA's "dismal performance" left her with "no doubt in my mind that the Federal government is simply not committed to worker safety."

Whatever the outcome of the trial, this case highlights the low priority worker safety has on federal, state, and local government agendas.

REFERENCES

1. Hearings Before a Subcommittee of the Committee on Labor. House of Representatives, 74th Congress, H.J. Res. 449, January 16, 17, 20, 21, 27, 28, 29. February 4, 1936. U.S. Government Printing Office, Washington, DC, 1936.

2. W. Pepper, The Medical Side of Dr. Franklin. *Univ. Penn. Med. Bull.* **23:** 547–585. (esp. pp. 562–565), 1910.

3. E. J. Gangerosa, Quoted in *The New York Times*, Sunday, June 2, 1985, p. E3.

4. J. W. Cook, C. L. Hewett, and I. Hieger, The Isolation of a Cancer Producing Hydrocarbon from Coal Tar. *J. Chem. Soc.* **1:** 395–398, 1933.

5. I. J. Selikoff, J. Churg, and E. C. Hammond, Asbestos Exposure and Neopasia. *J. Am. Med. Assoc.* **188** (1): 142–146, 1964.

6. I. J. Selikoff, E. C. Hammond, and H. Seidman, Mortality Experience of Insulation Workers in the United States and Canada, 1943–1976. *Ann. N.Y. Acad. Sci.* **xx:** xx–xx, 19xx.

7. K. M. Lynch and W. A. Smith, Pulmonary Asbestosis III: Carcinoma of Lung in Asbestos-Silicosis. *Am. J. Cancer* **24:** 56–64, 1935.

8. E. E. Sigurdson. Observations of Cancer Incidence Surveillance in Duluth, Minnesota, *Environ. Health Perspect.* **53:** 61–67, 1983.

9. M. R. Cullin, Controversies in Asbestos-Related Lung Cancer. In *Occupational Medicine—State of the Art Reviews*, Vol. 2, *Occupational Pulmonary Disease*, Linda Rosenstock (Ed.), Hanley & Belfies, Philadelphia, 1987, Chap. 4, pp. 259–272.

10. J. C. Wagner, C. A. Sleggs, and P. Marchand, Diffuse Pleural Mesothelioma in the North Western Cape Providence. *Br. J. Indust. Med.* **17:** 260–271, 1960.

11. J. C. McDonald, Health Implications of Environmental Exposure to Asbestos. *Environ. Health Perspect*, **62:** 319–328, 1985.

12. E. C. Hammond and L. Garfinkel, Mortality Experience of Residents in the Neighborhood of an Asbestos Factory. *Ann. N.Y. Acad. Sci.* **330:** 417–423, 1979.

13. J. C. Wagner, G. Berry, and F. D. Pooley, Mesotheliomas and Asbestos Type in Asbestos Textile Workers: A Study of Lung Contents. *Br. Med. J.* **285:** 603–606, 1982.

14. A. Churg, Lung Asbestos Content in Long-Term Residents of a Chrysotile Mining Town. *Am. Rev. Respir. Dis.* **134:** 125–127, 1986.

15. J. M. Hughes and H. Weill, Asbestos Exposure—Quantitative Assessment of Risk. *Am. Rev. Respir. Dis.* **133:** 5–13, 1986.

16. H. Weil and J. H. Hughes, Asbestos as a Public Health Risk: Disease and Policy. *Annu. Rev. Public Health* **7:** 171–192, 1986.

17. R. Peto and M. Schneiderman (Eds.), *Quantification of Occupational Cancer: Banbury Report 9*, Cold Spring Harbor Laboratory, Cold Spring Harbor, NY, 1981.

18. W. F. Harlan, A. R. Sharret, H. Weill, G. M. Turino, N. O. Berhani, and L. Resenkov, Impact of the Environment on Cardiovascular Disease, Report of the American Heart Association Task Force on Environment and the Cardiovascular System. *Circulation* **63:** 243–246A, 1981.

19. C. D. Jenkins, Psychosocial Risk Factors for Coronary Heart Disease. *Acta Med. Scand.* *(Suppl.)* **660:** 123–136, 1982.

20. S. G. Minter, Are Prosecutors Stepping in Where OSHA Fears to Tread. *Occup. Hazards* **49** (9): 101–103, 1987.

21. Getting Away with Murder: Federal OSHA Pre-emption of State Criminal Prosecutions for Industrial Accidents. *Harvard Law Rev.* **101:** 535–556, 1987.

SUGGESTED READINGS

Doll, R. Occupational Cancer: Problems in Interpreting Human Evidence. *Ann. Occup. Hygiene* **23** (3): 291–305, 1984

Kinney, T. A., and Holzhauer, J. D. *Safety at Bay: The Failure of the Department of Justice to Enforce Federal Occupational Safety Laws,* The National Safe Workplace Institute, Chicago, 1987.

Levy, B. S., and Wegman, D. H. (Eds.). *Occupational Health: Recognizing and Preventing Work Related Diseases,* Little, Brown & Co., Boston, 1983.

Walter, B., Jr. The Occupational Safety and Health Act. The National Policy for the Prevention of Work-related Illness. *Environ. Health* **47** (1): 2–23, 1984.

Sinclair, Upton. *The Jungle,* Heritage Press, New York, Special Edition, 1965 (dedicated to the Workingmen of America). Originally published in 1906.

9

Radiation and Nuclear Power Generation

Nothing in life is to be feared, it is only to be understood.
Now is the time to understand more, so that we may fear less.
—Marie Curie

Radiation is as natural as "little green apples in the summertime," as much a part of our world as sun, wind, and rain. Radiation was an integral part of the earth, the planet, long before we evolved from our primitive ancestors. And with the appearence of *homo erectus* 500,000 years ago, humandkind has been living side by side, as it were, with natural radiation. It was not until the closing years of the 19th century that the presence of radiation was revealed. For example, Wilhelm Konrad Roentgen, professor of physics at the University of Wurtzburg (Germany), discovered x-rays. It was Roentgen who found that these invisible rays possessed the energy to pass through paper, wood, and aluminum, as well as his hand. And it was in France that his contemporary, Henri Becquerel, inadvertently discovered that uranium emitted radiation. It was also in France that the Curies, Pierre and Marie, discovered radium in pitchblend—uranium oxide.

We are exposed to radiation daily from the sun and outer space; naturally occurring radioactive minerals are present in the earth, in the buildings we inhabit, and in the food and water we consume. There are radioactive gases in the air we breathe, and our bodies are radioactive.

Figure 1 displays the variety of sources contributing to our exposure levels. The natural background contributes 82% of the 360 millirem. Synthetic sources—medical x-rays, nuclear power production, nuclear

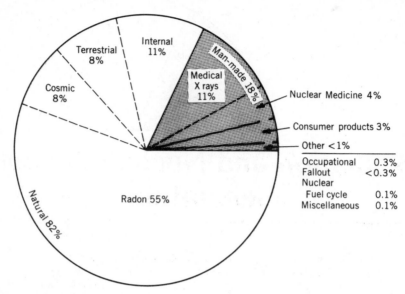

Figure 1. The percent contribution of various radiation sources to the total average effective dose equivalent in the U.S. population.

medical, agricultural, and industrial devices—at 18% are relatively negligible contributors to the doss most people receive. As averages, some people actually receive more than others.

Cosmic rays, a major contributor of radiation to the natural background, has long been a source of mystery to scientists. At a recent meeting of the American Astronomical Society, data were presented indicating that Cygnus X-3, a two-star system lying at the outer fringes of the Milky Way, may be the source of these high-energy particles. Cosmic rays include the most powerful form of radiation known to scientists. In space they consist primarily of high-energy protons, nuclei of hydrogen atoms, that generate showers of secondary particles which reach the earth. They contribute significantly to natural background. People in Denver the "mile-high city" and Flagstaff, Arizona, receive more cosmic radiation compared to people in New Jersey or New York. On the other hand, Michiganders, Pennsylvanians, and Tennesseeans may have greater contact with radon. Nevertheless, the 360 mrem level is far below the 5000 mrem federal standard allowed as safe occupational exposure.

I.   TYPES OF RADIATION

Although the term "radiation" is very broad and includes such things as light and radio waves, it most often refers to "ionizing" radiation, which is radiation that can produce charged particles or ions in materials that it strikes. This holds for both living and inanimate matters. Consequently, ionizing radiation can be hazardous; that is, it has the *potential* for harm.

A. Alpha, Beta, Gamma Radiation

Alpha radiation consists of positively charged particles and is emitted from naturally occurring elements such as uranium and radium as well as synthetic elements. As noted in Figure 2, alpha radiation will just penetrate the skin; it can be stopped completely by a thin sheet of notebook paper, or a thin film of water.

Beta radiation, is also particulate, consisting of electrons with a plus one charge. Much more penetrating, they can pass through 1–2 cm of water or human skin. A sheet of aluminum 2–3 mm thick can stop beta radiation, as can most clothing.

A third type of particle is the gamma ray, high-energy pulses of electromagnetic radiation emitted from nuclei as a result of nuclear reactions or decay. It is the most penetrating and therefore the most potentially harmful. This radiation can easily pass through the human body, clothing included. It penetrates tissue, organs, and bone, and accordingly, it is the gamma ray against which shielding must protect. Dense materials such as concrete and lead are thus the materials of choice.

The seriousness of radiation effects depends on two factors: the dose received and the time exposed to the dose. When ionizing radiation penetrates and interacts with a material, it imparts energy to it. For a given mass of material this energy is referred to as the absorbed dose. The unit internationally used to measure the absorbed dose is the rad, derived from

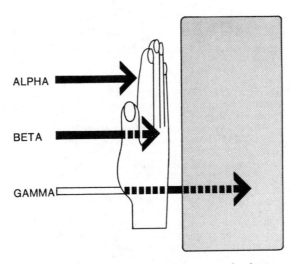

Figure 2. The penetrating power of radiation.

radiation absorbed dose. The rad is equal to 100 ergs* deposited in 1 g of material.

An important question, however, is how biological systems react. If a person is exposed to ionizing radiation, a different quantity, the dose equivalent, is used because the same dose of different types of radiation elicits different biological effects in the body. The dose equivalent is obtained by multiplying the absorbed dose by a value that varies with the type of radiation. As each type of radiation affects tissue differently, values must be assigned based on those effects. The value currently used is the QF (quality factor), where

$$QF = 1 \text{ for beta, gamma, and x-radiation}$$
$$= 10 \text{ for neutron radiation}$$
$$= 20 \text{ for alpha radiation}$$

Multiplying the QF and absorbed dose (rads) obtains the dose equivalent or rad-equivalent-man, the rem. The rem and millirem (mrem) indicate the dose equivalent to a population or individual. Millirem is 1/1000 of a rem. One chest x-ray, for example, is equivalent to approximately 20 mrem.

Currently, great pressure from Europe is being exerted to accept the SI system of units—the Systéme Internationale. It is absolutely vital in evaluating radiation from radioactive sources that definitions of and distinction between *activity* and *dose* be understood. However, comprehension is being complicated by the fact that since 1975 the traditional ("old") units, curies, rad, rem, and roentgen are being replaced by the SI units gray (gy)* and sievert (Sv).* In this system, the unit for expressing the amount og absorbed (radiation) dose is the gray—equal to 1 joule† deposited in 1 kilogram of material. And in 1977, the International Commission on Radiological Protection proposed the sievert as the SI unit of radioactive dose equivalent. It is in fact the joule-kilogram. It equals

---

* Scientists have used the erg (from the Greek meaning to work) as a basic unit of work as a means of measuring the energy levels of various activities. In the cgs or centimeter–gram–second system of physical units, it equals the work done by a force of 1 dyne acting over a distance of 1 meter. A dyne is the force necessary to move 1 gram of material 1 centimeter per second per second. An erg therefore is very small. For example, pressing a typewriter key requires 1 million ergs. It takes 10,000 ergs for a bee to flutter its wings, while the burning of a match requires 10 billion ergs. To lift a 1-lb weight 1 ft requires 1.356×10^{12} ergs.
* The gray is named for Louis Harold Gray, 1905–1965, a British physicist and radiobiologist who was an international leader in radiation research. The sievert is named in honor of Rolf Maximillian Sievert, 1896–1966, a Swedish mathematical biophysicist and outstanding pioneer in the development of radiation protection standards.
† A joule, named for the English physicist James Prescott Joule, is the amount of (mechanical) work required to produce a unit of heat. Thus, the current definition is the amount of heat required to raise the temperature of 1 lb of water 1°. In 1849, working with water, Joule determined the value of 772 ft-lb, remarkably close to the 778 ft-lb accepted today. One joule equals 10 million ergs.

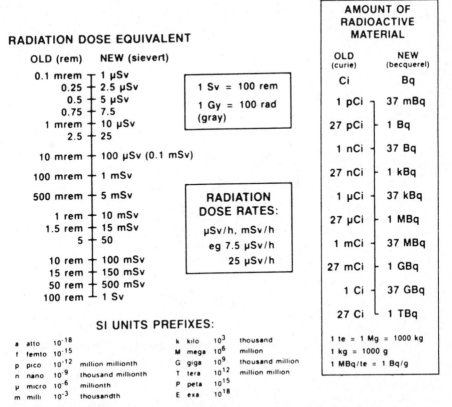

Figure 3. Guide to SI units. Reproduced with permission of Health Physics Dosimetry, Albuquerque, NM.

100 rems. Figure 3 lists the "old" and "new" dose equivalent as well as the old and new amounts of radioactive material. In the ensuing discussion, both old and new units will be used to encourage understanding and use.

Given that the natural background is the major source of radiation exposure, where do potential hazards lie? Certainly, radon requires scrutiny. And large amounts of radioactive waste containing isotopes with extremely long half-lives and gamma radiation are produced by nuclear reactors. These wastes have the potential for harm during transport and storage. Accidental releases of radioactive materials, as occurred at Windscale, England, in October 1957, and Three Mile Island in 1979, are other potential sources of hazardous radiation.

Radioactive materials lose energy as they radiate until they become stable or nonreactive. This loss of energy over time, in which half the radioactive nuclei disintegrate, is called radioactive decay. At some point, a sample or unit of radioactive material will lose half its radioactivity.

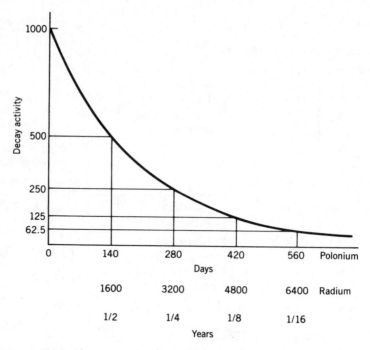

Figure 4. Half-life decay rate. Decay activity indicates nuclei that have not decayed.

Each radioactive element has a specific period for this loss, known as its *half-life*. Figure 4 displays a typical trend line of radioactive decay. Activity, the number of nuclei not yet decayed, is plotted against time. This can be seconds, minutes, hours, days, or years, even hundreds and millions of years. All half-life curves or trend lines will have a similar appearance. Only the time line will differ. Table 1 lists a number of radioactive isotopes and their half-lives.

For example, the curve for polonium depicted in Figure 4 can be transformed into the curve for radium by changing 140 days to 1600 years,

Table 1
Half-Lives of Radioactive Elements

Element	Half-Life
Nitrogen-16	8 s
Bromine-85	3 min
Sodium-24	14.8 h
Iodine-131	8 days
Cesium-137	33 years
Radium-226	1600 years
Carbon-14	5600 years
Thorium-232	10 billion years

250 days to 3200 years, 420 days to 4800 years, . . . ;, until 700 days when only 1/32 of the polonium will still be available or 8000 years at which time 1/32 of the radium will be available. Similar determinations can be made for each of the other isotopes in the table. Note that the other side of this coin is persistence. Persistence implies continuing hazard. Thus, shortness of half-life is a facet of hazard. The longer the half-life, the period over which the isotope emits radiation, the greater the opportunity for radiation injury.

Early in the 19th century, John Dalton, an English chemist, wrote that all elements are composed of small, indivisible particles called atoms. He noted too that all atoms of a given element are identical in mass, size, and shape and thus exhibit the same physical and chemical characteristics. That was fine until the discovery of radioactivity. It had been established that all atoms of the same element contain the same numbers of electrons and therefore must contain the same number of protons. However, it has since been determined that the number of neutrons in two atoms of the same element can be different. Atoms that have the same number of protons but a different number of neutrons in their nuclei are known as *isotopes*. Many elements have one or more isotopes that are unstable; that is, the composition of their nuclei changes with time. Isotopes with unstable nuclei (different masses) emit radiation. These are called radioactive isotopes, radioisotopes, or radionuclides.

B. Energy from Nuclear Power

While there are many ways to produce energy, some are not fully developed, while others are too expensive or are of limited potential. As shown in Figure 5, most of the energy currently available is supplied by three major sources: oil, natural gas, and coal. Oil and gas provide close to three-fourths of our energy requirements. However, if we continue to use these fuels at current rates, we will have consumed approximately 25% of our total resources by 1992. If no new fields are discovered and opened within the next 20 years, 50% of our remaining resources will be gone. It is vital that new sources of energy be developed.

The current U.S. population of 245 million will become 270 million by the year 2000. This alone will increase need and demand. By the year 2000, personal computers will be in 50–60% of all homes in this country. In addition, electronic banking, teleshopping, and entertainment are reshaping our lives. Given the increasing energy demand, a highly reliable supply of electric energy assumes paramount importance. Figure 5 suggests the sources that can be counted on to provide for the increasing needs. Solar, geothermal, and oil and gas are not capable of contributing significantly.

Uranium, a nuclear fuel rather than a fossil fuel, could provide as much energy as either oil or natural gas. If used in a new generation of advanced

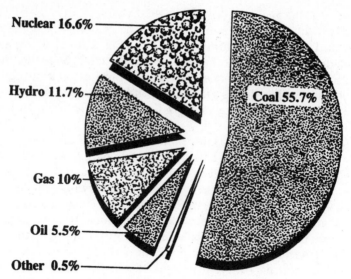

Figure 5. Sources of electricity in 1986.

nuclear reactors, the amount of energy obtained form uranium could be multiplied by a factor of 50 or 60.

The energy available within the nucleus of an atom is staggering. According to Einstein's penetrating equation, $E = mc^2$, the energy available in any atom of material is equal to its mass multiplied by the square of the velocity of light. In terms of energy and resources that means that on splitting (fissioning) 0.4 hg of uranium, energy equivalent to burning 2,500,000 lb of coal would be obtained.

Before exploring the state of nuclear power generation in the United States, it may be useful to review briefly the level of nuclear power supplied around the world.

During the 1960s, generation of electricity via nuclear power plants became a proven and commercially sound source of energy. By 1970, 90 nuclear stations were on-line in 15 countries with a capacity of 16,500 MWe. The number of plants fairly trebled by the 1980s, with 253 generating over 135,000 MWe in 22 countries. Nuclear power appeared on its way to becoming not only a reality, but the preferred method of generating electric power for the 21st century.

If the accident at Three Mile Island (TMI) near Harrisburg, Pennsylvania, in 1979 did not change that, the explosion at Chernobyl in the Soviet Ukraine in April 1986 seemed to do so—at least in the United States. However, the chronic uncertain conditions in the Middle East and the continued high price of crude oil spurred European and Asiatic countries to seek independence from oil. It is anticipated that by 1990, 350,000 MWe will be generated by some 500 operating stations. Many countries have made the commitment. Figure 6 shows the proportion of electricity generated by nuclear power around the world.

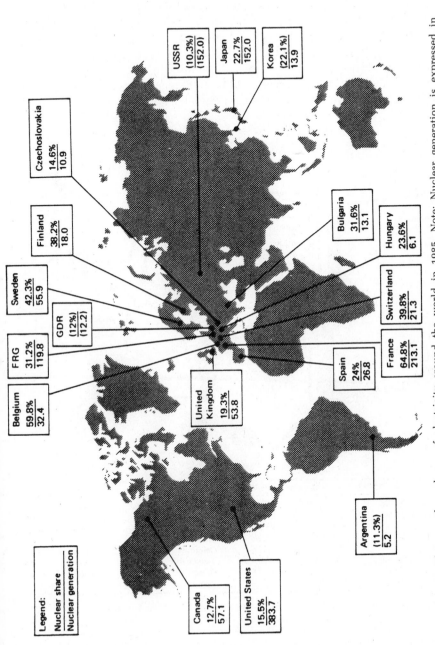

Figure 6. Highest nuclear shares of electricity around the world in 1985. Note: Nuclear generation is expressed in terawatt-hours and the nuclear share as a percentage of total electricity produced. FRG is the Federal Republic of Germany; GDR is the German Democratic Republic. In Taiwan, China, electricity generated by nuclear reactors was 27.3 terawatt-hours, or 52% of total electricity. Figures in parenthesis are IAEA estimates. Reproduced with permission of IAEA, Vienna.

Legend:
Nuclear share
Nuclear generation

USSR
(10.3%)
(152.0)

Japan
22.7%
152.0

Korea
(22.1%)
13.9

Czechoslovakia
14.6%
10.9

Finland
38.2%
18.0

Sweden
42.3%
55.9

GDR
(12%)
(12.2)

FRG
31.2%
119.8

Belgium
59.8%
32.4

United
Kingdom
19.3%
53.8

Bulgaria
31.6%
13.1

Hungary
23.6%
6.1

Switzerland
39.8%
21.3

Spain
24%
26.8

France
64.8%
213.1

Argentina
(11.3%)
5.2

Canada
12.7%
57.1

United States
15.5%
383.7

334   Our Precarious Habitat

Given the wide geographic variation in the United States, the figure of 15.5% can be misleading. A more representative rendition is seen in Figure 7. The range, 71.7% in Vermont to 1.8% in Ohio, highlights the wide disparity. Looking at the states collectively, there may be a suggestion that without openly acknowledging it, we may also have made a commitment. Perhaps the next stage will be failsafe stations.

Interestingly enough, 8 years after TMI, 3500 reactor years of operation without a fatality have been amassed: an enviable and creditable statistic.

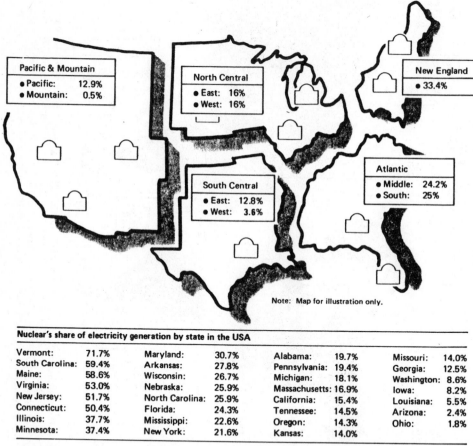

Nuclear's share of electricity generation by state in the USA

Vermont:	71.7%	Maryland:	30.7%	Alabama:	19.7%	Missouri:	14.0%
South Carolina:	59.4%	Arkansas:	27.8%	Pennsylvania:	19.4%	Georgia:	12.5%
Maine:	58.6%	Wisconsin:	26.7%	Michigan:	18.1%	Washington:	8.6%
Virginia:	53.0%	Nebraska:	25.9%	Massachusetts:	16.9%	Iowa:	8.2%
New Jersey:	51.7%	North Carolina:	25.9%	California:	15.4%	Louisiana:	5.5%
Connecticut:	50.4%	Florida:	24.3%	Tennessee:	14.5%	Arizona:	2.4%
Illinois:	37.7%	Mississippi:	22.6%	Oregon:	14.3%	Ohio:	1.8%
Minnesota:	37.4%	New York:	21.6%	Kansas:	14.0%		

Figure 7. Nuclear power generation in the United States, 1985. In the United States, 31 states have nuclear generating capability. The table indicates the amount of electricity generated by nuclear power in 1985 in those states; the map shows the nuclear share by region. Data are expressed as a percentage of the state's or region's total kilowatt-hours. They do not, however, necessarily indicate the amount of electricity actually consumed in a state or region, since in some instances a portion of the electricity generated by a utility's power plants in one state is sold to and used by customers in another state. Reproduced with permission of INFO Data.

C. Nuclear Reactors

To split atoms and release the tremendous amounts of energy they contain, specially designed vessels or reactors are required. In most of these conventional containment vessels, water is circulated through the fuel core. The water extracts heat from the core, keeping it from overheating, while transferring the heat outside where it produces steam that drives the turbines which ultimately generate electricity. Since water boils at 212°F (100°C), it must be kept under intense pressure in a vessel/reactor, or it will boil away and expose the core. This means that the reactor must be contained in a larger vessel. This vessel within a vessel concept is the basis for all existing plants generating power from atomic nuclei. In the following sections, a variety of reactor types are discussed.

The history of nuclear energy or power is relatively brief. The first nuclear reactor, a power plant (Chicago Pile—CP-1), was built under the bleachers of Stagg Field at the University of Chicago. The experiments performed there showed conclusively that a controlled fission or chain reaction could not only occur but could be maintained. The conclusion was evident. Tremendous amounts of energy could be obtained from small amounts of uranium.

In a conventional steam-generating plant, coal or oil is used to boil water, which is then converted to steam. A turbine, connected by a drive shaft to a generator, is driven at high speed as the steam strikes its blades and rotates them rapidly, enabling the generators to produce electrical energy. For the moment, a nuclear power plant can be thought of in much the same way, except for the method of heating water to steam.*

The heat in a nuclear plant is produced by the process of fission, which can occur in certain types of nuclear fuel. Figures 8 and 9 show, in a general way, the commonality of the processes.

Essentially, a nuclear power generating plant consists of six major components:

- Reactor
- Core
- Fuel elements
- Coolant
- Moderator
- Control rods

---

* This is only for purposes of understanding the process. In fact, nuclear power plants are different from traditional coal- or oil-fired generating plants in that they require special handling, siting, and construction because of the enormous power they can generate. It was, unfortunately, this "similar" attitude that has been responsible for many of the problems of nuclear reactors.

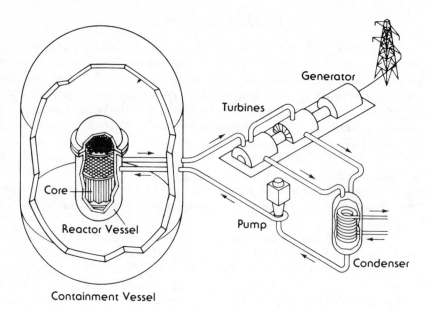

Figure 8. Schematic diagram of a nuclear plant.

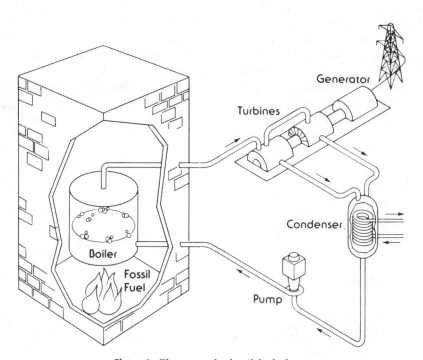

Figure 9. Elements of a fossil fuel plant.

A reactor is the part of a plant where the chain reaction occurs and where heat is generated. At the center of every nuclear reactor is the core, composed of nuclear fuel. When a neutron strikes an atom of fuel it can be absorbed (Fig. 10). This produces instability in the now heavier atom, which proceeds to split into two lighter atoms, as fission products. Heat (energy) and two or three neutrons are also produced. At this point the reaction becomes chainlike in that huge numbers of neutrons begin striking other atoms, producing a shower of fissions.

An optimum chain reaction can be sustained in uranium if the neutrons are slowed prior to striking the fissionable fuel. This is accomplished by surrounding the fuel by a moderator, which absorbs a portion of the energy of the neutrons as they are released during the process. Several different materials are suitable as moderators, including water, heavy water,* and graphite. In fact, each of these moderators is the basis for a specific class of reactor.

The heat from the fission process is removed from the core by a continuous stream of fluid called the coolant. The heat in the coolant can be used directly to produce electricity.

Uranium-235 is the isotope most commonly used. A neutron striking an atom of U-235 is absorbed. The U-235 becomes unstable and splits, yielding two lighter fission products, heat and neutrons.

Only a few elements fission easily enough to be used in a nuclear power plant. Uranium is probably the most suitable. But it must be processed before it can be used. After refining, it is shaped into small cylinders or fuel pellets. As shown in Figure 11, the pellets are less than half an inch (12 mm) in diameter but each can produce energy equivalent to 120 gal of oil.

Fuel pellets are stacked in hollow tubes some 12 ft long. When full these "pins" or "rods" are grouped together in bundles, the so-called fuel assembly. Within the assembly, rods must be carefully spaced to allow a liquid coolant to circulate freely between them. Approximately 200 nuclear assemblies are grouped together to make up the core of a reactor. It is this nuclear fuel in the core that generates the heat, just as oil heats water in a boiler.

Interspersed among the fuel assemblies are movable control rods (Figure 12). These are made of neutron-absorbing materials. It is upon the control rods that the speed of the chain reaction depends. Heat production is therefore moderated or augmented by inserting or withdrawing the control rods.

The entire reactor core, containing fuel assemblies and control rods, is

---

* Hydrogen has three isotopes: the common isotope contains one proton in its nucleus, another isotope has a nucleus containing one proton and one neutron. This isotope has twice the mass of the common isotope and is referred to as heavy hydrogen, or deuterium, meaning second. Water molecules consist of two hydrogen atoms and one oxygen atom. Water in which the molecule contains two deuterium atoms, D_2O, is called heavy water.

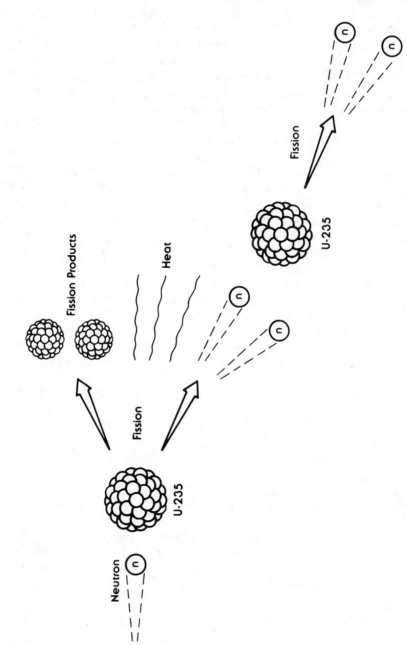

Figure 10. Process of nuclear fission. Courtesy of the U.S. Department of Energy.

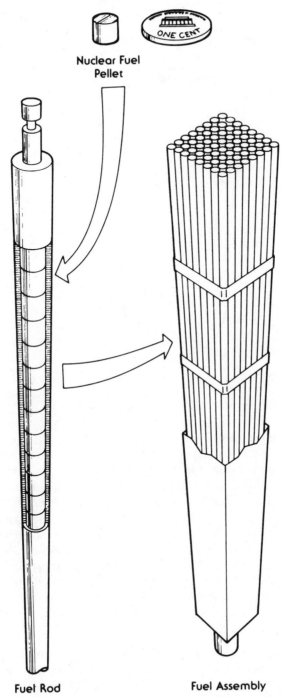

Nuclear Fuel Pellet

Fuel Rod   **Fuel Assembly**

Figure 11. Nuclear fuel elements. Courtesy of the U.S. Department of Energy.

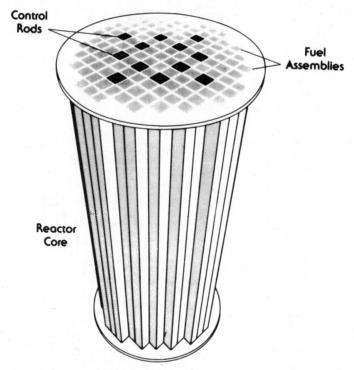

Figure 12. A reactor core. Courtesy of the U.S. Department of Energy.

enclosed in a stainless steel reactor vessel. And to further ensure safety, the steel-enclosed vessel is housed in a reinforced concrete structure.

Liquid coolant is pumped into the reactor through the core to remove heat. It is then pumped out of the reactor and used to produce steam. Those nuclear power plants using water as the coolant are known as light water reactors (LWRs). In this case, light water means ordinary water and it serves a dual function of moderating neutrons and transferring the heat of the reaction. Two distinct types of LWR are currently in use: boiling water reactors (BWRs) and pressurized water reactors (PWRs). In the United States, 84 nuclear reactors with operating licenses, 66% are PWRs, 27% are BWRs, and 7% are HWRs (heavy water reactors). Figure 13 shows the location of all the currently operating nuclear power plants (NPPs), as well as those under construction and planned.

1. Boiling Water Reactors

In the boiling water reactor, Figure 14 pressure is controlled to ensure that the water boils as it passes through the core. Consequently, steam is generated directly from the heat of the core. These are "direct cycle" systems with no intermediate steps. The steam produced drives a conventional turbine that runs a generator. In the PWR, Figure 15, pressure is

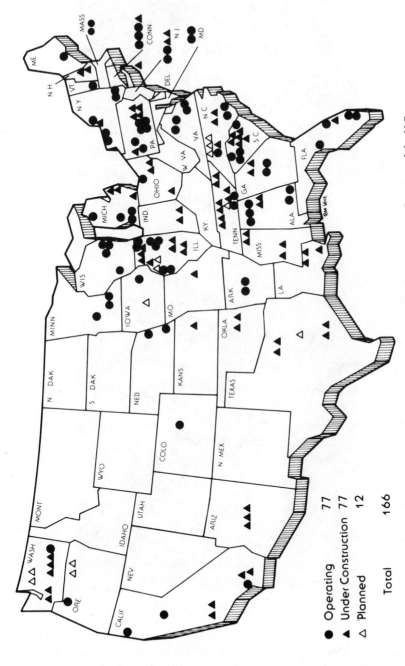

Figure 13. Nuclear reactor sites in the United States as of January 1982. Courtesy of the U.S. Department of Energy.

●	Operating	77
▲	Under Construction	77
△	Planned	12
	Total	166

341

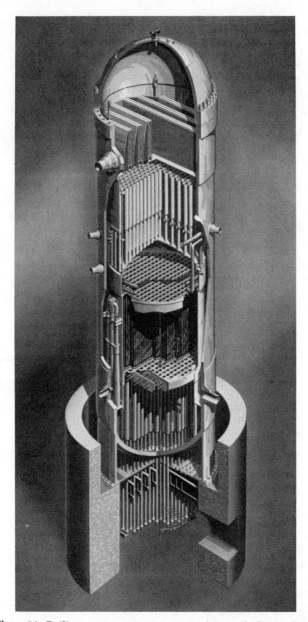

Figure 14. Boiling water reactor. Courtesy of General Electric Company.

maintained high enough to prevent boiling. It is pump. d under pressure and high temperature through a heat exchanger that transfers the heat to water in a secondary coolant system, which subsequently boils and forms steam. Here again, the steam drives a conventional turbine generator system. Water from the core is repeatedly circulated through the primary loop without being converted to steam.

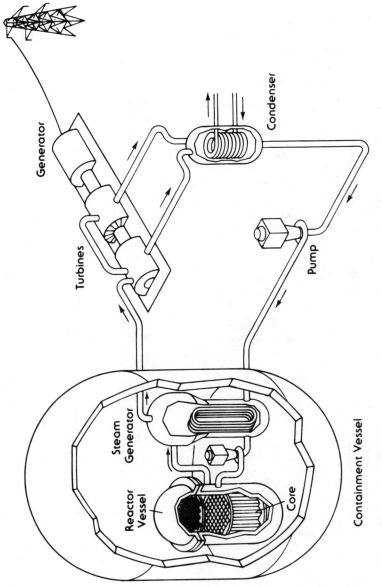

Figure 15. Schematic view of a pressurized water reactor. Courtesy of the U.S. Department of Energy.

During operation, the radioactive materials are contained within the fuel rods and pose no threat to public safety. The concern here is that should an accident occur, the fuel could become overheated to the point of melting, with the consequent release of fission products that accumulate during normal operation. Rapid interruption of the nuclear chain reaction can occur by inserting the control rods containing neutron-absorbing boron into the core.

2. Heavy Water Reactors

Heavy water reactors, Figure 16, have not been developed in the United States. Canada has used them most extensively where the CANDU (Canada deuterium uranium) reactors produce all their nuclear-generated electricity. Interest in the HWR stems from the fuel efficient, economic way heavy water moderates or slows down neutrons. Cost, combined with safety of operation, can be an overriding determinant. The fact is that worldwide HWRs appear to have demonstrated the most impressive performance records, especially for reliability, of all reactor types in use.

The design of a HWR is similar to a PWR insofar as the primary coolant

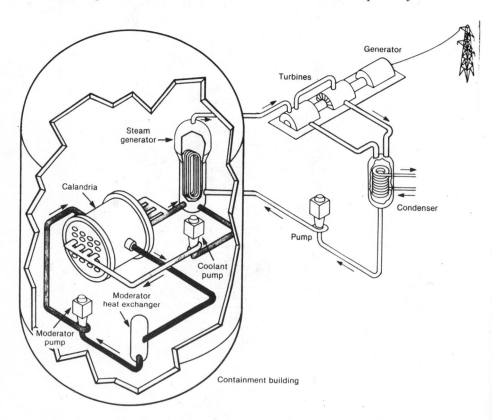

Figure 16. Heavy water reactor. Courtesy of the Office of Technology Assessment.

transfers heat from the core to a secondary coolant system by way of a steam generator. The similarity ends there. Heavy water is used to moderate the neutrons generated in the fission process. It is more effective at this than ordinary water, so that the core can operate with less concentrated fissionable material. The result is that the HWR can use unenriched or natural uranium. It is thus understandable why such countries as India, Korea, and Argentina have opted for this process. If U.S. utility companies decided in favor of HWRs, the readily available enrichment facilities would mean that HWRs would require only 60% of the uranium used in a LWR to produce an equal amount of electricity. Considering, however, that heavy water is far more expensive than ordinary water, economy is not all on the side of HWRs.

3. Gas Cooled Reactors

Reactors can be cooled and moderated by substances other than water. Both helium and liquid sodium have proved efficient. The high-temperature gas-cooled reactor (HTGR), Figure 17, is cooled by helium and moderated by graphite. Because helium is used in place of water, the reactor can operate at higher temperatures and lower pressure than the LWRs, which can be more efficient in generating electricity. In addition, it has several outstanding safety features. By not absorbing neutrons, helium remains nonradioactive as it conveys heat from the core. The use of graphite with its inherently high heat capacity reduces risk of core damage in case of an accident. Yet another compelling feature is the prestressed concrete reactor vessel which contains the entire primary coolant system. Concrete provides far more shielding from radioactive materials than does steel.

Since the development of the HTGR, only two have been built in the United States. Peach Bottom I, constructed by Philadelphia Electric Company, was the first. It was also the first nuclear plant to produce steam at 1000°F. Unusually high cost overruns forced the plant to shut down in 1974. In 1976, the Public Service Company of Colorado constructed the Fort St. Vrain reactor, which is still operating and by all accounts has functioned exceptionally well.

Modular High Temperature Gas-Cooled Reactors. Although the use of liquid sodium as a coolant avoids the multiple barrier systems required for water-cooled reactors, liquid sodium has the inherent disadvantage of reacting violently on contact with air or water.* To avoid this problem, a gas-cooled reactor with a radically different fuel element has been developed by the General Atomics Company of San Diego.

The unique features of their modular high temperature gas Cooled Reactor (MHTGR) is its ability to use helium, an inert gas, as a coolant

---

* Anyone who has poured a can of Drano pellets into a clogged drain knows this reaction well.

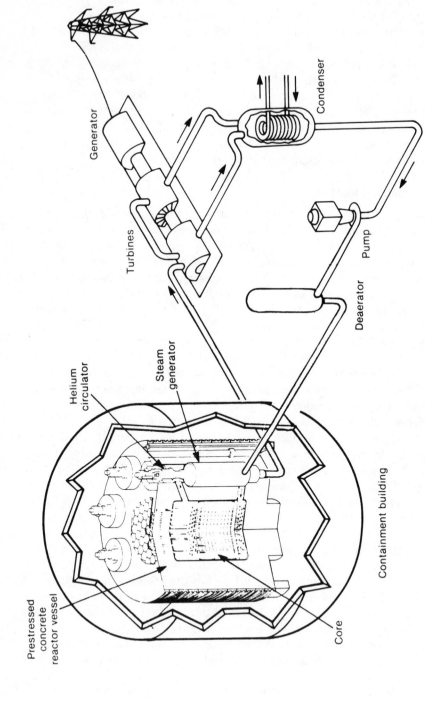

Figure 17. High-temperature gas-cooled reactor. Courtesy of the Office of Technology Assessment.

Helium circulator

Steam generator

Prestressed concrete reactor vessel

Core

Containment building

Generator

Turbines

Condenser

Deaerator

Pump

along with ceramic-coated uranium fuel pellets. General Atomics has designed these inherently safe features to avoid dependence upon engineered devices and human action, protecting the public while supplying needed electrical energy. Although the pellets' ceramic layers cannot melt, they can ultimately vaporize at temperatures above 3300°F (1800°C); well above the level encountered when a reactor loses all coolant. The fuel is actually composed of billions of tiny 1 mm diameter) grains of uranium oxycarbide–thorium oxide overlaid with a four-ply ceramic shell, effectively sealing the radioactive kernel from the environment. Figure 18a offers a cross-sectional view of a pellet. In this configuration, the uranium–thorium fuel is in a highly diluted form, which adds another dimension of safety.

Figure 18b shows one of the most recent designs for a 35-megawatt reactor. In this unit, helium flows downward toward the core, where it is heated by the ongoing chain reaction. The hot helium then passes through the inner cross duct and downward over the steam generator bundle, where its heat is transferred to water to make steam. The cooled gas then flows upward in an annulus between the steam generator and the vessel, is recompressed by the circulator, and driven into the annulus between the inner and outer cross duct. The cool gas entering the reactor vessel flows up through the annulus between the core and the reactor to the top of the core to complete the circuit.

The exceedingly compelling feature of the MHTGR is its relative lack of sensitivity to changes in fuel temperature, which should win it public confidence and acceptance.

4. Fast Breeder Reactors

Breeder reactors obtained their name because they were able to yield more energy from uranium than a LWR. Figure 19 shows the process of "breeding." There is a substantial difference between this and the chain reaction noted earlier in Figure 10. Take another look at the chain reaction. Where the LWRs depend on U-235 for fission, the breeding process depends on plutonium, an element that does not exist in nature: it cannot be mined. Plutonium is formed in all nuclear reactors employing uranium. When neutrons strike U-238, some neutrons are absorbed. When this occurs, a new atomic species, plutonium-239, is created. When neutrons strike Pu-239, a cascade of fission reactions occurs. Breeder reactors are designed to enhance the process, creating in the reaction more plutonium than is consumed.

Breeders also use liquid sodium as a coolant. Accordingly, this type of reactor is known as a liquid metal fast breeder reactor (LMFBR). Liquid sodium appears to have optimum heat transfer characteristics, thereby permitting the reactor to be operated at high temperatures and low pressures. This permits a more efficient conversion of heat to electricity. More efficient usually means less costly and better yields.

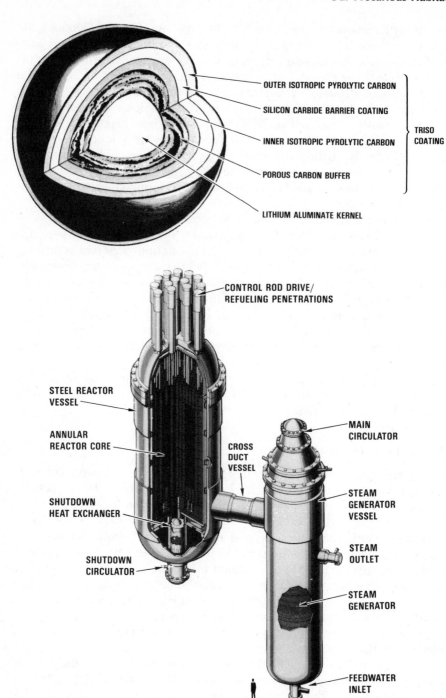

OUTER ISOTROPIC PYROLYTIC CARBON

SILICON CARBIDE BARRIER COATING

INNER ISOTROPIC PYROLYTIC CARBON

TRISO COATING

POROUS CARBON BUFFER

LITHIUM ALUMINATE KERNEL

CONTROL ROD DRIVE/ REFUELING PENETRATIONS

STEEL REACTOR VESSEL

ANNULAR REACTOR CORE

CROSS DUCT VESSEL

MAIN CIRCULATOR

STEAM GENERATOR VESSEL

STEAM OUTLET

SHUTDOWN HEAT EXCHANGER

SHUTDOWN CIRCULATOR

STEAM GENERATOR

FEEDWATER INLET

Figure 18. (a) Cross sectional view of a ceramic coated uranium pellet. (b) Recent design of a modular gas-cooled reactor.

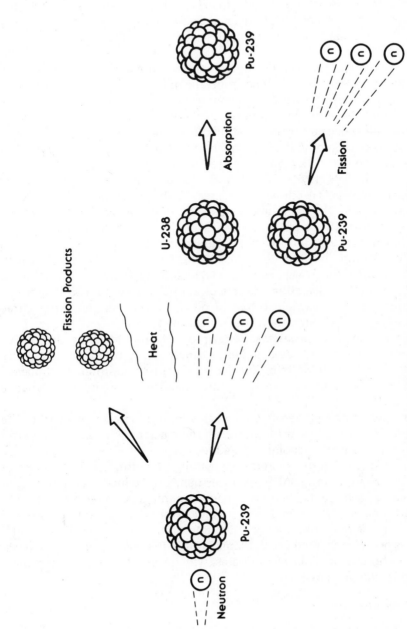

Figure 19. The breeding process of nuclear energy production.

Liquid sodium is pumped through the core and into a heat exchanger, where the heat from the core is transferred to a second sodium coolant system. It is this second system that is used to generate steam. As the sodium in the second system does not pass through the reactor's core, it does not become radioactive. In these features, it is reminiscent of the HTGR.

Given the fact that fast breeders do not require enriched uranium and that fuel can easily be obtained from stockpiles produced by enrichment plants, the federal government encouraged their development. Here was a reactor that abets the extraction from uranium of the considerable energy content left untapped by conventional reactors. Consequently, the major test of FBRs was to be the Clinch River breeder reactor, a prototype 375 MW power plant. Although authorized in 1970, funds were not forthcoming until the late 1970s. By 1983, $1.7 billion had been spent on site preparation near Oak Ridge, Tennessee. With the economic recession in full swing and the lame-duck status of its foremost backer, Senator Howard Baker (R., Tenn.), the death of the Clinch River reactor was foretold on May 12, 1983, when the House of Representatives voted 388 to 1 to deny funds beyond September 30th. On October 26th, the Senate virtually killed the project by voting 56 to 40 against providing additional funds. Although escalation of costs of nuclear power appeared to be the primary cause of the defeat, loss of public confidence following Three Mile Island and its aftermath was in fact the reason.

Reactors do have problems. Better products are indeed needed. Existing reactors are complicated and exceedingly expensive to build, far more so than comparable units in Europe and Japan. In addition, a number of the components are prone to failure, requiring lengthy maintenance shutdowns.

Among the design modifications being built into new plans will be the placement of the nuclear fuel lower in the reactor vessel, below the level of the pipes carrying coolant, to reduce the risk of pipe breakage which would result in an uncovered core leading to melting of the fuel. Pipes also are being redesigned with fewer bends to reduce accumulation of radioactive sediment. The new reactors are intended to be available 90% of the time. In the United States reactors are "down" for repairs and refueling over 30% of the time. Bear in mind that the life expectancy of a NPP is no more than 40 years. Plants built in the 1950s and 1960s are fast approaching their replacement dates, which can mean billions of dollars in replacement costs.

5. PIUS Reactors

In his book, *Atomic Energy: A New Start* (1), David Lilienthal[*] called on nuclear engineers to design reactors that were intrinsically or inherently

[*] Lilienthal was founding director and head of the Tennessee Valley Authority (TVA) and first chairman of the Atomic Energy Commission, the forerunner of the Nuclear Regulatory Commission (NRC).

safe—not as tall an order as might be suspected. He wrote, "I propose then a new start in nuclear research and development, aimed at finding and proving out safer and better methods . . . to provide the maximum possible protection against radiation hazards in the operation of the badly needed existing plants and those additional reactors that necessity and good sense may dictate that we complete." For Lilienthal nothing short of a fail-safe unit would restore the public's faith in energy derived from atomic nuclei. Three Mile Island had dealt nuclear energy an almost fatal blow, at least in the United States. Was it possible to design and construct a reactor that was absolutely safe? At first the idea was not taken seriously (2).

By 1982, however, the idea had been moved from design criteria to development by the Swedish nuclear firm ASEA/ATOM. They were building a PIUS (process inherent ultimately safe) reactor. The United States and Germany followed closely with a concept modeled on the HTGR concept. The goal is to ensure safety even if the reactor is subjected to human error, equipment failure, or natural disaster. As shown in Figure 20, PIUS reactor is a 500-MW PWR (500 megawatt pressurized water reactor) in which inherent safety is obtained by immersing the reactor core in a pool of cold borated water.

Two design objectives were set: ensure safe shutdown and adequate heat removal under any credible circumstance, and eliminate or avoid uncertainties associated with totally new designs. Thus, the LWR was used as a model.

The PIUS reactor is designed to shut down automatically as soon as the pool water begins to flow through the core. Added security is established by maintaining a high concentration of boron in the pool water. Boron is a strong neutron absorber and automatically interrupts a chain reaction. The entire system relies completely on thermohydraulic principles which makes it independent of electrical, mechanical, or human intervention. In addition, because the PIUS reactor does not require either earthquake proofing or a high-pressure containment, it can be built at a cost similar to a conventional PWR. However, for the foreseeable future, the benefits of a PIUS reactor may accrue only to European countries.

However, an "inherently safe" reactor may be available in the United States. Recent tests of an experimental system were successfully performed at the Department of Energy's Idaho National Engineering Laboratory at Idaho Falls.

According to American engineers, an inherently safe reactor would shut itself down without human intervention or back-up systems in the event of a breakdown similar to what occurred at TMI and Chernobyl. This would make the reactor invulnerable to the most feared of nuclear accidents involving runaway releases of radioactive material and, along with it, melting of the uranium fuel core.

The inherently safe reactor tested in Idaho is based on a wholly new concept. Called an integral fast reactor (IFR), it requires the use of liquid sodium in place of water as the cooling liquid. Liquid sodium boils at

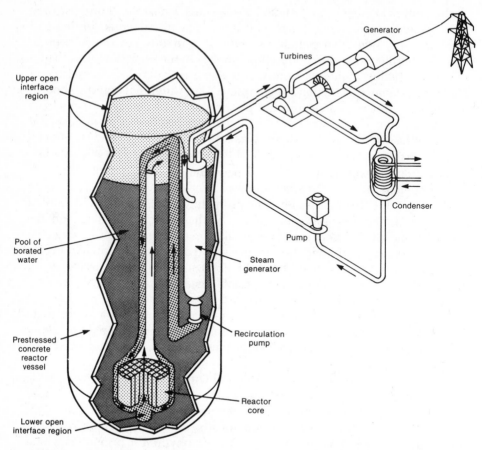

Figure 20. Cutaway rendition of a PIUS reactor. Courtesy of the Office of Technology Assessment.

1600°F (857°C), which permits its use at normal pressures. Since the vessel containing the reactor can therefore be much smaller, the entire system is submerged in a pool of liquid sodium (Figure 21a,b). Thus, all the potential cooling fluid, not just that flowing through the core, is available in the event of an accident. The liquid sodium carries the heat to water outside, where the steam to drive the turbines is produced—as in conventional systems.

Another change is the fuel itself. Where conventional reactors use uranium oxide, the IFR uses an alloy composed of 75% uranium, 15% plutonium, and 10% zirconium. And it utilizes a new fuel-processing technique that relies on electrorefining rather than chemical extraction. Since electrorefining requires smaller and simpler machinery, it will reduce the amount of radioactive waste. With a simpler processing system in the plant along with the reactor, risk of accident or theft during transportation is also reduced.

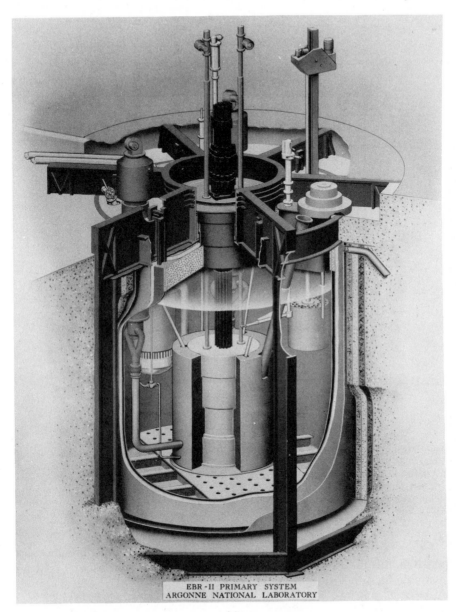

EBR-II PRIMARY SYSTEM
ARGONNE NATIONAL LABORATORY

(a)

Figure 21(a). The EBR-II reactor vessel is shown submerged under liquid sodium. The liquid metal concept exploits the inherent properties of liquid metal cooling and metallic (uranium) fuel to achieve both lower cost and inherent safety.

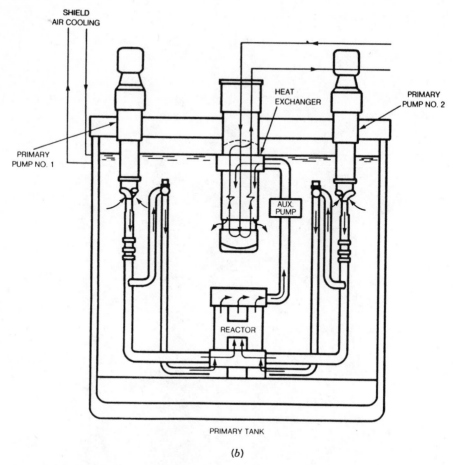

(b)

Figure 21(b). This cross-sectional sketch shows the internal mechanism of the ERB-II reactor vessel. Because of liquid sodium's high boiling point, the vessel does not need to be pressurized. Consequently large reactors are no longer necessary. Small reactors can be entirely submerged in the coolant thereby providing additional safety. The use of metallic fuel provides another safety feature, and reduces the amount of radioactive waste. (Courtesy of Argonne National Laboratory, Argonne, Ill.).

Other designs are being put forward and tested. When the generally negative attitude toward nuclear power dissipates, safe reactors will be available for generation of electricity.

II. DISPOSAL OF NUCLEAR POWER PLANT WASTES

In addition to heat, NPPs produce waste, both solid and liquid. Not only are radioactive wastes generated by reactor operations, but the entire cycle shown in Figure 22, from mining and milling of uranium ores, to uranium fuel fabrication plants, to fuel reprocessing plants and produc-

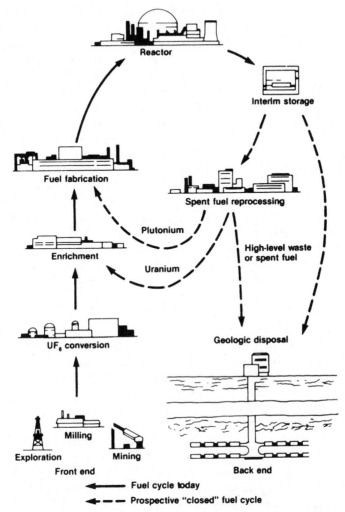

Reactor

Interim storage

Fuel fabrication

Spent fuel reprocessing

Plutonium

Enrichment

High-level waste
or spent fuel

Uranium

UF$_6$ conversion

Geologic disposal

Milling

Exploration   Mining

Front end   **Back end**

◀━━━━ Fuel cycle today
◀ ━ ━ ━ Prospective "closed" fuel cycle

Figure 22. Courtesy of the Office of Technology Assessment.

tion of isotopes, is a source of wastes. Because of their inherent capacity for harm, these wastes must be strictly controlled if people and the environment are to be appropriately protected. However, given the extraordinary life span of some of these radioactive materials, they are not typical wastes. Special care and concern must attend their disposal.

A. Low- and High-Level Wastes

Wastes are generally classified as low, medium, and high level, in relation to their radionuclide content, heat generation rates, and methods of treatment. Two distinct types of material are involved: (1) fission products, which emit penetrating radiation (see Figure 2) but which are relatively short-lived, and (2) the fuel materials, which are very long-lived and emit nonpenetrating alpha radiation.

Low-level wastes include a broad category of waste from operation and maintenance of nuclear power facilities, such as cleaning and cooling water, and wastes from use of radioisotopes in medical, agricultural, and industrial settings. Much of this waste is barely radioactive, while other wastes can require heavy shielding. Regardless of their radiation levels, locating disposal sites has proved difficult.

Originally, groups of states agreed to join compact coalitions—with the idea of reducing each of their chances of being selected as the disposal state. One such coalition had 11 states—New York, New Jersey, the six New England states, Pennsylvania, Delaware, and Maryland—joining to establish a single joint LLW disposal site. But when the July 1, 1984, deadline passed, only New Jersey, Connecticut, Maryland, and Delaware had signed, and they could not agree on which of the four wold be "it" and where the site would be. Obviously, foot dragging will be the rule and it will remain for the federal government to make the selection. Given the difficulty of siting incinerators, nursing homes, and drug rehabilitation programs, it will take more than a few meetings of governors to establish nuclear waste disposal sites.

Currently, three states—Washington (Hanford), South Carolina (Barnwell), and Nevada (Beatty)—maintain LLW disposal sites. But each has stated firmly that they will not receive waste from other states indefinitely. They have threatened to close their existing facilities, unless other sites are selected. In 1980, Congress mandated that new sites be opened by 1985. That has not occurred.

In August 1985, the state of Tennessee filed suit in Federal District Court seeking to enjoin the U.S. Energy Department from asking Congress for approval to store spent nuclear fuel in the state temporarily. The Energy Department told Governor Alexander that it intended to set up a temporary storage facility at one of three sites. Under the proposal, spent fuel rods from all nonmilitary nuclear power plants would be stored in concrete casks, with shipment to begin by 1995.

The state's response was that the Energy Department had violated federal law, requiring consultation with a potential host state in selecting a site for temporary storage of low-level waste (see Chapter 13), now being referred to as a monitored retrievable storage (MRS) facility.

Public resistance to disposal of radioactive waste recently received another setback. The citizens of South Dakota voted overwhelmingly to reject a plan that would have joined their state and North Dakota in an agreement permitting creation of a site to store low-level wastes from other states. The proposal would have allowed the two states to build a large disposal facility to handle waste from other states, or a smaller one to store waste generated from within their own borders. Both failed at the polls.

NIMBY, not in my backyard, is fast becoming a national slogan. However, a decision on storage of wastes cannot be put off indefinitely. A recent technological breakthrough may ease the decisionmaking process.

Leaking, decaying tanks of radioactive waste are being processed in a system the DOE and Westinghouse Electric Corporation anticipate will be a model for stabilizing and storing LLW. As shown in Figure 23, solid–liquid mixtures are separated at a site in West Valley, New York, 30 mi southeast of Buffalo, and sealed and isolated in cement blocks and borosilicate glass logs. In these cemented glass repositories, the radioactive waste cannot become environmental contaminants. These blocks to be stored locally have a radiation level of 35 mrem/h (350 μSv/h or 0.35 mSv/h) on their exterior—an amount the average person contacts from the natural background in 2 months. The glass logs are far "hotter." Their radiation level is approximately 5000 rem/h (50 Sv/h). This means that a lethal dose can be obtained in less than 10 min of contact. Once the logs are prepared, they will be transported to the national repository for deep storage. Although no one can offer guarantees of absolute safety, safer practices and procedures may ease the location problem somewhat.

Three principles have governed the management of radioactive waste:

1. Dilute and dispense radionuclide-containing wastes to the environment in amounts below authorized protection limits.

2. Delay and decay those wastes that contain only short-lived radionuclides.

3. Concentrate and confine those wastes that contain significant amounts of long-lived radionuclides.

Figure 24 shows the concept of relative levels of underground burial for the three classes of waste. The depth for high-level wastes (HLW) in both the United States and Europe varies from 1500 to 3000 ft (500–1000 m).

Although HLWs refer primarily to the spent fuel rods, any waste that is highly radioactive and generates large amounts of heat is high-level waste. LLWs are those materials that contain less than 1 curie of activity per cubic foot, or less than 10 nonocuries of TRU (transuranic waste) per gram (about 30 μCi/ft^3). Low-level wastes include such things as contaminated tools, machinery, and paper.

University research facilitates, hospitals, and industrial settings produce substantial amounts of LLWs. The diagnosis of heart disease, treatments for cancer, biomedical research, testing bridge supports and ship hulls, monitoring air pollution, and, of course, generating electricity all produce LLWs.

Under the Low Level Radioactive Waste Policy Act of 1980, each state was to assume responsibility by January 1, 1986, for the civilian (as opposed to military) radioactive wastes produced within its borders. That

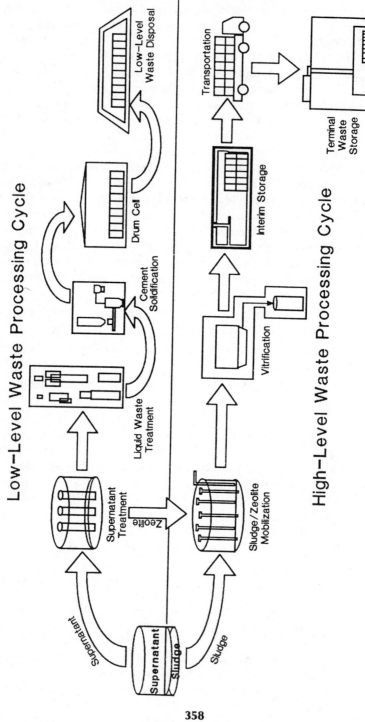

Low–Level Waste Processing Cycle

Supernatant Treatment

Liquid Waste Treatment

Cement Solidification

Drum Cell

Low–Level Waste Disposal

Supernatant

Supernatant

Sludge

Zeolite

Sludge

Sludge/Zeolite Mobilization

Vitrification

Interim Storage

Transportation

Terminal Waste Storage

High–Level Waste Processing Cycle

Figure 23. Process overview of both low-level and high-level waste processing cycles.

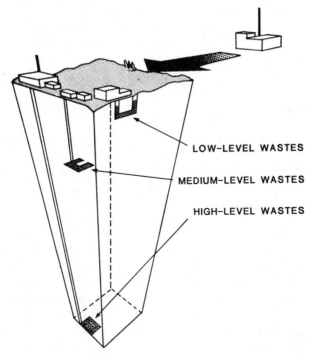

LOW–LEVEL WASTES

MEDIUM–LEVEL WASTES

HIGH–LEVEL WASTES

Figure 24. Diagrammatic representation of the containment of the three classes of radioactive waste.

policy has still to be implemented. Low-level wastes have been dealt with via dilution and dispersion, burial at sea, and burial in underground pits.* Recall that LLWs usually consist of short-lived isotopes and are thus of little real concern to human health. The problem most difficult to deal with is the HLWs with their extremely long-lived radionuclides.

As stated earlier, the primary purpose of any radioactive waste management program is to protect current and future generations from radiation exposure. The basic waste management scheme is: (1) spent fuel rods are held on the reactor site for about 6 months to allow the short-lived nuclides to decay and heat to dissipate; (2) fuel rods are sent for reprocessing; (3) solidification and consolidation of the high-level liquid wastes is accomplished; and (4) final disposal is achieved.

B. Near Field Considerations

In assessing a potential geological structure for its waste disposal capabilities, it is necessary to consider the characteristics of all components involved. One of the major considerations is the "near field."

* See Chapter 13, Hazardous and Solid Wastes, for a more detailed discussion of treatment of LLWs.

The near field is currently understood (by the international community concerned with waste disposal) to mean the excavated repository, including the waste package, filling or sealing materials, and those parts of the "host" or final site whose characteristics could possibly be altered by reaction with the resident contents.

The waste package includes the form of the waste, its container, and any additional barriers so long as they are an integral part of the "package."

The repository is the underground facility in which the waste packages will be stored. It includes any engineered barriers that separate the waste from host rock, as well as structural materials such as the lining of all tunnels and galleries along with materials used for backfilling the final sealing.

Concern for the host rock is actually directed at the physical–chemical properties of the medium that may be or have been modified by the repository. The actual extent of this potentially alterable area can vary with the type of waste, design of the site, and the characteristic of the rock itself. The type of rock currently under consideration as potential "hosts" for long-lived radioactive waste include granite, basalt, rock salt, and such argillaceous rocks as clays, shales, mud, and silt stones along with marls.

The waste package, the next element of the near field, involves complex methods of concentrating nuclear wastes. In the late 1970s President Carter indefinitely suspended the reprocessing of nuclear fuel. This means that spent fuel rods are piling up in onsite spent fuel pools. Rods that have been reprocessed have left large volumes of liquid and acidic waste. These liquids must be solidified. Currently, several processes are being studied. Calcination, vitrification, ceramic and supercalcine ceramics, and synthetic rocks (synroc) all appear promising.

Calcination is a process that atomizes and dries the liquid waste, leaving a highly radioactive granular substance. This calcined material may be stored for future processing. Virtrification is a process by which the calcine is mixed with a molten borosilicate glass and recooled. In the molten state the glass may be molded or poured into a canister for final storage. This technology is the most advanced at this time. Although there is some question about the integrity of the glass at high temperatures and pressures, it still remains a very promising technology. Crystalline and supercalcine ceramics are two additional alternatives being researched. These processes involve combining the waste with ceramics and storing the resulting mix. It has been found that some of the radioactive materials will leach into brine. As the radioactive material decays, it could disrupt the crystal structure of the ceramic and leave the "mix." Synroc is a process by which the radioactive materials are combined with certain minerals, melted, and cooled to form a synthetic rock. This process requires that the waste be partitioned into its constituents so that the final "rock" will be geochemically stable. Synroc is stable in fresh water but has problems when immersed in brine.

Once the wastes are solidified and packaged, they are ready for transport. The wastes are placed in canisters; the typical canister will hold about 20 million curies. the canister is similar in size and shape to compressed gas cylinders except that the canisters have much thicker stainless steel walls. Several of these canisters are placed in a cask, which will weigh anywhere from 20 to 100 tons. The cask is heavily shielded and extremely strong. Each has cooling fins and is clad with 1.5–2.5 in. of stainless steel. The casks are then put through a tortuous series of tests. There is extremely little risk in the transport of these wastes.

C. Deep Storage

Throughout the history of radioactive waste management, several ideas for final deposition have been suggested. Surface storage, ice-bed storage, and space disposal, all of which have major difficulties. The current thrust of research is on long-term geologic storage. This entails storage in stable geological formations. Areas such as salt domes, bedded salt, basalt, tuff, and crystalline rocks have been considered. Research started with bedded salt in the 1950s, salt domes in the 1970s, tuff in 1977, and basalt and crystalline rocks in 1978. Geologic disposal requires that certain conditions be met. First, limited or no seismic activity; second, protection of the groundwater; and third, protection from human intrusion, either deliberate or accidental. Thus, an arid area with a low water table, seismically stable, and free of any possible mineral deposits is required. The depository must be able to last longer than present human watchfulness.

The above is a brief and by no means complete description of radioactive waste management. It is a brief summary to familiarize the reader with the issue of radioactive waste disposal only.

Clearly, the northeastern states are facing the most difficult problem in locating suitable disposal sites. NIMBY is at work here and appears to be the attitude of most communities. Since 1957, 10,000 metric tons of spent fuel have accumulated at nuclear power facilities. By the year 2000 that is expected to double. The Nuclear Waste Policy Act of 1982 required that the Department of Energy (DOE) select nine possible disposal sites, rank the top three, and present them to the President. The nine sites selected (Figure 25) were in six states. The site in Louisiana is a salt dome; two sites in Mississippi are salt domes; one site in Nevada is a geological medium called tuff, a compacted volcanic ash; the two Texas sites are imbedded salt, as are the two Utah sites. The Washington site is in basalt, a fine-grained rock formed by the solidification of lava.

In January 1988, Congress chose Nevada's Yucca Mountain as the site of the nation's first high-level nuclear waste repository. Yucca Mountain is a 6-mi long ridge that rises 1000 ft above a remote Nevada desert. Over the next 7 years, the DOE and the U.S. Geological Survey will study the site. If they conclude that it is suitable, drilling the 1500-acre grid of

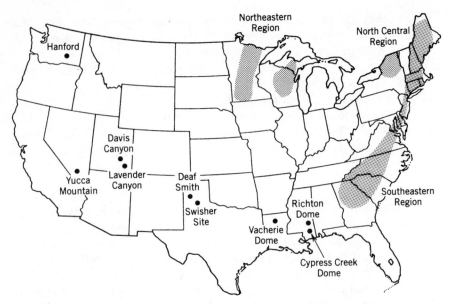

Figure 25. Places studied are now being considered for permanent storage of high-level nuclear waste. Courtesy of the Office of Civilian Radioactive Waste Management, U.S. Department of Energy.

tunnels into the mountain will begin in 1998. Five years later, waste-filled canisters will be placed in the tunnels. Once the repository is filled with some 70,000 metric tons, plans call for it to be permanently sealed by the year 2038.

Interestingly enough, and seemingly with far less political strife, the country's first deep burial site for spent military radioactive waste was to open in 1989 at Carlsbad, New Mexico. Called the Waste Isolation Pilot Plant, it was to receive transuranic waste from reprocessing facilities, clothing, tools, glass, and plutonium-contaminated sludges that have been stacked aboveground for years at the Idaho National Engineering Laboratory.

The Carlsbad site is a bedded salt basin laid down 200 million years ago, during the Permian period. Guidelines for selecting a site, in addition to those of the near field, include geologic capability of storing at least 70,000 metric tons of waste for 10,000 years—the time it takes for several longer-lived radioactive species to decay to harmless levels.

Suddenly, in September 1988, the Department of Energy announced that it could not prove that the $700 million facility could be operated safely. Four months before the scheduled opening, they announced an indefinite postponement. federal officials had hoped that the WIPP would show that they had the capability of managing a large radioactive waste site. The site, 2150 ft below the desert floor, and 26 mi east of Carlsbad, had been chosen after extensive surveys and test drillings. However, scientists at the University of New Mexico questioned the amount of moisture seeping into the underground chambers and galleries—which might mix with salt and form brine which could eventually corrode the waste storage drums. Representative Synar (Dem. Oklahoma) noted that the Department of Energy had an enormous amount of work to do before Congress would approve a bill authorizing use of the pilot plant.

D. The Legacy of Three Mile Island

Fear, ignorance, and loss of confidence are at the heart of the controversy over nuclear energy: fear of the type of damaging effects associated with Hiroshima and Nagasaki, the results of exploding atomic bombs over these cities; ignorance of the fundamentals of science and technology, especially radioactivity; and loss of confidence in political leaders, as well as an almost total loss of credibility in any statements emanating from corporate boardrooms or governmental regulatory agencies. As a consequence, the question posed in Figure 26 is as pertinent and germane today as it was when it first appeared 15 years ago. In fact, the accident at TMI has made matters a good deal worse. Never mind that by 1988 we had over 30 years of experience without a melt down and without a single death attributable to radiation. These powerful, positive facts, for all the good they do, simply do not exist for those who fear "the bomb." Indeed, NPPs are viewed as part of an arsenal of atomic weapons in armories around the world. Until these two concepts are separated, "good" statistics are as effective as sand castles against an incoming tide.

The most spectacular nuclear reactor accident did not occur at TMI. It did not occur in the United States, and it was not all that recent. Actually, the accident at Windscale, in the north of England, in October 1957, was far more hazardous. Over the course of 3 days a large amount of radioactivity in the form of the fission product iodine-131 was deposited over much of England and part of Europe (3). At TMI, there was only the possibility of a release of radioactivity.[*] Yet the reactions were far different, primarily because of the forthrightness of officials at all levels.

---

[*] When operators lost control of the unit 2 reactor, it was feared that Pennsylvania and possibly the United States generally might be blanketed by radioactive material. This did not occur. The reactor remained intact, even though as we now know the fuel was so hot it melted through an inner containment wall. Unknown to the operators, the water in the reactor drained away, exposing the upper half of the 177 uranium-dioxide-filled tubes, which then melted but did not breach the final containment. Perhaps the most important finding was the fate of the estimated 66 million curies of ^{131}I known to be in the reactor when it went out of control.

Unlike Chernobyl and Windscale, at TMI there was no major release of ^{131}I. Only 14–15 Ci (550 Gbq) escaped. Virtually all the ^{131}I had decayed away. Switching to the new system of SI units carries with it another problem. The numbers are higher—15 Ci becomes 550 Gbq, and numbers have a mystique all their own. For most people, the higher the numbers, the bigger the problem.

Of particular interest is the fact that the Japanese government has contributed $20,000,000 toward the clean-up of TMI in order to learn as much as they can about operations and management, now that they have made a commitment to build nuclear power plants.

'Well, do you or don't you want me ?'

Figure 26. Courtesy of Borgstedt, from the Philadelphia *Evening Bulletin*.

Still, Americans have lost trust in their leaders. In the Kemeny Report (4), following the accident at TMI, it was not the reactor design or construction that came under attack. Rather, the report scorched, and rightly so, all levels of management for the cavalier attitude with which they treated a nuclear plant, believing that it was no more than an ordinary manufacturing facility. That will have to change, and the message of such change will have to filter down to the grass roots before the public will give the question, "Well, do you want me?", the affirmative answer it deserves.

E. TMI—A Final Report: Nearly

On September 3, 1985, 6 years after the accident at Three Mile Island, State (of Pennsylvania) Health Secretary H. Arnold Muller announced that the Pennsylvania Department of Health, after a detailed epidemiologic study, had found no evidence of increased cancer among residents living within a 20-mi radius of the nuclear power plant.

He also iterated that there was no indication of increased new cases of leukemia, which has a shorter latency period than cancers of many other organ systems.

Actually, the study covered cancer deaths recorded from January 1974 through December 1983, including cases found in four areas "downwind of the TMI plant." Remarking on allegations of increased deaths since the accident, Muller noted that the Health Department found cancer deaths that had been diagnosed prior to the accident, as well as among long-term heavy smokers who had died of lung cancer, and among people with cancer who had moved into the area after the accident, or who were mistakenly counted as having lived in the area when in fact they did not.

The Health Department study identified a total of 154 cancer deaths since the accident in an area of 25,000 residents downwind of the plant. Calculations arrived at an estimated expected number (if TMI had never occurred) of 152.5.

For the area within a 10-mi radius of TMI, they found that 2892 cancer deaths occurred since 1979—slightly *less* than the 2909 expected. In a 20-mi radius, 7924 deaths were recorded where 8177 would have been expected. "The Department," Muller said, "focused its survey of new cases on four downwind communities nearest the facility, because of the potential importance of wind direction and radioactive plume dispersion during the early days of the accident."

Of perhaps overriding importance was the additional finding of four cancers among 3582 women who were pregnant at the time of the accident, or who became pregnant shortly thereafter. These women were living within a 10-mi radius of the plant at the time. The number diagnosed, four, was almost identical to the 3.9 expected.

As the first scientific analysis of cancer among area residents, it will be a milestone as well as a touchstone or lightening rod, drawing all manner of criticism. Be that as it may, substantive data are not available for examination and discussion. Given the wide discussion it needs and most assuredly deserves, a number of unfounded dismal anecdotes can be laid to rest.

F. Man and Mouse

The XVI International Congress of Genetics may mark a turning point in our understanding of the adverse effects of radiation to human health. The operative term here is human.

At that meeting held in Toronto during August 1988, James Neel of the University of Michigan's School of Medicine told the audience of scientists that the genetic consequences of radiation may not be as extensive as it was thought to be.

Neel has studied the survivors—and their children—of Hiroshima and Nagasake for over 40 years. His most recent analysis compared indicators of genetic damage in children of men and women exposed to radiation from bomb blasts, with children of comparable, unexposed individuals. The indicators measured were congenital malformations, stillbirths, and newborn deaths, pediatric cancers, chromosomal defects, and chemical alterations that could have occurred as a consequence of radiation-induced gene mutation. Neel reported no significant differences between the two groups for any of these markers.

Perhaps his most pertinent finding was that the "doubling dose"—the level of radiation required to produce numbers of mutations equal to those arising spontaneously in people, which they often do—is in the range of 145–255 rem, four times higher than the doubling dose projected from studies in mice. This can have enormous implications for us. Neel now believes that human beings are not more sensitive to radiation than mice and may well be less sensitive and may have developed more effective protective mechanisms than other mammals.

In 1982, the Nuclear Regulatory Commission estimated that people living in the vicinity of radioactive waste disposal sites might be exposed to levels of radiation on the order of 0.003 mrem/yr. Public pressure moved several states to enact more stringent requirements—lower levels—carrying with it greater economic burdens to achieve. Given Neel's new data, the NRC's estimates may be more than adequate protection.

G. Chernobyl

On April 26, 1986, the unit 4 reactor at the Chernobyl nuclear power plant in the Ukranian Soviet Socialist Republic exploded. The accident did more than destroy the reactor. It significantly damaged public confidence in the use of nuclear power. Figure 27 is a photograph from afar of the power plant at Chernobyl, and Figure 28 presents diagrammatic rendi- tions of units 3 and 4 before and after the explosion.

Commenting on the aftermath, Marvin Goldman (Laboratory for Energy-related Health Research, University of California) stated it suc- cinctly. "The cloud of radioactive material that was ejected from the reactor blanketed much of the Northern Hemisphere and precipitated a fallout of small radiation doses and large societal apprehensions" (5). With all that has been written about Chernobyl that pithy remark sums it up admirably: small amounts of radiation, but large doses of public suspicion and unease.

Writing in The New York Times, Serge Schmemann asserted that "the worst fears that seized Europeans after the Chernobyl power plant in the Ukraine ruptured and sent a radioactive plume swirling across the Continent have not been realized." And he continued, "a study by the European Community (EC) estimated that over the next 50 years, Cher- nobyl was likely to be responsible for about 1,000 extra cancer deaths, a

Figure 27. Power plant at Chernobyl.

statistically negligible addition to the 30 million deaths from 'natural' cancers projected over that same time in the 12-member Community's population of 315 million. The study also concluded that the average individual radiation dose caused by Chernobyl in the entire EC would be considerably less over 50 years than the effective average dose in one year from natural background radiation" (6).

The complex meteorological conditions, the local dispersion of rain clouds, and the varying characteristics of the release led to an intricate pattern of atmospheric transport and deposition of radioactive releases on the ground, both within the Soviet Union and in other countries. Figure 29 pinpoints the location of the power plant and indicates when increased levels of radioactivity were reported across Europe.

Shortly after the accident, concern developed to avoid the uptake of radioiodine—specifically ^{131}I—by the thyroid gland, where it concentrates, primarily from consumption of milk and leafy vegetables. Because of its short half-life—8 days—protective measures are required for a brief period. Strontium-89 with a half-life of 50 days requires additional concern.

However, from a strict radiation protection point of view, radiocesium and radiostrontium are the most difficult contaminants. Cesium-137 has a half-life of approximately 30 years, and strontium-90, 29 years. They contribute to human radiation exposure (dose) in two ways: externally from contaminated soil and surfaces, and internally by consumption of contaminated food. Taken internally, cesium distributes itself throughout the soft tissues of the body, while strontium seeks out bone. Figure 30 displays the major pathways by which radionuclides reach people as a consequence of uncontrolled releases of radioactivity.

In the United States, the Environmental Radiation Ambient Monitoring

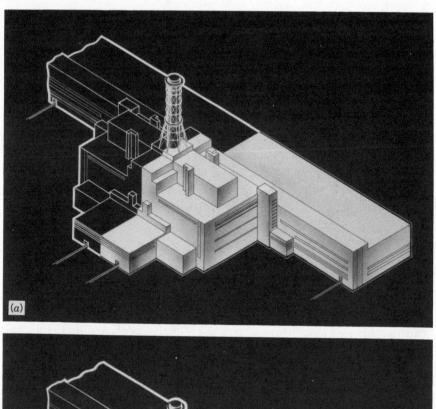

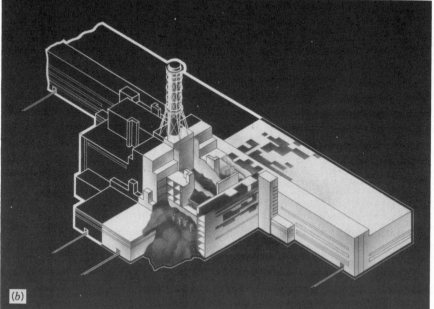

Figure 28. (a) General view of the Chernobyl nuclear power plant, units 3 and 4, before the accident. (b) General view of units 3 and 4 after the accident.

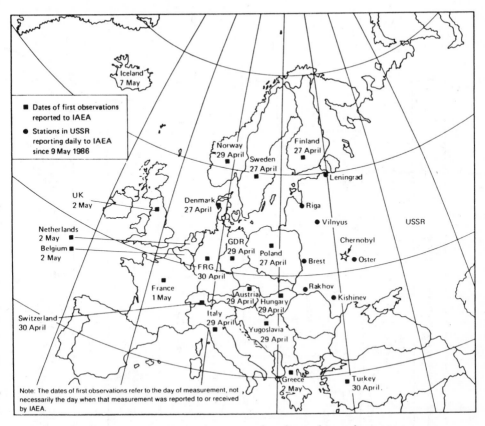

Figure 29. Dates when radiation was first detected in each country.

System (ERAMS) was organized in 1973 by the U.S. EPA. One of its mandates was to respond to accidental releases of radioactivity to the environment. Samples of milk are collected at some 268 locations in the United States and its territories and forwarded for analysis to the Eastern Environmental Radiation Facility in Montgomery, Alabama.

Since the inception of the nationwide monitoring program in the 1960s (preceding ERAMS), several events of international consequence have been closely attended. Several of these events included episodes of radioactive fallout during atmospheric nuclear weapons testing by both the Chinese and Americans in the 1950s, 1960s, and 1970s, continuing through the accidents at Three Mile Island and Chernobyl.

Figure 31, 32, and 33 show the levels (in becquerels per liter, which have replaced the curie)* of strontium-90, cesium-137, and iodine-131 in our pasteurized milk supply over the 25-year period 1963–1987. Several

---

* 1 pCi (picocurie) = 37 milliBequerels (mB); 27 pCi = 1 Bq.

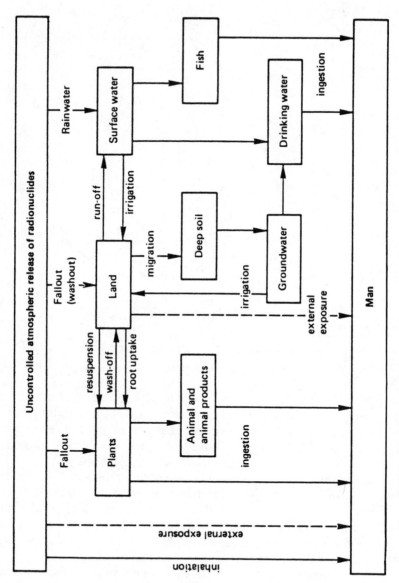

Figure 30. Major pathways of radionuclides to humans due to uncontrolled release of radioactivity. Reproduced from *IAEA Bulletin*, Autumn 1986.

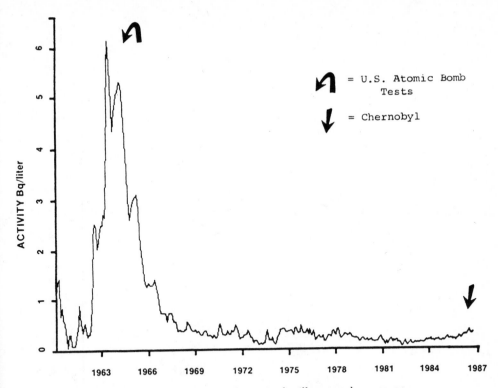

Figure 31. Cesium-137 in pasteurized milk: network averages.

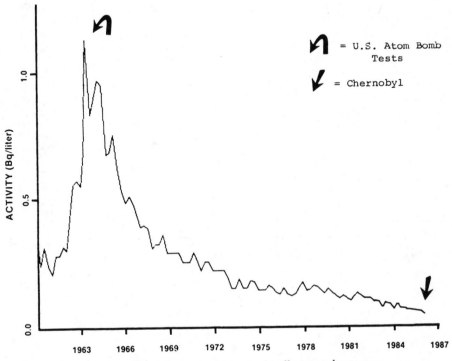

Figure 32. Strontium-90 in pasteurized milk: network averages.

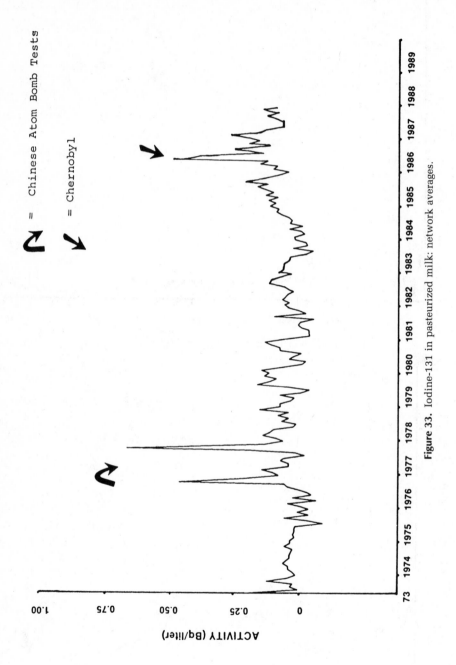

Figure 33. Iodine-131 in pasteurized milk: network averages.

372

facts are clear. Strontium and cesium have declined steadily, and iodine has remained stable at low levels, since the atmospheric tests of the 1960s. Milk has contributed as much as 50% of the total dietary intake of strontium during the periods of high fallout.

Charles R. Porter of the Office of Environmental Programs, EERA, reported that following the Chernobyl episode, the average values of ^{137}Cs in milk increased from 0.1 to 0.3 Bq/L and that Chernobyl added very little to the total. With respect to Three Mile Island, there appears to have been no detectable increased activity, and Chernobyl's radioactive cloud had no adverse effect on either the milk or food supply (7, 8). Note the far right portion of the trend line in Figure 30. It tells us that between May 1986 and April 1987 in the United States the integrated cesium-137 activity was estimated at 50 TBq of 6.2 Bq/m^2. Similarly, with strontium-90 the level is at the lower limits of detection. And, whereas iodine-131 rose to 0.5 Bq/L it rapidly declined to 0.25.

The capability displayed by ERAM is to be applauded. They demonstrated that they could mobilize their extensive network within 24 h to monitor specific sources. Their data should be a source of confidence for those anxious about the safety of the food supply. More about Chernobyl in Chapter 15.

III.   Isotopes for Humanity

Nuclear reactors do more than produce heat and neutrons. Unfortunately, most people are barely aware of the benefits reaped daily from the radioactive materials, the isotopes, generated in nuclear reactors. Over the past 30 years, radioisotopes and controlled radiation have been used, and continue to be used, to:

- Improve the yield of farm crops
- Preserve food
- Determine groundwater resources
- Sterilize medical supplies
- Monitor hormonal activity
- X-ray pipelines
- Control industrial processes
- Track environmental pollution
- Increase meat production
- Prevent animal diseases
- Determine the fate of pesticides
- Protect citrus crops from insects

- Diagnose human illness
- Treat human and animal diseases
- Enhance biological nitrogen fixation

These activities and benefits are largely taken for granted, if they are realized at all. Although some appear aware of the use of medically related diagnostic and therapeutic procedures containing radionuclides, these uses are not associated with nuclear reactors. Nevertheless, it is the reactor that makes these uses possible.

With the advent of nuclear reactors, it became possible to make large amounts of radioactive material at low cost. However, radioactive isotopes were available and being used at the dawn of the 20th century.

Hungarian-born, Nobel Prize laureate George Charles de Hevesy was the first to make use of radioactive material as a "tracer." When Hevesy was a young assistant in Ernest Rutherford's laboratory at the University of Manchester (England), Rutherford assigned him the onerous task of separating radium-D from lead. He failed. Isotopes are not chemically separable. Radium-D is the fourth decay product of radium and as it turned out is in fact lead-210. In the process, however, he became skilled in using radioactive materials. He used lead-210 as a tracer, studying the process of lead uptake and distribution in bean plants.

At the time, 1913, he had little money. Most of his meals were eaten with his testy landlady. The story is told that he became appalled by the repetitiveness of the food she served and suspected that leftovers were being served week after week. In order to test his suspicion, Hevesy incorporated a trace of his radioactive radium-D into the leftovers on his plate. When the same dish reappeared several days later, he used a gold leaf electroscope (which measures electric charge) and obtained a positive response. His landlady did not appreciate his detective work and invited him to find new lodgings.

Today, radioisotopes were still used as tracers but now scientists are able to determine where atoms go in a biological process, and what role they play in living tissue. The spectrum of uses to which radioactive materials have been put goes well beyond the 15 areas noted above. Hardly an aspect of life has not benefited from their use.

A. Medical Devices and Diagnostic Tools

The commercial use of gamma radiation to sterilize* medical devices began in the 1960s. Earlier, the accepted means of sterilization employed either steam under pressure or chemical fumigants such as ethylene oxide

---

* Sterilization is an absolute term. It means the complete removal of all living things, in this instance, microbes: bacteria, fungi, and viruses. Sterilization and disinfection are not synonymous. Disinfection refers to a reduction in the number of microbes, not total removal.

Table 2
Radiation-Sterilized Single-Use Medical Devices

Hyperdermic syringes	Burn dressings
Needles	Intravenous solution tubing sets
Scalpels	Rubber and plastic sheets
Surgeon's gloves	Absorbable sutures
Gowns	Irrigation sets
Blood collection tubes	Medicated tapes and powders
Cotton	Heart valves and other prostheses
Drainage bags	

(ETO). More recently, irradiation sterilization of single-use medical devices has developed into a major commercial application of gamma rays. Thirty percent of all single-use items used in the United States and Canada are so sterilized.

The growth of radiation sterilization has come at the expense of ethylene oxide, which has come under increasing criticism by both the EPA and FDA as being mutagenic and possibly carcinogenic. The growth of radiation sterilization is attributable to such advantages as speed and nontoxicity. Unlike chemicals, gamma radiation is insensitive to temperature, pressure, and humidity so that the only variable requiring control is the time of exposure of the product to the cobalt-60 source.[†] In addition, gamma rays can penetrate all forms of packaging material, including sealed glass and metal containers, and is able to sterilize the product uniformly without leaving chemical residues or raising the temperature of the product. Finally, gamma radiation, like x-rays, is not itself radioactive. Substances exposed to it cannot become radioactive. Table 2 lists a sampling of products sterilized by gamma rays.

Radioisotopes are an essential part of current clinical diagnostic procedures. Improved instrumentation and technology, as well as new radiopharmaceuticals, have broadened the range of possible applications. Notable among these are improved imaging devices coupled to computers. With these devices, not only can the spatial distribution of the radionuclide in an organ be assessed, but also the dynamic process occurring in the organ itself.

Another example of the use of nuclear medicine diagnostic procedures it the thallium-201 scan used in the diagnosis of chest pain. The problem with certain types of chest pain is to determine whether a person has disease of the coronary arteries. The procedure has several roles: selecting appropriate patients for coronary artery surgery, assessing the adequacy of

---

[†] Cobalt-60 is produced by placing cobalt-59 (the naturally occurring nonradioactive isotope) in a nuclear reactor to absorb a neutron and become cobalt-60. The strength of ^{60}Co is reduced through decay at a determinable rate so that it must be replaced from time to time to maintain its activity. ^{60}Co has a useful life of approximately 15 years.

medical treatment, and helping to increase understanding of the causes of the problem.

Using a scintillation camera, both regional ventilation and perfusion of the lungs can be quantified. Patients inhale xenon-133 gas and from the computerized data ventilation is assessed. For perfusion studies, patients receive an intravenous injection of technetium-99-labeled albumin microspheres. These are transported through the heart and lungs depending on the condition of flow. Both ventilation and perfusion images can be combined into a single image which expresses the ratio between ventilation and perfusion. Using a color system, multiple functions and spatial relations can be seen in the same image. Figure 34, a brain scan, displays the concentration of a radioisotope. This scan, superimposed on an x-ray taken at the same time, is of a normal brain. The distribution and level of the isotope can then be compared with a similar scan of a patient with a dysfunction. The thyroid, liver, kidneys, and lungs have all profited from application of radiation techniques.

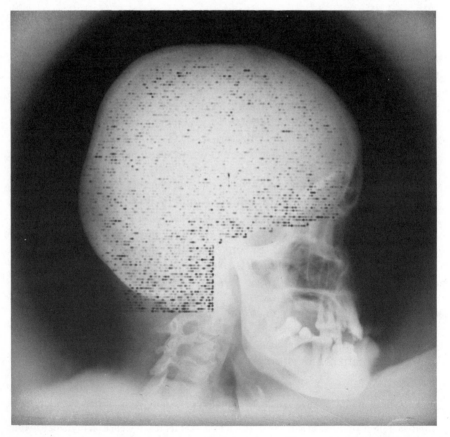

Figure 34. A brain scan is superimposed on an x-ray. The spatial distribution of the radionuclide is evident.

B. Pest Control

Insects cause losses estimated at between 10 and 20% of the total production of crops and livestock around the world. To substantially reduce a portion of this loss, emphasis is being placed on the sterile male or sterile insect technique (SIT). This involves the radiation sterilization and release of large numbers of insects of the target species into the area where control is desired. There, the sterile males mate with the fertile wild insects. The results of this mating produce no progency. The technique is therefore a highly specific form of birth control.

The Mediterranean fruit fly, *Ceratitis capitata*, commonly known as the "medfly," is one of the most devastating pests of citrus, stonefruits, and certain vegetables around the world. Countries where this insect pest does not exist refuse to import produce from areas where it does. Japan's recent refusal to accept citrus from California paralyzed the industry and all but created an international incident.

The development of the SIT has reaped large benefits for a number of countries. When the medfly invaded Mexico and Guatemala, the Medfly Eradication Program was activated. A mass rearing factory at Tapachula currently produces over 500 million medflies per week. About half of these are male. Eradication is proceeding and has prevented losses estimated at $550 million.

The program is currently under way in Guatemala, and a major effort is in progress in Egypt. The objective there is to construct a mass rearing facility capable of irradiating and sterilizing a billion flies per week.

Nagana, cattle trypanasomiasis, a tsetse-fly-transmitted disease, has been a major deterrent to agricultural development in Nigeria. In a cooperative project between the International Atomic Energy Agency, a WHO affiliate, and FAO (Food and Agriculture Organization of the United Nations), the SIT was employed to help eliminate *Glossina palpales palpalis.*

Raising millions of tsetse flies to maturity had never been done before. This meant that researchers had to learn not only the diet preferred by the flies, but how they eat. Of particular importance is their feeding on animals and people. Which meant that scientists had to simulate skin with appropriate texture and odor.

Over the past 10 years these problems have been solved and 1500 km^2 on north-central Nigeria, heretofore closed because of the tsetse fly, have been opened to agriculture.

C. Rusitec the Cow

The rumen is an important part of the digestive tract of ruminant animals such as cows, sheep, goats, and buffalo. It contains large numbers of bacteria whose function is to metabolize such fibrous foods as grass and straw and convert them to nutrients that can be used by the animal to produce meat, milk, wool, or energy for work.

J. W. Czerkawski of the Hannah Research Institute in Scotland has developed an artificial cow—Rusitec, rumen *simulation* technique—to analyze the contribution of different feed stuffs. As Rusitec "chews" its way through different fields, radioactive isotope tracings are used to compare digestibility in the artificial rumen, which consists of flasks, bottles, and jars. Data from these experiments provide information that suggests dietary improvements.

Tagging antibodies unique to the parasitic water-borne schistosome enables detection of antigens in either the serum or urine of infected individuals. Having this ability permits greater flexibility in field studies as urine specimens are so much easier to obtain then blood samples.

D. Bore Hole Logging

In oil and mineral exploration, drilling is required to determine if the ore content is worth mining. Drilling thousands of holds is expensive. To cut costs of shipment of samples to laboratories as well as their analysis, x-ray fluorescence analyzers can be used on the spot. A particularly useful approach obtains an analysis of the surrounding rock by running a probe through the hole. The advantage of using nuclear radiation is that it is highly penetrating and information can be obtained in water-filled boreholes and a large volume of rock can be analyzed quickly.

Engineers are using isotopes to solve hydrological problems dealing with water resources, sediments, transport, and infiltration.

The uses of radioactive materials are limited only by the imagination.

E. Food Preservation

For hundreds of years great effort has been expended on ways of protecting food supplies from the depredation of insects, microorganisms, and inherent chemical changes which produce food unfit to eat.

Realizing that an "army travels on its stomach," Napoleon offered a rich prize to anyone who found a method of preserving food quickly, wholesomely, and conveniently. At the time, smoking, pickling, and drying were the traditional methods. Nicholas Appert won the prize in 1810 for his canning* technique.

Over the ensuing 175 years, frozen foods have been the only new addition to commercial food processing. That is changing. Fresh strawberries in Belgium, potatoes in Japan, rice pudding in the Soviet Union, papaya in South Africa, shrimp in Australia, rye bread in Holland, onions in France, and cod fillets in Canada are all part of an expanding

---

* Actually, the first canisters were bottles. Appert used lightly corked bottles that were immersed in boiling water. After a period of cooking they were removed, tightly sealed, and cooled. It was not until the 1840s that tin-coated, double-seamed steel containers came into use.

worldwide menu of foods preserved by gamma radiation. In the United States, the unrestricted irradiation of food products has not yet been approved by regulatory authorities. However, in March 1981, the FDA, charged with regulating the sale of foods, drugs, and cosmetics, issued an advance notice dealing with proposed regulations which, if adopted, would approve irradiation processing for the preservation of specific foods and dry ingredients.

On July 5, 1983, the FDA issued a rule amending its regulations to permit the use of gamma radiation to reduce or control microbial contamination in spices and vegetable seasonings. In its release the FDA noted that it intends to propose comprehensive regulations for food irradiation in the near future. Among the more promising applications proposed would be the treatment of citrus fruits for fruit fly control, irradiation of dry products such as cocoa, nut meats, and starches for the reduction of microbial activity, as well as irradiation of fruits and fish to extend their edible shelf life. Table 3 lists the foods that have received clearances in 21 countries around the world. Clearly, this is a procedure whose time has come.

Radiation preservation of foods is considered a "cold" process as there is only a slight temperature rise induced in the food during the irradiation period. This makes the procedure especially attractive for such heat-sensitive considerations as nutrient retention. With only the small rise in temperature, adverse changes such as altered flavor, odor, color, texture, and nutrient quality are significantly minimized. Consequently, the treated food retains more of the appearance, taste, and quality of fresh* raw food.

Another advantage of this type of treatment is flexibility. Irradiation can be used to preserve a variety of foods in a range of sizes and shapes: crates of potatoes, flour in 50- or 100-lb sacks, entire roasts of meat, whole turkeys, and/or sandwiches of sliced meat, fish, or chicken.

Four treatment levels have been developed:

1. Doses sufficient to reduce the number of viable microorganisms of public health significance or capable of causing food spoilage (the two are not the same) so that few, if any, are detectable. The term "radappertization" (after Nicholas Appert) has been suggested for this level of treatment. The resulting processed food items can be stored at room temperature as are the more common processed foods (canned tuna, vegetables, soups, etc.). Doses are typically greater than 1 Mrad (10 kGy).

---

* Fresh is a relative term. Apples plucked from a tree and eaten must be fresh. What about "fresh" fish caught, iced, and stored at sea for days or weeks? Fruits and vegetables brought to refrigeration plants and frozen within hours of harvesting seem to be in this category. Is time for picking, harvesting, or catching the crucial point? Or does fresh mean unprocessed in any manner?

Table 3
Unconditional and Provisional Clearances of Food Products
in Different Countries, 1984[a]

Foods	Argentina	Bangladesh	Belgium	Canada	Chile	Denmark	France	Hungary	Israel	Italy	Japan	Netherlands	Norway	Philippines	Poland	South Africa	Spain	Thailand	Uruguay	United States	USSR
Potatoes	x	x	x	x	x	x	x		x	x	x	x			x	x	x	x		x	x
Onions		x	x	x			x	x	x	x		x				x	x	x	x		x
Garlic			x				x			x						x					
Shallots			x				x														
Wheat, flour, whole wheat flour	x			x	x															x	
Spices		x	x				x		x	x		x	x							x	
Chicken	x		x							x		x				x					
Fish and fish products (chilled, frozen)	x		x									x									
Frozen shrimps	x											x									
Frog legs	x											x									
Rice and ground rice products	x					x						x									
Rye bread												x									
Egg powder												x									
Blood proteins												x									
Cocoa beans				x								x									
Dates				x																	
Pulses	x			x																	
Papaya	x			x												x					
Mango	x			x												x					
Strawberries			x	x								x				x					
Paprika			x																		
Mango achar																x					
Bananas (fresh, dried)																x					
Litchis																x					
Dry food concentrates																					x
Grain																					x
Dried fruits																					x
Mushrooms												x									
Endive												x									
Asparagus												x									
Battermix												x									

[a] Data provided by the International Atomic Energy Agency Bulletin, June 1984.

2. Doses sufficient to kill or render harmless disease-producing organisms such as *Salmonella*. Viruses and spore-forming bacteria are not affected at this level. This radicidation occurs at doses typically below 1 Mrad.

3. Doses sufficient to enhance keeping quality by killing substantial numbers of microbes. Similar to pasteurization, this process is being referred to as radurization. Here too processing takes place below 1 Mrad.

4. Doses sufficient to interfere with physiological processes rather than microbial. Sprout inhibition of potatoes and delayed maturing of fruit are in this category. Levels of radiation at 0.003–0.015 Mrad (30–150 Gy) or less are used.

Table 4 shows the range of gamma doses used to achieve specific effects.

The irradiation process occurs in a mazelike chamber. The chamber or irradiation room, as well as all the people in it, is protected from the radiating source by concrete walls at least 6-ft thick. The cobalt-60[*] is inserted into 18-in.-long stainless steel tubes called pencils. They are not unlike the rods shown previously in Figure 11, only smaller. The cobalt pencils are loaded into a ladder at the bottom of a 25–30-ft-deep pool. Because of the gamma emissions, the water vibrating at the frequency of light takes on an aquamarine hue.

As the products to be treated move through the chamber, the cobalt-60 rack is raised above the water. Gamma rays strike everything in the chamber. Depending on the food being processed, it can spend 2–6 h being bombarded.

Among the variety of irradiators being designed and developed are the batch carrier and pallet units. In the batch carrier type, Figure 35, the product is loaded into aluminum "carriers," which are suspended from and driven through the irradiation chamber by an overhead monorail system. Loaded on tall carriers—up to 9 ft high—the food is passed along an overlapping source plaque, which permits uniformity of dose.

Table 4
Dose Range for Various Kinds of Effects in Food Irradiation

Effect	Dose (Gy)
Destruction of bacterial spores	30,000–50,000
Destruction of vegetative bacteria	1,000–10,000
Destruction of food-borne pathogens	2,000– 8,000
Insect disinfestation	150–500
Inhibition of maturation of raw fruits and vegetables	50–500
Chemical change	variable

[*] Cesium-137 is also used in the irradiation process.

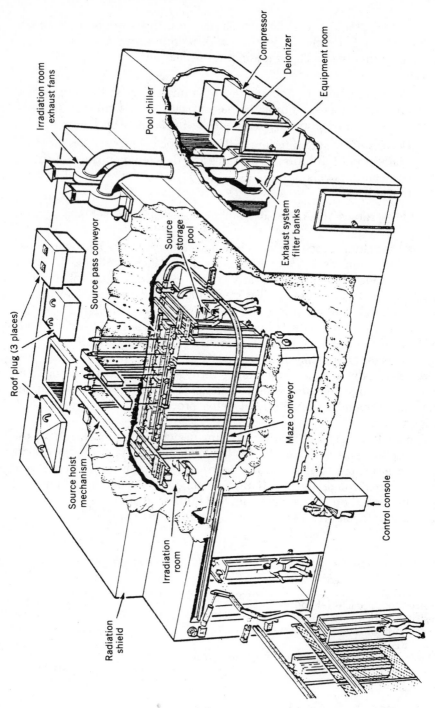

Figure 35. A batch carrier irradiator. Courtesy of Atomic Energy of Canada, Ltd.

Irradiation room exhaust fans

Compressor

Deionizer

Equipment room

Pool chiller

Source pass conveyor

Source storage pool

Exhaust system filter banks

Roof plug (3 places)

Source hoist mechanism

Maze conveyor

Radiation shield

Irradiation room

Control console

Table 5
Dose Levels Used to Irradiate a Selection of Food Products

Different doses of ionizing energy can be used for different purposes in food preservation:

1. **Low,** up to 1 kGy:
   (a) Inhibition of sprouting
   (b) Insect disinfestation
   (c) Delay of ripening
2. **Medium,** 1–10 kGy:
   (a) Extension of food shelf life
   (b) Reduction of microbial load
   (c) Improvements in technological properties of food
3. **High,** 10–50 kGy:
   (a) Commercial sterilization
   (b) Elimination of viruses

The following are a few of the more important uses:

1. **Fish**
   (a) To control insect infestation of dried fish during storage and marketing (0.5 kGy)
   (b) To reduce the microbial load of the packaged fish and fish products (2–3 kGy)
   (c) To eliminate certain pathogenic microorganisms in packaged frozen fish and fish products (3–5 kGy)
2. **Potatoes and onions**
   To inhibit sprouting during storage (0.05–0.15 kGy)
3. **Food grains**
   To control insect infestation during storage (0.5–1.0 kGy)
4. **Cocoa beans**
   (a) To control insect infestation during storage (0.5–1.0 kGy)
   (b) To reduce the microbial load of fermented beans with or without heat treatment (2–3 kGy)
5. **Dates**
   To control insect infestation in dried dates (0.5–1.0 kGy)
6. **Mangoes**
   (a) To control insect infestation (0.25–0.5 kGy)
   (b) To improve keeping quality by delaying ripening (0.5–1.0 kGy)
   (c) To reduce the microbial load through combined irradiation and heat treatment (0.5–1.0 kGy)
7. **Spices and condiments**
   (a) To control insect infestation (0.5–1.0 kGy)
   (b) To reduce the microbial load (5–10 kGy)
   (c) To reduce the number of pathogenic microorganisms (5 kGy)
8. **Meat and chicken**
   To extend shelf life and eliminate pathogenic microorganisms (2–4 kGy)

Among the products that can be processed in this irradiator, with consequent benefits, are:

- Fresh poultry (shelf-life extension and salmonella control)
- Spices (disinfection, total microbial count control)
- Food waste (elimination of viruses and pathogens)
- Animal feed (elimination of pathogens)
- Potatoes (sprout inhibition)
- Starch powder (reduction of microbial contamination)

The pallet irradiator is designed to process packaged goods stacked on 100-cm × 120-cm pallets as shown in Figure 36. These are transported through the irradiators and loaded into steel carriers suspended from overhead rails. Two pallets can be loaded in each carrier. Each side of the pallet is exposed to the source twice, thus ensuring uniformity considering the large size of the stack.

By 1990, processing units may be as different as conventional food plants are today. In fact, one of the most unique designs is shown in Figure 37. This circular unit conveys potatoes on a trolleylike track. It

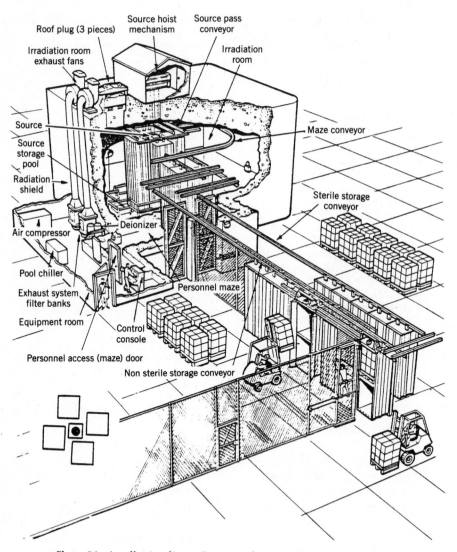

Figure 36. A pallet irradiator. Courtesy of Atomic Energy of Canada, Ltd.

(a)

(b)

Figure 37. Shihoro potato irradiator in Japan. Courtesy of IAEA, Vienna.

Table 6
Foods Approved by FDA for Irradiation Treatment

Food	Purpose	Dose Limit	Date Approved
Fruits and vegetables	To slow growth and ripening and to control insects	Up to 1 kGy	April 18, 1986
Dry or dehydrated herbs, spices, seeds, teas, and vegetable seasonings	To kill insects and control microorganisms	Up to 30 kGy	April 18, 1986
Pork	To control *Trichinella spiralis* (the parasite that causes trichinosis)	Minimum 0.3 kGy to maximum of 1 kGy	July 22, 1985
White potatoes	To inhibit sprout development	50–150 Gy	August 8, 1964
Wheat, wheat flour	To control insects	200–500 Gy	August 21, 1963

requires relatively little space and is highly efficient. The brain child of the Shihoro Company in Hokkaido, Japan, It may be the wave of the future.

Are these foods radioactive? No. It is a physical impossibility because the levels of radiation are incapable of activating the foods. Sophisticated analytic techniques have isolated and identified new substances now called "unique radiolytic products (URPs) but these have also been found not to be human health hazards (8–11).

The three preconditions necessary for the widespread use of irradiated foods appear well on their way to being fulfilled. These are:

1. Proof of safety for human consumption.

2. Technological feasibility.

3. Economic competitiveness.

Given the lack of irradiation facilities in the United States, it would take at least 3 years after the FDA gave final approval before consumers would find irradiated foods on their supermarket shelves.

Approval of this food-processing method has been slow in coming primarily because of concern for the possibility of induced radiation. Over the past 25 years, we have learned that irradiation will not make a treated food radioactive. Consequently, consumers cannot be exposed to radiation. At the levels established for the various foods, irradiation is not a hazard and does not present a risk to health.

Ari Byrnjolfsson of the Department of Applied Biological Sciences, MIT, recently had this to say: "The data on the animal feeding studies conducted at a great many laboratories as well as the radiation chemistry data indicate that irradiation of foods does not introduce harmful effects and the process is ready for practical applications for the benefit of the consumer" (9).

Approval for irradiation of pork to destroy the parasite of trichinosis, *Trichinella spiralis,* was granted by the FDA in July 1985. In April 1986, Frank E. Young, Commissioner of the FDA, announced at a Congressional hearing that permission to irradiate fresh fruits and vegetables would shortly be permitted. Approval was announced in the *Federal Register* on Friday, April 18th.

Unfortunately, the strident voices protesting the FDA's action has stayed the approval, and with the food industry's inability to state the case for electronic preservation, it may be some time before products appear on market shelves. Richard Hickey and his colleagues at the University of Pennsylvania have stated the problem precisely (12): "Looking back at the Three Mile Island incident it appears," they said, "that no serious health effects have resulted from radiation or from radionuclide releases. The main problem has been reported to have been the stress and anxiety generated among the nearby population." They went on to note that, "as Meinhold observed [13] . . . just as no one should be exposed to ionizing radiation for trivial reasons, so no one should be denied the benefits . . . by trivial objections."

Figures 38 and 39 show the beneficial effects of radiation on potatoes and strawberries.

IV. LOW-DOSE IRRADIATION

Because ionizing radiation can inactivate deoxyribose nucleic acid (DNA), the constituent chemicals of the genes which provide the code for protein synthesis, it can be lethal to all living things in sufficiently high doses. But is all nuclear radiation harmful? What of low doses? And what constitutes a low dose?

To the first question, conventional wisdom would respond with a resounding yes. There is, however, a growing body of knowledge that suggests that such a response may be incorrect. Substantial evidence indicates that ionizing radiation may be classified with those substances which at high doses are toxic but which at low or lower doses are innocuous or even beneficial. The dual nature of aspirin, vitamin D, sunlight, alcoholic beverages, caffeine, and anesthetic gases, to note a

Figure 38. Inhibition of sprouting. These potatoes were photographed 8½ months after exposure to gamma rays. The potato at the upper left was not exposed; it sprouted and became soft in the normal fashion. The other potatoes were exposed to varying dosages of gamma rays. The following dosages were administered: in the top row, left to right, non; 0.12 kGy; 0.5 kGy. In the bottom row, left to right, 2 kGy; 8kGy; 10 kGy. This photo suggests that the middle range of radiation best controls sprouting of potatoes, thus extending storage life.

sampling, are well established. The process whereby low doses of an otherwise harmful substance may result in stimulatory or beneficial effects is called hormesis and is commonly found in nature.

Radiation protection has generally predicted low-dose responses based on extrapolation from effects observed at high doses. However, increasing numbers of studies report beneficial effect from low-dose exposure. These beneficial effects challenge the linear, no-threshold theory of radiation exposure.

It was noted earlier that some types of radiation can penetrate the body, producing highly reactive, electrically charged atoms and molecules (ions) and free electrons in tissue cells. This ionizing radiation when received in small doses is referred to as low-level radiation (LLR). The contributions of various natural sources—the natural background—are foremost examples of LLR. Emissions of LLR also occur from such synthetic sources as TV sets, smoke detectors, nuclear power plants, and medical devices and diagnoses. At sea level none of these sources, natural

Figure 39. Radiation inhibition of mold growth.

or synthetic, exceeds 100 mrem (1 mSv). This level and below is often referred to as low level.

Radioactive effects are often described by mathematical models. A model of carcinogenesis is a set of assumptions about the mechanism by which cancer is induced along with deductions of the consequences of those assumptions for cancer risk. Because of limited knowledge and understanding of the processes by which cancer occurs, assumptions themselves cannot be directly verified. Consequently, the level of built-in uncertainty can be large. Different models (four or five are in current use; cf. Chapter 15) can vary in their estimates of risk by as much as a factor of 1,000,000.

Within the range of low-level radiation, some researchers potulate a threshold below which the risk is effectively zero. Figure 40 indicates the area and level of radiation dose for which concern is being expressed. The range is well below 10 mrem. This is the area highlighted in the figure.

Attempts to prove conclusively that any one theory is the correct one are confounded by the fact that effects predicted by each are exceedingly small, and the difficulties in identifying these effects are exceedingly great. As the dose becomes smaller, the effects become fewer. Therefore, to identify conclusively that certain health effects have been caused by a given dose, the number of people required to be studied (examined) becomes larger and larger as the dose decreases. For example, to detect

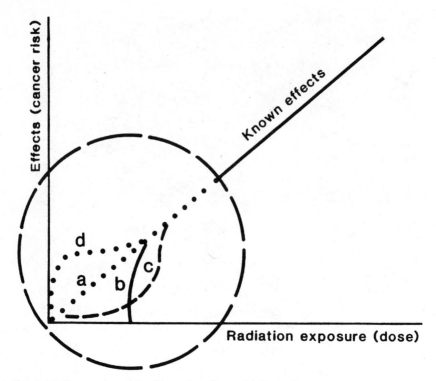

Figure 40. Proposed models of how the effects of radiation may vary at low dose levels.

any change in the number of genetic defects that could be produced by an increase in radiation dose of 3 mrem/yr would require the close observation of a population of 700 million for three generations. Indian has a population of that size! Is it any wonder then that models are used, or even animals with short life spans? But which species should be used? Of course, this requires the translation of animal data to man which injects yet another variable and level of uncertainty into the models.

There is, however, another side of the low-level radiation coin. Rather than attempt to deduce from models what may or may not occur, scientists have been studying effects of low doses directly in a variety of animal species. Among the effects studied are growth and development, cancer induction, wound healing, life span, and immunity.

As early as 1961, Hugh F. Henry, professor of physics at DePauw University, asked, "Is all nuclear radiation harmful?" (14). Writing in the *Journal of the American Medical Association*, he noted that "a significant and growing amount of experimental information indicates that the overall effects of chronic exposure (to low levels) are not harmful." He was concerned primarily with life span and somatic and genetic effects.

By 1980, Thomas D. Luckey of the Department of Biochemistry, University of Missouri Medical School, was able to catalog over 1000 articles supporting beneficial effects of LLR (15). He called this effect "hormesis" and he defined it as a natural response, "probably homeostatic, evoked in an organism by low-levels of insult* that would be lethal or seriously incapacitating at high exposure levels" (16).

Low-levels of insult or occurrence have been described as inducing higher growth rates in irradiated *Daphnia*, flies, moths, silkworms, and blue crabs, compared with nonirradiated controls. Vertebrates such as dogs, mice, chickens, and fishes have also been reported as responding beneficially with more litters and longer fertile life.

Although no one questions the harmfulness and lethality of excessive radiation, ample evidence supports the idea that low doses of whole-body radiation reduce tumor induction. It has been shown that low doses of radiation administered to mice resulted in a fivefold reduction in the incidence of leukemia (17).

At the Los Alamos National Laboratory, Dr. Gary L. Tietjen surveyed workers exposed to plutonium to measure adverse health effects of inhalation of the highly insoluble forms of plutonium oxide. Studies in animals predicted excess lung cancers. He reported that "exposures of 370 Bq (10 mCi) seem to be safe as far as lung cancer is concerned. From animal experiments we expected an excess in lung cancer, but lung cancer is conspicuous by its absence. Gofman . . . gave rather alarming predictions of what would happen to Pu workers who smoke. His results, when used in the workers in the study, are shown in the table" (18).

Average length of exposure	26 years
Lung cancer dose at age 38	285 Bq (7.7 mCi)
Average total Pu deposition	666 Bq (18 mCi)
Lifetime excess lung cancers	48–65
Expected cases by 1980	9–13
Observed cases by 1980	0

Similar benefits have been described for wound healing. Rabbits exposed to x-rays and dogs exposed to low doses of radium exhibited faster healing to incised skin. And irradiated animals also responded to low levels of radiation with heightened antibody production and enhanced bactericidal activity. Improved defense mechanisms would be expected to manifest themselves in increased longevity.

Dr. Bernard Cohen, professor of nuclear engineering at the University of Pittsburgh, recently noted that "the area of southeastern Finland, and Skaraborg County, Sweden have radon levels similar to those of Cumberland County, Pa., and their lung cancer rates are similarly below

---

* The term insult, taken from the medical literature, is used in its broadest sense to mean risk, hazard, and/or experience.

average." He noted too that "the evidence argues against the linear, no threshold relationship, suggesting that it probably overestimates effects at low levels by at least something like a factor of 4." And finally, "failure of the linear, non-threshold theory is a necessary condition before hormesis can be accepted" (19).

Because of the increasing concern about radiation hormesis and its implications, an international conference was held in August 1985. The Health Physics Society devoted an entire issue (Vol. 52, No. 5, May 1987) to papers presented at the conference. It is worthy of study.

After reviewing 40 years of research on human populations that were exposed to excessive radiation either by medical treatment or by fallout from atomic weapons, Lyon further commented that "the mixture of positive, negative and equivocal results leads us to postulate that ionizing radiation is itself neither safe nor dangerous. The safety or danger involved varies from setting to setting, depending on type of radiation, dose or dose rate, tissue exposed and age at exposure" (20).

Obviously, mathematical models obtain "results" faster than experiments on humans or animals, but they can neither replace nor substitute for them. Ignoring the hormesis phenomenon could be a serious error. Limiting attention to models could compound it.

Is all nuclear radiation harmful? The data suggest there are levels for which there is not only no injury, but benefit. This is a phenomenon that bears watching.

V. THE FUTURE OF NUCLEAR GENERATED ELECTRICAL POWER

Over the next 50 years, demand for electricity will surely increase, especially if the elevated temperatures currently being forecast for our hemisphere are borne out. Greatly increased countrywide needs can probably only be realized by the establishment of a national nuclear power network. Solar, tidal, wind, and geothermal energy sources, individually or in combination, are simply inadequate to the task.

Generation of electricity via nuclear power is to be encouraged because it alone can meet the heavy demands, but especially because carbon dioxide, a major contributor to the greenhouse effect (cf. Chapter 10) is not a by-product of the process.

Nevertheless, nuclear power plants have serious shortcomings that need to be addressed. In the United States, generation of electric power by the release of energy from the nucleus of atoms is treated much like any product offered for sale to the public. In our free-market economy, anyone with sufficient financial backing can build a plant and generate the steam necessary to drive turbines. And the stubborn belief is that steam is steam no matter the source. Unfortunately, this laissez-faire attitude often results in poor design and construction, astronomical cost overruns, and mismanagement, with loss of public confidence and trust—terribly important ingredients for a national commitment to nuclear power.

For all its vaunted lack of government intervention, the United States is not the only country wherein market forces are unfettered by government. Canada, France, England, Japan, Italy, and the Federeal Republic of Germany all claim the engine of private enterprise as the driving force of their economic systems. But not where nuclear power is concerned. Government control is implacable.

Until the United States takes nuclear power seriously, which means our elected officials and the electric power industry understand its potential for harm as well as its benefits, and develops stringent standards for construction and operation, and imposes strict measures of accountability, operating licenses must be withheld.

It may be in everyone's best interest that until the electric power industry can formulate an acceptable code of operation, our nuclear power needs be supplied by French or Canadian engineering and construction firms, under contract and control of a new federal agency mandated by Congress to assure appropriate compliance. Simultaneously, an American company, chosen by lottery, should be allowed to build a power plant under close supervision of the new agency, to ascertain if the required technical capability exists. Then and only then can foreign firms be replaced by American initiatives. If public confidence is to be restored, nothing less will do.

REFERENCES

1. D. Lilienthal, *Atomic Energy: A New Start*. Harper & Row, New York, 1980.
2. A. M. Weinberg and I. Spiewak, Inherently Safe Reactors and the Second Nulcear Era. *Science* **224**: 1398–1402, 1984.
3. Resurrecting a nuclear accident. *Nature* **32**: 207–209, 1983.
4. Kemeny Commission findings. *Hearing Before the Sub-Committee on Energy Research and Production*. 96th Congress, 1st Session, November 14, 1979. U.S. Government Printing Office, Washington, DC, 1980.
5. M. Goldman, Chernobyl: A Radiobiological Perspective. *Science* **238**: 622–623, 1987.
6. S. Schmemann, Chernogyl and the Europeans. *The New York Times* June 12, 1988.
7. C. R. Porter, J. A. Broadway, and B. Kahn. *Methodology of Procedures for Surveillance of the Food Chain as Conducted by the United States*. U.S. EPA, Office of Radiation Programs, EERF. Montgomery, AL, 1988.
8. W. C. Cunningham, W. B. Stroube, Jr., and E. J. Baratta, Radionuclides in Foods 1953–1986. *J. Assoc. Off. Agric. Chem.* (in press).
9. A. Byrnjolfsson, Wholesomeness of Irradiated Foods: A Review. *J. Food Safety* **7**: 107–126, 1985.
10. W. M. Urbain, Irradiated Foods: A Giant Step Beyond Appert. *Nutr. Today* **19** (4): 6–11, 1984.

11. WHO, 1981. Wholesomeness of irradiated Food. Report of a Joint FAO/IAEA/WHO Expert Committee. Geneva, Oct 27–Nov. 3, 1980 WHO Tech. Rept. Ser. #659, Geneva.

12. R. J. Hickey, E. J. Bowers, and R. C. Clelland, Radiation Hormesis, Public Health and Public Policy: A Commentary. *Health Phys.* **44** (3): 207–219, 1983.

13. C. B. Meinhold, President's Message—1981. *Health Phys.* **41:** 1, 1981.

14. H. I. Henry, Is All Nuclear Radiation Harmful? *J. Am. Med. Assoc.* **176:** 671–675, 1961.

15. T. D. Luckey, *Hormesis with Ionizing Radiation.* CRC Press, Boca Rato, FL, 1980.

16. T. D. Luckey, Physiological Benefits from Low-Levels of Ionizing Radiation. *Health Phys.* **43** (6): 771–789, 1982.

17. A. P. Jacobsen, F. A. Plato, and N. A. Frigerio, The Role of Natural Radiations in Human Leukemogenesis. *Am. J. Public Health* **66:** 31–37, 1976.

18. G. L. Tietjen, Plutonium and Lung Cancer. *Health Phys.* **52** (5): 625–628, 1987.

19. B. L. Cohen, Tests of the Linear, No-Threshold Dose–Response Relationship for High-LET Radiation. *Health Phys.* **52** (5): 629–636, 1987.

20. J. F. Lyon, Radiation Exposure and Cancer. *Hosp. Prac.* 159–173, July 1984.

SUGGESTED READINGS

Anspaugh, L. R., Catlin, R. J., Goldman, M. The Global Impact of the Chernobyl Reactor Accident. *Science* **242:** 1513–1519, 1988.

Boice, J. D., and Fraumeni, J. F., Jr. (Eds.), *Radiation Carcinogenesis: Epidemiology and Biological Significance.* Raven Press, New York, 1984.

Committee on the Biological Effects of Ionizing Radiation (BEIR III), *The Effects on Populations of Exposure to Low Levels of Ionizing Radiation.* National Academy of Sciences, Washington, DC, 1980.

Jablon, S., and Bailar, J. C., III. The Contribution of Ionizing Radiation to Cancer Mortality in the United States. *Prev. Med.* **9:** 219–226, 1980.

McCracken, S. *The War Against the Atom.* Basic Books, New York, 1982.

Report on the Safety and Wholesomeness of Irradiated Foods. Advisory Committee on Irradiated and Novel Foods, Department of Health and Social Security, Ministry of Agriculture, Fisheries and Food, Scottish Home and Health Department, HMSO, London, 1986.

Symposium on the Perception of the Risk of Radiation. *Am. J. Roentgen.* **140** (3): 595–610, 1983.

Truswell, A. S. Food Irradiation. *Br. Med. J.* **294** (6585): 1437–1438, 1987.

Upton, A. C. The Biological Effects of Low-Level Ionizing Radiation. *Sci. Am.* **246:** 2, 1982.

10

Air Pollution

It is unhistorical to believe that a society can somehow return to the simple past while retaining the complex advantages of the technological present.

—R. J. Forbes

A young gentleman who had inked himself by accident, addressed me from the pavement and said, "I am from Kenge and Carboy's, miss, of Lincoln's Inn."

"If you please, sir," said I.

He was very obliging; and as he handed me into a fly, after superintending the removal of my boxes, I asked him whether there was a great fire anywhere? For the streets were so full of dense, brown smoke that scarcely anything was to be seen.

"O dear no, miss," he said. "this is a London particular. A fog, miss," said the young gentleman.

"O indeed," said I.

I. THE AIR RESOURCES

A. The Setting

Although the above dialogue from Charles Dickens' *Bleak House* makes it clear that mid-19th century London was no stranger to air pollution, it was as long ago as the year 1273 that a law was passed to control the burning of soft coal, in order to curtail local air pollution. Complaints against such pollution were frequent in 13th and 14th century England. In 1306 a proclamation was issued by Parliament, requiring the burning of wood rather than coal by the artisans and manufacturers of London during sessions of Parliament. Shortly thereafter, a man was executed for

395

violating this early smog ordinance. Apparently, air pollution is not a modern invention.

A landmark in the history of the subject was the pamphlet entitled *Fumifugium*, addressed by John Evelyn to Charles II in 1661. Evelyn described the "Evil" as "epidemicall: indeangering as well the Health of Your Subjects, as it sullies the Glory of this Your Imperial Seat." Evelyn suggested that factories using coal be moved farther down the Thames valley and that a green belt of trees and flowers be put around the heart of the city. "But I hear it now objected by some," Evelyn wrote, "that in publishing this Invective against the smoake of London, I hazard the engaging of a whole Faculty against me, and particularly, that the College of Physicians esteem it rather a Preservation against Infections, than otherwise any cause of the sad effects which I have enumerated." One is reminded of the French aphorism, "the more things change, the more they remain the same."

The famous 17th century physician Thomas Sydenham, however, had no doubt about the ill effects of London's air. "The fumes that arise," he wrote, "from the several trades managed here, but especially sulphur and fumes of sea coals with which the air is polluted, and these, being sucked into our lungs and insinuating into the blood itself, give occasion for a cough."

Yet more than 300 years later there is still uncertainty about the precise effect the fouling of the atmosphere has on health. This lack of understanding is chiefly due to the complexity of the problem. Anyone who has experienced the dense yellow fog that used to be known as the "London particular" will not be surprised by the suggestion that it is bad for the lungs. But a doctor faced with a patient from an area where air pollution has long been severe may find it difficult to determine the extent to which the patient's health has been affected by it. A person with chronic bronchitis may be a heavy smoker, a worker in heavy industry, may have lived in a crowded home, or have had repeated attacks of pneumonia. It is difficult therefore to assess the additional contribution that exposure to polluted air may have on a person's health. That, however, cannot be construed as license for air pollution to continue unabated.

B. Clean Air Acts

On December 17, 1963, President Johnson signed into law the first Clean Air Act; Public Law 88-206 was a historic milestone in the control of community air pollution. It established a national program to meet the steadily growing demands for cleaner air.

Less than 2 years later, President Johnson signed an amendment to the act. In doing so he indicated that the federal government would assume an even greater role in guiding and planning for pollution prevention and control in the air over our cities. On signing, the president remarked: "We have now reached the point where our factories, our automobiles, our furnaces, and our municipal dumps are spewing out more than 150 million tons of pollutants annually into the air we breathe—almost one-half million tons a day."

Continuing to fire its broadsides, Congress enacted the Clean Air Act of 1970. This required all states to submit to the Environmental Protection Agency a plan to achieve the act's ambient air quality standards. The primary intention of the act was to control air pollutants that adversely affected human health. It specified that not only should National Ambient Air Quality Standards (NAAQS) be set, but that they be set at levels sufficient to protect human health with an adequate margin of safety.

That was 25 years ago, and we still do not know what the dangers are; consequently, we have little real idea of what constitutes a safe level of any specific pollutant. This creates perplexing problems for regulatory agencies.

The question that arises almost immediately is: What is health? Unfortunately, the World Health Organization advanced an unchallenged, abstract definition: "health," it said, "was a state of complete physical, mental, and social well being, and not merely the absence of disease or infirmity." How was anyone going to be able to meet so esoteric a goal. What did Congress intend? Over the past 25 years, what have we learned about the health effects of air pollutants?

C. What Do We Breathe?

Stretching from the surface of the earth toward outer space is a relatively thin layer of air. This troposphere, some 5–11 mi deep, contains the air we breathe and the air we foul. From the day we are born we contaminate the air around us with every breath, cough, and sneeze. As the years roll by, we add the by-products of cigarette, cigar, and pipe smoking, cooking, driving our cars, heating our homes, and industrial effluent.

Air pollution is the presence in the air of substances in amounts great enough to interfere directly or indirectly with our comfort, safety, and health. Public and governmental concern with air pollution is generally considered to spring from the fact that the average person inhales approximately half a liter of air with each breath. As most of us breathe some 22,000 times every 24 h, we inhale about 2000 gal of air per day. Each liter (approximately 1 quart) of air in our urban centers has been estimated to contain several million particles of foreign matter. Thus, in a day a city dweller can inhale some 20 billion particles of . . . what? These particles, it is claimed, must certainly impair human health. After all, we can clearly see the ill effects on ornamental plants and crops, as well as the corrosion of buildings and building materials. How can human tissue withstand the onslaught? Does it?

Each breath consists of about half a liter of air, 20% of which is molecular oxygen. The air swirls briefly through a maze of branching ducts leading to tiny sacs (alveoli), a gas-exchange apparatus in which some of the gaseous oxygen is dissolved in the bloodstream. I am, of course, describing the lungs. Do the billions of foreign particles actually enter the alveoli? What protects the lungs and the air ducts leading to them from contamination?

Recall that in most acute episodes only a small fraction of the population succumbs. The great majority do not become ill. This implies that those who do are more susceptible or their defenses are over-whelmed. In either case, does the evidence suggests the presence of physiological defense mechanisms? Indeed, the human animal is neither an alabaster statue nor a passive receptor. It was "made" to fight back.

D. The Bronchial Tree of Life

Figure 1 is a cross-section of the respiratory tract. The nose can be compared to an air-conditioning unit because it controls the temperature and humidity of the air entering the lungs and filters out foreign particles. External respiration begins and ends with the nose. It filters, warms, and moistens the air. The interior of the nose is divided by a wall of bone and cartilage, the septum. On both sides of the septum are a series of scroll-like bones, the turbinates or conchae. The purpose of the turbinates is to increase the amount of tissue surface so that inhaled air will be further conditioned before continuing toward the lungs. The surface of the turbinates is covered with a mucous membrane secreting a continuous supply of mucus, which drains slowly into the throat. The mucus gives up heat and moisture to incoming air. It also helps trap irritating substances contained in the inhaled air. The inhaled air passes through the nasal cavity into the pharynx (a common passage for both air and food), which leads to the esophagus (the food tube) and the larynx (the voice box or Adam's apple). When food is swallowed, a flap of cartilage, the epiglottis, folds over the opening; at the same time, the larynx moves up to help seal the opening. The trachea continues down the neck into the chest and branches into the right and left bronchi.

Each bronchus divides and subdivides; between 20 and 22 bronchial subdivisions have been counted. The smallest bronchi in the depths of the lung are called bronchioles. They end in some 300–400 million air sacs, called the alveoli, which are 75–300 μm* in diameter. The alveoli are balloonlike structures that give the lungs their spongy quality.

---

* The micrometer μm, a unit of the metric system, is 0.001 mm, or 1/25,400 of an inch. A ¾-in.-long firefly is to the 1472-ft Empire State Building as 1 μm is to 1 in.—25,000 times smaller.

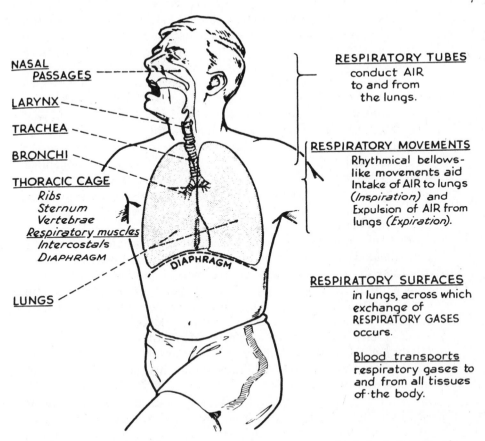

NASAL PASSAGES

LARYNX

TRACHEA

BRONCHI

THORACIC CAGE
Ribs
Sternum
Vertebrae
Respiratory muscles
Intercostals
DIAPHRAGM

LUNGS

DIAPHRAGM

RESPIRATORY TUBES
conduct AIR
to and from
the lungs.

RESPIRATORY MOVEMENTS
Rhythmical bellows-
like movements aid
Intake of AIR to lungs
(*Inspiration*) and
Expulsion of AIR from
lungs (*Expiration*).

RESPIRATORY SURFACES
in lungs, across which
exchange of
RESPIRATORY GASES
occurs.

Blood transports
respiratory gases to
and from all tissues
of the body.

Figure 1. The respiratory system. From *Illustrated Physiology*, 4th ed., by A. B. McNaught and R. Callander. By permission of Churchill Livingstone, Edinburgh, New York, 1983.

It is here in the alveoli that the business of the lung occurs. The lungs exist for one purpose, gas exchange or transpiration. Inhaled air brings oxygen to the blood for distribution to tissues and cells throughout the body, and waste carbon dioxide is brought from the cells to be exhaled. The walls of the alveoli, less than 1 μm thick and surrounded by a complex of capillaries, are the site of this two-way diffusion.

The hairs of the nose filter out large particles. Beyond the hairs, the involuted contours of the turbinates force the fast-moving airstream to impact on the mucus surfaces again trapping suspended particles. In this way, the filtering mechanism almost completely removes particles larger than 10 μm in diameter. Particles from 2 to 10 μm usually settle on the walls of the trachea, bronchi, and bronchioles. Particles between 0.3 and 2.0 μm may reach the alveoli. Particles smaller than 0.3 μm are likely to remain in suspension as aerosols to be blown out of the lungs with exhaled air. Foreign matter that settles on the walls of the nose, pharynx, trachea, bronchi, and bronchioles may also be expelled by the explosive blast of air generated by a sneeze or cough; more often it is removed by ciliary action. Each cilium makes a fast, forceful forward stroke, followed by a slower return stroke that brings it into starting position again. The strokes of the rows of cilia are so coordinated that the hairs move together as a wave approximately 12 times per second. The cilia are embedded within a protective sheet of mucus secreted by glands in the trachea and bronchi. The effect of their wavelike motion is to move the entire sheet, and anything trapped on it, up through the tract of the pharynx, where it can be expectorated or swallowed.

In spite of all these preventive measures, inhaled particles, particularly those suspended in fluid droplets, may pass into the alveoli. Before arriving there, however, another defense comes into play. Macrophages, ameboid white blood cells, surround foreign particles, engulf them, and often destroy them or carry them away. Some particles do remain permanently attached to lung tissues, as the darkened lungs of coal miners and city dwellers demonstrate. Many of these particles are essentially harmless; others, such as silica, asbestos, and beryllium, can induce the formation of tough, fibrous tissue that causes serious pulmonary disease (see Chapter 8).

One of the most difficult facts to uncover about the many chemicals released into the air has been which, or how many, are associated with or directly responsible for respiratory dysfunction. Another question is whether the chemical is synthetic. This implies an ability to control it. A third is question is whether the effect is dose related.

II. TYPES OF POLLUTION

A. The Process

Air pollutants can be gases mixed in the air, or solid, or liquid particles dispersed in the air (aerosols). The most widely recognized pollutants are the familiar products of combustion: carbon monoxide, oxides of sulfur and nitrogen, hydrocarbons, and particulates. Table 1 indicates the many and varied sources and tonnages of pollutants released into the air in 1982. Table 2 presents the estimates, again in metric tons,* for the period

---

* A metric ton is 2205 lb. The standard American ton is 2000 lb.

Table 1
National Air Pollutant Emissions[a] Estimates, by Pollutant and Source, 1982

Source	Particulates	Sulfur oxides	Nitrogen oxides	Volatile organic compounds	Carbon monoxide
Transportation	1.3	0.9	9.7	6.1	53.3
Highway vehicles	1.1	0.5	7.8	4.8	46.3
Aircraft	0.1	0.0	0.1	0.2	1.0
Railroads	0.0	0.1	0.7	0.2	0.2
Vessels	0.0	0.2	0.2	0.4	1.4
Other off-highway vehicles	0.1	0.1	0.9	0.5	4.4
Stationary-source fuel combustion	2.4	17.4	9.6	2.0	6.6
Electric utilities	1.0	14.3	6.2	0.0	0.3
Industrial	0.4	2.3	2.7	0.1	0.5
Commercial-institutional	0.1	0.6	0.3	0.0	0.1
Residential	0.9	0.2	0.4	1.9	5.7
Industrial processes	2.4	3.1	0.6	7.1	4.8
Solid waste disposal	0.4	0.0	0.1	0.6	2.1
Incineration	0.2	0.0	0.0	0.3	1.2
Open burning	0.2	0.0	0.1	0.3	0.9
Miscellaneous	1.0	0.0	0.2	2.4	6.8
Forest fires	0.9	0.0	0.2	0.8	6.2
Other burning	0.1	0.0	0.0	0.1	0.6
Miscellaneous organic solvent	0.0	0.0	0.0	1.5	0.0
Total	7.5	21.4	20.2	18.2	73.6

[a] In millions of metric tons per year.

Source: U.S. Environmental Protection Agency, *National Air Pollution Emission Estiamtes, 1940–1982,* February 1984.

1940–1982. Note that 1974 is considered the base year. Clearly, since 1974 the trend has been distinctly downward.

The figures readily show that particulates, sulfur oxides, and nitrogen oxides are all generated by burning fuel for home and office heating. On the other hand, carbon monoxide and hydrocarbons are the result of operating automobiles and trucks. This suggests the most appropriate sources where preventive and control measures might be applied.

The number of suspended particles in urban areas is subject to daily, weekly, and seasonal variations associated with the rhythm of human activity.

Table 2

National Air Pollutant Emissions Estimates, by pollutant, 1940–1982

Year	Particulates	Percentage of 1974	Sulfur Oxides	Percentage of 1974	Nitrogen Oxides	Percentage of 1974	Volatile Organic Compounds	Percentage of 1974	Carbon Monoxide	Percentage of 1974
1940	22.4	182	18.1	67	6.7	34	17.1	76	79.8	92
1950	24.2	197	20.4	76	9.2	47	19.3	86	85.3	99
1960	20.9	170	20.1	74	12.7	64	21.9	97	87.9	102
1970	18.0	146	28.4	105	18.1	92	25.3	112	100.2	116
1971	16.8	137	26.9	100	18.5	94	24.5	109	98.6	114
1972	15.0	122	27.6	102	19.7	100	24.5	109	95.7	111
1973	13.9	113	28.9	107	20.2	103	24.0	107	91.6	106
1974	12.3	100	27.0	100	19.7	100	22.5	100	86.5	100
1975	10.3	84	25.7	95	19.2	97	21.0	93	82.4	95
1976	9.6	78	26.3	97	20.4	104	22.1	98	87.2	101
1977	9.0	73	26.3	97	21.0	107	21.9	97	83.0	96
1978	8.9	72	24.6	91	21.2	108	22.4	100	82.3	95
1979	9.0	73	24.6	91	21.3	108	21.9	97	79.5	92
1980	8.6	70	23.3	86	20.7	105	20.8	92	77.6	90
1981	8.1	66	22.5	83	20.9	106	19.4	86	75.3	87
1982	7.5	61	21.4	79	20.2	103	18.2	81	73.6	85

[a] In millions of metric tons per year.

Source: U.S. Environmental Protection Agency, *Natioal Air Pollutant Emission Estimates, 1940–1982*, February 1984.

B. Acute and Chronic Exposure and Levels of Emission

1. Acute Exposure

a. **Los Angeles Versus New York.** We have noted two major sources of air pollution: fuel for heating and electricity, and fuel for automotive transportation. These two fuels produce dissimilar types of air pollutants and, depending on prevailing geographical and meteorological conditions, produce two dissimilar types of air pollution. The Los Angeles or smog type of pollution is a characteristic of automotive effluents acting in concert with the location of the city. The New York or London type of air pollution, on the other hand, results from burning huge quantities of fossil fuels—coal and oil. The effects of both types of air pollution are notably worse during the meteorological mixup called an inversion.

Pilots and mountain climbers are quite aware of the reduction in temperature that occurs the higher they climb. With each 1000-ft rise, the temperature drops about 5.5°F (1.8°C). When this normal condition is reversed—that is, when there is a temperature *increase* with increasing height—the condition is called an inversion (Figure 2). Inversions occur most often in the autumn and winter months and during the early morning hours. On a clear, calm night, the surface of the earth cools

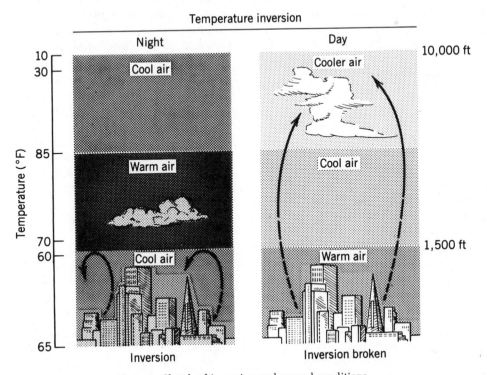

Figure 2. Sketch of inversion and normal conditions.

quickly. In response to this surface cooling, the air in contact with the earth's surface is also cooled. During the hours between evening and morning, this air has cooled considerably, while the air in the upper atmosphere has remained relatively unchanged. The result is a layer of warm air sandwiched between two cold layers—an inversion.

The feature that all inversions have in common is the damping action of the warm layer on the cool bottom layer. This acts like a lid on a pot. It prevents or severely reduces the air's natural upward movement. With such a condition prevailing, the air over a city or region is literally trapped. Pollutants cannot be blown away (diluted), but remain fixed in the area and can build up to dangerous levels.

The Los Angeles basin is an area of about 400 square miles with the Pacific Ocean to the west and hills on three sides. This unique topographic feature inhibits horizontal air movements. Thus, air masses have little opportunity to escape from the basin. In addition, temperature inversions, prohibiting vertical air movement, occur 260–270 days a year. During prolonged inversions, pollutant concentrations increase precipitously as the air mass shifts back and forth—from the city to the ocean during the night, when land breezes prevail, and back over the city in the morning, when sea breezes prevail. The predominantly sunny weather of Los Angeles adds yet another dimension to the problem. In the presence of nitrogen oxides and the sun's energy (in the form of ultraviolet irradiation), organic chemicals undergo a photochemical reaction that can produce entirely new compounds, reduce visibility, and produce eye irritation, crop damage, and characteristic noxious odors. This condition was first noticed in Los Angeles and is therefore referred to as Los Angeles or photochemical smog. An important part of the photochemical problem is our transportation system's almost total dependence on gasoline and on the internal combustion engine, which is the principal source of the hydrocarbons (some 130 have already been identified), carbon monoxide, and oxides of nitrogen present in the atmosphere. The concentrations of these compounds emitted by an individual car seem insignificant. However, when multiplied by thousands of cars in city traffic, the emissions can be measured in tons. Even these large amounts appear harmless until the air masses over a community stagnate as they do during an inversion.

Besides its engine, the automobile has two additional emission sources: evaporation from the fuel tank and the carburetor. It has been estimated that for every 1000 gal of gasoline put into gas tanks, five leave unburned, two leave the crankcase ventilation pipe, and one is lost through evaporation. It is these evaporative losses that contribute hydrocarbons.

In the New York and London type of air pollution, most often found in large industrialized communities, vertical inversions are not as frequent as in the Los Angeles type, nor is horizontal air passage usually impeded. Dispersal of pollutants is not restricted. Although vertical inversions do occur, they are relatively infrequent. Restrictive terrain, however, is not a necessary condition for extreme pollution levels if contaminants are being discharged into the atmosphere at a sufficiently high rate.

b. Three Examples. Much of the available knowledge of the effects of air pollution on human health has come from acute exposures. Two of these acute episodes occurred in areas where large amounts of fossil fuels were used and where heavy industry also contributed its special pollutants. These two areas were further handicapped by restrictive topographical characteristics. A third acute episode occurred in a large city with no restrictive air flow. Only under the extreme conditions described below have fairly strong causal links between air pollution and acute upper respiratory illness been established.

The Meuse Valley of Belgium, scene of some of the bloodiest battles of World War I, is a heavily industrialized area. Blast furnaces, glass factories, lime furnaces, and sulfuric acid and artificial fertilizer plants spew a variety of contaminant chemicals into the atmosphere. During the first week of December 1930, a thick fog blanketed most of Belgium. The air was especially stagnant in the river valleys, particularly along a 15-mi stretch of the meuse. Three days after this abnormal weather condition began, residents began to report shortness of breath, coughing, and nausea. Thousands became ill; the exact number was never ascertained. About 60 people died. Again, deaths were primarily among the elderly and those with chronic illnesses of the heart and lungs. Once the fog lifted, no new cases occurred.

Because this was the first notable acute air pollution episode of the 20th century, public health scientists were unprepared. A study of the area after the incident suggested that the effects on health had been caused by a mixture of the sulfur oxides, sulfur dioxide gas, and an aerosol of sulfur trioxide. This has never been fully substantiated, and such an episode has not recurred in the Meuse Valley over the ensuing 50 plus years.

The episode in Donora, Pennsylvania, occurred during the last week of October 1948. Donora is located some 30 mi south of Pittsburgh in a highly industrialized valley along the Monongahela River. On the morning of October 27 the air over Donora became very still and fog enveloped the city. The air was trapped, and it remained so for 4 days. In addition to sulfur dioxide, nitrogen dioxide, and hydrocarbons from the burning of coal for heating and electricity, the air contained the effluents from a large steel mill and a large zinc reduction plant, where ores of high sulfur content were roasted. During the period of the inversion these pollutants piled up. As they did, severe respiratory distress occurred in the older members of the population. Eye, nose, and throat irritations were common. Twenty people died in a period in which only two deaths were expected. Autopsies of many of those who died showed chronic cardiovascular disease. This finding confirmed the opinion that preexisting heart disease increased the chances of serious illness during an acute air pollution episode. Before the weather changed and broke the inversion, 5910 of the 12,000 inhabitants of Donora had become ill.

The highly inefficient burning of soft coal in open grates by the citizens of London was primarily responsible for the fog that blotted out their capital on December 5, 1952. The city, located on a gently sloping plain, is not hemmed in by hills, as are Donora and the Meuse Valley communities. The flow of air over London is not impeded by topographical barriers. But for 5 days a strong inversion and fog enveloped the city to such a degree that the "ceiling" was some 150 ft high. Within 12 h after the fog had settled over London, residents began complaining of respiratory ailments. By the time the inversion lifted on December 9, 4000 deaths in excess of the normal rate for a 4-day period had been recorded in the Greater London area. Some striking differences between this episode and earlier ones were noted. The increase in mortality was not confined to the very old. Although the highest increment was among those over 45, deaths occurred in all age groups. Another difference was in the rapid onset of illness: 12 h as compared to 48 and 72 h in the earlier episodes. A third difference was the increased death rate in London compared with Donora and the Meuse Valley.

Yet another difference was the fact that London was to experience an acute episode a second and a third time. In 1956, 1000 people died, and in 1962, 700 deaths were recorded. The 30% reduction is believed due to the strenuous preparations made to reduce concentrations of sulfur oxides.

Accidents can also yield valuable information. In November 1950, the Mexican village of Poza Rica was subjected to an unusual pollution experience. Tanks of waste hydrogen sulfide accidentally released their contents and blanketed the entire village in a yellow haze. People of all ages were hospitalized, and 22 died.

It becomes increasingly clear that air pollution is an extremely complicated process: the degree and type are influenced by climate, weather, industry, traffic, density, heating practices, topography, and atmospheric chemistry. Yet there is another, perhaps more critical, problem that requires elucidation.

2. Chronic Exposure

I noted earlier that acute episodes and accidents have strengthened the relationship between air pollution, illness, and death. As a result of these dramatic periods of abnormal pollution, it has become mandatory to investigate whether higher death and illness rates occur during periods of less intense pollution, that is, in the long-term, low-level, "normal" conditions of pollution to which we are often subjected. This recalls a question posed earlier: Can air be lethal at one time when contaminated with chemicals at high concentration, and benign at another when exposure is continuous but the concentration low?

This question strikes at the heart of the Clean Air Act, as well as the tentative efforts of those mandated by law to establish standards for the protection of the public. The following may offer a partial response.

III. ACID RAIN

A. The Process

Acid deposition, or acid rain as it is commonly referred to, is a recent addition to our language.* As of 1984, the term had not yet been added to any dictionary. Acid rain refers to both wet and dry acidic deposits which occur both within and downwind of areas of major industrial emission of sulfur dioxide (SO_2) and the oxides of nitrogen (NO_x). After SO_2 and NO_x are emitted into the atmosphere, they are transformed into sulfate or nitrate particles and by the subsequent contact and combination with water vapor are transformed into sulfuric and nitric acids. The process is shown in Figure 3.

In eastern North America, synthetic emissions of SO_2 account for approximately 90% of the total sulfur moving through the atmosphere. Table 3 lists the major sources: electric utilities, industrial, commercial,

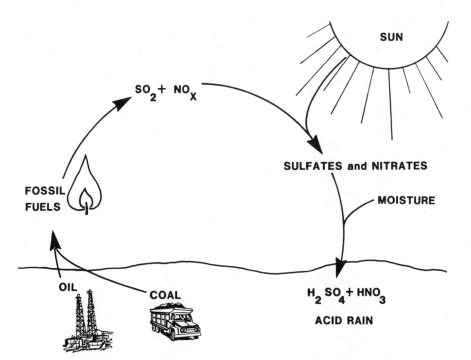

Figure 3. Schematic representation of the formation of acid rain deposition.

* In fact, Robert Angus Smith described the condition in Manchester, England, in 1872.

Table 3
SO$_2$ Emissions for 1980 (tons/year)

Electric utilities	15,800,000
Heating	3,200,000
Industrial processes	2,900,000
Smelters	1,400,000
Transportation	800,000

and residential heating, and smelters. Diesel engines and marine and rail transport are additional significant sources.

About 60% of the total emission occurs in the eastern portion of the United States, particularly in the highly industrialized central states.

Emissions of synthetic NO$_x$ follow a similar pattern. The north-central states account for most of the nitrogen oxides moving across the United States and Canada. The largest sources of these effluents result from vehicular transportation (cars, trucks, buses, and planes), electric utilities, and nonutility fuel combustion facilities. Table 4 lists the yearly tonnage of NO$_x$ and its primary emitting sources.

Some pollutants are deposited as dry fallout. Others react with water vapor to form H$_2$SO$_4$ and HNO$_3$ and return to earth as acidic dew, drizzle, fog, sleet, snow, and rain. Pollutants are often carried hundreds or even thousands of miles by prevailing winds. Consequently, political or geographic boundaries pose no obstacle. This phenomenon is referred to as long-range transport of airborne pollutants (LRTAP). Canadian scientists estimate that as much as 50% of the acid rain falling in Canada comes from U.S. sources. Upward of 10% of the acid precipitation falling in the northeastern United States comes from Canadian sources. However, a growing body of data clearly shows that acid rain is developing into a national problem; no longer a matter of concern just to New England and New York's Adirondack Mountains.

The most vulnerable areas of the continent are those in which the soils are thin and the bedrock granite—which cannot adequately neutralize the action of the acids. Acidification of lakes with the loss of all biological

Table 4
NO$_x$ Emissions for 1980 (tons/year)

Transportation	8,000,000
Electric utilities	5,000,000
Industrial boilers and process heaters	3,500,000
Resident and commercial	700,000
Other industrial processes	700,000
Miscellaneous	300,000

forms is extensive. Over 200 lakes in the Adirondacks no longer support
fish and thousands more are slowly losing their capacity to buffer acid
rain (1, 2).

B. pH

The acidity of aqueous solutions is most often indicated by pH, which is a
logarithmic scale from 0 to 14. Seven is neutrality. Values below 7 are
acidic while those about 7 are basic or alkaline. Being a "log" scale, each
unit represents a change in value 10 times that of either the preceding or
following value. And a change from 7 to 5 represents an increase in
acidity of 100 times. Accordingly, the lower the number, the greater the
acidity. The higher the number, the greater the alkalinity. Figure 4 shows
a pH scale with values for a selected number of familiar solutions. At pH
5.6–6.0, "clean rain" is somewhat acidic. Below this level precipitation is
considered abnormally acidic. By way of comparison, lemon juice and
battery acid are 1000 and 10,000 times more acidic. But it is not necessary
to be all that acidic to produce harmful effects. Most biological species
live within narrowly defined pH levels. Slight shifts can be harmful to
fatal.

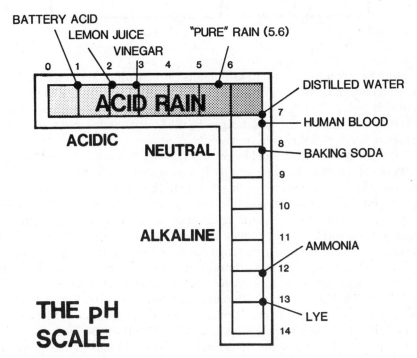

Figure 4. Scale of pH values for acid rain and a selected number of common liquids.

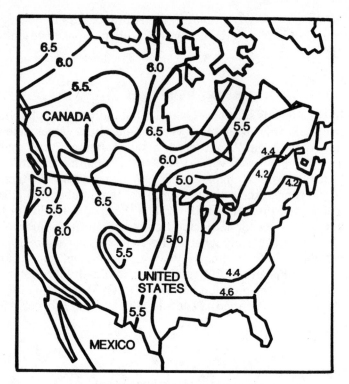

Figure 5. Patterns of precipitation.

More than 2 million square kilometers of the United States and Canada now receive rain with a pH of 4.6 or lower, that is, ten or more times as acidic as "clean" or normal rain. Figure 5 shows the current pattern of precipitation. East of the Mississippi in the United States and Canada pH values of 4.5 and below are prevalent.

Lakes and rivers with a pH below 4.5 will not support fish populations. Even at pH 5.0 only limited populations can survive: frogs, salamanders, insects, algae, and other species through the ecologic food chain are decimated. Fish taken from acidic lakes have excessively high mercury, aluminum, lead, copper, and cadmium in their tissues, presumably leached from soil by acidic moisture percolating through. Of course, these same lakes are sources of drinking water for surrounding communities.

Currently, it appears to be emissions of SO₂ that produce the most extensive and measurable environmental damage. A 1980 study completed by the University of Wyoming for the U.S. EPA found that the dollar damage to natural and synthetic resources from acid rain in the eastern third of the United States exceeded $5 billion per year.

But eastern United States is no longer the sole focus of the problem. Recent studies have shown the Rocky Mountains to be threatened with the same devastation that has degraded the Appalachian Mountain ecosystems. The commanding features that lend such character to the West also make it vulnerable to acid deposition. High mountain watersheds with their thin soil, rocky outcroppings, and sparse flora lack the requisite chemical capability for neutralizing the acidity.

In November 1984, a team of scientists from the State College of Environmental Science and Forestry at Syracuse found, after a year-long study of hundreds of reports on acid rain, that, "based on the available evidence, we cannot attribute the decline of forests solely to acid rain." Robert L. Burgess, who headed the study team, remarked that "everywhere we look we can document forest decline. What we haven't been able to do is unequivocally tie that decline to acid rain" (3).

To further muddy the waters, a team from the University of New Hampshire tested the acidity of ice core samples from the Antarctic and Himalayan Mountains. They found the samples, one of them 350 years old, laden with acid. It seems acid rain was around prior to the industrial revolution. If Mother Nature has been creating and coping, more or less, with acid rain for centuries, it will not be easy to make a case, certainly not an urgent case, for regulatory action, especially with a conservative administration, to keep human activities from tipping a delicate balance.

C. Health Effects

And what of acid rain's effect on the health of people? Here too the issue is far from clear. For example, in highly acidic water (below pH 5.5) female fish fail to reproduce. The acid triggers a depletion of calcium from bone producing skeletal deformities, and metals such as aluminum, normally tightly bound in soils, leach into water where they pile up in fish gills, smothering them. (Curiously enough, the more acidic a lake becomes, the more "crystal clear" and visually appealing it becomes. The lakes have a healthy appearance.) Effects in lower forms of life suggest human consequences; but again these are not always translatable.

1. Ocular Damage

The possibility that acidified water may cause eye injury to swimmers and others using recreational areas has kept many people from them. P. K. Basu and colleagues at the University of Toronto's Department of Ophthalmology and Preventive Medicine bathed the eyes of human volunteers and rabbits with water from lakes whose pH was as low as 4.5, equal to the level of most acidified lakes. He found lack of ocular congestion, corneal damage, or irritation. Morphological examination using both light and scanning electron microscopy confirmed the lack of ocular effects as a result of short-term contact (4, 5). In a second study using 190 rabbits, one eye of each animal was exposed to samples of water with pH values of

5.18 to 4.5. The other eye was simultaneously exposed to a sample having a pH of 6.40 or 6.21. The water was continuously instilled for 15 min every day for 7 days. Again, no significant differences were found between pH of lake water and eye irritation in the animals tested.

2. Indirect Effects

Every year hundreds of thousands of people lose their ability to remember names, places, and events. Most of them eventually are diagnosed as having organic dementia. The frontal lobes of the brain become atrophic and senile plaques and tangles of neurofibrils abound. In 1907 a German neurologist, Alois Alzheimer, recognized these signs as a specific clinical and pathological entity. The fact that this condition antedates the human-induced rise of acid deposition and is seen in all areas of the world, many free of such deposition, is a factor to bear in mind. Nevertheless, in the past few years a great deal of research activity has given the study of Alzheimer's disease new prominence.

3. Alzheimer's Disease and Aluminum

Evidence that aluminum deposition in brain tissue may contribute to this condition comes from several directions. Scientists at the National Institute of Neurological Diseases and stroke found that injecting aluminum salts into rabbits and cats (but not monkeys, rats, or mice) generated neurofibrillary tangles. Interestingly enough, Allen Alfrey at the University of Colorado School of Medicine described an irreversible dementia in people who had repeated kidney dialysis with dialysate high in aluminum salts (6).

Aluminum intoxication or the dialysis encephalopathy syndrome was suggested as early as 1972. Balance studies performed on a group of uremic patients receiving aluminum-containing phosphate-binding gels found positive aluminum balance (7). With the advent of flameless atomic absorption techniques, a sensitive procedure for determining tissue aluminum content down to ppb became available and has been able to demonstrate that tissue stores of aluminum are markedly increased in individuals on dialysis (8).

The association of aluminum with memory loss was first demonstrated in 1921 in a metalworker. In 1926, the potential health hazards of aluminum in drinking water and certain medications was reported (9). In 1983, Wills and Savory remarked that "the increased tissue content of aluminum appears to be an important factor in the pathogenesis of a progressive encephalopathy" (8).

These reports indicate that aluminum does have an affinity for brain tissue. It is also well known that aluminum, abundantly distributed in the earth, is insoluble in neutral and alkaline water and therefore biologically unavailable. However, as a consequence of "wet" acid deposition, the concentration of aluminum in water supplies can be greatly increased. It is being suggested that the continued ingestion of aluminum-containing water may be responsible for the unusually high aluminum concentrations found in brain tissue of some individuals with Alzheimer's disease. Support for this association comes from studies in Japan, Guam, and New Guinea (10).

Amyotrophic lateral sclerosis (ALS), or Lou Gehrig's disease, is known to be substantially higher on the Island of Guam compared to the continental United States. Analysis of brain tissue of ALS victims has shown both increased numbers of neurofibrillary tangles and aluminum deposits. Similar findings have been reported among Japanese living on the Kii Peninsula.

All the above notwithstanding, the evidence implicating aluminum remains circumstantial. Recall Chapter 3 and what has been shown about ALS on Guam. Currently, in addition to the toxin theory, which includes aluminum, five additional theories purport to account for Alzheimer's disease. Originally, Alzheimer ascribed the condition to "hardening of the arteries." He may not have been far from the mark. The current *blood-flow theory* associates the disease with a substantial reduction in the amount of blood reaching the brain, in the amount of oxygen and glucose extracted, and energy generated from both oxygen and glucose.

The genetic model posits an inheritable factor. There are families in which the incidence of Alzheimer's disease is unusually high. Is a faulty gene at work? Does abnormal DNA render an individual susceptible to some environmental factor? Support for a genetic etiology comes from the well-known finding that upward of 80% of all people with Down's syndrome develop Alzheimer's by the age of 40.

The *abnormal-protein theory* proceeds from the clear presence of neurofibrillary tangles, the amyloid which surrounds and invades cerebral blood vessels as well as the amyloid-rich plaques that replace degenerating nerve terminals. Each of these represents an accumulation of protein not normally found in the brain.

There is also an *infectious-agent theory* which postulates an agent of the slow-virus group reminiscent of Scrapie and Creutzfeldt–Jakob disease. If Alzheimer's disease is the result of an infectious agent, it should be transmissible. Thus far attempts to do just that have been unsuccessful.

In 1976 a report of biochemical abnormality associated with Alzheimer's disease appeared. It indicated that the level of the enzyme choline acetyltransferase was reduced by as much as 90% in Alzheimer's patients. This suggests an explanation for the disease's cardinal symptom, loss of memory. Thus, challenges to the evidence implicating aluminum come from a number of directions. It is, of course, possible that aluminum cannot by itself produce the clinical and pathologic indicators of the disease. Its presence may contribute to their appearance in those exposed to other risk factors. It is also possible that aluminum has an affinity for brain but that its presence is of no adverse consequence. Unraveling this intricate and complex web will take time. Meanwhile, any conclusion on aluminum would be premature.

Aluminum is not the only heavy metal mobilized by wet acid deposition. Lead, mercury, cadmium, and cooper are also found in increasing concentrations.

4. Mercury

Evidence is accumulating suggesting that acid deposition alters the biogeochemical cycle of metals. Mercury appears to be among those primarily affected. The toxicity of methyl mercury has been well documented since the poisonings in Minimata, Japan, due to consumption of methyl-mercury-contaminated fish. In this instance and in others, organic mercury proved to be a neurological toxin.

Recent reviews (11, 12) establish fish and fish products as the primary mercury risk factor for human populations. Increasing evidence suggests a strong correlation between pH levels of lakes and mercury levels in fish. A model developed by Hakanson (13) supports this contention. Metallic mercury appears also to be mobilized at a greater rate by low pH levels. Its availability for methylation is also increased, as is its solubility. Decreased pH appears to be related to greater retention of inorganic mercury in waterways, thus increasing exposure of fish to mercury. However, most authorities agree that the chemistry and appropriate sampling of mercury are highly complex. The many variables involved have either not been controlled or measured in many studies to date. Nevertheless, a recent study noted that 30% of total mercury in samples taken from Canadian and Japanese rivers could be accounted for by methyl mercury. Mercuric mercury (Hg^{2+}) accounted for approximately 50%. Two important conclusions can be drawn from these data: (1) precipitation is an important source of mercury in fresh water, and (2) the level of mercury in drinking water is too low to be a health threat. Accordingly, additional mobilization of mercury into water by acidic deposition should not pose a threat to health.

5. Lead

Although it appears clear that acid rain contributes to the distribution of lead in air, water, and food, it is not (yet) possible to quantitate the potential increase in lead exposure attributable to acid rain. Acidification of drinking water increases the solubility of lead in water. Individuals who obtain potable supplies from sources other than large distribution systems may be at increased risk.

A number of biological effects of lead accumulation have been documented: kidney damage, alteration of vitamin D metabolism, central and peripheral nervous system toxicity, and depressed biosynthesis of protein, nerve, and red-blood-cell formation.

In addition, it was recently demonstrated that "low" lead concentrations (less than 10 ppm lower than doses that are clearly symptomatic) are neurotoxic. The investigators maintain that their data "may have relevance to the current regulatory and legislative initiatives directed at the control of airborne lead" (14).

6. Copper

Copper is an essential element for plants and animals. A limit of 1 mg/L of copper in domestic drinking water has been set by the EPA on the basis of taste rather than toxicity. There is no legislated limit for copper in food. For otherwise healthy people, copper deficiency is more widespread than copper toxicity (15). As with aluminum, lead, and mercury, decreasing pH increases the availability of copper. However, current levels do not appear to warrant serious concern.

7. Cadmium

Elevated cadmium levels have been recorded in both rainfall and drinking water supplies in regions receiving acidic deposition (16). Cadmium is the most mobile of the common heavy metals. This mobility is highly dependent on the pH of soil moisture. Acid water holds between 10 and 100 times more cadmium than neutral water.

Absorption of cadmium by plants (crops) increases with declining pH and increasing cadmium concentrations in soil. For the human population, the most serious concern is the increase in its accumulation in agricultural lands.

The major adverse human health effect associated with long-term low-level ingestion of cadmium is renal disease. Although inhalation of cadmium is known to produce lung disease, and possibly cancer of the lung, the primary contribution of cadmium to the total body burden is from ingestion not inhalation.

Accumulation is greatest in the kidney's renal cortex. The level that produces disease is not yet clear. However, because there is no acceptable method for removing this metal from the body, prudence dictates concern for activities which may add to exposure.

IV. GENERAL HEALTH EFFECTS OF AIR POLLUTANTS: EPIDEMIOLOGIC CONSIDERATIONS

The human health effects of low-level air pollution have been extremely difficult to ascertain. To set appropriate air quality standards adequate to protect the health of the public generally, regulatory agencies require reliable data on the consequences of various pollutants, Tom Lehrer's clever lyrics notwithstanding.*

For standards to be developed, data from human and animal toxicological studies, epidemiological studies, atmospheric chemistry, physiology, clinical medicine, and air monitoring are needed. Scientists obtain these data in two ways. They can create a "polluted environment" within a laboratory setting which approximates the ambient environment. Volunteers, who may or may not be representative of the groups at highest risk within the target community, are invited for testing. These "subjects" must then be exposed to well-controlled conditions and they must be tested for responses to the exposures. Perhaps most essential, the test subjects must be compared to unexposed controls.

Another, more realistic approach, but fraught with difficulties and uncertainties, is to choose two communities, one known or believed to be exposed to regular and continuous pollution, the other free of or with far less of the pollutant in question. Individuals from both communities must be tested and compared to determine differences or similarities that may be related to the specific exposure.

The effect that an air pollutant has on health is related to the absorbed dose. The concentration of the pollutant depends on a variety of factors, only one of which is the amount emitted (this must be known). The location of the source and the meteorological conditions are others. The amount deposited in the lungs, the absorbed dose, is dependent on the concentration to which an individual is exposed, the amount of time exposed, as well as an individual's unique physiological characteristics. Consequently, the effects of a specific absorbed dose can be expected to vary considerably in a population, further complicating interpretation. Theoretically, the following spectrum of effects may occur:

- A few may die.
- Some may become ill.
- A larger number will indicate annoyance or discomfort.
- A still larger number will exhibit physiological changes of uncertain significance.
- Most will show no reaction at all to the exposure.

---

* If you visit an American city
  You'll find it very pretty
  Just two things of which you must be aware,
  Don't drink the water and don't breathe the air.
  Pollution, Pollution—
  Wear a gas mask and a veil
  Then you can breathe, long as you don't inhale.

From *That Was the Year That Was*. Music and lyrics by Tom Lehrer, 1965.

For the most part, scientists searching for relationships between exposure and health effects do not have the full spectrum with which to deal. Most often death and evident illness are criteria unavailable to them, as such severe effects rarely occur. The must struggle with comparisons between annoyance and no reaction at all. Consider for a moment the range of subjectivity to be encountered in descriptors such as "annoyance" and "discomfort."

To establish a relationship between an air pollutant and a health effect, the first obstacle that must be surmounted is a definition of what constitutes an "adverse health effect." Then one must ascertain if in fact there are differences in health effects in a matched community. P. R. Sherwin (17) recently proposed an approach to this problem.

He suggested that an adverse health effect has the potential for lowering the quality of life, causing a disabling illness or leading to premature death. This definition focuses on the earliest state of disease, and to detect deviations from "normal" it requires the search for minor adverse effects on cells and tissues. This poses complications and raises barriers which current state-of-the-art tests are unable to meet. The available tools simply are not sensitive enough. Symptoms far more evident have been undiagnosed or misdiagnosed.

Two recent reports focused on this problem (18, 19). Both were concerned with the need for increased use of the autopsy to better clarify causes of death. Misdiagnoses are increasing in the United States. Death certificates are wrong as to underlying cause of death from 25 to 42% of the time. Not only does this cast doubt on the diagnostic acumen of our physicians, but the error is extremely critical because control measures instituted on the basis of an erroneous cause of death may be aimed at the wrong target.

The autopsy, of course, provides the only way one can positively identify what a person died of. Another important concern is autopsy frequency. More autopsies are performed on individuals in the 30–40-year age range than on older people. The ratios are 1 in 6 in the range 60–70 years; 1 in 10 from 70 to 80; and 1 in 16 beyond 80. This situation is inherently bad for the study of adverse reactions because it is in the older members of our population that the diseases most likely due to air contaminants would have the opportunity to become developed sufficiently to be detectable.

In addition, there are a multiplicity of confounding factors which, most often in combination, make epidemiologic and toxicologic studies increasingly difficult to evaluate. Some of these are:

- Estimation of cumulative exposure
- Cigarette smoking
- Population density

- Socioeconomic status
- Chemicals in the indoor environment
- Age, sex, and race
- Mobility of the population
- Occupational factors
- Lack of consensus about which pollutants to monitor

In either of the two approaches to air pollution studies, one of the most complex tasks is to adjust for or control these various influences as they affect either mortality or morbidity. Then, of course, the regulatory agency is faced with interpreting studies conducted on animals or people exposed to a single pollutant at a constant concentration when in reality the body is subjected to mixtures of pollutants whose concentration varies continually with time.

As if these were not problems enough, experimental studies with human subjects have ethical limitations. Studies with animals have the inherent difficulty of having their effects translated to people. Doses given to animals are often far higher than those to which human populations would be subject. Not only have these considerations placed formidable obstacles to setting of standards, they have over the past 30 years been responsible for the inability of scientists to make definitive statements about the effects of air pollution on health. Nevertheless, the quality of air over most cities has improved. This has further complicated the problem. As the level of pollutants decreases (the dose), it becomes all the more difficult to relate exposure to effect.

Recently, Dr. Alice S. Whittmore of the Department of Family, Community and Preventive Medicine, Stanford University School of Medicine, reviewed the studies attempting to relate air pollution to specific health effects (20). She remarked that "prognosis is poor for reliable quantitative data relating chronic pollutant exposure to respiratory damage." She went on to say that "the existence of respiratory damage due to air pollution has been established beyond reasonable doubt, nevertheless, little of the work that has been reviewed can be used to obtain reliable estimates of effects at specified levels. None of the data discussed," she continued, "indicate that current or even moderately less stringent standards are inadequate to protect the public health. On the other hand, there is a dearth of reliable data indicating that standards can be realized."

Most scientists dislike this state of affairs and, being inherently conservative, would tend to agree with her estimation. There is, however, a frustrating quality about it all. It must be there so why can't we find it? Yet when one looks at the leading causes of death (Figure 10, Chapter 14), it is exceedingly difficult to relate any of them to air pollution. One can find reports of both positive and negative effects. As Hackney properly noted, "statistically significant changes with exposure may not always have medical significance or imply risk to public health. They may represent an appropriate homeostatic response to an entirely tolerable stress, or may reflect a minor annoyance rather than a meaningful threat to health" (21). But that will not, should not, deter continuing study. Failure to find statistically significant changes does not necessarily rule out risk.

The most ambitious study to date has been mounted by a group at the School of Public Health, Harvard University (22). Theirs was a six-city study designed to test the adequacy of the current SO_2 and particulate matter standards. The question is, should they be relaxed or made more stringent? The study examined the chronic effects of pollutants in six cities: two clean—Topeka, Kansas, and Portage, Wisconsin; two with levels slightly below the primary standards—Watertown, Massachusetts, and Kingston-Harrison, Tennessee; and two dirty cities, Steubenville, Ohio, and the southern tip of St. Louis, Missouri, Some 1800 people age 25–74 have been enrolled and will be followed prospectively for 14 years. Final results should be available by 1990. Ten years into the study little if any differences in levels of respiratory damage between the six cities have been found.

V. THE GREENHOUSE EFFECT

Fossil fuels, coal, oil, gas, and wood have been the primary sources of energy for centuries. Energy, especially economic electrical, has been closely linked with the Western world's enviable economic growth. The burning of fossil fuel (combustion) is a process in which these hydrocarbons break down to yield their energy content along with carbon dioxide and water. Combustion is the process that powers our cars, heats our homes, and literally drives our industry. Altogether, coal, oil, and natural gas provide some 20 quadrillion Btu annually.

But there is a price for all this energy. That price is the production of carbon dioxide. Because of its natural presence in the atmosphere and its fundamental role in photosynthesis and respiration, carbon dioxide (CO_2) has not been seen as an air pollutant. The availability, circulation, and interaction of carbon, nitrogen, and sulfur have been essential for the development of life as we know it.

The availability and interaction of these elements is often referred to as biogeochemical cycles. Five primary cycles are recognized.

- The hydrologic cycle, which describes the movement of all water, including water vapor
- The nitrogen cycle
- The sulfur cycle
- The phosphorus cycle
- The carbon dioxide cycle

A. Sources and Levels of CO_2

Ordinarily, the primary sources of CO_2 to the atmosphere would be the microbial decomposition of organic matter, respiration by all animals, and gas exchange in the oceans. However, the advent of modern industry with its widespread combustion of fossil fuels has released far larger amounts of carbon dioxide into the atmosphere. Figure 6 represents the carbon dioxide cycle.

According to the accumulating evidence, combinations of gases—carbon dioxide, chlorofluorocarbons, and methane—have been pouring into the atmosphere and stratosphere encircling the planet like a band of insulation. This insulating barrier acts much like a greenhouse, and like a greenhouse, temperatures within rise. But that is only part of the problem. Several adverse effects are occurring simultaneously. We shall look at each separately, then at the entire problem.

1. The Greenhouse

In wide use in Europe by the 17th century, greenhouses were devices for keeping plants from freezing to death in winter. They were equipped with glass roofs (and in many instances siding as well) that prevented passage or removal of heat from the house and temperatures rose. Without proper ventilation, especially in summer, the plants would burn up.

2. CO_2

The accumulated insulating ring of CO_2 now encircling the earth acts as a greenhouse without a ventilator. Energy from the sun is trapped by the CO_2 as it radiates upward from the earth's surface.

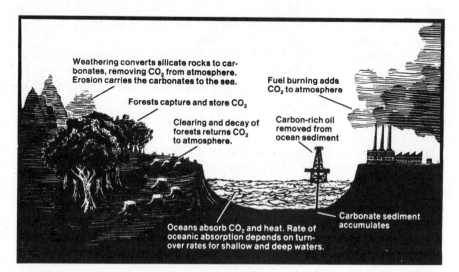

Figure 6. The carbon dioxide cycle: how CO_2 is discharged into and removed from the atmosphere.

John Tyndall appears to have been the first to raise the carbon dioxide alarm. In 1861 he noted that sunlight beaming earthward consists of ultraviolet (UV), gamma, and visible radiation. Ozone absorbs high-intensity UV and gamma radiation, keeping these from the earth's surface. That is a key, and a second problem. Keep it in mind as we proceed. The unabsorbed UV and visible radiation proceed downward, providing the warmth we associate with sunlight.

Tyndall correctly hypothesized that as the earth was not a perfect emitter, some radiation was absorbed. It was the absorption of radiation that caused the surface of the planet to warm.

At night this absorbed radiation is reemitted to the atmosphere. The emitted radiation is of a lower intensity and is easily trapped by certain atmospheric gases. Carbon dioxide is a very good absorber of low-intensity radiation. The presence of carbon dioxide in the atmosphere means that some radiation will remain in the lower atmosphere, instead of being lost into space. A small fraction of the radiation is trapped, causing a net effect of keeping the earth warmer than if the radiation was not present. This "greenhouse effect" is shown in Figure 7.

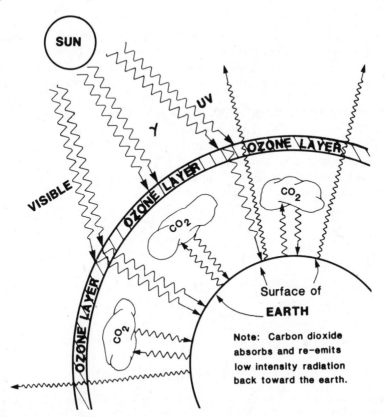

Figure 7. Diagrammatic portrayal of the greenhouse or warming effect of carbon dioxide on the earth.

Table 5
Carbon Dioxide Concentrations[a] in Air, 1958–1982

Year	Mauna Loa Hawaii	South Pole Antarctica	Year	Mauna Loa Hawaii	South Pole Antarctica	Year	Mauna Loa Hawaii	South Pole Antarctica
1958	NA[b]	315	1966	321	320	1974	330	328
1959	316	316	1967	321	321	1975	331	329
1960	317	316	1968	322	322	1976	332	330
1961	317	317	1969	324	322	1977	334	332
1962	318	318	1970	325	324	1978	335	333
1963	319	318	1971	326	325	1979	337	335
1964	319	319	1972	327	326	1980	338	337
1965	320	319	1973	329	327	1981	339	338
						1982	341	339

[a] Concentrations are average annual concentrations in parts per million (ppm).
[b] NA—Not available.

Sources: 1958–1980: R. B. Bacastow and C. D. Keeling. Atmospheric Carbon Dioxide Concentration, the Observed Airborne Fraction. Courtesy of the Council on Environmental Quality.

In 1896, the Swedish chemist Svanté Arrhenius advanced the hypothesis that an increase in atmospheric concentrations of CO_2 would raise the earth's temperature and lead to changes in rainfall and patterns of soil moisture. But it was not until the more recent detailed calculations of Charles Keeling of the Scripps Oceanographic Institute that the steadily increasing concentrations of CO_2 became a subject of heated discussion. Table 5 shows the CO_2 concentrations in air between 1958 and 1982. Figure 8 makes these increases readily apparent. Over the 25 years there has been an increase of approximately 0.3% per year.

Using estimates of atmospheric CO_2 at the inception of the Industrial Revolution, scientists calculate that an 18.6% increase has occurred. The

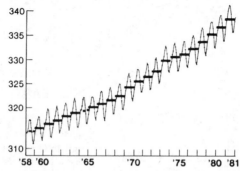

Figure 8. Variations in CO_2 levels. Horizontal bars represent mean concentrations in CO_2; peaks and valleys show seasonal oscillation. CO_2 concentration is given in parts per million. Readings were taken at the observatory at Mauna Loa in Hawaii. Reproduced with permission of the National Research Council.

possibility that a significant warming is under way is serious enough to have persuaded many people to call for reducing the hazard by sharply restricting the use of coal and other fossil fuels and reconsidering nuclear energy. By doing so, it is believed the current level of 340 ppm would be held to 415 ppm by the year 2050 (23, 24).

If the current rate of buildup continues, it is believed that between the years 2025 and 2050, the temperature of the earth's surface may increase by an average of 3–9°F. This could produce a rise in sea level of 1–4 ft. At the same time, the additional heat is expected to cause inland waters such as the Great Lakes to recede.

B. Ozone

A second look at Figure 7 is in order. The ozone layer between the earth and the stratosphere plays a most important role in permitting life on earth to occur. Without it, or with a substantially depleted layer, gamma and UV radiation would strike the earth incessantly. Anything that affects ozone reduction could result in reduced crop yields along with genetic changes in plants. The incidence of skin cancer would increase sharply as would cataracts and blindness. Damage to DNA could be reflected in increased birth defects. Most serious of all could be the reduction of photosynthesis in plankton, the basis of the oceans' food chain. How might a decrease in ozone occur?

Fluorocarbons released by refrigeration coolants, spray cans, and foams are a major source of damage to the ozone layer. Despite the virtual elimination of freon propellants from American spray cans, freon remains unrestricted on a worldwide basis, and freon emissions from air conditioners and other sources continue.

Freon, a refrigerant also referred to as fluorocarbon or fluorochloromethane, is a synthetic compound of fluorine, chlorine, and carbon. When freon molecules reach the stratosphere they are split by sunlight, releasing chlorine that forms chlorine monoxide. This then acts, over and over again, to break up ozone molecules, an unstable form of oxygen. The reactions appear to be

$$(1) \quad CFCl_3 + UV \longrightarrow CFCl + Cl$$

$$(2) \quad Cl + O_3 \longrightarrow ClO + O_2$$

$$(3) \quad ClO + O \longrightarrow Cl + O_2$$

$$(4) \quad Cl + O_3 \longrightarrow ClO + O_2$$

In reaction (1) a free chlorine atom is released. This atom reacts with a molecule of ozone, producing chlorine monoxide and an oxygen molecule (2). Chlorine monoxide reacts with atomic oxygen, producing oxygen gas and a free chlorine atom (3). In (4) we see the free chlorine atom

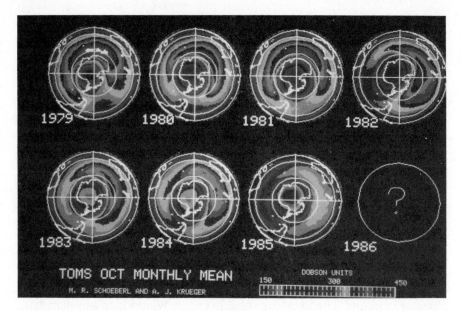

Figure 9.

combining with another ozone molecule and the chain reaction continues with the reduction of ozone and its protective layer.

It is the stratospheric ozone belt some 15–20 mi above the earth that has prevented lethal wavelengths of UV light from reaching the earth.

In March 1985, British scientists working in Antarctica reported that the amount of ozone over Halley's Bay had declined from 300 Dobson units to 180, a reduction of over 40% in the 8 years between 1977 and 1984. Later that year, NASA scientists produced satellite data confirming the British findings. There was a hole in the ozone layer over the Antarctic larger than the size of the continental United States. Figure 9, a series of computer-generated plots of data obtained by a total ozone mapping spectrometer (TOMS) aboard NASA's Nimbus 7 satellite, shows the growth of the hole. Note particularly the changes between 1982 and 1985. The views are from a point above the South Pole. The southern edges of Africa, Australia, and South America are visible at the sides. The depletion of the protective ozone layer from 8 to 16 mi up is extreme at the center where the levels have fallen since 1979. This discovery is cause for concern. Alarm is not inappropriate. Although ozone makes up less than 1 ppm of the atmospheric gases, it absorbs most of the sun's UV rays. It is, in effect, the earth's UV shield. Should UV light become more intense, the problems noted earlier could readily occur.

C. Methane

If one person can be said to have his fingers on the pulse of atmospheric pollutants, it is Sherwood Rowlands, along with his colleague Mario F.

Molina, both of the University of California at Irvine. In 1974, they reported on the increasing use of chlorofluorocarbons (CFCs) and before that were concerned about potentially harmful effects to the atmosphere of emissions of nitrogen oxides and water vapor from a fleet of supersonic transports (SSTs) that Boeing Aircraft was planning to build. These 1800 mi/h, high-altitude planes were a storm center of controversy and never made it off the drawing boards.

More recently, Rowland and Donald R. Blake have warned about the significant increases in atmospheric methane and its contribution to the greenhouse effect.

Although methane, or marsh gas, is the major component of natural gas, about 80% of current atmospheric methane comes from decomposition in rice paddies, swamps, garbage, and the intestines of ruminants, especially cows. According to Rowland, each cow produces about a half-pound of methane daily. Considering there are upward of a billion and a quarter cows around the world, there may be enough methane to raise the earth's temperature 5°F by the year 2050.

However, they also raise a curious paradox. On the one hand, methane may increase the size of the hole over the Antarctic but it may also reduce the ozone depletion over the rest of the world.

As methane rises into the polar stratosphere it breaks down, releasing hydrogen which combines with free hydroxyl groups (OH^-) to form water which freezes into clouds. Cloud formation abets widening of the ozone hole. Outside Antarctica, wind and higher temperatures prevent stratospheric cloud formation. And by attacking chlorine, methane may reduce the amount available to destroy ozone.

D. Ocean Sediments

Warming of the atmosphere clearly suggests a concomitant rise in global sea levels. However, recent deep drilling along both Antarctica's eastern and western coasts tells a different story. Sediment cores taken from over 4000 ft beneath the ocean floor reveal a history dating back 60 million years. Not only do these cores, shown in Figure 10, confirm theories that the Antarctic continent had a temperature climate, but more importantly, it contradicts the belief that west Antarctica occasionally sheds its ice and rapidly raises global sea levels. Sediments taken from the Weddel Sea show no evidence of ice loss over the past 5 million years. The evidence does indicate that though the climate at times was almost subtropical, the west Antarctic ice remained intact.*

An additional consideration must be introduced into the discussion of a warming trend. Remember El Niño? (See Chapter 1.) El Niño made its

* Disaster preachers led by Richard Kieninger of Adelphi, Texas, predict that on May 5, 2000, earthquakes will rumble across the globe. They believe that a massive buildup of ice at the South Pole will trigger a series of earthquakes, tidal waves, and volcanic eruptions—to welcome the millennium.

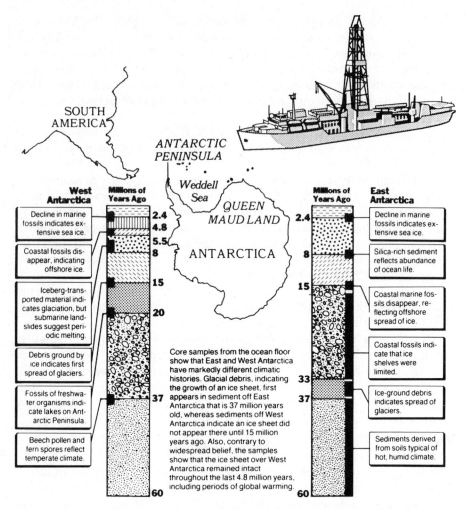

Figure 10. Ocean sediments reveal Antarctica's climatic history. Reproduced with permission of *The New York Times*.

last appearance in January 1986 and lasted until March 1988. During those 27 months, it wrought climatic havoc and undoubtedly contributed to an already overly warm decade. The fact of a warming trend seems conclusive. Only the cause remains uncertain. Should 1988 and 1989 remain unusually warm, without the presence of El Niño, strength of association would tilt toward an advance of the greenhouse effect. The year 1990 could be the year of decision.

E. Recapitulation

New and startling data indicate a worldwide warming trend as a consequence of years of polluting the atmosphere with a variety of chemicals.

This warming trend is seen as bringing with it drought in our grain-producing states and increased sea levels around our southeastern port cities, with all the consequent dislocations.

On the other hand, new and startling discoveries indicate that the Antarctic ice has remained stable through eons of warming and cooling without loss of ice.

Decisionmakers are now faced with a conundrum. Where do we place our efforts and our money? This will be the eye of the coming storm. Nevertheless, swift action can be taken. Planning must begin for major reductions in the use of fossil fuel. And because trees absorb and store CO_2, a vigorous worldwide program of reforestation must get underway. More efficient use of energy needs to be instituted to reduce fuel use, and like France, we must make a national commitment to nuclear power.

Of all our environmentally related problems, this one is potentially the most consequential. I am not alone in believing this. Nevertheless, political concern is lacking. Our elected officials are not talking about it. This may well be the "hot potato" no one wants to touch—at least not until it reaches crisis proportions. That would fit the political mentality that prefers not to be associated with causes or needs requiring action or expense. One is bad enough; the two are deadly, politically speaking, especially if the need is seen as upsetting, threatening, or alarming to the electorate. John Dunster, a British politician, stated it with the brevity that remains the soul of wit: "Any politician," he said, "would prefer a dead body to a frightened voter."

Remember, most elected offices are for periods of 4 years or less; consequently the effort must be on those issues which can produce a payoff before the next election. Long-term problems such as acid rain or the greenhouse effect do not really have a chance.

Incinerators, landfills, and hazardous wastes are middle-distance-type problems. Accordingly, they will get some attention, but not enough to disturb the community's equanimity. We are currently moving into a period of multiple community needs, particularly as it relates to waste disposal. At the moment we are in the foot-dragging stage. Because concern implies turning down thermostats in homes, offices, and factories, that means less available heat (read fuel) for warmth at home, but most especially less energy for production at the office and in the factory: perhaps less work and fewer jobs, unless the nuclear power option is reviewed. Is it any wonder then that our elected officials turn a blind eye to the problem. Nothing about it contains votes. On the contrary, it could be unseating.

F. Indoor Air Pollution

Since 1815, when the first air pollution ordinance was enacted in the United States, air pollution has meant outdoor air over our cities, or closely related to industrial sites. Whatever has been gleaned about the association between respiratory or other illnesses has come from this setting. Although many millions of hard to come by tax dollars have been spent and enormous quantities of data generated, the relationship between indoor pollution and illness remains tenuous and elusive.

During the 1960s and 1970s data began to emerge suggesting that many people spend as much as 90% of their time indoors—infants, the aged, and infirm. Those with full-time employment can spend from 40 to 60% of their time at home, while homemakers can be indoors more than 80% of the time. This house-bound condition may persist for years on end.

Between school, jobs, and home, most people are in contact with a variety of indoor environments most of their lives. This must raise questions as to the possible effects on health of the many chemicals released in the home from the following sources:

- Heating
- Cooking (gas)—frying, broiling, baking and boiling
- Hobbies—solvents, cements, dusts
- Dander—pets
- Fibers and dust—rugs, carpets, and furniture
- Pesticides
- Household cleaners—ammonia, chlorine, and floor, tile, and wood polishes
- Paint
- Powders, moth crystals
- Air disinfectants
- Solvents—turpentine, alcohol, paint thinners
- Radon
- Smoking—cigarettes, cigars, and pipes

There is little doubt but that indoor air requires investigation for its possible or potential contribution to illness. Nevertheless, expectations should not run too high. Given the concentrations of pollutants required to produce illness in Donora, London, and the Meuse Valley, it is reasonable to expect that for the most part air may not be the major contributor to disability or death that many believe it to be. Radon may not fit this pattern.

1. Radon

Ernest Rutherford and Stanley Watras have one thing in common. While experimenting with radium, Rutherford discovered radon in his laboratory in 1899. In 1985, Watras discovered that his house was perched atop an outcropping of low-grade uranium ore. And in the year he had lived there it had accumulated enough radon to set off the alarms of the radiation monitors at the Limerick nuclear power plant where he worked.

Radon is a radioactive gas formed from the decay of uranium. Given uranium's radioactive half-life of 4.5 billion years, radon has been with us a long time. Uranium miners in Czechoslovakia and the United States are known to have had cases of lung cancer far in excess of unexposed workers. But until Stanley Watras set off the alarm, our homes were never suspected.

Radon exists in the form of two radioactive isotopes ^{220}Rn with a half-life of 54 s, and ^{222}Rn with a half-life of 3.8 days. Each of these decays away through the series of stages shown in Table 6. With decay, there is emission of energy in the form of alpha and beta particles into the air. As we breathe, some of the radioactive particles, primarily ^{222}Rn, attached to dust are trapped on the moist mucosal surfaces of the bronchus and bronchioles. These packets of energy damage lung tissue, inducing tumor development (25).

As with Watras' house, radon seeps into the premises through cracks in the foundation, the water supply, as well as building and construction materials. As with uranium, the original source of radon is found in soils and rock containing granite, shale, pitchblend, and phosphate; those area of the country with formations of these materials are at increased risk. The uranium-rich Reading Prong, for example, slices through Pennsylvania,

Table 6
Process of Decay of Uranium-238 to Lead-206

Isotope[a]	Half-Life		Emitted Radioactive Particle
Uranium-238	4.5	billion years	Alpha
Thorium-234	24.1	days	Beta
Proactinium-234	1.18	min	Beta, gamma
Uranium-234	250,000	years	Alpha, gamma
Thorium-230	80,000	years	Alpha
Radium-225	1622	years	Alpha, gamma
Radon-222[b]	3.8	days	Alpha
Polonium-218	3.05	min	Alpha
Lead-214	26.8	min	Beta, gamma
Bismuth-214	19.7	min	Beta, gamma
Polonium-214	.0016		Alpha
Lead-210	19.4	years	Beta, gamma
Bismuth-210	5.0	days	Beta
Polonium-210	138.4	days	Gamma
Lead-206	Stable		Stable

[a] Approximately one-third the way through its decay sequence, uranium is transformed to the gas radon. Radon then undergoes a series of four disintegrations, yielding polonium-218, bismuth-214, polonium-214, and lead-214, all with brief half-lives.
[b] Radon-220 is often called thoron because it arises from the decay of thorium and is an alpha emitter.

New Jersey, and New York State from Reading, Pennsylvania, through Boyertown, Watras' hometown, up into Morristown, New Jersey, and Suffern and Peekskill, New York. These are areas which until recently had been countryside, with few homes. The rapid development of these areas brought this problem to light.

Being odorless, tasteless, and nonirritating, radon was never suspected. Note again Figure 11, Chapter 8, with its estimate of the contribution of ionizing radiation to cancer mortality. Clearly, radon, a part of the natural background, may contribute as much as 1% to the overall cancer rate. Estimates indicate a range of from 0.1 to 0.9. Smoking complicates the problem. Those who smoke and live in radon-containing areas are at additional risk. As much as 85% of the presumed radon-related deaths are among moderate to heavy smokers. Consequently, a primary preventive measure for people living in radon-containing homes would be cessation of smoking. Table 7 and Figure 11 indicate the level of risk anticipated from 70 years of continued exposure to radon at various concentrations.

Testing for radon is a relatively simple process. two methods are most common and least expensive. Charcoal canisters provide accurate information in a short time. Sealed charcoal-filled canisters are exposed for 2–3 days. They must then be sent to EPA-approved laboratories for analysis of their level of gamma radiation—which can be correlated to the radon level. Alpha track detectors, or track etch detectors as they are often called, are small cuplike affairs containing plastic strips that are hung about the house like fly paper and on which alpha particles etch a track as they strike. These strips are left in place for 60–90 days. Analysis requires the counting of the microscopic tracks.

Radon is measured in terms of picocuries per liter of air (pCi/L)* as well as working levels (WL).† The EPA has set 4 pCi/L (0.02 WL) as its action level. And they have estimated that regular exposure to 10 pCi/L has a similar risk of lung cancer as smoking one pack of cigarettes per day. Recall, however, that not everyone who smokes gets lung cancer. Radon is no different. And bear in mind that estimated (predicted) adverse effects are based on 70 years of continued exposure and remaining indoors 70% of the time.

2. Indoor Concentrations

High(er) concentrations of chemicals accrue indoors simply because they are contained within a structure. In too many instances these buildings—

---

* The picocurie is equal to 10^{-12} curies.

† Working level month (WLM) is the product of exposure to Rn daughters in working level (WL) multiplied by duration of exposure in "working months" of 170 h. One working level is a combination of Rn daughters that will yield an emission of 1.3×10^5 MeV of alpha-particle energy per liter in air. One WL is assumed to be equivalent to 200 pCi/L of radon.

WL = 0.02
pCi/l = 4

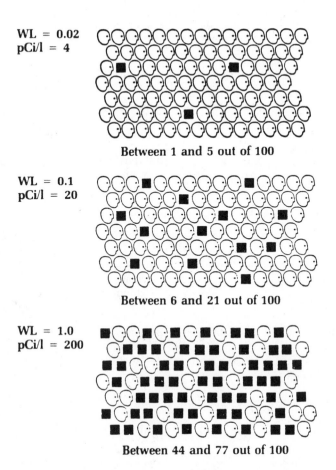

Between 1 and 5 out of 100

WL = 0.1
pCi/l = 20

Between 6 and 21 out of 100

WL = 1.0
pCi/l = 200

Between 44 and 77 out of 100

If these same 100 individuals had lived only 10 years (instead of 70) in houses with radon levels of about 1.0 WL, the number of lung cancer deaths expected would be:

WL = 1.0
pCi/l = 200

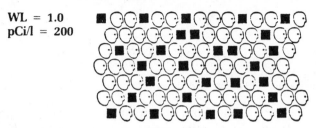

Between 14 and 42 out of 100

Figure 11. Lung cancer deaths associated with exposure to various radon levels over 70 years. Courtesy of the U.S. Environmental Protection Agency.

Table 7
Radon Risk Evaluation Chart

pCi/L	WL	Estimated Number of Lung Cancer Deaths Due to Radon Exposure (out of 1000)	Comparable Exposure Levels	Comparable Risk
200	1	440–770	1000 times average outdoor level	More than 60 times nonsmoker risk
				4 pack-a-day smoker
100	0.5	270–630	100 times average indoor level	
				20,000 chest x-rays per year
40	0.2	120–380		
				2 pack-a-day smoker
20	0.1	60–210	100 times average outdoor level	
				1 pack-a-day smoker
10	0.05	30–120	10 times average indoor level	
				5 times nonsmoker risk
4	0.02	13–50		
				200 chest x-rays per year
2	0.01	7–30	10 times average outdoor level	
				Nonsmoker risk of dying from lung cancer
1	0.005	3–13	Average indoor level	
				20 chest x-rays per year
0.2	0.0001	1–3	Average outdoor level	

Source: U.S. Environmental Protection Agency.

homes—are simply too tight, too well insulated for purposes of heating to allow the free dispersion of other gases. Accordingly, modification of buildup of chemical vapors lies with building occupants. Assuming that rooms are ventilated at appropriate intervals is their responsibility.

Employing an industrial hygienist to determine the rate of air flow as well as the number of air changes per hour can be helpful in establishing a baseline prior to making the required modifications. Of course, the most notorious conditions exist in houses of inveterate cigarette smokers, as well as those who smoke and are among the 2 or 3% with indoor radon levels of some consequence. In either case, increased ventilation is the prescription of choice.

Recently, Burr and colleagues of Cardiff, Wales, examined 72 asthmatic individuals and 72 matched controls aged 15–60, for evidence of allergy to molds in their homes. Positive skin tests to any five molds were found in 38 asthmatics (53%) and 7 (10%) of the controls. Visible mold was found in the homes of 19 of the asthmatics and 9 controls. Penicillium was the mold most often found and to which antibodies occurred most often (26). Exposure to indoor mold may be more of a contributing factor to asthma than previously believed. This is the type of study that needs to be done for a variety of chemicals.

3. House Plants and Pollution

Indoor air pollution has spawned the "sick building syndrome." The key contributors to this contemporary malady are overly tight building construction—a by-product of the energy crisis of the 1970s—with greatly reduced rates of ventilation and major changes in the nature of building materials and household furnishings.

Given their concern for completely closed buildings in outer space, the National Aeronautics and Space Administration (NASA) has been studying the problem for over a decade. In anticipation of long-term manned space flight during the 1990s and into the 2000s, NASA is evaluating the potential contribution of common, and no so common, house plants as systems for purifying air in space stations (27). If laboratory results thus far obtained are borne out in field trials, indoor air pollutants may be substantially reduced economically, effectively, and satisfying.

4. Photosynthesis

The absorption of gases, chemicals, by plants is a fundamental natural process. To live and grow, plants continuously exchange gases with their surrounding environment. Photosynthesis is a uniquely complex process (not fully understood) that produces the relatively elemental overall reaction in which carbon dioxide and water are absorbed and converted into energy-rich carbohydrate in the presence of solar energy. The process is described by the following overly simplified equation:

$$6CO_2 + 6H_2O + 688 \text{ kcal} \longrightarrow C_6H_{12}O_6 + 6O_2 \uparrow$$

carbon dioxide   water   (sunlight and   glucose   oxygen
                                    chlorophyll)

But carbon dioxide* is not the only gas that plant leaves can absorb. According to B. C. Wolverton of NASA's John C. Stennis Space Center, at Bay St. Louis, Mississippi, household plants in laboratory tests have markedly reduced air concentrations of formaldehyde, carbon monoxide, benzene, and trichloroethylene. For example, the common green spider plant, Chlorophytum elatum, and Scindapsus aureus, the golden pothos, proved superior to other plants in removing carbon monoxide and formaldehyde, while the Philodendrons domesticum (elephant ears) and oxycadium (heart leaf) appear to cleanse closed environments of benzene.

Similarly effective in removing benzene are the Gerbera daisy (Gerbera jamesonii) and the Chrysanthemum (Chrysanthemum morifolium). All of these, either in the form of broad leaves (Philodendron) or long and narrow leaves (Chlorophytum) present huge amounts of surface area with their profusion of stomates for efficient gas absorption.

In addition, spider plants effectively reduced levels of nitrogen dioxide. As Wolverton described it, "one spider plant per room where combustion is occurring in conjunction with normal air exchange, would help alleviate the build up of NO_2" (28).

There is a potential downside to the use of plants for purification that must also be investigated. Since plants take in CO_2 and oxygen, the question that must be answered is whether plants placed in environments containing benzene, formaldehyde, and trichloroethylene could give off trace levels of chemicals that could have adverse effects in closed or poorly vented rooms. NASA expects substantive answers to the question of biological purification of closed structures within 2 years.

Nevertheless, to the question, would placing more plants in buildings improve air quality, Wolverton is unequivocal. The more foliage, the healthier the environment is going to be.

VI.   AIR QUALITY POLICY

Air quality policy is not new to the United States. The first emission control ordinance was adopted in Pittsburgh in 1815. Antismoke and antiodor ordinances were later enacted in a number of other industrial cities. The objective of achieving clean air generally finds universal support. What controversy there is centers on the trade-off between costs and benefits, the degree or level of air purity necessary to ensure health.

The thrust of the Clean Air Act is that economic considerations must not dominate. To this end, the act provides a basic structure for the administration of its objectives.

---

* It is the significant removal of trees worldwide that is also contributing to the greenhouse effect. Reforestation can be a major preventive measure for removal of large quantities of carbon dioxide from the air.

Table 8
National Ambient Air Quality (Primary) Standards (NAA QSs)

Pollutant	Standard	Period of Measurement
Ozone (O_3)	0.12 ppm	1 h
Sulfur dioxide (SO_2)	0.14 ppm	24 h
Carbon monoxide (CO)	9 ppm	8-h average
Nitrogen doxide (NO_2)	0.05 ppm	1 year
Total suspended particulates		
H_2SO_4		
$(NH_4)_2SO_4$	260 mg/m³	24 h
NH_4NO_3		
$Fe_2(SO_4)_3$		
Peroxyacetylnitate	None	

A. Standard Setting

Title I of the Clean Air Act provides for "establishing national primary and secondary ambient air quality standards." The U.S. EPA, mandated by Congress to carry out its intent, issued national ambient air quality standards for six pollutants. Table 8 lists each pollutant along with the standard and its average period of measurement. Shown in Table 9 are the alert, warning, and emergency levels for the four major pollutants.

Establishing national ambient air quality standards is a lengthy and complex process. By law the EPA must prepare a criteria document which provides the scientific basis for each standard. Given, however, the less than successful efforts over the past 20 years to relate low-level exposure with illness, these standards must be understood to be more in the nature of guidelines rather than scientific verities. Little if any direction exists to ascertain just how pure air must be if it is to protect human life. William D. Ruckelshaus, former administrator of the EPA, understood this. In testimony before the committee on Environment and Public Works of the U.S. Senate, he stated that "our inability to precisely measure pollution discharges, quantify them, relate them to given concentrations in air and in turn translate those concentrations into known health or environmental effects has resulted in severe enforcement complexities." He then went on to ask: "Is the Clean Air Act causing our country to devote substantial resources for little public benefit?" The question remains to be answered.

Table 9
U.S. Alert, Warning, and Emergency Air Pollutant Level Criteria

Pollutant	Alert	Warning	Emergency
SO_2	0.3	0.6	0.8
CO	15.	30.	40.
NO_2	0.6	1.2	1.6
O_3	0.1	0.4	0.5

B. Innovative Approaches

Responding to the Clean Air Act's directive tht it set nationwide ambient air standards, EPA in its wisdom developed incentives to induce compliance. The act also directs individual states to develop and enforce EPA-approved plans. Thus, it is the states that are primarily responsible for attaining air of high quality.

Four incentive-based reforms (of course they are controversial) have been promoted by EPA for state implementation:

1. The bubble
2. Offsets
3. Netting
4. Emissions banking

The bubble allows emitters to treat their existing sources of air pollution as if they were enclosed by a great dome or bubble. In this way the emitting company can trade reduced control on some of its stacks for extra, compensating reductions on other stacks.

The offset permits new plants, or new additions to established facilities, to operate in dirty air areas if they secure more reductions from their existing sources than the new one will add. Netting, on the other hand, exempts factory modernization plans, from time-consuming scrutiny if reductions elsewhere in the facility can assure that the net emission for the entire plant does not increase.

Finally, emission banking does just what its name implies. Companies can obtain "credits" for surplus reductions they can achieve which can be used or stored (banked) for future use or even sold to other companies if the credit is not deemed necessary for their own use in the near future. Thus, International Harvester, for example, had banked or been credited with hundreds of tons of emission reductions it was able to achieve on its own facility and was therefore able to "sell" those banked tons (at a handsome profit of course) to General Electric, who was unable to meet specific state emission standards (in the same state) in its plant. By buying those credited tons from International Harvester, it allowed its new plant to come on-line, rather than being forced to shut down by the restrictive standards it could not meet at the time. Nor did it have to invest in expensive equipment which would quickly have become obsolete when new processing procedures went into effect.

Apparently, these real-world solutions are being credited with significant environmental improvement. A problem of some consequence is the question of who "keeps the books." Be that as it may, those people who have been walking around with gas masks can safely remove them.

Another innovation currently in its early demonstration stage promises to have greater impact and benefit for communities generally.

Everyday, 6000 buses lumber up New York's avenues and across its streets. Twenty-four hours a day the fleet of local buses spew out the smoke, carbon dioxide, and ozone that makes eyes tear and noses wrinkle. Those are the buses that run on diesel fuel, the use of which is one of the major reasons why the city is unable to meet EPA's air quality standards. But New York is not alone. Cities across the nation, over 50 of them, have similar problems. But there is a remedy: the natural gas bus. By substituting natural gas—which we have in almost unlimited supply—for diesel fuel that generates the polluting fumes, much of the low-level air pollution problems can be eliminated. The Brooklyn Union Gas Company has built two demonstration natural-gas-powered buses for New York City.

Because natural gas burns clearly and completely, buses will run without plumes of smoke trailing behind. Furthermore, natural gas is neither corrosive nor toxic, and the engines are quiet—another plus. As shown in Figure 12, these 40-ft-long buses are similar in configuration to existing buses. Only the engine and fuel tanks, shown in Figure 13, are different, and the buses can be converted at reasonable cost.

Figure 12. Brooklyn Union Gas Company's newly developed natural-gas-fueled bus.

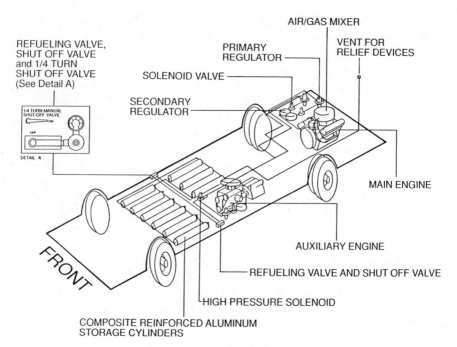

Figure 13. NGV bus fuel system.

Fuel stations can be built quickly. It takes only tapping into the gas main and running a service line to the fueling station. New York could prove to be the magnet that attracts other cities to natural gas. Clearly, this is the type of innovation needed by some of our grimy cities. Unfortunately, bureaucracies such as city governments are notoriously slow in adopting new ideas and technologies. This may be an issue on which the voice of the people should be heard in the land.

REFERENCES

1. C. L. Schofield, Acid Precipitation: Effects on Fish. *Ambio* **S:** 5–6, 228–230, 1976.
2. C. L. Schofield, Lake Acidification in the Adirondack Mountains of New York: Causes and Consequences. In *Proceedings of the 1st International Symposium on Acid Precipitation and Forest Ecosystems,* USDA FOrest Service General Technical Report NE-23, 477, 1976.
3. R. L. Burgess, Acid Rain and the Decline of Forests. *The New York Times,* Sunday, November 4, 1984, Sec. I, p. 84.
4. P. K. Basu, M. Avaria, A. Cutz, and M. Chipman. Ocular Effects of Water From Acidic Lakes: An Experimental Study. *Can. J. Ophthalmol.* **19** (3): 134–141, 1984.
5. P. K. Basu, M. Avaria, and S. M. Hasany, Effects of Acidic Lake Water on the Eye. *Can. J. Ophthalmol.* **17** (4): 74–78, 1982.

6. A. C. Alfrey, G. R. LeGendre, and W. D. Kaehn, The Dialysis Encephalopathy Syndrome. *N. Engl. J. Med.* **294** (4): 184–188, 1976.

7. E. M. Clarkson, V. A. Luck, W. V. Hynson, and R. R. Bailey, The Effect of Aluminum Hydroxide on Calcium, Phosphorus and Aluminum Balances. *Clin. Sci.* **43**: 519–531, 1972.

8. M. R. Wills, and J. Savory, Aluminum Poisoning: Dialysis Encephalopathy, Osteomalacia, and Anemia. *Lancet* **1983**: 29–33, July 2, 1983.

9. C. T. Betts, *Aluminum Poisoning*. Research Publishing, Toledo, OH, 1926.

10. R. J. Wurtman, Alzheimer's Disease. *Sci. Am.* **252**: 62–74, 1985.

11. *Acidic Deposition: Trace Contaminants and Their Indirect Human Health Effects* (Scott Q. Quinn, Ed.). Research Needs Workshop, June 19–22, 1984. Top Bridge Conference Center Paul Smith's New York. Corvallis Environmental Research Laboratory, Corvallis, OR, 1984.

12. A. Kudo, H. Nagase, and Y. Ose, Proportion of Methyl Mercury to the Total Amount of Mercury in River Waters in Canada and Japan. *Water Res.* **16**: 1011–1015, 1982.

13. L. Hakanson, Quantitative Impact of pH, Bioproduction and Hg Contamination in the Mercury Content of Fish (Pike). *Environ. Pollut. Ser. B* **1**: 285–304, 1980.

14. H. L. Needleman, S. K. Geiger, and R. Frank, Lead and IQ Scores: A re-analysis. *Science* **227**: 701–704, 1985.

15. S. Quinn, and T. W. Clarkson (Eds.), *Research Needs Regarding Trace Contaminants and Acid Depositin as Related to Increased Human Exposure*. Report to NY State Department of Environmental Conservation, Albany, NY, and Office of Research and Development, U.S. EPA, Washington, DC, 1984.

16. R. J. P. Paolini, and E. Somers, The Potential Health Effects of Acid Rain—The Canadian Experience. Environmental Health Directorate, Health Protection Branch, National Health and Welfare, Ottawa, Canada, July 1984.

17. P. R. Sherwin, What Is an Adverse Health Effect? *Environ. Health Perspect.* **52**: 177–182, 1983.

18. C. F. Enloe, Leaning on a Weak Reed. *Nutr. Today*, Jan./Feb. 1984.

19. Study Attests Value of Autopsy Despite Diagnostic Innovations. *Med. World News*, Apr. 9, 1984.

20. A. S. Whittemore, Air Pollution and Respiratory Disease. *Annu. Rev. Public Health* **2**: 397–429, 1981.

21. J. D. Hackney, W. S. Linn, and E. L. Avol, Assessing Health Effects of Air Pollution. *Environ. Sci. Technol.* **18** (4): 115A–121A, 1984.

22. F. E. Speizer, Epidemiology of Environmentally Induced Chronic Respiratory Disease. *Chest* **80** (1): 215–235, 1981.

23. *Changing Climate*. Carbon Dioxide Assessment Committee, National Academy of Sciences, National Academy Publications, Washington, DC, 1983.

24. S. Seidel, Can We Delay a Greenhouse Warming. Environmental Protection Agency, Office of Policy Analysis, September 1983, DEA 4900-198.

25. J. B. Hursh, D. A. Morken, T. P. Davis, and A. Lovass, The Fate of Radon Ingested by Man. *Health Phys.* **11**: 465–476, 1965.

26. M. L. Burr, J. Mullins, T. G. Merrett, and N. C. H. Stott, Indoor Molds and Asthma. *J. R. Soc. Health* **108** (3): 99–101, 1988.

27. B. C. Wolverton, Foliage Plants for Improving Indoor Air Quality. Presented at Nationl Foliage Foundation, Interiorscape Seminar, Hollywood, FL, July 19, 1988.

28. B. C. Wolverton, R. C. McDonald, and E. A. Watkins, Jr., Foliage Plants for Removing Indoor Alr Pollutants from Energy-Efficient Homes. *Econ. Botany* **38** (2), 224–228, 1983.

SUGGESTED READINGS

M. R. Becklake, Asbestos-Related Diseases of the Lungs and Pleura. *Annu. Rev. Respir. Dis.* **126:** 187–194, 1983.

Can We Delay a Greenhouse Warming? EPA, Office of Policy Analysis, Washington, DC, 1983.

Mathews, J. T. Global Climate Change: Toward a Greenhouse Policy. *Issues Sci. Technol.* **3** (3): 57–68, 1987.

Nero, A. V., Jr. Controlling Indoor Air Pollution. *Sci. Am.* **258** (5): 42–48, 1988.

Samuelsson, C. Retrospective Determination of Radon in Houses. *Nature (London)* **334:** 338–340, 1988.

Spengler, J. D., and Sextori K. Indoor Air Pollution: A Public Health Perspective. *Science* **221:** 9–16, 1983.

Stolarski, R. S. The Antarctic Ozone Hole. *Sci. Am.* **258** (1): 30–36, 1988.

11

Water Pollution

*Strongly held opinions often determine what kind of facts
people are able or willing to perceive.*

—Robert Waelder

Water and its pollution command
attention and wide discussion. Considering, however, that most discussions generate more heat than light, they are often more cathartic than productive.

Among the critical issues competing for public attention is the need to maintain water resources of high quality and generous quantity. A growing population and simultaneously expanding industrial and agricultural bases, as well as concomitantly increasing energy needs, with their waste by-products, require immense quantities of water.

I. THE WATER SUPPLY AND THE COURSE OF POLLUTION

The increasing importance of groundwater is reflected in the staggering amounts withdrawn each day. Withdrawals have increased from about 20 billion gallons per day (bgd) in 1945 to approximately 95 bgd in 1985. Although this represents an increase (almost 12% per year), the preponderant upturn in usage more than likely began in the early 1960s. This means that demands for water have been rising far faster than is generally conceded or perceived.

The types of use to which groundwater is put, along with estimates of the quantities withdrawn, are as follows:

Irrigation	63	bgd
Industry	13	
Drinking	12.5	
Rural, home, and stock watering	3.5	
	92.0 bgd	

Projections suggest that by the year 2020, 135 bgd will be withdrawn. Will it be there? Perhaps the question should be posed with a qualification. Will it be there in usable form? Or will it be so contaminated as to be unfit for consumption?

Before discussing groundwater pollution, such essential questions as What is it? and Where does it come from? require consideration.

A. The Hydrologic Cycle

The earth's water cycle is a complex closed system of heroic proportions that maintains a balance between water in the atmosphere and water in and on the earth. Most of the earth's water is stored in the saline waters of the world's oceans. Fresh water for the most part is frozen in glaciers, with smaller amounts contained in lakes or underground. Every molecule of this water will move through the hydrologic cycle. Figure 1 shows this is a generalized global proportion, while Figure 2 describes it in greater "local" detail.

Each year some 95,000 cubic miles of water evaporate.* Approximately

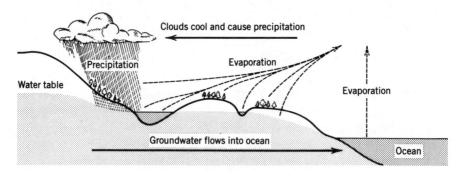

Figure 1. The hydrologic cycle: a global view.

* To give some idea of the vast quantities involved, 1 cubic foot of water equals 7.5 gallons. Thus, it is necessary to multiply 95,000 by $(5280)^3$ then by 7.5 to convert to gallons.

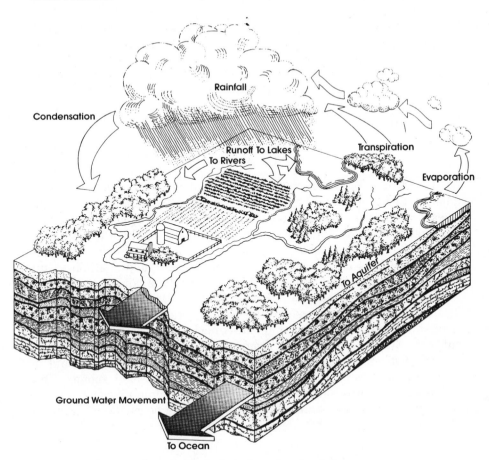

Figure 2. The hydrologic cycle: a local view.

75,000 cubic miles fall back as precipitation—rain, snow, sleet, and hail—into the oceans, lakes, and rivers, and onto the land. Responding to the complex of mighty forces from the rotation of the earth, the radiant heat of the sun, and the gravitational effects of the earth and the other eight planets of the solar system, we have then an endless cycle of evaporation and precipitation.

At any instant, about 0.005% of the total supply moves through the cycle. In the United States, a drop of water spends an average of 12 days passing through the air, then may remain in a glacier for 40 years, in a lake for 100 years, or in the ground from 200 to 10,000 years. Eventually, every drop moves through the cycle. The hydrologic cycle uses more energy in 24 h than humans have generated throughout their total existence. The awesomeness of this energy use should not be quickly dismissed or forgotten. Nor must it be forgotten that the *total supply of water never grows or diminishes.* That is crucial to comprehension of the entire issue.

The water we use and drink today was doubtless contained in barrels carried on the Santa Maria, Pinta, and Nina as Columbus sailed west in 1492. This endlessly recycled water was the same consumed by Caesar and Cleopatra as they sailed serenely down the Nile. And the hunter–gatherer–artists of the upper Paleolithic period more than 18,000 years ago, who lived in and painted the caves at Altimira in Spain and Lascaux in France, drank the same water. In fact, we are all drinking "used" water.

B. Groundwater

Potable (drinkable) water exists in the ground at some depth and at some quantity nearly everywhere on earth. Gravity attracts the water from the skies, pulls it beneath the surface, distributes it among peremeable layers (shown in Figure 3), and influences the direction in which it flows.

The amount of drinkable water stored within the first half mile of the surface is at least 20 times greater than the amount held in all our rivers and streams. The major source of this vast supply is rain.

Water from rain-soaked surfaces slowly percolates through the earth's layers at a rate of from several inches to several feet per day. Figure 3 suggest this movement. A common misconception has it that groundwater consists of large underground rivers and lakes similar to those above-ground. Instead, the seeping or percolating water may reach rock, sand, or gravel formations where it collects as if in a saturated sponge. This collection, at various depths, is referred to as groundwater, and the collection area is called an aquifer. Aquifers may be small and extend a

Figure 3. Groundwater location and movement.

few miles, or, as in the case of the Ogallala aquifer, they can spread over many states.

The great advantage of groundwater as a drinking water source has always been its comparative purity. Unlike surface water, groundwater ordinarily needed little or no treatment to safeguard its quality. There are other advantages: the large storage potential of aquifers, replenished by rainfall, mitigates the need for expensive reservoirs and long-distance transmission lines.

Unfortunately, industrial and agricultural expansion has created monumental problems of waste disposal. Each year some 70 billion gallons of waste are disposed of from both accidental and purposeful "spills" and waste chemicals are being leached into the soil and then into major aquifers.

Nearly 50% of the population of the United States depends on groundwater as their primary source of drinking water. The quality of this water, a national resource, can be adversely affected by the chemicals and microbes scattered throughout our environment, which it contacts as it migrates across and through the surface of the land. Consequently, the level of pollution is related to the degree of contamination of the soil, as well as to the waste effluents pouring into streams, lakes, and rivers. Accordingly, groundwater contamination has become one of the nation's most pressing environmental concerns.

Not only do many sources of contamination exist, but they vary from area to area. For example, agriculturally related pollution in the western United States is extensive and likely to become increasingly serious. A widespread misconception is that groundwater pollution is singularly the consequence of the chemical or "high-tech" industries producing the new class of hazardous and toxic wastes. Recall for a moment the enormous number of cattle raised in the United States (Chapter 4) and recall too that growing our food crops—corn, rice, wheat, fruits, nuts, potatoes, vegetables, and soybeans—requires extensive use of pesticides. Obviously, the contributions of agriculture and farming are substantial and significant, in addition to industrial wastes. Furthermore, municipal landfills are a continuing source of a conglomerate mixture of substances, while septic tanks contaminate the groundwater for millions of people. Currently, the total effect is the pollution of about 1% of the country's aquifers, but because these aquifers are close to large population centers, the impact can be disproportionately large.

Another, and long forgotten, contributing source of pollution was recently discovered in the form of underground gasoline storage tanks leaking their contents. Estimates place the number of tanks currently in use at close to 2 million (in all 50 states). This does not include abandoned tanks or tanks used to store other volatile chemicals. Most of these tanks were made of steel and few were protected against corrosion. Again, going by estimates, the number of leaking tanks is on the order of

75,000–100,000 and the number is rising. Although the leaks may be slow, one gallon of gasoline per day leaking into groundwater can pollute the water of a 50,000-person community to a level of 100 ppb. Depending on the chemicals involved, that may or may not be excessive. However, and this is crucial to policy decisionmaking, comprehensive knowledge and understanding of groundwater quality, and by extension assessment of risks, are hampered because the quantities are so vast, so slow moving, and largely inaccessible. Therefore, far less is known about the quality of groundwater than is known about surface waters.

C. Computer Modeling

Given the exceptional characteristics of aquifers, Promethean approaches are called for to determine their pollution. Recently, a Boston-based consulting engineering firm, Camp, Dresser and McKee (CDM), won a national award for the analysis of a complex groundwater contamination problem. Although their solution accelerated the removal of the pollutants, it did much more. Past operations at a manufacturing plant had produced several onsite hazardous waste lagoons and storage areas. In time, chemical wastes from these lagoons seeped into the ground and contaminated an aquifer underlying the site as well as two drinking water supply wells southwest of the site. Because it was known that the groundwater flowed southward from the site toward the well field, it was believed that the glacial aquifer material was the primary route of the well field's contamination.

A 90-day pump test and over 1000 samples of groundwater revealed that most of the water entering the wells came from two proximate uncontaminated surface ponds. This ruled out the aquifer material as the primary pathway for both water and contaminants. A comprehensive computer modeling program was needed to track the pollution plume from the surface, into the ground, and on to the wells.

After developing computer models for water flow and contaminant transport, they constructed a finite-element grid of the site. This grid, shown in Figure 4, consisted of more than 3000 tetrahedral monitoring segments for which the computer calculated water level and flow.* This model was singularly unique because of its ability to track pollutants both vertically and horizontally. In doing so, it was found that the highest concentrations flowed downward through the aquifer, into and through a layer of fractured bedrock covered by a layer of till, and up into the well field where the till layer terminated. Figure 5 shows, from left to right, the three-dimensional computer graphics which CDM generated of the bedrock and site surface elevations, contaminant concentrations, and piezometric contours.

* The large gray areas are natural ponds. The rectilinear shapes are buildings on the site, and the arclike configuration of particles on the right are contaminants.

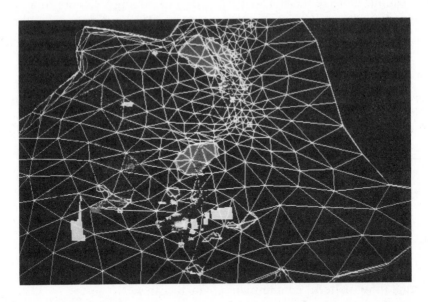

Figure 4. A finite-element grid with some 3000 tetrahedral monitoring segments. Courtesy of Camp, Dresser and McKee.

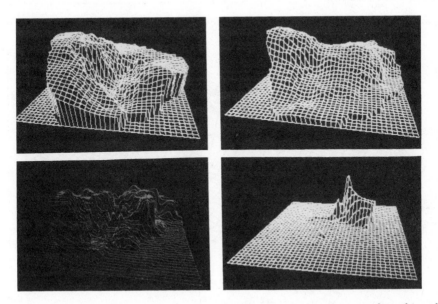

Figure 5. From left to right, a series of three-dimensional computer-generated graphics of bedrock, surface elevations, contaminant concentrations, and piezometric contours. Courtesy of Camp, Dresser and McKee.

Contour maps of the site are prepared by digitizing XYZ coordinates on an equilateral grid. A computer program fits the points to a three-dimensional surface. Contour maps and perspective surface projections can be drawn from the three-dimensional surface computer file. Cross sections can also be extracted and used for volume computations.

Five test wells that were subsequently drilled into the fractured bedrock confirmed the model's prediction. Based on the data, a pumping and treatment system was set up that intercepted the pollution plume in the fractured bedrock. The discovery by computer of this high-concentration, low-volume plume resulted in an efficient aquifer restoration system.

D. Electronic Sensing of Leachate Plumes

Over the past 10 years, development of remote sensing devices and associated computer modeling and graphics have permitted the location of buried waste sites and allowed the determination of hydrological features that influence leachate migration as well as identification of plume (flow pattern) contaminant characteristics.

For example, Roy Evans and Glen Schweitzer of the U.S. EPA (1), used ground-penetrating radar, electromagnetic induction (EM), and magnetic surveys to locate and estimate the volume of buried waste. Figure 6 shows

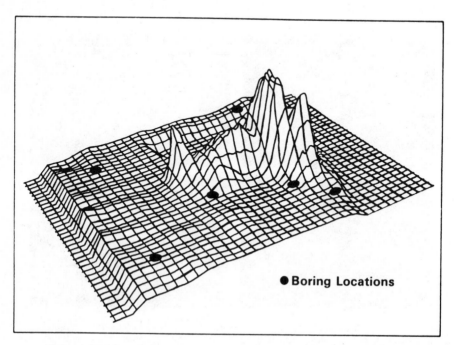

● **Boring Locations**

Figure 6. EM data showing extent and magnitude of conductivity anomaly.

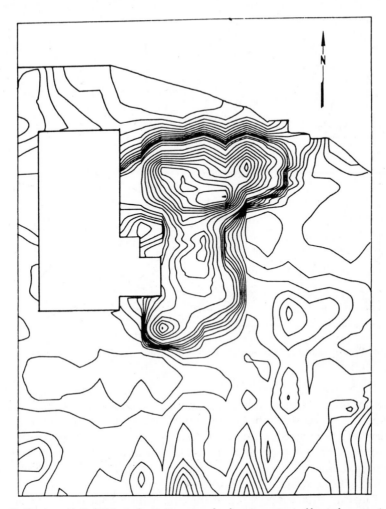

Figure 7. Contour plot of EM conductivity anomaly showing extent of buried contaminants.

in three-dimensional perspective the results of such plotting. The black dots were six well boring sites that missed the burial site. Figure 7 is a contour plot of the same site. In this, the extent of the buried contaminants become clearly defined.

In another instance a resistivity study was conducted to map a landfill leachate plume. Although chemical analysis of the groundwater had identified both organic and inorganic contaminants, the plume was electrically conductive because of the presence of sodium, potassium, magnesium, and chloride ions. Had these inorganic ions been absent, resistivity techniques would have been inappropriate. Figure 8 shows the type of plume patterns obtained, as well as the extent of migration between 1977 and 1981 (1).

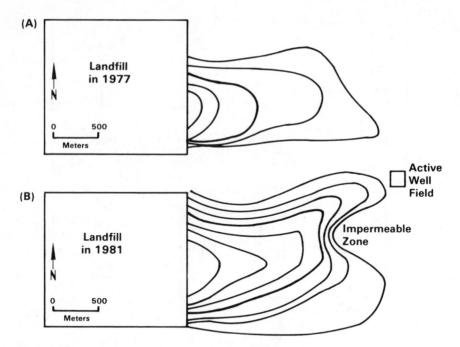

Figure 8. Plume patterns. Courtesy of Dr. Roy B. Evans, Environmental Research Center, University of Nevada.

We are only at the beginning of gaining knowledge about underground transportation of substances and about the movement and migration patterns of these subterranean "pools." New, highly sensitive analytical instruments of the type described will be crucial for identifying and quantifying the multiplicity of contaminants that are water soluble and can reach the underground reservoirs.

While our rate of contaminating the water appears to be slowed, residuals from earlier carelessness remain to be found and purged. Furthermore, health officials need to learn where and why synthetic organic chemicals occur in groundwater, and they need to ascertain the long-term effects on health of exposure to such compounds as trichloro-ethylene, trichloroethane, chloroform, and carbon tetrachloride. There is much to be done, little time in which to do it, and perhaps most important of all, diminishing federal funds to support the investigations.

II. LEVEL AND TYPE OF POLLUTION

The question of ultimate concern and most often asked is: Is our drinking water safe? Few individuals would be satisfied with anything less than a clear, affirmative response. "Yes" or "maybe" would be terribly uncomfortable. Interestingly enough, many are convinced that it is not safe. Before rendering my own opinion, or better still, allowing you to formulate your own, a look at the available evidence seems reasonable. Also reasonable at this juncture is the question, what do we mean by "safe"?

A. What Is Safe?

This seemingly simplistic question is hardly that. But neither is it so convoluted or abstruse as to be beyond consideration. Much of the confusion and misconception about the nature of "safe" and "safety" in decision making at all levels would be dispelled if their meaning were understood and agreed upon.

Before "safe" or "safety" can be defined, the specter of risk must be raised, and along with it the question of whether we are willing to take chances. Of course we are. We do it every day. Often many times a day. Inherent in choice is concern for risk. Are we willing to take risks? Again, of course we are, but assuredly not all. What about acceptable risks?

If we can go that far, we can define "safety" as a judgment. That's the key—a personal and individual judgment about the acceptability of risk; and in turn we can define "risk" as a measure of the probability of harm to our health. Safety then is a value judgment of how much risk we as individuals, or collectively as a community or society, will accept, and risk is a quantitative assessment of the degree of harm to health that may be anticipated.

Consequently, a chemical, a process, a drive to the beach, or a plane trip from Chicago to Miami is safe if its risks are judged to be acceptable, or unsafe if they are judged to be unacceptable. For many people leaving the ground is completely out of the question. Obviously, this definition differs markedly from that in most dictionaries where "safe" is defined as "free from risk"—a definition both misleading and unrealistic. Nothing we do can be totally without risk, neither taking a bath, crossing the street, eating a meal, having a baby, playing a round of golf, nor jogging through the park. There are only degrees of risk and, thereby, degrees of safety.

This concept emphasizes the relativity and judgmental nature of the concept of safety. In this system, there can be no absolutes; a thing can never be either all black or all white. That is a basic concept. Shades of gray are in fact nature's way.

This definition implies that there are two very different aspects to determining how safe the environment, including our drinking water, is: (1) measuring risk, a technical or scientific problem, and (2) judging the acceptability of that risk, a matter of personal and social values.

Failure to appreciate this dichotomy may just be the shoal on which understanding founders. Failure to appreciate this duality gives rise to the false expectation that scientists can measure whether something is safe.

They cannot. They are limited to the first function, to measurement of probabilities and consequences, deaths, injuries, or disease. The second, judging the acceptability of the risk, is not a decision scientists should make or be asked to make. Deciding whether people might or should be willing to bear the estimated risks is a judgment call that scientists are no more qualified to make than any other citizen.

Safety, as we have come to realize, is a relative attribute that changes from time to time and place to place. People clearly show their acceptance or rejection of risks, though they may not think of their behavior in these terms.

Deaths from automobile accidents, as noted in Chapter 6, have taken 45,000–50,000 lives each year for the past 20 years. And there is no reason to expect or anticipate a decrease in the carnage in the years ahead. It has also been estimated that 2000 additional deaths over the next 20 years, 100 extra deaths per year, may occur as a consequence of radiation exposure from all the active nuclear power plants. If the public outcry against nuclear power plants is any bellwether, a large segment of the public is unalterably opposed to such a risk. Choices are indeed made and risks taken. Fifty thousand actual deaths per year from motor vehicles are quite obviously acceptable; 100 possible or potential deaths from radiation are not. "Safe" then is anything people decide is safe. On what basis have some people decided our drinking water is unsafe?

B. Pollution and Oxygen Debt

Having thus suggested a view of safe and safety, the concept of pollution as it relates to water deserves fuller consideration.

Pollution is not an all or nothing concept. Pure water does not exist outside analytical laboratories. Rivers, streams, and lakes are natural habitats for a wide variety of plants and animals. When they die, the organic and inorganic chemicals of which they are constituted pollute the water. Runoff from land as a result of rain or flood and leaf fall during the year add additional contaminants. Consequently, natural bodies of water cannot be thought of as "pure" in either a biological or chemical sense.

Microbiologists and sanitary engineers consider water pollution as a problem of oxygen deficit; the lower the dissolved oxygen, the greater the pollution. The public often takes a broader view. Anything added to water is a pollutant. Thus, discarded bottles, tires, shoes, and egg crates, along with the rest of the trash that careless members of our society choose to toss away, are considered pollutants. Figure 9 typifies this view.

These things are unsightly and aesthetically unappealing, but they have little or no effect on the dissolved oxygen in the water. To a smaller but highly vocal segment of the population, pollution refers to anything other than chemically pure water, which, of course, has never existed on this planet.

Figure 9. "It's pollution all right, but it's pollution of rather a high order." Drawing by Stevenson; © 1965 The New Yorker Magazine, Inc.

If pollution is taken to mean the depletion of oxygen with consequent septic conditions such as offensive odors, floating masses of sludge, and death of fish and other aquatic life, then pollution is undoubtedly older than recorded history. As people cleared the land and cultivated it, surface water runoff increased and brought with it large amounts of organic material, causing depletion of oxygen. As soon as settlements arose in river valleys, human wastes began to be emptied into streams. The course of history leads directly from the Stone Age tanner who scraped his furs into a stream to the river-polluting oil refinery, tannery, or plating plant of the 20th century. There is, however, an important difference. As recently as 30 years ago little was known of the consequences of dumping industrial wastes into rivers. Current knowledge suggests that other forms of disposal require development.

1. Biological Oxygen Demand

One of the most widely used measures of the concentration of biologically oxidizable matter in a river or stream is the biochemical oxygen demand (BOD). Although used as a chemical test, it is in fact a biological procedure, depending for its activity on the microorganisms in the biological material. The BOD test is a quantitative measure of the oxygen used up by a sample of sewage effluent or river water during a 5-day incubation period at 20°C. The sample is diluted with fully oxygenated water and the initial amount of oxygen determined. After 5 days in the dark, the oxygen content is determined again; from the difference between the initial value and the 5-day value, the BOD or strength of the original sample can be calculated. Several state health departments have set their water conservation policies by limiting the BOD to a stipulated maximum, while others attempt to control added pollution by limiting the BOD of effluents. In a word, the purpose of sewage treatment is to consume most of the BOD before the effluent is discharged.

Several random BOD values may be helpful: the BOD of the sewage of a residential community generally varies between 150 and 250 milligrams per liter (mg/L). (The BOD of the sewage produced in the five boroughs of New York City averages about 100–200.) Milk processing and cannery waste range from 5000 to 6000 mg/L, while the liquid discharged from pulping operations, which contains large amounts of wood sugar, casein, and starches, often has 5-day BOD of 10,000–15,000 mg/L. Among the strongest wastes going into rivers and streams are those from the wool scouring industry; its waste can have a BOD above 20,000 mg/L.

Another useful indicator of the overall condition of a river or stream is its content of dissolved oxygen (DO). This value, more than any other, determines whether self-purification is likely to proceed satisfactorily and whether the stream community or ecosystem is adequately balanced. University, government, and industrial laboratories are currently pressing research investigations dealing with induced reaeration by mechanical aerators and flow augmentation schemes as a means of maintaining DO contents compatible with a biologically healthy waterway.

Theoretically, under conditions of standard temperature and pressure (20°C, 760 mm Hg), 9 mg of oxygen can be dissolved in 1 L of water; this is the saturation concentration for this particular temperature. Healthy streams and rivers generally yield DOs of 5–7 mg/L. Values consistently below 4 mg/L tend to indicate organic overloading, developing sepsis, and migration of fish species downstream.

One aspect of these chemical tests seems a bit perverse. Chemical effluents notwithstanding, water pollution is essentially a biological phenomenon, but biological indicators of adverse effects of organic (and inorganic) loading are rarely employed. Yet it is change in or destruction of the fauna and flora of the stream or river that is central to the problem.

2. Self-Purification and Reuse

When a river receives organic matter, it tends to purify itself; that is, it can overcome or adequately handle the organic load in time. "Running water purifies itself" is an old adage that had much truth to it, but one that may no longer be true. The magnitude and types of waste presently being discharged frequently overwhelm the recuperative capacities of the stream and thus stall recovery. Self-purification, which leads to the stabilization of the added organic matter, depends primarily on the metabolic activities of bacteria, which, given sufficient dissolved oxygen, utilize the organic matter as nutrients and break down (neutralize, stabilize, digest) organic compounds to innocuous end-products. The process also requires substantial growth of aquatic vegetation whose photosynthetic processes return oxygen to the water.

This cycle is interrupted by deficient dissolved oxygen levels: turbidity can limit light penetration, which in turn limits photosynthesis. Should self-purification of a watercourse become impossible, metabolism of organic matter proceeds via the anaerobic route, characterized by the production of foul-smelling by-products and incomplete stabilization of the organic load.

Some waterways undergo self-purification in fairly short distances from point of sewage outfall, while others require 25, 50, or more miles in which to accomplish this. The process is complex, and each stream or river is a unique ecosystem with its own specific capacity for purification and recovery. If oxygen is removed faster than it can be replaced by natural aeration or by photosynthesis, conditions worsen and pollution of virtually an entire river may result.

Therefore, the function of sewage treatment is removal of enough organic matter to produce an acceptable effluent, whose oxygen demand is low or virtually nonexistent. Thus, water pollution, the availability of acceptable drinking water, and sewage disposal are intimately related.

Not uncommonly, a city locates its intake water pumping stations upstream on a river and discharges its treated sewage downstream. The city downriver repeats the process, as do each of the subsequent towns. However, distances between discharges of sewage effluents and of intakes of raw water are being compressed as a result of the constant expansion of communities due to increased population. Figure 10 depicts the progressive intake–outlet reuse pattern currently prevalent on many water courses.

Today's communities and modern industry are both elephantine users of water. The average city-dwelling family of four uses about 600 gal of water each day. Industry requires some 5 gal of water to produce 1 gal of gasoline, 10 gal to produce each can of vegetables, 25,000 gal to process 1 ton of steel, and 50,000 gal to produce 1 ton of paper. After use, the spent water, loaded with organic waste, is often discharged directly into a stream or river with the generally mistaken idea that dilution with fresh water will prevent pollution locally or farther downstream. Unfortunately, however, the available water resources are unevenly distributed; many sections of the country are forced to reuse the available supply long before the river from which they draw it has had a chance to purify itself. It has been estimated, for example, that the 1300-mile Ohio River is used around five times before it joins the Mississippi at Cairo, Illinois.

The present state of our technical capability requires that we accept a certain degree of pollution. We are still far from developing alternative methods of liquid-waste disposal; consequently, the use of rivers for the disposal of waste is essential. The fact is that industry and residential communities alike would be forced to shut down if unified public opinion

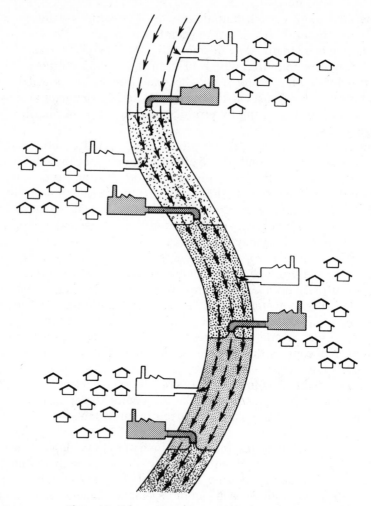

Figure 10. Schematic representation of water reuse.

demanded pollution-free waterways under the broad definition of pollution. It is therefore in the best interests of the community to support research investigations aimed at discovering the degree of waste a stream can tolerate. Decisions as to what constitutes a tolerable degree of pollution must be guided by considerations of the water's natural purification capacity and the purpose for which it is to be used.

C. Types of Pollutant

Scientists concerned with water pollution have categorized pollutants as chemical, biological, physiological, and physical. The chemical pollutants include both inorganic and organic compounds, such as dairy, textile, cannery, brewery, and paper-mill wastes, ensilage, laundry wastes, manure, and slaughterhouses wastes. These essentially contain proteins, carbohydrates, fats, oils, resins, tars, and soaps. If these pollutants are not excessive, they will be stabilized by the self-purification process. If they are excessive, death of fish and offensive odors can result. In addition, such minerals as phosphates, nitrates, and potassium, released in the stabilization process, encourage growth of aquatic plants. As the plants decay, further depletion of oxygen occurs. Metals, such as chromium, cadmium, lead, and mercury, and cyanides, primarily from engineering industries, can be toxic to fish and inhibit the biological waste stabilization process by killing the necessary microflora.

Biological pollutants include the many types of microscopic animal and plant forms, such as bacteria, protozoa, and viruses, that are associated with disease transmission. These come from domestic sewage, farms, and tanneries.

Physiological pollution manifests itself as objectionable tastes and odors. These may be imparted to the flesh of fish, making it inedible; or water itself may become unfit to drink owing to odors and tastes caused by the presence of inorganic chemicals, such as hydrogen sulfide, or the extensive growth of certain species of algae. Some impart musty odors to the water, while others give it a fishy, pigpen, spicy, or chemical taste.

Various physical effects, such as foam, color, turbidity, and increased temperature, are also considered forms of pollution. Water from a nearby stream or river is often pumped into a plant to cool a machine or process that normally generates heat. The transfer of heat to the cooling water raises its temperature several degrees. When this heated water is discharged back into the stream or river, it may disrupt ecological relationships. A rise in temperature of only a few degrees may be lethal to a variety of aquatic plants and animals that, like most living things, are sustained only within a narrow temperature range. The death of certain species can remove the food supply of the species that prey on them; without food they in turn will die or be forced to move downstream. Furthermore, the warmer the water, the less oxygen it will contain. At elevated temperatures all chemical and biological activity proceeds at a more rapid rate. This further depletes the sensitive oxygen balance of the stream, as well as adversely affecting respiration and metabolism of a wide variety of aquatic plants and animals. This series of events can destroy a stream's capacity for self-purification by altering or eliminating vital links of the stream's ecosystem.

Thermal pollution requires additional comment. Each aquatic species adapts to the seasonal variations in water temperature of its particular habitat. Species cannot, however, adapt to abrupt thermal shocks. Because of the growth of industrial cooling requirements, ecologists are concerned that discharge of heated effluents may destabilize aquatic ecosystems. As a result, heat has been added to the list of potential pollutants.

On the Green River in Kentucky, the Tennessee Valley Authority has built three cooling towers (437 ft high, 320 ft in diameter, with a capacity of 280,000 gal/min) that enables the world's largest coal-fueled electric plant to discharge its waste water harmlessly. Fortunately, the addition of heat to water can readily be corrected.

1. California's Central Valley

Water pollution can occur in ways never suspected, primarily because our knowledge of the biosphere and local ecosystems are inadequate.

For years California farmers in and around Fresno had been warned of problems for the lush and bountiful Central Valley. Few people expected the day of reckoning to arrive so soon. Farmers in the valley's west side were stunned in March 1985 when the Interior Department said it would cut off irrigation water to 42,000 acres of land. The move was necessary, Bureau of Reclamation officials told a Congressional hearing, to halt the drainage of dangerous wastes into the Kesterton Wildlife Refuge, which has become a death trap for migrating birds. The land used for growing cotton, fruits, and vegetables will be taken out of production at a cost of millions of dollars. Some warned it would revert to a desert.

The pollution is caused by the lack of natural outlets for the water after it is used for irrigation and by geologic conditions that prevent it from seeping down into the earth. The water becomes toxic when it leaches natural salts and chemicals from the irrigated soil.

In February, California water officials warned that the selenium-contaminated water that was killing the water fowl could eventually pose a threat to people, including millions in Los Angeles, a part of whose water supply is channeled through farming areas.

The Bureau of Reclamation, which helped build the dams and canals that turned the valley into an "agricultural paradise," has been under fire by environmental groups (more concerned about the water supply than the available food supply), claiming the Bureau has been ignoring the problem. They may be right. Clearly, water pollution cannot be considered a simple problem.

D. Biological Monitoring

Although chemical monitoring can be accomplished relatively rapidly and is generally uncomplicated, it has deficiencies. A sample may not contain enough contaminant to measure by available chemical techniques even though it is present. In addition, an analytical procedure is often designed to identify a specific compound. More importantly, the sample is usually destroyed during analysis and prolonged or improper storage can produce changes in the sample.

In contrast, the advantages of biological monitoring and specimen banking are that organisms and tissues can be preferentially selected for their specific accumulation of material over a prolonged period. This can substantially reduce monitoring frequency while increasing the sensitivity of the procedure.

These characteristics can be used as "sentinels" or early warning indicator systems for protection of the environment and of people.

Specific "bioconcentrators" can be selected for their sedentary nature, because they are easily mapped or because they are highly mobile and sample a large area. Often the biological effects of contaminating chemicals can be correlated with their tissue concentrations, and information obtained from biological monitoring provides for effective evaluation and management of chemicals in watercourses because only living systems can directly measure biological effects (2). Measurement of biological effects and attainment of specific effluent levels (the goal of chemical testing) are both worthy goals, but they are totally different concepts and serve different needs.

For over 150 years, limnologists the world over have developed biological pollution assessment methods based on the presence or absence of certain characteristic species in rivers and streams. One of the most widely known of these methods is that developed by Ruth Patrick of The Philadelphia Academy of Natural Sciences. Patrick classifies animals and plants in seven groups. On the basis of their occurrence (or absence) she divides watercourses into zones:

Healthy: Abundance of species from all seven groups.

Semihealthy: Species from several groups overabundant, others reduced in number.

Polluted: Worms particularly abundant and great reduction in fish and insects.

Very Polluted: Most groups eliminated?

Septic: Biologically dead.

While a number of variations on this system have been proposed, essentially they all deal with the alteration of biota. In effect, each stream and river can be "fingerprinted" by its unique pattern.

Thus, the toxicity of chemicals emptied into a stream and moving along can quickly be shown by analyzing samples of water. A wipeout of expected species quickly indicates the presence of a hazardous substance. For reasons that are difficult to discern, these biological indicators are rarely used even though they are more reliable and more economical measures than chemical tests.

1. Mussel Watch

Edward D. Goldberg of the Scripps Institute of Oceanography proposed a monitoring system in 1975, which would use data obtained from bivalves (mussels) to assess changes in organic and heavy metal loadings of coastal waters. With support from the EPA, 106 sampling stations were set up around the United States. Over the ensuing years, mollusk tissue was analyzed for trace metals, transuranics (radioactive decay products), and hydrocarbons.

The "mussel watch" program has located "hot spots" for PCBs and lead and has shown that heavy organic loading of coastal waters has readily measurable effects on these sensitive ecosystems.

2. Mysid Shrimp

Biomonitoring is being scrutinized both regionally and nationally as a means of meeting the objectives of the Clean Water Act, because it can be effective in assessment of risk as well as setting limits on potentially toxic effluents.

Indeed, the country's first official industrial biomonitoring application is on-line at the Ciba-Geigy chemical plant in New Jersey. The Department of Environmental Protection required the plant to continually monitor a colony of mysid shrimp, which are unusually sensitive to pollutants.

3. Paper Mill Waste and Hydropsychids

Effluents of pulp and paper mills are known to be toxic to a wide variety of aquatic species. Recently, L. B. M. Petersen and R. C. Petersen of the University of Lund in Sweden (3) found that net anomalies spun by caddis fly larvae could be valuable indicators of pulp mill effluents at sublethal levels.

From their studies, they found that the effluent from paper mills induced the insects to weave nets in grotesque patterns, compared with nets woven by caddis flies in unpolluted streams. They also discovered that when the caddis fly larvae in polluted waters were allowed to pupate and emerge as flying insects, the pupation periods were longer and more variable than flies from unpolluted water.

These results demonstrate that the distorted nets, which can provide a new tool for monitoring toxic substances, are also indicators of physiological damage to insects exposed to low levels of chemical contaminants. Because these nets are produced and remain in the water for weeks on end, they are in effect recording stream conditions over a relatively long period. Sufficient numbers of these could evolve into an "index of disruption," which could relate to chemical concentration or even type of chemical.

Normal caddis fly nets consist of silklike strands that form a series of connected rectangles, a web, to the left and right of a central line. Nets spun in polluted water have oddly shaped openings, distorted center lines, and extra "wandering" strands. Figure 11 shows the normal geometric patterns, while Figure 12 shows the misshapen webs.

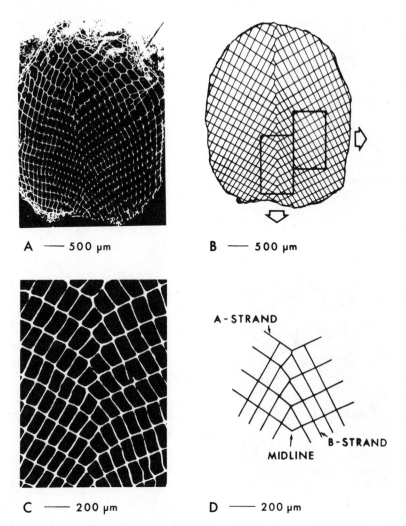

A ——— 500 μm   B ——— 500 μm

C ——— 200 μm   D ——— 200 μm

Figure 11. Normal character of caddis fly nets. Courtesy of Dr. Robert C. Petersen, Jr., Department of Limnology, University of Lund, Lund, Sweden.

4. Unexplained Fish Kills

Unknown toxic chemicals, often from clandestine releases, are assumed responsible for all sudden, unexplained deaths of massive numbers of fish and other types of marine life. Two such events recently occurred in the greater New York area.

Early in July (1988) during the protracted heat wave, some 30,000 menhaden (a nonedible, oily cousin of the herring) died in creeks and shallow river beds of the northeast Bronx. Given their high natural oil content, the decaying fish created an unbearable stench. Apparently, the

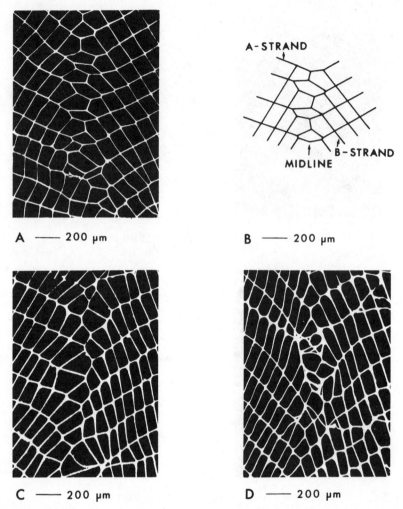

Figure 12. Abnormal nets of caddis flies resulting from polluted waterways. Courtesy of Dr. Robert C. Petersen, Jr., Department of Limnology, University of Lund, Lund, Sweden.

huge school of fish were chased into the brackish water of Eastchester Bay by bluefish, their natural enemy. Packed together in a watery cul-de-sac, the menhaden quickly depleted the available supply of oxygen. Simply stated, they suffocated. And as noted earlier in the discussion of BOD, the amount of dissolved oxygen declines as water temperature increases.

Thus, the summer heat was responsible for the warming of the shallow bay, and the sudden influx of literally thousands of highly stressed fish quickly depleted the oxygen, which resulted in their death. With their deterioration, whatever oxygen remained was quickly lost.

Two weeks earlier, thousands of flounder, bass, eels, and crustacea washed up on a beach in northern New Jersey. The first accounts on TV and in the newspapers held that gross chemical pollution was continuing to kill our natural resources. Upon further investigation it appeared that warm water and low oxygen levels were the likely causes.

Unfortunately for public understanding, first reports are usually accompanied by intense emotion and allegations of chemical spills. Follow-up reports are either nonexistent, or in the case of newspapers buried deep inside. This is not meant to absolve the polluters, of which there are far too many, but it does not help convey the urgent message that natural forces—ecological balances—can also be involved in water pollution episodes.

III. WATER-BORNE ILLNESSES

Historically, the primary reason for water pollution control was prevention of waterborne disease (see Table 1). If public health authorities in the United States and western Europe were asked to point to their single greatest accomplishment, it would undoubtedly be eradication of the classical waterborne diseases. Nevertheless, many infectious waterborne diseases remain endemic in other areas of the world, particularly the nonindustrialized nations of Asia, Africa, and Latin America.

In addition to cholera, typhoid, and dysentery, the developing coun-

Table 1
Diseases Transmissible by Water and Their Etiological Agents (Worldwide)

Ascariasis	Hydatidosis
Ascaris lumbricoides	*Echinococcus*
Amoebic dysentery	Leptospirosis
Entamoeba histolytica	*Leptospira interrogans*
Primary amebic meningoencephalitis	Melioidosis
Naegleria and *Acanthamoeba*	*Pseudomonas pseudomallei*
Hepatic capillariasis	Paratyphoid fever
Capillaria hepatica	*Salmonella paratyphi* A
Cholera	*S. schottmülleri*
Vibrio cholerae	Other salmonella infections
Diarrheal disease,	Typhoid fever
enteropathogenic	*S. typhi*
Escherichia coli	Schistosomiasis
Dracunculiasis	*Schistosoma mansoni*
Dracunculus medinensis	*S. haematobium*
Giardiasis	*S. japonicum*
Giardia lamblia	Shigellosis
Viral hepatitis	*Shigella dysenteriae*
Hepatitis A	*S. flexneri*
	S. boydii
	S. sonnei

tries contend with bilharziasis, commonly contracted when bathing in fecal- and urine-polluted streams and canals; urban filariasis, transmitted by the bite of an insect vector that breeds in polluted water; and infectious hepatitis, which was responsible for a recent major epidemic that included approximately 30,000 cases in New Delhi, India.

Because of the rapidity of human migration from endemic areas, coupled with the possibility that certain disease agents may adapt to new habitats in new environments, constant surveillance of water supplies must be maintained. It is necessary, however, to bear in mind that current water-borne microbial hazards are trivial in the United States, Canada, England, and western Europe.

Figure 14 shows that proper treatment of raw water can substantially reduce illness resulting from the ingestion of water containing pathogenic microorganisms. In Philadelphia, there were from 6000 to 10,000 typhoid cases each year between 1890 and 1907. With the introduction in 1907 of so simple a device as sand filtration of the untreated water, a precipitous reduction in illness was achieved. Six years later, the introduction of chlorination, over the strident protests of those opposed to the addition of synthetic chemicals to water, produced another abrupt drop in typhoid

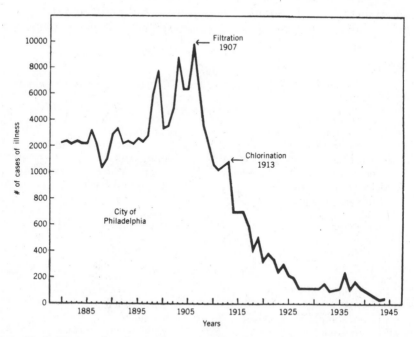

Figure 13. Reduction of typhoid fever in Philadelphia following treatment of the water supply. Adapted with permission of The Macmillan Company from *Preventive Medicine and Public Health*. 3rd. ed., by W. G. Smillie and E. D. Kilbourne. © The Macmillan Company. 1962.

cases. Clearly, mechanical and chemical treatment could control water-borne disease.

In a recent report, the Council on Environmental Quality remarked that "the water quality problem most often reported nationwide is excessive fecal bacteria levels that interfere with the recreational uses of water bodies." It went on to note further that "although not harmful by themselves, coliform bacteria have traditionally been used as indicators of sewage contamination, and the possible presence of microbial pathogens that are hazardous to human health." In fact, many rivers, lakes, and streams are widely believed to be major contributors to the burden of community illness, particularly so as fecal bacterial levels (coliform counts) have become synonymous with pathogens and, by extension, illness. This propinquity supports regulatory standards at federal, state, and local levels as well as enabling legislation and judicial decisions. But is the relationship more cordial than correct? Has it, like "topsy," just grown up?

There is little doubt that the waterborne epidemics of the 19th and early 20th centuries still influence attitudes. There is also little doubt that, in the public's mind, there exists an irrational fear that water supplies are suspect, a major contributing source of illness in the country today. It would be foolish to argue that water is not a potential vehicle for the transmission of enteric diseases. Water containing untreated fecal waste may be considered a health hazard. But bear in mind that fecal matter without accompanying pathogenic organisms may be no more than an aesthetic problem.

A. Chlorine Disinfection and Trihalomethane

Chlorine was added to drinking water supplies to destroy microbial pathogens. It proved to be a highly effective public health measure. In 1974, some 60 years later, J. J. Roock reported that the addition of chlorine to water containing organic matter produced trihalomethanes (THMs): chloroform, bromoform, bromodichloromethane, and chlorodibromomethane (4). Animal studies indicated that chloroform, the most abundant and simplest to detect, could be a human carcinogen. By the 1980s, additional halogenated compounds were identified. These included halo-acetonitriles and halogenated phenols and alcohols.

How should the presence of these THMs be interpreted? Certainly with a degree of caution. Considering the length of time chlorine has been used in water supplies in almost every country in the world, and under widely varying conditions, especially with reference to concentrations of chlorine, the likelihood of THMs being a risk to health appears minimal.

The National Academy of Sciences (NAS) has indicated that toxicological studies using laboratory animals indicated that exposure to chloroform posed a risk to human health. An epidemiological review by NAS arrived at the following conclusion. "The results suggest that higher concentrations of THMs in drinking water may be associated with an increased frequency of cancer of the bladder. The results do not establish causality, and the quantitative estimates of increased or decreased risk are extremely crude." The positive association found for bladder cancer was small and had a large margin of error, not only statistical but much more because of the very nature of the studies. (See Chapter 15 for a detailed discussion of source of error in laboratory studies.)

Responding to the NAS hint of possible human health effects, the EPA developed a standard for drinking water of 0.01 mg/L. This was determined by balancing public health needs and feasibility of achieving this level in public water supplies. It also reflected the proven use of chlorine as an established primary preventive health measure, with an outstanding track record. Of surpassing importance must be the steady decline in both stomach and liver cancer over the past 50 years. Figure 11 and 12 in Chapter 14 shows this clearly.

All the above notwithstanding, waterborne infectious disease remains the health risk of greatest concern to consumers of public drinking water supplies. It is suggested that the magnitude of these be given assiduous scrutiny. Guessing and intuition are wholly inappropriate activities where waterborne illness is concerned. The numbers tell a significant story. On the other hand, a caveat seems appropriate at this juncture. It should not be allowed to slip from working memory as it suffuses all environmental issues.

Decreases in mortality from infectious disease and certain cardiovascular diseases have led to increases in the proportion of deaths from cancer. Comprehension of this is of the utmost importance. Proportion is the operational term. Cancer is increasing in the population because deaths from other diseases are declining. This is not one of those issues from which "cold comfort" can be derived. Indeed, more than a modicum of warmth should be drawn from this verity.

B. Types of Illness

At this juncture at least two questions require elucidation. What diseases are known to be waterborne, and what levels of morbidity and mortality can reasonably be attributed to the water supply—both from infectious and noninfectious (chemical) agents? Table 2 lists the sum of microbial and parasitic diseases generally understood to be transmitted by the water route, around the world. However, this does not preclude their transmissibility by other means as well. Such overlapping is confounding.

Although the illnesses commonly seen in developing countries cannot be ascribed in the United States, discussions of waterborne diseases in the United States conveniently overlook this, creating misleading impressions about the safety of our water supply. An estimate of the level of new cases occurring during the period 1981–1983 is shown in Table 3.

Table 2
Diseases Transmissible by Water in the United States

Giardiasis
Hepatitis
Salmonellosis
Shigellosis
Typhoid fever
Campylobacteriosis
Viral diarrheas

Interestingly enough, these approximately 30,000 cases are the largest number that have occurred since 1940 (5). As such, they require some discussion.

The reporting, as weak as it is,* indicates that for the 36-month period, 28,791 of the total 29,185 were due to water from potable sources. There were 394 cases due to water from nonpotable sources. But more importantly, it would be misleading to average the total for the 3 years. For example, 4450 cases were reported in 1981 and 3611 in 1982. They are reasonably close. With 21,124 cases reported, 1983 must be considered highly unusual. Of the 21,124 cases, 11,400 became ill in Pennsylvania during a religious festival after consuming contaminated well water.

Another 871 individuals contracted campylobacteriosis as a result of drinking water from a system contaminated by wild birds. Investigation indicated that birds deposited the organism *Campylobacter* into an open settling tank during water treatment. Furthermore, 50% of all the cases

Table 3
Etiology of Waterborne Outbreaks in United States 1981–1983

Disease	Outbreaks	Cases of Illness
Gastroenteritis, undetermined etiology	50	20,346
Giardiasis	42	3,169
Chemical poisoning	8	149
Shigellosis	7	532
Hepatitis A	6	267
Viral gastroenteritis, Norwalk agent	6	826
Salmonellosis	2	1150
Campylobacteriosis	2	952
Viral gastroenteritis, rotavirus	1	1761
Cholera	1	17
Yersiniosis	1	16
Total	126	29,185

Source: Adapted from Craun (6).

* Reporting of illness is believed to be low because few people report transient bellyaches, and because it is extremely difficult to determine the route of infection days or weeks after the fact.

occurred in noncommunity water systems. These have a far greater potential for conveying waterborne diseases than community systems, since as Gunther Craun, of the U.S. EPA's Health Effects Laboratory, estimated, community water systems serve 180 million people while noncommunity systems serve approximately 20 million (thirty million are served by individual systems). Thus, the incidence rates for community-based systems are approximately 26 cases per million persons at risk, compared with 243 per million people drinking from noncommunity systems (6).

As Table 3 indicates, over 20,000 of the cases of gastroenteritis were undetermined as to etiologic agent. This suggests that although a report was filed at the state level, attempts to determine the inciting causes were not made. For those in which appropriate determinations were made, Giardia, a protozoan, and Salmonella, a bacterium, were the major risk factors. More than likely, a substantial portion of the undetermined 20,346 were due to both of these pathogens.

As for the contributions of chemicals, consideration must be given to acute and chronic involvement. Table 3 suggests that over the 3 years, 149 cases of acute poisoning occurred. By tracing these down, lead, copper, fluoride, nitrate, and benzene were found to be involved.

What is the evidence for more serious chronic effects? Recently, Crump and Guess (9) reviewed the epidemiologic evidence and assessed the risk of cancer, especially as it is related to chlorination. They found that "the associations between chlorination and cancer . . . are weak by traditional epidemiological standards." In addition, they noted that "no clear trend of increasing risk with increasing exposure has been demonstrated by the studies conducted to date." They concluded their assessment by reporting that "estimates based on animal data of human cancer risks from lifetime consumption of water from some highly polluted wells are small enough that they would probably not contribute noticeably to existing human cancer rates." (See Chapter 10, p. 413–415, for health effects of heavy metals.)

Even for the most suspicious, it is evident that water is not a major risk factor for disease in the United States, nor is it even a minor risk factor. Using the total number of cases for 1983, an unusually high year, the number of reported incidents from all causes approximates the number of people murdered in all 50 states each year. Much of the waterborne illness, however, can easily be prevented. What is required is increased surveillance of the smaller community water systems and noncommunity systems. Table 4 directs attention to those areas in which preventive adjustments would yield the greatest benefits.

Figure 14 shows the pattern of typhoid fever in the United States over the period 1955 to 1981. Note that these are rates per 100,000 people. Clearly, the disease is no longer the threat it was. In the 1940s and 1950s, for example, 2000–6000 cases per year could be expected down to about 300 cases in 1980. Trend lines similar to that of typhoid can be drawn for

Table 4
Primary Causes of Waterborne Illness in the United States, 1981–1983

Cause	Outbreaks	Cases of Illness
UNTREATED WATER		
Contaminated surface water	6	743
Contaminated springs	9	264
Contaminated wells	28	2,119
TREATMENT DEFICIENCIES		
Surface water sources	31	4,004
Spring water sources	5	659
Well water sources	17	18,004
DISTRIBUTION DEFICIENCIES		
Corrosion of plumbing	5	101
Contamination during water main breaks	5	1,600
Cross-connection	3	102
Storage	3	1,282
New plumbing	1	750
MISCELLANEOUS		
Ingestion during swimming	8	313
Water not intended for drinking	4	68
Insufficient data	2	47
Totals	126	29,185

Source: Adapted from Craun (6).

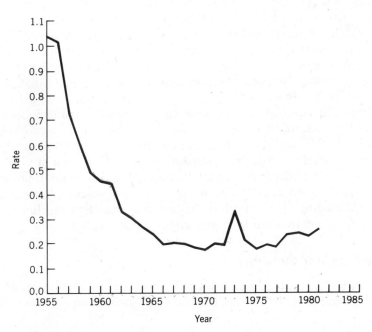

Figure 14. Reported cases of typhoid fever per 100,000 population, by year, for the United States, 1955–1981.

most diseases known to be carried by the water route. The trend is clearly downward.

C. Fecal Contamination

Although water has been established as an appropriate vehicle for transmission of human disease, the evidence for a cause–effect relationship is not all that clear for more than a few diseases. Fecal matter contains pathogens only when an individual is ill or for a short period during recuperation.* As a rule, fecal matter of itself does not a priori contain pathogenic microorganisms. However, as fecal matter can contain pathogens, it is assumed that the greater the concentration of feces in water, the greater the risk of illness to consumers of such water. Unfortunately, many of the diseases that can be waterborne are also transmissible by other routes. Until the necessary epidemiologic studies have demonstrated a positive relationship between pathogenic microbes in water and frank human illness, the microbiological tests for water quality must be considered to be no more than untested assumptions. What then do they do?

The tests for water quality seek primarily to verify the presence or absence of recent fecal contamination. If intestinal discharges have contaminated the water, large numbers of coliform bacteria are certain to be present. These organisms are benign types, living a saprophytic existence as part of the natural flora of the large intestine (the colon). They incite no illness but are always present in feces; thus, their presence in water testifies to the fact of fecal discharge. As natural inhabitants of the human bowel, they do not find environmental conditions in natural waters suitable for multiplication and begin to die off rapidly. Their presence in water samples therefore indicates recent contamination.

If fecal matter is of recent origin it is assumed that, along with the harmless coliforms, there may also be pathogenic organisms such as *Salmonella typhosa,* the organism associated with typhoid fever; *Shigella dysenteriae,* the organism associated with bacterial dysentery. *Vibrio comma,* the cholera organism; and *Entamoeba histolytica,* a protozoan associated with amoebic dysentery. Pathogens such as these, if present in water supplies, are usually few in number and exceedingly difficult to detect. Accordingly, no effort is made to test for them during routine examinations; instead, an indicator organism is used.

1. The Indicator Organism

Although the scientific underpinning for equating fecal coliform density with pollution is far from firm, the choice of such organisms is critical if

---

* In addition, there are instances of asymptomatic individuals (carriers) who shed organisms but are themselves healthy.

they are to be used. Suitable organisms must be associated with feces and/or pathogens. Such microbes should be easy to grow and differentiate. A relatively simple, accurate, rapid, and standard test to enumerate the indicator should be available.

In addition, the organism must not multiply rapidly (at water storage temperature), otherwise the degree of contamination may be obscured. If, on the other hand, it dies rapidly, its value as an index of contamination is reduced.

If it is less stable, it may disappear before the pathogen. If it is more stable, it may persist long after the pathogen. Either way misinterpretations of conditions may occur. Clearly, the choice is frought with difficulties. The real question is whether the energy spent developing such an indicator is commensurate with the quality of the data obtained.

Here again mature, professional judgment, combined with standardized criteria, is needed to determine when a water supply is unfit to drink or bathe in. Such decisions cannot be made lightly or be based on political expediency. Nonetheless, public opinion does allocate priorities, which may have little to do with the prevalence of disease.

Although several groups of bacteria have served as indicators, no group is as well established as the coliforms. The routine test for determination of coliform density requires that samples of water be inoculated into a series of tubes containing a medium specific for coliform growth, followed by incubation at $35 \pm 0.5°C$ for 48 ± 3 h. Gas production, in the form of bubbles of CO_2 rising in the tubes of liquid, is positive evidence of the presence of coliform organisms. A coliform index, or a count of the most probable number of organisms present in the original water sample, can be made on the basis of the amount of sample inoculated into a stated number of tubes. Criteria for the acceptance or rejection of a water supply are based on a long record of correlation of this parameter with the occurrence of fecal pollution.

Often, after prolonged or torrential rains, the coliform counts of water samples rise precipitously. Does this reflect the fact that human fecal matter has been washed into waterways as surface runoff? In the United States or western Europe, this could hardly be the case. How then can this rise in coliforms be interpreted?

By definition, coliforms include all aerobic and facultative anaerobic rod-shaped, non-spore-forming, gram-negative bacteria that can ferment lactose with gas production within 48 h at 35°C. Unfortunately, this encompasses organisms that are nonfecal in origin as well. That is, although they are called coliforms, some of them do not live in the colon of human or animal species. This group is widely distributed in nature, being found in grasses, in soil, and on plants. Accordingly, since these organisms are carried into water sources that are intended for drinking or bathing, and since they are not indicators of fecal contamination or pathogens, a means of distinguishing between fecal and nonfecal coliforms is necessary. Biochemical tests suitable to make these distinctions are available and well standardized. All can be found in *Standard Methods for the Examination of Water and Waste Water* (7).

The membrane filter (MF) technique, which involves a direct plating for detection and estimation of coliform densities, is as effective as the multiple-tube fermentation test* for detecting bacteria of the coliform group.

Two additional points are noteworthy although not directly related to indicator organisms. The Atlantic Ocean has a salt content of approximately 3.5%. This is not conducive to the survival of disease-transmitting pathogens, which are among the most fragile and sensitive of all microbes. Furthermore, the ocean has never been incriminated as a vehicle for any waterborne disease. French scientists studying the problem of waterborne poliomyelitis found that the prevalence of clinically manifest paralytic poliomyelitis was not higher in bathers who swam regularly in the Mediterranean Sea near Marseilles, France, than in nonbathers. Speaking at a symposium, "Transmissions of Viruses by the Water Route," Joseph Melnick of Baylor University Medical School noted that small doses of virus in water supplies may serve to immunize against rather than produce disease. He noted that sewage workers had the lowest rate of absenteeism among all occupational groups studied. "It appeared that sewage workers were regularly immunized by their exposure to small amounts of infected material."

With the coming of summer, the Atlantic Ocean off the beaches of New York and New Jersey is once again a subject of some concern: this time because of AIDS. Can swimmers or bathers generally become infected? The many people staying out of the water think so. Discarded vials of AIDS-containing serum were believed to have been washed up on the beaches along with other flotsom and jetsom. A number of questions immediately come to mind. For example, is AIDS virus transmissible via water? By swallowing or via a cut on an arm or leg? Can any amount of virus produce disease? Should people be warned to stay out of the water?

Although the coliform index is an indicator or sentinel of pathogenic organisms whose source is the human bowel, it is unlikely to be of help with organisms whose habitat is elsewhere. Legionnaires' disease and its associated organisms *Legionella pneumophila* is the most recent addition to the list of illnesses transmissible by the water route.

With legionnaires' disease, however, epidemiologic evidence has made reasonable connections between the organism, water, and human illness. It should be noted that the connections are not of cause and effect, but one that is "strongly circumstantial."

---

* Reporting of results by the multiple-tube fermentation procedure as a most probable number (MPN) is an index of the number of coliform bacteria that "more probably than any other number, would give the results shown by the laboratory exam." It is not an actual enumeration. The MF, however, permits a direct count. Either test is used to appraise efficacy of a sewage treatment process.

From almost the day it opened in 1977, Wadsworth Medical Center in Los Angeles was beset by legionnaires' disease (Kegionellosis) on a regular basis. Between 1977 and 1982, some 250 cases among its hospitalized patients were recorded. The epidemic was finally terminated by the process of hyperchlorination of the potable water supply. So far, however, the source of contamination has eluded efforts to find it, as has the question of how the organism survives (8).

D. Schistosomiasis

Schistosomiasis or bilharziasis* (snail fever) is another example of the ability of water to convey infectious material. Schistosomiasis is considered to be the world's second most prevalent disease. Only malaria accounts for more cases. Indications are that some 250 million cases of snail fever occur worldwide. It is known to be endemic in 74 developing countries.

It is primarily a disease of tropical and subtropical areas, but because of the ease and rapidity of world travel, it is also found in other regions. For example, New York City has thousands of cases imported from Puerto Rico and Cuba. There is little danger that the disease will spread in the United States, as it is not directly communicable and our unique ecological conditions hinder its establishment.

Bilharziasis, while seldom directly fatal, is debilitating and eventually contributes to the death of its victims. While the afflicted endure the illness for years, their ability to work is severely restricted.

Over 100 years passed from Bilharz's discovery of the flatworm in the blood of a patient to the complete untangling of the complex life cycle of the human illness, which includes a snail, three growth stages in the development of the minuscule worm, and humans.

Figure 15 shows this complex transmission cycle. Considering the period, the area of the world, and the intricacy of the cycle, it is surprising that it did not take longer for the connection between the worm and the illness to be made.

A crop of eggs laid by the worms gains entrance to irrigation canals and streams via urine or feces. The eggs hatch out as pear-shaped, ciliated miracidia which must find a snail host within 24 h or die. Those that locate a suitable snail host burrow into it and undergo a 2-month period of incubation. During this period, the miracidia are transformed. They become the fork-tailed cercariae that leave the snail, as shown in Figure 16, and seek another host, usually a warm-blooded one, human or animal. In countries where streams, canals, ponds, or any other watery area such

---

* In 1851 Theodor Bilharz discoverd the parasite in the blood of victims. The disease is often called bilharziasis in his honor. Schistosomiasis, another name for the disease, refers to the type of parasite that causes the disease, a schistosome or flatworm.

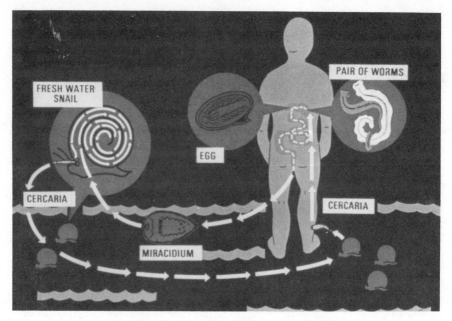

Figure 15. The life cycle of *Schistosoma mansoni*.

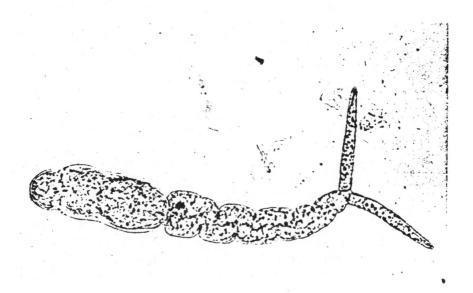

Figure 16. Free-swimming cercaria that can penetrate skin.

as rice paddies serve as latrines, laundries, and recreational areas, it is the rare cercaria that cannot locate a human host. Thus, the cycle is completed—and often repeated.

These free-swimming cercariae appear to fasten themselves to human skin by a sticky secretion, then penetrate with the aid of proteolytic enzymes, which dissolve a path for them. Upon reaching a blood vessel they pass to their ultimate sites, the veins of the upper and lower intestinal tract, or the veins of the bladder and rectum. After several months the cercariae develop into adult schistosome worms. At the height of the disease hundreds of worms can be found in a victim; each lays as many as 3000 eggs per day. The number and size (approximately 2 cm) of the adult worms causes mechanical blocking of hepatic circulation, which produces the fluid-filled, swollen abdomen characteristic of bilharziasis. The young boy shown in Figure 17 is an excellent example of the characteristic distended belly with its enlarged liver and spleen—pathognomonic of the disease.

After 3–6 weeks, an explosive onset of fever occurs, along with generalized aches, diarrhea, and extreme weakness. During the chronic stage the victim grows emaciated and steadily weaker. Until 1982, when the Bayer Company of West Germany developed praziquantil, drug therapy had been worse than the treatment of populations; attempts continue to rid streams and canals of the host snail. However, the most

Figure 17. Young boy with distended belly characteristic of schistosomiasis.

vigorous efforts are being given to education of children and adults in an attempt to change their sanitary habits.

Although snail fever was described in the Kahun papyrus (1900 B.C.), it has been given added impetus by humans in the name of progress. The construction of huge irrigation projects in the underdeveloped countries of the world has spread the disease into areas where it was previously unknown. Construction of the Aswan High Dam in Egypt, the Gezira Irrigation Project in the Sudan, Lake Kariba in Zambia, and Lake Volta in Ghana have been responsible for increasing the number of new cases by several million. While these dams and water management projects have brought heretofore thousands of unavailable acres into cultivation, the economic burden of the new cases may offset the gains.

As was noted above, although there are known cases in New York City and some additional cases spread around the country by New Asian immigrants and tourists, bilharziasis is not expected to establish itself here. However, two of our native snails, both of the genus *Pomatiopsis*, are potential hosts of the fluke. One is limited to Louisiana; the other is more widespread—its range is from the Great Lakes to Kentucky and from Iowa eastward to the coast. Public health authorities appear fully convinced that our system of sewage disposal and our personal hygiene habits do not offer the necessary ecological conditions for the infection to become established. Again, vigilance and regular surveillance are expensive but necessary precautions.

E. Cercarial Dermatitis

It is of interest to note that cercarial dermatitis, also known as swimmer's itch and clam digger's dermatitis, is the result of an infection by an avian form of schistosome worm. The eggs reach the water in bird droppings, and the miracidia locate suitable host snails in which to develop into fork-tailed cercariae. The cercariae invade human skin, producing pustules that itch intensely. The itch is well known in such diverse states as Michigan, Minnesota, New York, Rhode Island, Florida, and California.

IV.   QUALITY AND REGULATION

Water should have only one quality—safety. All legislative and public health authorities have that as their goal. However, aspiration and attainment are one thing; for the public to believe their water supply is safe is another. In fact, the public's perception may be the more difficult to ensure.

Congressional concern for the quality of the nation's water supply was spelled out in Public Law 93-523, The Safe Drinking Water Act of 1974. Congressional intent was clear: to provide the public with safe drinking water and protect them for waterborne illness.* The act required the U.S. EPA to develop primary drinking water regulations that apply to public water systems,† and "specify(s) contaminants which in the judgment of the Administrator may have any adverse effect on the health of persons." They were further directed to establish for each contaminant either a maximum contaminant level (MCL) or a specified treatment if it is not feasible to monitor the concentration of that contaminant. In 1975, as a first step, the EPA promulgated National Interim Primary Drinking Water Regulations, with amendments occurring in 1976, 1979, 1980, and 1986 for organic and inorganic chemicals, radionuclides, and microbiological contaminants.

A. Recommended Maximum Contaminant Levels

Looking to protect the American people, Congress directed the EPA to devise a set of even more comprehensive standards, The Revised Primary Drinking Water Regulations, which were based on data obtained by the National Academy of Sciences¶ from a 2-year study of contaminants in drinking water. The revised standards introduced the concept of recommended maximum contaminant levels (RMCLs). RMCLs are and must be understood to be ultimate goals, levels at which no adverse health effects can be expected to occur. Here too, sound judgment prevailed. The RMCLs were established as *nonenforceable* health goals, or desirable objectives to be striven for. They were set at levels which in the administrator's best judgment would assure that "no known or antici- pated adverse effect on the health of persons occurs and which allow an adequate margin of safety."

As already noted, the primary regulations required the EPA to set MCLs. These would be the enforceable working standards, and they could not be set so far below the RMCLs as to make the RMCLs appear unattainable.

According to the EPA, development of the regulations will proceed in five phases:

- Phase 1—volatile synthetic organic chemicals (VOCs).
- Phase 2—synthetic organic chemicals (SOCs), inorganic chemicals (IOCs), and microbial contaminants.

---

* Clearly this was a commendable attitude. However, in the early 1970s the level of water-borne illness, or disease, was miniscule compared to other routes. Obviously, perception played a major part in their actions.
† A system is considered to be public if it has at least 15 service connections or regularly serves at least 25 people.
¶ The initial toxicological assessments are contained in a series of volumes titled *Drinking Water and Health*. These have since been argumented by additional volumes in 1980 and 1983.

- Phase 3—radionuclides.
- Phase 4—disinfectant by-products, including the trihalomethanes (THMs).
- Phase 5—other SOCs, pesticides, and IOCs not previously considered.

The approach will be that of an open public hearing with four successive stages.

An advance notice of proposed rule-making will be published. This will be followed by a period for public comment and a meeting open to all interested and concerned parties. RCMLs are expected to be developed from these comments and meetings. Finally, both RMCLs and MCLs will be proposed.

B. Volatile Synthetic Organic Chemicals

After further public comment and hearings, an MCL will be promulgated. This process is estimated to require 3–5 years. An advance notice of proposed rule-making for VOCs was issued on March 4, 1982. A proposal for RMCLs for the nine VOCs listed in Table 5 was published on June 12, 1984. Final MCLs were expected in April 1986. Given the number of chemicals involved in the five phases, the process may be interminable.

Two EPA sampling surveys of potential drinking water are currently underway. Approximately 1000 public water systems are being tested for inorganic and radionuclide levels. Completion was anticipated in 1986, but 1989 may be more realistic.

Table 5
Recommended Maximum Contaminant Levels (RMCls)[a] for VOCs

VOC	Proposed RMCL
Tetrachloroethylene	0
Trichloroethylene	0
1,2-Dichloroethane	0
Carbon tetrachloride	0
Vinyl chloride	0
Benzene	0
1,1-Dichloroethylene	0
1,1,1-Trichloroethane	200 μg/L
para-Dichlorobenzene	750 μg/L

[a] The 1986 amendments required the EPA to consult with the National Academy of Sciences and propose a nonmandatory maximum containment level goal (MCLG) which is to replace the RMCL. Simultaneously, however, the EPA must also propose MCLs, set as close to the MCLGs as technically and economically feasible.

Obviously an effort to provide a safe water supply is being made. Regulations and standards are one way of approaching the goal. Standards, however, generally derive from a series of calculations, most of which are based on mathematical models with very weak biological bases. What biological data enter the model are usually animal data, and that can be orders of magnitude above the range of environmental or "real" exposure. Thus, caution in interpretation is required. Nevertheless, many of the calculations, or "extrapolations" as they are often called, are all the decisionmakers have to work with. Consequently, the decision as to which level to peg a standard is often a judgment call. The implications of this will be discussed at greater length in Chapter 15.

C. Standards

These standards set limits on the amounts of specified chemical and biological materials permitted in water supplies. This means that every community supply serving 15 or more connections or 25 people must ensure that its water meets these minimum standards. Even noncommunity supplies such as trailer parks, camping sites, and roadside motels with their own sources are included in the regulations. Most public drinking supplies, particularly the larger ones, already meet these standards.

To help states comply, Congress directed the EPA to provide the states with both financial and technical assistance. As of June 24, 1977, every water supply would require periodic testing. From then on, "Notice of Violation" has become a regular occurrence. Just what does this mean in terms of the safety of a supply?

The limits or standards set for each chemical, as well as bacterial level, were based on the amount of each substance planners calculated most individuals would consume over a lifetime without adverse health effects. The limit was based on consumption of 2 L of water (or water-based beverages such as tea, coffee, and soft drinks) every day for a lifetime. Of course, this allows for a large margin of safety. The fact that a violation is noted for a water supply is not necessarily cause for alarm. Rather, it is a device to call attention to deficiencies in the system. This was Congress' way of giving anyone who wants it "the right to know" about their particular supply. Unfortunately, neither Congress nor the EPA authorized an education campaign to provide an understanding of the meaning of the "standards" and "violations." Nevertheless, two sets of standards have been established: the National Interim Primary Drinking Water Regulations (NIPDWRs) along with the National Secondary Drinking Water Regulations (NSDWRs).

The NIPDWRs set what the EPA believes are achievable levels of quality to protect health. They are referred to as "interim" because

Table 6
Secondary Maximum Contaminant Levels for Public Water Systems

Contaminant	Level
Chloride	250 mg/L
Color	15 color units
Copper	1 mg/L
Corrosivity	Noncorrosive
Foaming agents	0.5 mg/L
Iron	0.3 mg/L
Manganese	0.05 mg/L
Odor	3 threshold odor number
pH	6.5–8.5
Sulfate	250 mg/L
Total dissolved solids (TDSs)	500 mg/L
Zinc	5 mg/L

research in ongoing and as new developments occur, changes may be made. Standards have been set for ten chemicals, six pesticides, bacteria, radioactivity, and turbidity.

The NSDWRs control contaminants in water that for the most part affect the aesthetic qualities relating to public acceptability. These regulations are not federally enforceable but are intended as guidelines for each state and by extension for individual localities. Table 6 lists secondary maximum contaminant levels (SMCLs) established by the regulation.

As we approach the waning years of the 20th century the weight of evidence indicates that biologically, chemically, and epidemiologically our water supply is safe (9). Psychologically, that may not be the case. Too many people prefer to believe otherwise.

D. What Do the Levels Mean?

With our ability to test for smaller and smaller concentrations of chemicals (we are not able to identify contaminants in the parts per trillion range) the recurring questions are: What do these levels mean? and What is their implication for human health?*

Perhaps the questions can be answered by asking another. If untoward health effects are not observable at the parts per million (ppm) level, will they become manifest at the parts per billion or the parts per trillion level—six orders of magnitude less? Let us restate that. Can chemicals

---

* Current analytic capability permits detection of dioxan levels of parts per quadrillion— that is 1 followed by 15 zeros. Other than telling us that the substance is present in these close to molecular quantities, it does little more for us. If we can find no adverse health affects at the ppm or ppb level, surely amounts in the ppq range can not be expected to be harmful.

undetectable at one level, become harmful when newly developed sophisticated analytical techniques are able to detect them at far lower levels? That is what has happened over the past 20 years. Chemists are closing in on detecting the presence of molecules, but does that make the environment or the water supply less safe? Hardly.

At the levels found in groundwater and surface waters, they do not induce pathologic, malignant, or mutagenic changes in laboratory animals or microbial and tissue culture test systems. They do, however, create headlines and doubt about the safety of the water supply.

On the other hand, to load test animals with concentrations five, six, or seven orders of magnitude above those found in some of the country's water supplies creates a dilemma of another sort. How does one translate those studies into human terms—human protection? These analytic innovations notwithstanding, as we approach the second millenium, and more into the 21st century, the health of the American people is at its highest level. Never before in the history of our country has the general well-being been better, and the future indicates still further improvement for more people. Unfortunately, too many of us prefer to believe otherwise.

There is yet another problem: the lack of understanding between a threat to health and the citation of a water supply for violation of noncompliance with standards of the Safe Drinking Water Act. A great breakthrough will have occurred when the American people understand that difference and also accept the fact of their generally high level of health.

REFERENCES

1. R. B. Evans and G. E. Schweitzer, Assessing Hazardous Waste Problems. *Environ. Sci. Technol.* **18**(11):330A–339A, 1984.

2. M. Rogul, An Assessment of Biomonitoring. EPA Draft Document, July 1983. U.S. EPA, Washington, DC.

3. L. B. M. Petersen and R. C. Petersen, Effect of Kraft Pulp Mill Effluent and 4,5,6-Trichloroguaiacol on the Net Spinning Behavior of *Hydropsyche angustipennis* (Trichoptera). *Ecol. Bull (Sweden)* **36**:1–7, 1984.

4. J. J. Rook, Formation of Haloforms During Chlorination of Natural Waters. *Water Treat. Exam.* **23**:234–243, 1974.

5. G. F. Craun, A Summary of Waterborne Illness Transmitted Through Contaminated Groundwater. *J. Environ. Health* **48**(3):122–127, 1985.

6. G. F. Craun (Ed.), *Waterborne Diseases in the United States.* CRC Press, Boca Raton, FL, 1986.

7. *Standard Methods for the Examination of Water and Wastewater*, 16th ed. American Public Health Association, Washington, DC, 1985.

8. K. N. Shands, J. L. Ho, R. D. Meyer, G. W. Gorman, P. H. Edelstein, G. F. Mallison, S. M.

Finegold, and D. W. Fraser, Potable Water as a Source of Legionnaire's Disease. *J. Am. Med. Assoc.* **253**(10):1412–1416, 1985.

9. K. S. Crump and H. A. Guess, Drinking Water and Cancer: Review of Recent Epidemiological Findings and Assessment of Risks. *Annu. Rev. Public Health* **2**:339–357, 1982.

SUGGESTED READINGS

Drinking Water and Health. Safe Drinking Water Committee Assembly of Life Sciences National Research Council. National Academy of Sciences, Washington, DC, 1977.

Lechevallier, M. W., and McFeters, G. A. Recent Advances in Coliform Methodology for Water Analysis. *J. Environ. Health* **47**(1):5–9, 1984.

Webbe, G. The St. Lucia Project and Control of Schistosomiasis. *Endeavor* **11**(13):122–126, 1987.

12

Sanitary Sewage/Waste Water Treatment

It is possible for a description of things to be completely accurate, without being at all truthful.

—Trevanian

"**T**hou shalt have a place also without the camp, whither thou shalt go forth abroad: and thou shalt have a paddle upon thy weapon; and it shall be, when thou wilt ease thyself abroad, thou shalt dig therewith and shalt turn back and cover that which cometh from thee" (Deut. 23:12–13).

Although the Israelites did not have the benefits of waste disposal by water carriage,* they seemed well aware that human excreta could be a vehicle of disease transmission. It was not until several thousand years later in the filthy, sewage-laden industrial cities of London, Boston, Cologne, Edinburgh, and New York, that social pressures forced the introduction of the water-carriage system of waste disposal.

After a rainfall in the city of London, Jonathan Swift wrote the following lines; they could apply to any community circa 1740:

> Now from all parts of the swelling kennels flow and bear their trophies with
> them as they go; filth of all hues and odors seems to tell what street they

---

* The water-carriage system of waste removal requires a supply of water under pressure that removes waste from dwellings. Its introduction in the 1840s involved engineering problems of considerable magnitude. However, it only shifted the problem away from the dwellings. At some point the combined discharges of the whole community had to be dealt with. The introduction of sewage-treatment plants was a logical extension of the waste-removal system.

sailed from by the sight and smell, sweeping from the butchers' stalls, dung, guts and blood, drowned puppies, stinking sprats, all drowned in mud; dead cats and turniptop come tumbling down the flood.

Descriptions of conditions in streets of European and American cities in the 18th and 19th centuries clearly show the appalling filth (by our standards) in which the people lived. One traveler wrote: "In the tenements of Glasgow dung was left lying in the courtyards as there were no lavatories in houses. This lack of lavatories led to the habit of house dwellers filling chamber pots with excreta and after some days, when completely full, and with a shout of, beware slops, emptying the contents out of the window into the street below." This practice quickly led to the pollution of wells and to the infestation of cities with rats; both gave rise to diseases of epidemic proportions, which regularly took a heavy toll in human lives. It was not until the latter part of the 19th century that disease transmission via human waste was actually demonstrated.

In fact, it was John Snow (1813–1858), a distinguished British anesthesiologist, famed for having administered chloroform to Queen Victoria at the birth of her two children, who established the waterborne transmission of cholera. His elegant epidemiologic studies, 30 years before Robert Koch isolated the cholera vibrio, clearly linked cholera with contaminated water.

Unfortunately, too many people who ought to know better continue the fiction that the outbreak of cholera in the Golden Square area of London during August and September of 1854 was halted by his having had the handle of the Broad Street pump removed—thereby preventing the residents from consuming the cholera-containing water. In fact, the epidemic peaked on September 1st and was well on its way out of the community by September 8th, the day the handle was removed.

I. THE NEED FOR SEWAGE TREATMENT

While the objective of disposal is transportation of sewage away from homes and offices to an appropriate treatment site, the primary goal, if community health is the concern, is treatment in a manner calculated to protect health, preserve natural resources, and prevent nuisance conditions.

With the aggregation of millions of people in large urban centers, there is an unprecedented volume of human waste to be treated. In this treatment, pathogenic microbes (bacteria, viruses, protozoa, fungi, spirochetes, and helminths) must be removed or reduced to harmless levels. It is well known that in any community, at any time, there are always small numbers of people who are either manifestly ill, in some stage of illness but not demonstratively so, or healthy carriers of disease. Together, they constitute a community's normal background or reservoir of pathogenic microoganisms whose elmination as feces means they may be present in sewage. As a result, sewage can be a threat to health, particularly to those who come in contact with water drawn from sewage-contaminated sources. Thus, disposal of sewage implies two entirely different propositions: (1) the removal of energy-laden organic matter from liquid waste and its conversion to an innocuous form in order to prevent or control pollution of lakes, streams, and rivers; and (2) the prevention of water borne disease.

A. Sewage Constituents

The term *sewage* usually connotes the liquid wastes from homes, schools, commercial buildings, hotels, hospitals, and industrial plants. In addition to human excretal material, it generally contains industrial wastes such as those from meat-packing operations, breweries, milk and food plants, and chemical processing plants. Domestic sewage is generally limited to household wastes from residential areas and contains wastes from water closets and wash water from baths and kitchens. Curiously enough, in the United States domestic sewage varies little in composition and strength from community to community across the country. Although we like to think we are uniquely individualistic, in fact we are remarkably similar. So much so in fact that most of us flush our toilets at about the same time. Figure 1 shows the hourly volume of sewage flow resulting from this national behavior pattern.

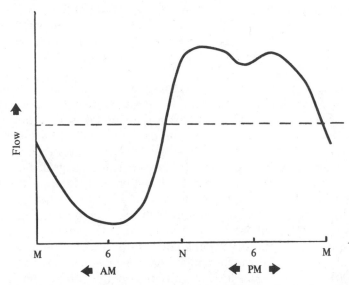

Figure 1. Nationwide pattern of sewage flow. From *Microbiology for Sanitary Engineers* by Ross E. McKinney. McGraw-Hill Book Company, New York. Used by permission of McGraw-Hill Book Company.

Sewage or liquid waste has the appearance of spent dishwater. By the time it reaches the main sewer trunk lines, combined with waste from other areas of the community, and arrives at a waste treatment facility, it usually contains paper, organic material from feces and urine, soap, and such exotic items as dead animals, fruit skins, old shoes, last week's fish stew, hosiery, tampons, drugs, and anything—repeat anything—anyone feels can be quickly flushed out of sight.

A useful measure of the strength of sewage is the proportion of suspended solids it contains. The suspended solids of interest to the chemist are not the dead animals, shoes, or fruit skins, but organic particles of carbohydrates, fats, and proteins. Generally, sewage consists of 99.9% water and 0.02–0.04% solids. Although the amount of suspended solids is small when evaluated on a percentage basis, the total daily amount from a large city can be considerable. For example, in the sewage of a city the size of Washington, DC, solids amount to approximately 200 tons/day. Of this, 40–50% is protein, another 40–50% consists of carbohydrates (sugars, starches, and cellulose), and the remaining 5–10% is fat. Discharging 200 tons/day of carbohydrates, fats, and proteins would provide more sumptuous meals for the microbes in lakes, streams, and rivers in and around Washington than they could digest. The result would undoubtedly be pollution of a magnitude as yet unknown in that area. Although such a discharge is only in the realm of conjecture, it does indicate the degree of treatment necessary.

B. Pathways to Treatment Sites

Before discussing sewage treatment, it may be helpful to delineate the path taken by sewage as it flows from home to treatment plant to final outfall in a watercourse. Treatment of sewage from homes with septic tanks, cesspools, or seepage pits will be discussed separately from community systems.

Suppose that while you are washing your hands a ring slips off your finger, and before it can be retrieved it has gone down the drain. Figure 2 is a diagrammatic summary of the path the ring takes as it moves along with the liquid waste. At top left, house and street are seen in cross-section. Each of the various utilities—gas, electricity, water, and sanitary sewers—leads from the street to the house. The largest of the conduits found under the street is the main or lateral branch of the sewage system that collects domestic waste from houses on the street. A cross-section of the inside of the house is depicted in Figure 2a at top right. Connections and flow from sink, tub, and flush toilet lead to the street sewer.

The sewer under the street joins large trunk sewers from other streets and neighborhoods, through which the waste stream flows toward a treatment facility.* Figure 3 shows a typical collection system gathering

---

* Many cities have combined sewers in which both sewage and storm water runoff from street washing and rain are collected and flow together to the treatment plant.

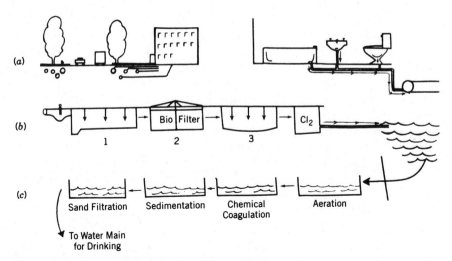

(a)

(b) Bio Filter → Cl₂

1   2   3

(c) Sand Filtration ← Sedimentation ← Chemical Coagulation ← Aeration

To Water Main
for Drinking

Figure 2. Diagrammatic representation of the route taken by domestic waste from an urban dwelling to its final outfall in a river.

waste from homes, commercial and industrial sites, and flowing on to a treatment plant. Figure 2b represents the trunk line arriving at a treatment plant, as well as the typical processes to which a waste stream will be subjected. Wherever practical, sewers are built to allow sewage to flow by gravity; when this is not possible, pumping stations are strategically placed to pump sewage through pressure conduits to points where they can discharge into a gravity sewer.

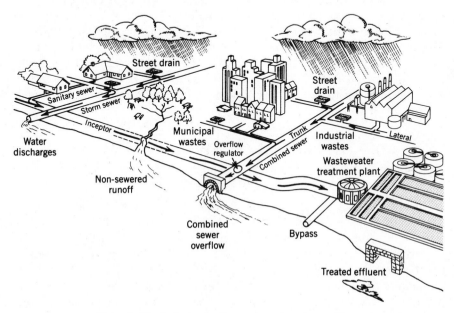

Figure 3. Typical sewage flow pattern in urban settings.

As the sewage flows along through the many miles of piping, biological activity engendered by the extensive microbial flora naturally present in human waste (together with additional microbial forms from soil, street washings, dead animals, fruit skins, and whatever else falls into sewers) progresses at an ever-increasing rate. The longer the sewage remains in the system, the greater will be the activity. If the sewers are extensive and the rate of flow is low, the biological activity will be so advanced that the waste can be considered as having received partial treatment before it reaches the plant. The key points here, and as the waste continues on, are *biological activity* and *treatment*. Both are concerned with rendering this highly putrescible organic material innocuous; that is, preventing water pollution, odors, and unsightly conditions.

At this point, a digression is in order.

C. Sulfate-Reducing Bacteria

Much of the piping used to transport liquid waste has been made of concrete. Nevertheless, it is not immune to microbial attack. These pipes are often corroded by the metabolic activity of sulfate-reducing bacteria (SRB). Their metabolism is largely dependent on the sulfate molecule being used as a primary oxidizing agent for the dissimilation of organic matter, analogous to oxygen in human respiration. In anaerobic respiration, the sulfate molecule is reduced to water and sulfide.

Under aerobic conditions, the free sulfide will become hydrogenated, forming hydrogen sulfide (H_2S), which is itself extremely corrosive. The presence of this dissolved gas will also tend to lower the redox potential of the micro (localized) environment, making it even more favorable for further SRB growth. However, sulfur-oxidizing bacteria are also present. They can convert H_2S to sulfuric acid, which in turn reacts with the lime in concrete, forming calcium sulfate. Calcium sulfate, having little structural integrity, crumbles. This type of problem requires ongoing materials research in order to prevent continued replacement and excessive costs.

II.   THE PROCESS OF SEWAGE TREATMENT

A. Primary Treatment

At the treatment plant, sewage flows through coarse and fine screens to trap floating objects such as fecal solids and other items that may impede treatment. The coarse screens consist of a bank of vertical bars set approximately 1 in. apart. The lost ring might not pass through unless it approached the bars sideways.

Backing up the bar screens is a comminutor, which grinds the remaining solids to a size that should prevent damage to the machinery. Sand, gravel, and stone (grit) are then removed. True grit usually consists of particles of stone larger than sand. Grit removal is simply a process of sedimentation or settling out as a function of gravity. Grit chambers are provided to prevent clogging and wearing of moving parts such as sludge-collector mechanisms, pumps, drains, and pipes. The waste water is slowed down by design to allow the heavier particles to settle out. The next step is the actual settling basin for lighter materials requiring more time—the primary settling tank—in which retention time is sufficiently long to give the lighter material an opportunity to fall to the bottom. The velocity of flow through the tank is adjusted to achieve a 30–90 min retention time, depending on the strength of the sewage (BOD_5)* and on whether additional treatment is to be given. It is quite likely that the ring originally lost down the bathroom sink will settle out in the primary tanks, as a result of the decreased velocity and gravitational effects, and will finally be lost in the accumulated sludge. With each of these steps, the waste water becomes increasingly clarified.

Although meeting increased resistance from the EPA, too many communities continue to discharge their waste into a river or stream without further treatment. Settling or primary treatment may be all a waste receives. Raw sewage is thus emptied into a watercourse with the idea that dilution with large volumes of water will prevent pollution or nuisance conditions. This may be wholly unjustified. The problem of course, is the distaste for raising taxes to obtain the funds needed to build a secondary treatment plant.

The waste water leaving the settling tank or clarifier has been relieved of much of its solids content; nevertheless, it still contains a high demand for oxygen (BOD) in the form of fats, proteins, and carbohydrates (sources of energy). It is this oxygen demand that can produce "pollution" of a waterway by having its natural oxygen content used up in the process of metabolism of the organics via mcirobial activity. To prevent this, secondary treatment of sewage is designed into the process to remove or stabilize (neutralize) the remaining organic compounds. Depending on the type of secondary treatment, upward of 90% of the suspended organics can be removed. This degree of stabilization can prevent pollution in most waterways.

B. Secondary Treatment

At this point, depending on the type of secondary treatment to be given, the sewage will pass to either a trickling filter (biofilter) or an activated sludge aeration tank. Both are based on the stabilization or neutralization of organic waste by biological action. That is, a host of biological species ranging from bacteria to snails and flies take part in the process that

---

* Five-day biological oxygen demand.

eventually produces an inoffensive effluent. It is in either the trickling filter or the aeration tank that the business of removing the remaining suspended matter occurs. From either the trickling filter or the aeration tank, the stabilized and clarified liquid flows or is pumped to a secondary settling tank for an additional short holding period.

Up to this point the second of the two major functions of waste treatment, the prevention of disease transmission by virus, bacteria, and protozoa, has been given only cursory attention. Although pathogenic mciroorganisms are partially removed in the settling process and others are mechanically removed by filtration and aeration, it is the process of chlorination that destroys the great majority of organisms. In addition, chlorine abets the stabilization of the remaining organic matter by its highly oxidizing character. After appropriate contact with chlorine, the clarified disinfected liquid is discharged into a watercourse.

This is a generalized description of the trip sewage takes from home to stream or river. It is at this point of sewage outfall in some watercourse that water pollution can begin, and it is at the point of water intake farther downstream that many of the problems of drinking water begin. Figure 2c also indicates the general path followed by water from point of intake downstream of a sewage outfall, through treatment, and back to the consumer.

The question that remains to be answered is: If sewage follows the path outlined, why aren't our waterways pollution free?

Figure 9 in Chapter 11 offers an appropriate point of departure. Pollution means different things to different people. Not all potential pollutants begin as sanitary waste and pass through primary, secondary, or tertiary treatment. A substantial amount of organic matter, with high BOD, is contained in runoff from agricultural lands. Storm water containing the washings of city streets can bypass treatment plants, and the treatment plants themselves are far from 100% efficient.

Then of course there is an entirely different class of waste—the chemicals from commercial and industrial processes that never see a treatment plant, and for which traditional treatment plants were not developed. Sending chemical wastes through these plants without adequate dilution wipes out the microbial flora that makes trickling filters and activated sludge processes the marvels they are.

Finally, watercourses have traditionally been the end-of-the-line for all manner of discarded waste, as Figure 2c shows. Indeed, there is good reason for the existence of pollution despite substantial treatment.

A sewage treatment plant employing secondary treatment is in fact utilizing aerobic biological processes to stabilize waste. Where biofilters or trickling filters are used, the liquid waste from primary settling tanks flows to the filter or percolation bed. Here rotary distributor arms spray the waste over a bed of rocks. Figure 4 shows a unit in operation.

Figure 4. Trickling filter in operation outside Paris.

Figure 5 is a close-up view of the liquid falling on the bed. As the liquid trickles over and between the stones, bacteria, protozoa, worms, snails, spiders, and flies utilize its protein, carbohydrates, and fats to make additional bacteria, protozoa, worms, and so on. In the process, the waste is transformed into energy-free particles that do not use the oxygen dissolved in the stream; with little or no oxygen demand, pollution cannot occur. Thus, the main purpose of secondary treatment is reduction of the amount of dissolved oxygen withdrawn from the stream.

Figure 5. Close-up view of a section of a distributor arm of a trickling filter.

1. Activated Sludge

The activated sludge process is a direct evolutionary decendant of the trickling filter. In place of rocks for creating surface area and a place for the organisms to fasten onto, experiments indicated that dissimilation of organics could proceed as well in a tank of water into which oxygen was bubbled. Although appropriate hydraulic designs required development, the activated sludge process was born.

Essentially, the process can be represented as follows:

organic waste + microbes + oxygen $\rightarrow$ CO_2 + H^2O +
$$energy + additional\ organisms$$

Underlying the process is the concept of continual reuse of the microorganisms to seed or "activate" incoming waste. Accordingly, it is a "continuous culture" process, as depicted in Figure 6.

Airjets churn the sewage as it flows into and through the tank, up to 8 h in some installations. Microbes grow and reproduce as they feed on the sewage particles. These particles and microbes begin to coalesce, producing aggregates that flow and settle out as a sediment or sludge. A portion of this "activated" flow is returned into the incoming tank to begin the process again. This aeration is shown in Figure 7.

As organisms absorb and metabolize waste, oxygen, initially high, begins to fall rapidly while the concentration of organisms increases appreciably and waste is degraded. The relationships are depicted in Figure 8.

The activated sludge contains a variety of microbes, each in effect waiting their turn to participate. One set of microbes attacks initially, breaking down the organics to a form others can attack, until the remnants are carbon dioxide and water. In these continuous flow systems, the attempt is made to prevent the depletion of oxygen and organism reduction, so that the process can continue efficiently.

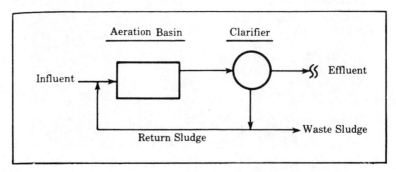

Figure 6. Flow diagram of the activated sludge process.

(a)

(b)

(c)

(d)

Figure 7. (a)–(c) Successive close-up views of the vigor of aeration given sanitary waste in activated sludge treatment plants. The draft tube, covered by the metal baffle in accompanying figures, is uniquely important. It provides the vigorous agitation of air and liquid. Without it, mixing would be limited to surface action. It is the continuous availability of large amounts of oxygen which permits the microbes to metabolize the waste to inert material. Chocolate-brown in color, the liquid waste is pumped up through a 20-ft high aerator unit. (d) The dimension of the unit in one section taken out of service for a brief period. Photos taken at the Sewage Operating Commission's Sewage Treatment Plant, Princeton, NJ.

Even greater reductions of generally refractory organic molecules such as DDT, dioxin, and other organohalides may be in the offing if recent investigations at Michigan State University prove fruitful. Researchers there have isolated a soil mold that, if added to activated sludge tanks to become an integral part of its active microflora, could readily deal with these otherwise refractory chemicals. (See Chapter 13 for a more detailed discussion of this.)

2. The Carousel System

The carousel system shown in actual use in Figure 9, and in schematic form in Figure 10, consists of a surface aerator mounted in a continuous channel to provide both oxygen and mixing to an activated sludge system.

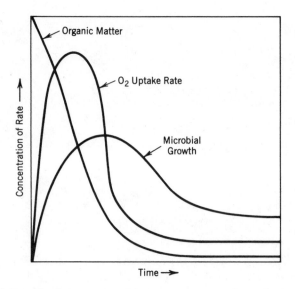

Figure 8. Relationships between organic matter, oxygen uptake, and microbial growth.

Figure 9. Eimco carousel biological oxidation system. Courtesy of Envirotech Corporation, Salt Lake City, Utah.

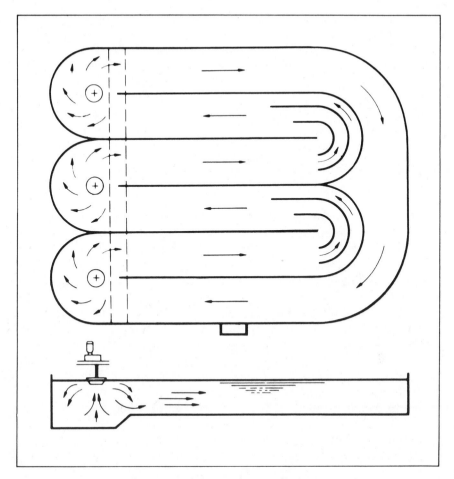

Figure 10. Schematic of carousel system.

Developed in Holland in the early 1960s and used in Europe since 1968, the carousel has only recently been gaining favor in the United States. Waste water and return sludge are mixed in the first aeration zone. A flow velocity of 1 ft/s is maintained in the channel as a function of the pumping action of the aerators. This action is achieved by lining up the partition with the aerator in such a manner that the aerator pushes the flow through the aeration zone and out into the channel. The flow passes through successive aeration zones and around a final outer loop where the effluent exits over a weir just before the first aeration zone.

The liquid waste in the channels completes its circuit ever 5–20 min, depending on channel length and loading. Carousels, especially in Holland, have normally been operated with extended aeration times of 18–28 h. This produces stable sludges. It is for this reason that the carousel process is favored when disposal takes the form of land dispersal for agricultural purposes.

Effluents from carousels can be expected to yield 5-day BODs (BOD_5) of less than 5 mg/L, which can represent reductions of 99% plus. Ammonia–nitrogen (or NH_3-N) levels of less than 1 mg/L, along with nitrogen removals of from 40 to 75%, are common.

In the oxygen-rich zones of the tank, ammonia and proteins are metabolized by bacteria to nitrates and nitrites. In the oxygen-poor zones, denitrifying microorganisms utilize the oxygen in nitrate/nitrite with release of gaseous nitrogen to the atmosphere.

3. Oxidation Ponds

A low-cost method for stabilizing sewage is the oxidation pond or lagoon. Like the other methods described, it is based on biological degradation. The sewage, which is not subjected to any preliminary treatment, is pumped directly from a collecting point to the pond, which is essentially a shallow ditch 2–6 ft deep. The area of the pond is usually dictated by the available land.

Because the pond is designed to provide a great deal of surface area as well as easy penetration of sunlight, oxygen is available in large supply for the algae, which play a major role in metabolism near the surface of the pond.

Anaerobic organisms working on their dietary requirements closer to the bottom, produce methane, ammonia, and carbon dioxide, which are vented to the atmosphere. Other organisms can further dissimilate these gases to water and oxygen.

After a rather long retention period, 30–40 days, the stabilized effluent is released into a nearby waterway. From time to time the accumulated sludge must be removed in order to keep the lagoon at peak efficiency.

Such industries as dairies, food canneries, and breweries often use oxidation ponds to handle their high-oxygen-demand wastes economically and to prevent overloading of already overburdened municipal treatment plants.

C. Septic Tanks

Another means of sewage disposal widely used in the United States, one which often contributes to pollution of streams and drinking water, is the septic tank. This is an individual system used where water is available to carry waste from a home not served by municipal sewer lines. The system is composed of three simple elements: the septic tank, a distribution box, and a disposal field.

Liquid waste is conveyed by pipe from a dwelling to a waterproof concrete or metal container. Figure 11 shows the general plan of the

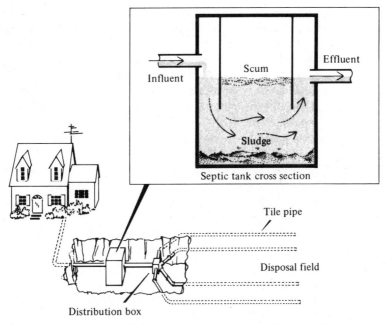

Figure 11. Schematic view of a septic tank system. Courtesy of John E. Kiker, Jr., University of Florida.

system and a cross-section of the primary container. The tank usually contains two baffles to prevent the heavier solids and the buoyant scum from passing out with the effluent to the distribution box and into the disposal field.

During the retention period in a properly operating system, from 50 to 70% of the suspended solids are removed by sedimentation, which forms a sludge that must be removed periodically. Failure to remove the accumulated solids often results in discharge of solids in the absorption field, with resultant clogging and ponding.

The principle is relatively simple. Because this is a closed system with little ventilation, anaerobic conditions prevail. This implies that the bacterial metabolism will be carried out by anaerobic species. The process is one of digestion, in which both floating scum and the settled solids are reduced in volume and transformed into liquids, gas, and an irreducible minimum of biologically inert solids. The gas diffuses into the atmosphere and the liquids containing organic acids pass into the fields and percolate through the soil, where aerobic bacterial species complete the stabilization.

The distribution box provides a means for even dispersion of the liquids to the perforated tiles. An integral part of the system is the disposal or absorption field, which should be carefully designed. Inadequate field size and too compact a soil can prevent percolation of the liquid sewage and result in ponding. As a consequence, public health nuisances as well as hazards to health may occur. In areas with sandy soil, contamination of groundwater may be an additional risk.

Although septic tanks are a direct descendant of cesspools, they are not the same. Cesspools have been described as wells in reverse. Untreated domestic waste is allowed to run into a hole or pit in the ground from which seepage or leaching through soil occurs. This stabilizes the waste through bacterial action. All too often the depth of the cesspool places the raw sewage in contact with the water table, with the result that gross contamination of the drinking-water source occurs.

The lack of any disinfection procedure in either the septic tank or the cesspool does not recommend them from the community health standpoint. Although some pathogenic microbes are removed in septic tanks during sedimentation and digestion, a significant number do enter the disposal fields. Cesspools are even more hazardous, as they place disease-carrying organisms in close proximity to wells and ground water aquifers. Public health agencies are quite correct in attempting to eliminate their use.

Although farmers and rural populations have used septic tanks since their development around the turn of the century, it is the rapid expansion of residential areas into formerly rural communities in recent years that has greatly accelerated the number of private sewage-disposal systems. Of course, they do require periodic emptying to remove the accumulated sludge.

Because sludge disposal sites are declining, it is becoming exceedingly difficult to simply dump sludge. Communities are requiring septic tank cleaners to obtain licenses. This aids in maintaining a modicum of control on clandestine dumping.

Of our 245 million people, three-fourths or 187 million are served by municipal sewers. The remainder use septic tanks. Of those using city sewer lines, about 5% live in communities that discharge untreated sewage directly into streams. Another 10% live in communities where raw sewage is discharged after a short retention in primary settling basins. Recall that primary treatment includes screens to remove large floating objects, a communitor for shredding, a grit chamber, and a sedimentation basin for settling of the heavier solids. Little stabilization occurs up to this point.

It has been estimated that the sewage discharges from communities that give primary treatment or none at all correspond to the raw waste of almost 30 million people.

Many treatment plants are old, inefficient, and overloaded, so that a substantial portion of the sewage is only partially stabilized. It is not difficult to see why water is polluted by domestic and industrial wastes. It was also noted earlier that many of the new industrial wastes are not fully metabolized by the organisms of the secondary treatment facilities and that some of these chemicals coming into the treatment plants in high concentration prove toxic to the microbes.

III. PROFIT IN SLUDGE AND COMMUNITY BENEFIT

A unique form of sewage treatment makes use of the many nutrients in sewage to increase food supply. Although the need for protein supplements to diet is greater in Asia, Africa, and South America than in the industrialized nations of western Europe, the huge sewage system in Munich, Germany, utilizes fish ponds both to purify sewage effluent and to increase the supply of fish. Instead of being discharged into a river, the effluent is sprayed into a pond stocked with rainbow trout or carp. The fish rapidly increase in weight on the luxurious diet and at the same time purify the waste, which is then discharged in the Yser River. The fish are sold for a smart profit.

Closer to home, we are beginning to give serious consideration to land disposal of municipal sewage and waste water effluents. Land disposal is really nothing more than employment of the soil as a huge filter. The soil microbial biomass consisting of soil bacteria, fungi, algae, protozoa, worms, and insects has been termed the "eye of the needle," through which all organic matter entering soil must pass as it is broken down to the inorganic components that plants can effectively incorporate.

Land use planners envision the discharge of treated effluents (secondary treatment plus chlorination) for crop irrigation, fertilization of gardens, public parks, and golf courses. This would free rivers and streams of a major organic burden, return nutrients to the soil, restore groundwater, and reduce or eliminate the need to fertilize crops chemically. With so many compelling features, it would be difficult not to consider it. One drawback may be the potential for infectious diseases as a consequence of broadcasting throughout our environment waste from human and animal feces.

Recovery and reuse of waste is not a new idea. Victor Hugo wrote passionately about the problem in 1863:

> Paris casts twenty-five millions of francs annually into the sea, and we assert this without any metaphor. How so, and in what way? By day and night. For what object? For no object. With what thought? Without thinking. What to do? Nothing. By means of what organ? Its intestines. What are its intestines? Its sewers . . . Science, after groping for a long time, knows not that the most fertilizing and effective of manures is human manure. The Chinese, let us say to our shame, knew this before we did. . . . Thanks to the human manure, the soil of China is still as youthful as the days of Abraham, and Chinese wheat yields just one hundred and twenty-fold the slowing. There is no guano comparable in fertility to the detritus of a capitol, and a large city

is the most important of dung heaps. To employ the town in manuring the plain would be certain success, for if gold be dross, on the other hand our dross is gold.[*]

A. The Odessa Project

Making his way to Texas and the School of Pulic Health, Geoffrey Stanford, a British-born and educated physician, was first heard from in 1973. He was about to put Odessa, Texas, "on the map." Set in the arid plains of West Texas, Odessa came into being for one reason only—to remove as much "black gold" as could be managed. By most calculations, the oil would run dry by 1990, and Odessa would become another ghost town. According to Stanford, that not only need not be, but Odessa could prosper on its own waste. Central to Stanford's plan was the fertilizing of the desert with sludge, a purified end-product of sewage treatment that "looks like gruel, smells like tar and is loaded with nutrients." That, together with waste water and cellulose-containing materials in garbage, would turn the desert into a vast garden. But Stanford was no idle dreamer. He had a plan, and a well-thought out one at that.

The Odessa Project was scheduled to start in the fall of 1974. Two hundred fifty tons of garbage, 20 cubic yards of sludge, and 400,000–500,000 gallons of sewage were to be sent to a 640-acre plot as an initial experiment. His objectives were laudable: beautify the city, make its environs more attractive, and restore the eroded rangeland back to its original prairie condition. The methods would provide immediate benefits:

- Recovery of metals from city refuse.
- Energy recovery from refuse and waste water.
- Water conservation from waste water and rain.

The project ran for 8 years but was bedeviled by lack of adequate financing. Nevertheless, the concept was correct: garbage makes fine manure. By 1981, the project had blossomed into the Landmix Program.

The city of Odessa had dedicated a significant portion of its refuse stream for agricultural soil enhancement. Landmix adds shredded, partially sorted, municipal refuse to digested, partially dewatered sewage sludge, in the top 6–12 in. of soil. Landmix approached waste disposal with a totally unique concept. Ordinarily, maximum amounts of wastes are added to minimum amounts of land. In this case, wastes were added in limited quantities to permit the particular soil type time to assimilate them beneficially. This has increased the nutrient content of the soil and

[*] *Les Miserables*, Vol. V, Book II, Chap. I. Routledge & Sons, New York, 1886.

provided a means for holding the nutrients in the root zone where they remain available to the crops.

During the period 1974–1981, Odessa landmixed 25%, or 53 tons, per day of its daily 205-ton refuse stream. The carrying capacity of the land increased over 12 times, from 3 cow-calf per acre mile to over 40 (1).

Under a grant from the U.S. EPA, Dr. Stanford initiated a 3-year soil and crop response test in a $5 \times 5 \times 3 \times 3$ multivariate trial in Houston. The trial compared the effects of various loadings of refuse and sludge, as well as the effects of three different inorganic fertilizer loadings. Refuse application rates from 16 to 256 tons/acre were utilized.

The results were impressive. All proportions of the refuse/sludge mix outproduced the highest inorganic fertilizer loading. In some plots the yield more than doubled. Landmix does not eliminate the need for other forms of disposal. Old furniture, worn-out washing machines, lamps, pots, and pans cannot be landmixed. It does, however, offer larger crop yields, fertile soil with improved water-holding capacity, reduced need for synthetic fertilizers, and smaller landfills. Hugo suggested it, and Stanford tested it. It may be time for wider application.

B. Health Effects of Land Disposal

1. Heavy Metals

Metals such as lead (Pb) or cadmium (Cd) have no known beneficial role in human metabolism. They can be toxic at high concentrations. The essential metals zinc (Zn) and copper (Cu) can also become toxic at concentrations in excess of their optimum for enzyme activity in metabolic pathways. When present in soil, these metals are available for uptake by plants and thereby enter human and animal food chains. Application of waste water and sludges to agricultural lands may produce potentially toxic quantities of Pb, Zn, Cd, Cr, and Ni, which accumulate with each application.

Recently, Brookes and his colleagues at the Rothamsted Experimental Station, Harpenden, England (2), investigated the effects of heavy metals on microbial activity and biomass in soils treated with sludge. They found that heavy metals did reduce soil microbial activity. In contrast, "phosphatase activity, viable and total counts of bacteria and protozoans were unaffected." They were unable to explain the discrepancy.

Given the potential for heavy metal accumulation and bioavailability, Davies and Houghton proposed the use of the radish (*Raphanus sativus* L.) as a more relevant test rather than conventional soil extractant analyses (3). The radish has been cultivated for some 2000 years. It takes up little space (in laboratories), its seeds germinate easily, and it is pH tolerant, growing bewteen pH 5.5 and 7.0. In recent tests of metal uptake in radish compared with wheat, rye, lettuce, and red fescue, they found that "radish metal levels correlate well with soil metal concentrations." Either the bulb or leaves can be used. Radish, it appears, will satisfactorily predict the uptake of metals by other plants grown in similar soils, and only in the case of zinc is radish a strong metal accumulator. The process of accumulation can lead to misinterpretation of actual or true bioavailability.

In common with heavy metals, the presence of organic contaminants in sludges has implications for animal and human consumption. Davis and co-workers of the Water Research Center, Medmenham, England, suggested recently that the use of sludge contaminated with organics (PCBs and benzene hexachloride-lindane) should be avoided on pasture land, but areas growing small grain crops represent a comparatively low risk. Organic contaminants such as PCB and lindane are lipophilic and could find their way into milk following application of sludge to pastures (4).

2. Virus Contamination

Because virus particles are more resistant than bacteria to chlorination, their presence in waste water effluents and sludges may be anticipated. Transmission of enterovirus from person to person via the fecal–oral route is well established. Since some 100 different viruses can be found in human fecal waste, and as virus particles can survive in favorable environments, the potential for further conveyance in sewage and night soil requires evaluation. Their presence is directly dependent on waste treatment practices, but given the current health status of most American communities, viral loading of sewage can be expected to be minimal.

Referring back to the route taken by sewage from home to treatment plants, it can be expected that a portion of the virus content adsorbed to solids will be removed by gravity in the primary settling tanks. As the waste passes to secondary treatment facilities, additional removal occurs. In trickling filters, the glutinous zoogleal mass covering individual rocks offers adsorption area for viruses to adhere. Experiments have indicated 10–15% removal can be expected this way. This is far less than occurs in the extended aeration tanks of the activated sludge process. Fifty percent removal is not unusual; 90% can be expected. In such tertiary processes as activated carbon and reverse osmosis, total removal is often the case.

Land treatment itself is highly effective in removing viral particles. Removal is a physicochemical adsorption phenomenon rather than antibiosis. Furthermore, if sludges are heated to 145°F (50°C), virus removal is total. In summary, if treatment processes are adequately controlled, there should be little, if any, health hazard from viruses (5).

Earlier, I noted the intimate relationship between water quality and waste disposal treatment. The federal government understood this and during Richard M. Nixon's presidency passed the Clean Water Act of 1971. Its intention was to prevent the discharge of pollutants into the nation's waterways and to make rivers, lakes, and streams swimmable and fishable. President Ronald Reagan extended the act in 1987 with an additional appropriation of $18 billion for the period 1987–1995.[*]

This money is to be used to finance the construction or upgrading of sewage treatment plants in 200 communities throughout the Northeast. Table 1 lists the 11 states and the amounts they can expect each year for the duration of the appropriation.

The act specifies that a revolving fund be established by each state as a mechanism to help wean them from further federal support. Accordingly, they must each begin contributing to the fund by 1990 and also begin providing low-interest loans to cities needing treatment plants. The loan repayments would be used to make new loans.

Although treatment of household and commercial/industrial sewage has a relatively long history, scientifically the process is only beginning to be unraveled. Waste water treatment is in fact the largest volume application of complex mixtures of microorganisms used to degrade (oxidize) a broad spectrum of organic chemicals while eliminating pathogenic organisms. The ability of microbes to metabolize potential pollutants is exploited in sewage systems. Thus, these systems must be properly conceived and constructed. Unfortunately, few people know what happens to their sewage.

Lee M. Thomas, Administrator of the EPA, announced with more than a degree of satisfaction that 87% of the country's municipal sewer systems met the July 1, 1988 deadline set by the Clean Water Act. These 87% handle 95% of the 35 million gallons of sewage treated daily. The goal of the CWA is the removal of human and household wastes from water, and secondary treatment requires removing 90% of the solids. As a conse-

Table 1
Funding Provided by the Federal Government to the
States for Sewage Treatment Facilities

State	Appropiation (millions of dollars)
Connecticut	$ 29.7
Delaware	11.9
Maine	18.8
Maryland	58.7
Massachusetts	82.4
New Hampshire	24.3
New Jersey	99.2
New York	267.9
Pennsylvania	96.1
Rhode Island	16.3
Vermont	11.9

[*] In fact, Congress in its wisdom overrode the president's veto.

quence of the CWA, once polluted rivers, the Potomac, Mohawk, Susquehanna, and Savannah, are again fishable and swimmable.

Since 1972, when the CWA was passed, $60 billion has been provided by federal, state, and local governments for additional sewage facilities. Another $80 billion will be needed by the year 2010, if the goals of the act are to be achieved.

IV. TERTIARY TREATMENT: NONBIOLOGICAL APPROACHES

An approximation of average individual water use is shown in Table 2. On the basis of 4 L/gal, 40–50 gal/person per day is not unrealistic for cities across the country. Multiplied by the total population, the amount of water needed is staggering. From the figures, however, it is evident that some 50% of the consumption is used for purposes that do not require water of drinking quality. Obviously then, much of our high-quality water is wasted. That is an appalling discovery. Clearly, we should be thinking about nonbiological advanced waste treatment to retrieve large quantities. Hundreds of millions of gallons could be available for irrigating parks and gardens, and if the quality is high enough, crop irrigation and even drinking water. Several treatment procedures follow.

A. Carbon Adsorption

After secondary treatment and chlorination to remove remaining pathogens, the clarified waste can be passed through a bed of activated carbon. As the liquid passes through the granular carbon particles, the waste

Table 2
Water Usuage of a Spectrum of Daily Activities

Activity	Water Use (L/inhabitant·day)
Bathing/showering	60
Toilets	50
Dish washing	10
Laundry	17
Personal hygiene	8
Watering of house plants and gardens	5
Drinking/cooking	3
Room cleaning	2.5
Car washing	2.5
Small trades	7
Total	165

particles cling to the carbon. The waste water changes from translucent to transparent instantaneously. Unfortunately, the cargbon granules become saturated fairly rapidly and must be removed and regenerated. As saturation proceeds, removal of waste declines. Constant vigilance is necessary if optimum removal is to be assured. The effluent discharged approaches the drinkable.

B. Reverse Osmosis

Reverse osmosis, obtaining wider acceptance at sewage plants around the country, is a direct descendant of water desalinization. It uses semipermeable membranes for removing both inorganic and organic waste.

Recall that the phenomenon of osmosis is the passage of liquids through a semipermeable (porous) membrane and that the rapidity of that diffusion is dependent on the concentration, or unequal concentrations, of solutes (solids) within the liquids on either side of the membrane. Thus, it is fluid from the less concentrated side which passes to the more concentrated side.

Since reverse osmosis (RO) must work against osmotic pressure, greater pressures are needed to make the process work. RO therefore requires the forcing of waste or brackish* water under pressure (from 20 to 2000 psig) through membranes so fine that a drinkable effluent freed of bacteria and viruses is produced. In addition, most of the calcium and magnesium (hardness) as well as other inorganic salts are left behind.

To achieve this level of removal RO employs cellulose acetate, triacetate, and polyamide polymers as selective barriers or filters. Bear in mind that incremental amounts of energy are needed to continually overcome the growing resistance to flow of the remaining solvent. Obviously, the membranes employed must be both fine enough to prevent solids from passing and strong enough to withstand the driving force of the hydrostatic pressures applied. Remember too that the application of driving force requires energy and energy adds cost to the process.

C. Ion Exchange and Electrodialysis

Solids dissolved in water consist primarily of the salts of sodium, magnesium, and calcium along with traces of iron and ammonia. Calcium and magnesium are the major problem ions because they contribute "hardness" and blockage of piping. One degree of hardness is equal to 1 g of calcium carbonate in 100 L of water. A soft water will usually have less than 10 degrees of hardness. A very hard water is always above 45. The harder the water, the more soap needed to produce a lather. In ion

---

* Although brackish is often taken to mean dirty or tainted, it actually refers to water that is slightly salty, up to 1.5%.

exchange and electrodialysis, inorganic salts are removed from liquids by "exchanging" the unwanted ions for more desirable ones, usually hydroxyl (OH^-), which combines with hydrogen (H^+) to form HOH or H_2O.

Ion-exchange resins are currently among the most effective means of removing these "hard" ions. Pure water can be produced by passing waste water through a mixed exchanger, containing negative sites coupled with hydrogen ions, and positive sites coupled with hydroxyl ions. Positive ions Na^+, Ca^+, and Mg^+ are then replaced by H^+, and negative ions such as Cl^- and SO_3^- are replaced by OH^+. The demineralized water produced is as pure as distilled.

Coupling an electric current to a selectively permeable membrane creates an electrodialytic process. Here an electric current is used to drive the inorganic ions out of solution to a charged electrode. In a saline solution consisting of NaCl in water, Na ions migrate to the negative electrode and Cl^- to a positive. The water, now freed of its unwanted salts, can be inoffensively discharged to a river or stream or can be recycled.

If these physicochemical treatment systems are to find their place in our communities, they will have to become appropriately economical and operable by moderately skilled workers—a difficult order at this time.

The discharge of waste water is subject to stringent controls. Consequently, many industries must employ costly measures to pretreat their waste. As a result, industrial effluents often contain high concentrations of chemicals, a high proportion of which are in dissolved form and are biodegradable. However, construction of conventional treatment plants are impractical because of lack of space.

For these concentrated wastes, deep tanks 50–75 ft high have been especially developed for biological treatment. Due to the depth of water in these covered aeration tanks, the utilization of oxygen in the injected air is extremely high, and because of high oxygen utilization, considerably less air is required. This significantly reduces the volume of exhaust air. With covered tanks the effect of wind is eliminated. This yields two benefits: the clarifier units operate more efficiently and odors to nearby residential areas are eliminated.

The problems are universal—and the solutions unlimited. A botanist at the University of Tel Aviv, Margalith Galun, has developed a fungal waste removal system that acts as an absorbent sponge, soaking up large amounts of metals often found in waste liquor from mining and manufacturing processes. The fungi can extract mercury, uranium, lead, nickel, silver, and zinc. As effluents pass through a filter containing the fungus, the metals are rapidly separated out. And the filters can be quickly regenerated when they reach saturation. The metals are extracted and the organism reused—a recycling that keeps costs low.

Cell-free enzyme systems are being tested at universities across the country. These offer highly selective, controlled, and efficient processes for use in the food industry. For example, amylases are used for treating starch-containing (rice, potatoes) waste waters, and proteases are being used to upgrade trash fish and fish-processing waste to fish meal for animal feed, thereby keeping these otherwise high BOD wastes out of watercourses.

The possibilities are limited only by our creativity.

REFERENCES

1. G. Stanford and D. Donohue, The Landmix Program. *Biocycle*, 48–51, May/June 1983

2. P. C. Brookes, S. P. McGrath, D. A. Klein, and E. T. Elliott, Effects of Heavy Metals on Microbial Activity and Biomass in Field Soils Treated with sewage Sludge. *Environmental Contamination International Conference*, London, July 1984. CEP Consultants Ltd., Edinburgh, UK.

3. B. E. Davies and N. J. Houghton, The Use of Radish as a Monitor Crop in Heavy Metal Polluted Soils. *Environmental Contamination International Conference*, London, July 1984. CEP Consultants Ltd., Edinburgh, UK.

4. R. D. Davis, K. Howell, R. J. Oaks, and P. Wilcox, Significance of Organic Contaminants in Sewage Sludges Used on Agricultural Land. *Environmental Contamination International Conference*, London, July 1984. CEP Consultants Ltd., Edinburgh, UK.

5. R. Feacham, H. Garelick, and J. Sade, Enteroviruses in the Environment. *World Health Forum* 3(2):170–180, 1982.

SUGGESTED READINGS

Dean, R. B., and Suess, M. J. The Risk to Health of Chemicals in Sewage Sludge Applied to Land. *Waste Management Res.* 3:251–278, 1985. (This is the report and conclusions of a WHO Working Group convened in Malta, October 23–26, 1984. It is a thorough review of the subject and well worth reading.)

Junkins, R., Deeny, K., and Echkoff, T. *The Activated Sludge Process: Fundamentals of Operation.* Ann Arbor Science Publisher, Ann Arbor, MI, 1983.

Sanks, R. L., and Asano, T. *Land Treatment and Disposal of Municipal Waste Water.* Ann Arbor Science Publishers, Ann Arbor, MI, 1976.

Sawyer, C. N. Milestones in the Development of the Activated Sludge Process. *J. Water Pollut. Control Fed.* 37(2):151–161, 1965.

Stanford, G. New Directions in Century Three: Strategies for land and Water Use. In Proceedings of the 32nd Annual Meeting, Soil Conservation Society of America, August 7–10, 1977, Richmond, VA. (This is well worth reading, and is available from Dr. G. Stanford, Greenhills Center and Experimental Station, Cedar Hill, TX 75104).

13

Hazardous and Solid Wastes

The public interest requires doing today those things that men of intelligence and good will wish, five or ten years hence, had been done.

—Edmund Burke

William T. Love had a noble idea: build a power-producing canal and offer free energy to attract industry for a model city to be built on the shores of Lake Ontario. His canal would tap the unrelenting force of Niagara Falls as it raced into the rapids before dropping 180 ft into the Niagara River.

With his remarkable powers of persuasion, he raised sufficient funds to organize a development company in 1892 (1). However, President Grover Cleveland's second administration was not 3 months old when a series of bank failures and industrial collapses inaugurated the panic of 1893. Love's canal was an early casualty. Three thousand feet long, 60 ft wide, and 10 ft deep, the canal (or ditch) remained fallow among the orchards and farms east of Niagara Falls.

Enter Elon Hooker. Also quick to see the benefits that proximity to the Falls bestowed in the way of cheap electric power, Hooker established the Electrochemical Company that bears his name for the production of chlorine and sodium hydroxide.

The company thrived and the city of Niagara Falls prospered. In 1942, permission was given Hooker Chemical to dispose of its chemical waste in Love's canal. In 1947, the "canal" was sold to Hooker. From 1942 to 1953, the canal was used as Hooker's primary waste chemical disposal site. Over 20,000 tons of chlorinated hydrocarbons, fatty acids, alkalis, and other chemicals (200 have since been catalogued) were buried there in metal and fiber drums or as sludge. By 1953, it was full. It was capped with clay, covered with soil, and shortly thereafter it became the neighborhood's grassy playing field.

Between 1957 and 1977 intimations of trouble were broadcast about. In May 1977, a survey showed that 12% of the homes adjacent to the canal had chemical residues in their basements, and the city's storm sewers contained PCBs (polychlorinated biphenyls). The area with homes and their relationship to the canal is shown in Figure 1.

Toward the end of April 1978, New York State Health Commissioner Robert Whalen stated that the conditions at Love Canal were a serious threat to health. Heavy rains caused flooding and leaking drums were popping to the surface. Samples of the chemical sludge showed the presence of dioxin. On August 2, Whalen issued the following order:

> Love Canal Chemical Waste Landfill constitutes a public nuisance and an extremely serious threat and danger to the health, safety and welfare of those using it, living near it, or exposed to the conditions emanating from it, consisting, among other things, of chemical wastes lying exposed on the surface in numerous places and pervasive, pernicious and obnoxious chemical vapors and fumes affecting both the ambient air and the homes of certain residents living near such sites.

He then declared "the existence of an emergency" and recommended the immediate evacuation of pregnant women from the area. Hazardous and toxic wastes became known to the world. A new pollution problem was born.* Love's canal had come to an ignoble end.

I. WHAT IS A HAZARDOUS WASTE?

Table 1 is a compilation (not exhaustive) of waste produced by a variety of industries and processes which the EPA has designated hazardous. According to the *Random House Dictionary of the English Language* (unabridged edition), *hazard*, a noun, is defined as something causing danger, peril, risk, or difficulty. Danger is exposure to risk (as in "there's danger lurking there"), and peril is also the exposure to risk, jeopardy, or danger. Risk appears central to the problem of hazard. It is generally understood to mean exposure to the chance of injury, harm, or less. Inherent in hazard is the concept of probability or uncertainty. There is nothing definite about it. It does, however, raise the flag of prudence.

On the other hand, the adjective *toxic* describes something that has the effect of a poison. It derives from the Latin *toxicum*, or late Latin *toxicus*, both meaning poisonous. Clearly, hazardous and toxic are neither the

---

* In fact, the term hazardous waste had an earlier conception. In 1976, Congress passed the Resource Conservation and Recovery Act (RCRA), P.L. 94-580, which directed the EPA to regulate hazardous waste. At the time, few people were aware of this development.

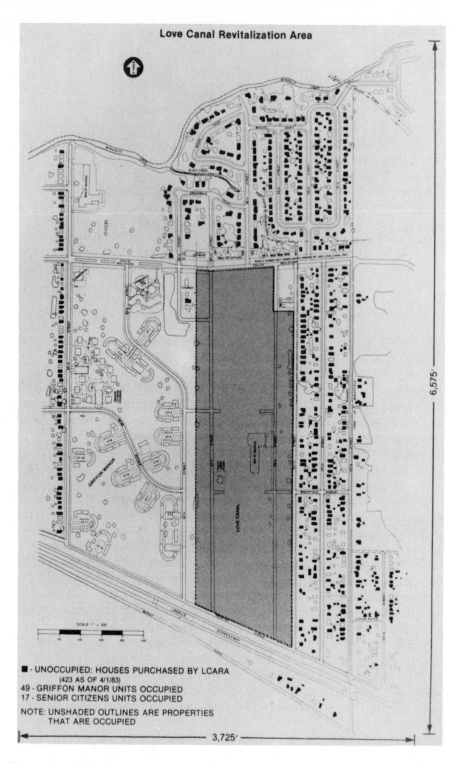

Figure 1. Love Canal revitalization area. Reproduced with permission of the Office of Technology Assessment.

Table 1
Wastes that May Be Considered Hazardous by the U.S. Environmental Protection Agency

Alloy production
  Electric furnace dust
  Electric furnace sludge
Purified 1-chloro,4-nitrobenzene production
  Vacuum distillation residue
Vinylchloride production
  Still bottoms
Ethylenedichloride production
  Still bottoms
Paint and pigment production
  Still bottoms from solvent reclaiming
  Ash from incinerated still bottoms
Chlorine production—mercury cell, Dow cell, and diaphragm processes
  Mercury-bearing sludges
  Chlorinated-hydrocarbon-bearing waste chlorine purification
  Sodium–calcium sludges
  wastewater treatment sludge
  Asbestos wastes
  Brine purification
Carbon tetrachloride production
  Still bottoms
Benzene and chlorobenzene recovery
  Still bottoms
Explosive propellants and initiating compounds
  waste water sludges
Oil drill muds
Ink formulation
  Sludges, wastes from ink tub washer
Pigment production
  Chromium-, zinc-, ferric ferrocyanide-, mercury-, or lead-bearing
  waste water treatment sludge
Boric acid production
  Arsenic-bearing waste water treatment sludges
Methyl methacrylate production
  Still bottoms from methanol recovery
Benzyl chloride distillation
  Still bottoms
Antimony oxide production
  Arsenic-bearing sludges from purification process
  Antimony-bearing waste water treatment sludge
Lead acid storage battery production
  Waste water treatment sludges
  Clean-up wastes from cathode and anode production
Electronic component production
  Immersion degreasing of ferrites
  Lacquer equipment clean-up
  Dip varnish equipment clean-up
Paint and pigment production
  Still bottoms from solvent reclaiming
  Ash from still bottom incineration

Table 1 (*Continued*)

Petroleum refining
  Fluid catalytic cracker catalysts fines
  Neutralization HF allcylation sludge
  Treatment of boiler feedwater
  DAF sludge
  Kerosene filter cakes
  Lube oil filteration clays
  Nonleaded tank bottoms
  Silt from storm water runoff
  Slop oil emulsion solids
  Leaded and crude tank bottoms
  Coke fines
Epichlorohydrin production
  Still bottoms from fractionator
Maleic anhydride production
  Vacuum still bottoms
Ethyl chloride production
  Still bottoms from fractionator
Trichloroethylene production
  Still bottoms
Coke by-product recovery
  Waste ammonia liquor
  Decanter tank tax
Baghouse operations
  Waste from gray iron cupola furnace
Foundry operations
  Pouring and shakeout dust
  Cleaning and finishing dust
Grinder operations
  Tank oil
Tanning operation—beamhouse/tanhouse
  Wastewater treatment sludge
Tanning—complete chrome tannery
  Wring, split, shave waste
  Dry, trim, condition, buff, finish waste
  Waste water treatment sludge
Tanning—leather finisher and retan/finisher
  Split, shave
  Dry, trim, condition, buff, finish waste (retan only)
  Waste water treatment sludge (retan only)
Tanning—sheepskin tannery
  Fleshing process
  Dry, trim, condition, buff
  Waste water treatment sludge
Tanning—split tannery
  Split, shave, trim
  Buff, trim
  Waste water treatment sludge

same, similar, synonymous, or interchangeable. Hazardous waste should be viewed in this context. For example, Table 2 lists the 127 organic and inorganic chemicals which the Office of Technology Assessment considered its priority pollutants to be assessed at Love Canal. According to their recent report, "review of available data on constituent levels within treated leachate indicates that the highest value recorded for any of the priority pollutants was 46 parts per billion. Even this low concentration has been detected only occasionally; most results of the analyses indicate no detection." Forty-six (46) ppb is the same as 0.046 ppm. The potential may indeed be there, but actual harm remains to be demonstrated.

There is another unfortunate aspect here. Inadvertently, the EPA sends mixed messages to the community. To check and sample for chemicals at waste sites, the EPA outfits technicians in fully encapsulating "moon" suits, hooked up via long lines (umbilical cords) to air supplies. What can people think? They must ask themselves and rightly so, "What do they know that I don't. If it's not toxic, why is he dressed that way?" A mixed message if ever there was one.

Because of the potential for harm, Congress enacted, in 1976, the Resource Conservation and Recovery Act (RCRA). This piece of legislation was a "cradle to grave" law that placed the burden of proof of safety on the waste generator. Under RCRA, the generating industry or company was to be responsible for the waste from its creation to such time as it no longer posed a threat. The RCRA stressed safe disposal rather than recycling or reductive methods and was poorly or weakly enforced. Most generators simply dumped their wastes in a pit or impoundment or paid to have it hauled from their premises unmindful of the manner of its disposal.

Although we talk about a waste as hazardous or toxic, a usable, clear working definition has yet to emerge from the EPA. We can say unequivocally that hazardous waste is a subset of solid waste, which encompasses garbage, refuse, and sludges of various kinds. The state of California, unwilling to wait interminably for the EPA to publish its definition, defines· hazardous waste as "any waste material or mixture of wastes which is toxic, corrosive, flammable and irritant; a strong sensitizer, which generates pressure through decomposition, heat or other means, if such waste or mixture of wastes may cause substantial personal injury, serious illness or harm to wildlife during or as a proximate result of any disposal of such waste or mixtures of waste." Although lawyers are obviously at the heart of this, it leaves us wondering at the meaning of such ideas as substantial and serious. Just how bad does it have to be? The definition, however, is all-inclusive, providing the state with power to control disposal of the greatest percentage of waste.

But it does not end there. Defining and identifying toxic chemicals can be singularly difficult because all substances become toxic at sufficiently

Table 2
Priority Pollutants

VOLATILE ORGANIC
COMPOUNDS

Acrolein
Acrylonitrile
Benzene
Carbon tetrachloride
Chlorobenzene
1,1-Dichloroethane
1,2-Dichloroethane
1,1,1-Trichloroethane
1,1,2-Trichloroethane
1,1,2-2-Tetrachloroethane
Chloroethane
2-Chloroethylvinyl ether
Chloroform
1,1-Dichloroethylene
1,2-trans-Dichloroethylene
1,2-Dichloropropane
1,3-Dichloropropene
Ethylbenzene
Methylene chloride
Methyl chloride
Methyl bromide
Bromoform
Dichlorobromomethane
Trichlorofluoromethane
Chlorodibromomethane
Tetrachloroethylene
Toluene
Trichloroethylene
Vinyl chloride
Bis(chloromethyl)ether

BASE-NEUTRAL EXTRACTABLE
ORGANIC COMPOUNDS

Acenaphthene
Benzidine
1,2,4-Trichlorobenzene
Hexachlorobenzene
Hexachloroethane
Bis(2-chloroethyl) ether
2-Chloronaphthalene
1,2-Dichlorobenzene
1,3-Dichlorobenzene
1,4-Dichlorobenzene
3,3'-Dichlorobenzidine
2,4-Dinitrotoluene
2,6-Dinitrotoluene

1,2-Diphenylhydrazine
Fluoranthene
4-Chlorophenyl phenyl
  ether
4-Bromophenyl phenyl
  ether
Bis(2-chloroisopropyl)ether
Bis(2-chloroethoxy)methane
Hexachlorobutadiene
Hexachlorocyclopentadiene
Isophorone
Naphthalene
Nitrobenzene
N-nitrosodimethylamine
N-nitrosodiphenylamine
N-nitrosodi-n-propylamine
Butyl benzyl phthalate
Di-n-butyl phthalate
Di-n-octyl phthalate
Diethyl phthalate
Dimethyl phthalate
Benzo[a]anthracene
Benzo[a]pyrene
3,4-Benzofluoranthene
Benzo[k]fluorathene
Chrysene
Acenaphthylene
Anthracene
Benzo[ghi]perylene
Fluorene
Phenanthrene
Dibenzo[a,h]anthracene
Ideno[1,2,3-cd]pyrene
Pyrene
Bis(2-ethylhexyl) phthalate

ACID EXTRACTABLE ORGANIC
COMPOUNDS

2,4,6-Trichlorophenol
Parachlorometa cresol
2-Chlorophenol
2-Nitrophenol
Pentachlorophenol
2,4-Dimethyphenol
4-Nitrophenol
2,4-Dinitrophenol
4,6-Dinitro-o-cresol
2,4-Dichlorophenol
Phenol

PESTICIDES AND PCBs

Aldrin
Dieldrin
Chlordane
4,4'-DDT
4,4'-DDE
4,4'-DDD
a-Endosulfan
b-Endosulfan
Endosulfan sulfate
Endrin
Endrin aldehyde
Heptachlor
Heptachlor epoxide
a-BHC
b-BHC
q-BHC
w-BHC
PCB-1242
PCB-1254
PCB-1221
PCB-1232
PCB-1248
PCB-1260
PCB-1016
Toxaphene
2,3,7,8-Tetrachlorodi-
  benzo-p-dioxin
  (TCDD)

METALS

Antimony
Arsenic
Beryllium
Cadmium
Chromium
Copper
Lead
Mercury
Nickel
Selenium
Silver
Thallium
Zinc

MISCELLANEOUS

Asbestos
Total cyanides

high doses. Recall, *dosis sola facit veneum*—what makes a poison is dose. Two aspirin four times a day is a therapeutic dose, but 20, four times a day, can be deadly. Thus, the question becomes: At what point does the shift from safe to hazardous to toxic occur? This concept is central to the problem. It may also be a reason why the EPA is so long in developing an appropriate definition.*

In addition, the mixtures of chemicals further confounds the problem. Pharmacologists are only beginning to understand the problem of adverse drug reactions posed by the simultaneous use of multiple medications. And toxicologists are only beginning to think about testing the effects of two and three individual chemicals in animal systems. Attempting to ascertain the effects of a single chemical posed difficult enough problems. To now add mixtures of chemicals, each at varying doses, will be staggering in its complexity.

A. Waste Characteristics

Under the RCRA's Part 261, Subpart c, three criteria appear to establish a waste as hazardous. A waste is deemed to be hazardous if it:

1. meets the statutory definition of hazardous waste,
2. is measurable by standardized test methods and is available and within the capacity of generators or private sector laboratories, or
3. can be reasonably detected/determined based on knowledge about the waste.

This type of language guarantees that other jurisdictions such as California will surely move to develop definitions with less ambiguity. However, as noted earlier, the act did provide specific adverbs by which hazardous wastes could reasonably be differentiated from other solid waste (2). Their inherent qualities were:

ignitability

corrosivity

reactivity

toxicity

---

* Although the U.S. EPA has not yet defined the term hazardous and toxic waste, the RCRA has. For Congress, a hazardous waste is defined as "a solid waste, or combination of solid wastes, which, because of its quantity, concentration, or physical, chemical or infectious characteristics may cause an increase in mortality or an increase in serious illness . . . or . . . pose a substantial present or potential hazard to human health or the environment when improperly treated, stored, transported, or disposed of, or otherwise managed." This does not require a second reading to realize its almost total uselessness in the real world.

1. *Ignitability* (Code I). A substance is considered to exhibit this quality if a representative sample

(a) has a flash point less than 140°C) as measured by the Pensky–Martin closed cup method. This applies to all liquids except those containing 24 percent alcohol by volume.

(b) is a nonliquid capable of sustained and spontaneous combustion.

(c) is an ignitable compressed gas or an oxidizing substance.

Wastes that meet these criteria are designatd "I" within the EPA hazard code and are identified by the designation D 001.

2. *Corrosivity* (Code C). Any substance, usually a liquid,* with a pH of less than 2 or greater than 12.5,† or if it can be shown to corrode steel at a rater greater than 0.25 in./yr.

3. *Reactivity* (Code R). A solid waste is considered reactive if it violently combines with water, forming explosive mixtures, or if it is chemically unstable, forming toxic gases, or fumes on contact with water. In addition, if the substance or mixture could detonate if heated at standard temperature or pressure or under confinement, it is considered reactive.

4. *Toxicity* (Code T). A solid waste is considered toxic if, using prescribed chemical extraction procedures, the extract of a representative sample contains any of eight heavy metals or six pesticides in the amounts noted in Table 3. Toxicity is based primarily on responses observed in laboratory animals, which, as noted earlier, has inherent uncertainties. Each hazardous waste of the types noted in Table 3 has been assigned an EPA identification number. For example, petroleum

Table 3
Extraction Procedure Toxicity Levels

Heavy Metals (mg/L)		Pesticides (mg/L)	
Arsenic	0.5	Endrin	0.02
Barium	100.	Lindane	0.4
Cadmium	1.0	Methoxychlor	10.0
Chromium	5.0	Toxephene	0.5
Lead	5.0	2,4-D	10.0
Mercury	0.2	2,4,5-T	1.0
Selenium	1.0		
Silver	5.0		

* Although hazardous wastes are specifically defined as solid wastes, the term "solid" also refers to solids within liquids and mixtures of wastes.
† If a refresher on pH is needed, refer to Chapter 10, specifically the section dealing with acid rain.

refinery waste, explosive propellants, and tannery waste are within the designation K 001–K 069. These numbers must be used in recordkeeping and with all notifications.

Provisions also exist within the RCRA for waste to be removed from the list of hazardous substances, and wastes can also be declared non-hazardous. But the onus is upon the generator to show beyond a reasonable doubt that the wastes are safe and qualify for "delisting" or nonhazardous status.

II. GENERATION OF HAZARDOUS WASTE

The intent of solid waste legislation is to ensure proper management of hazardous wastes. Though there are many problems associated with their generation, storage, and transportation, disposal elicits the greatest concern. In implementing hazardous waste legislation, certain chemicals, waste streams, and types of industry have been especially selected as targets for appropriate management. According to U.S. EPA estimates, American industry generates some 30 million tons of waste per year. Approximately 10% of that is deemed hazardous and/or toxic.

A. Magnitude

Figure 2, shows the tonnage of waste generated, as well as the percentage of the total for each of the ten EPA regions. It is apparent that the Atlantic seaboard and Texas are the primary sources of the more than 245,000,000 tons generated. Over 70% comes from these areas. What is not so apparent is the difficulty gathering this type of data, along with the uncertainty inherent in it. It appears that the reporting process for hazardous waste is far worse than the reporting of the incidence of food poisoning in the U.S. (cf. Chapter 2)

The number of active generators within each region is shown in Figure 3. The number of known generators—and for the most part, these are the large ones—has grown from an estimated 14,000 in 1980, to over 40,000, by 1985. Again, the numbers are only rough estimates. The industries that are the major sources of waste, and the type of waste each contributes, is shown in Figure 4.

Table 4 lists eight prominent classes of consumer products, along with their inherently hazardous by-products. A moment's reflection will reveal that every item and object on the planet is composed of chemicals, including our food, and will leave waste and residues. That is an integral and inescapable fact of the human condition. Problems arise when waste becomes excessive (and/or harmful) and out of control. That is what the solid waste problem is about: excess, potential harm, and preventing loss of control.

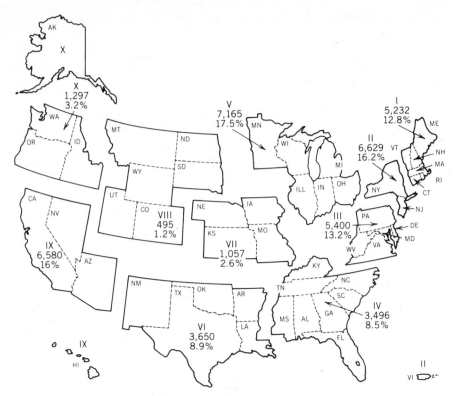

Figure 2. Regional distribution of hazardous waste generators, 1985. These numbers are reported to EPA by each state and includes small generators as some states do not make distinctions. Estimated total number of generators active in 1985: 41,000; tons of waste generated: 245,155,000. (Source: USEPA, Office of Solid Waste and Emergency Response. Draft Report. These numbers represents only large quantities as defined by RCRA. In any single calendar month a source generating greater than 1000 kilograms of RCRA defined hazardous waste, or generating greater than one kilogram of acutely hazardous waste.)

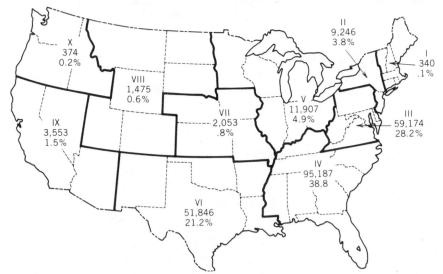

Figure 3. Federally regulated hazardous waste generation within each EPA region (in tons, 1985) and percentage managed within the RCRA permitting system. Numbers are based on a Draft Report for 1985. Information compiled from State Biennial Program Reports. (Source: USEPA, Office of Solid Waste and Emergency Response.)

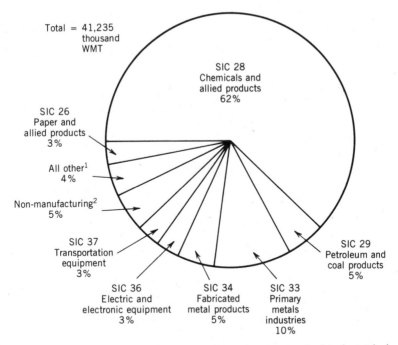

Figure 4. Percentage of 1985 hazardous waste generation by standard industrial classification (SIC) code. Total waste generated is estimated at 245,000,000 tons.

B. Who Generates and Where?

As of 1984, the EPA had documented the fact that some 15,000 plus generators were responsible for 98% of all hazardous waste. The remaining 2% appears to be contributed by the half-million or so who produce less than 1 metric ton per month. The map in Figure 5 indicates the sites

Table 4
Consumer Goods and Some of Their Hazardous By-Products

Plastics—organic chlorine compounds
Pesticides—organic chlorine and phosphate compounds
Medicines—organic solvents and residues
Paints—heavy metals, pigments, solvents, organic residues
Petroleum products (gas, oil, etc.)—oil polynuclear aromatics, phenols, other organic compounds, heavy metals, ammonia salts, acid, caustics
Metals—heavy metals, fluorides, cyanides, acid and alkaline cleaners, solvents, pigments, abrasives, plating salts, oil, phenols
Leather—heavy metals, organic solvents, sulfides
Textiles—heavy metals, dyes, organochlorines, solvents

Figure 5. Geographical locations of the major identified commercial hazardous waste management facilities, 1985. (Source: USEPA, Office of Solid Waste and Emergency Response).

of the major commercial disposal dump sites around the country. Only Nevada and Alaska are free of hazardous dump sites. The unfortunate thing about many of these sites is their placement on or near major aquifers whose groundwater is the main source of drinking water for millions of people. Eleven states contribute over 60% of the country's known and registered sites and the largest number of generating sources are within these same 11 states.

C. Laws and Clean-up

Regulatory agencies at both the state and federal levels are mandated to protect and ensure that the public and the environment are protected from untoward effects of waste chemicals. With the Resource Conservation and Recovery Act, Congress handed the EPA another colossal responsibility. Establish, they in fact said, a comprehensive program to control hazardous wastes. In the process, protect the health of the public and the environment and conserve valuable material and energy sources.

Congress was not content to sit on that laurel. It passed in 1980 the Comprehensive Environmental Response, Compensation, and Liability Act (CERCLA), which has come to be known as Superfund—Superfund because $9.2 billion dollars were appropriated to clean up existing hazardous waste sites. EPA supposedly had a weapon with which to set liability for spills, injuries, and abandoned sites. A fund was established, financed primarily by levies on producers of chemicals and supplemented in small measure by general tax dollars.

President Reagan's budget requested $900 million for Superfund in the fiscal year starting October 1, 1985, and $1.1 billion annually for the 1986–1990 period. Authorization came with SARA, the Superfund Amendments and Reauthorization Act of 1986. An important section of this act, 104(k), threatens a cutoff of federal funds to those states that do not assure the availability of future treatment capacity including procedures to override local vetoes. Although meant to encourage state involvement, it could also see them dig in their heels in opposition to any proposal.

In an attempt to prevent wastes from becoming hazardous in the first instance, Congress went to the source. In enacting TOSCA, the Toxic Substances Control Act (Public Law 94-469, 1976), manufacturers of chemicals were required to notify the EPA prior to offering for sale any substance not previously included in the EPA's inventory of toxic substances. In addition, responsibility for ascertaining toxicity now rightly fell on the producer.

It would appear that the EPA has been given an impressive array of tools with which to pursue its mandate. It remains only for governmental policy and the administrators to use these tools appropriately. Given the intimate ties between industry and the Reagan administration, expectations should not be excessive.

Nevertheless, on May 20, 1985, the U.S. Department of Justice and the U.S. EPA announced that the Westinghouse Electric Company had been found guilty of violating the RCRA. Under the consent decree, the company would clean up six hazardous waste sites near Bloomington, Indiana, where it had dumped PCBs for some 18 years.

Lee M. Thomas, administrator of the EPA, said the agreement was 'the largest hazardous waste settlement in the history of the Agency." Under the settlement, Westinghouse would build an incinerator to destroy the wastes. Apparently, Westinghouse would spend from $75 to $100 million to clean up the sites. The company responded by noting that its costs for building an incinerator would be offset by collecting fees from the city for burning its municipal solid wastes, as well as selling steam and electricity to other nearby industries.

In addition, the decree required Westinghouse to monitor underground water supplies in the area for the next 30 years and provide water to residents whose wells may become contaminated from leaking dump sites.

The important point here is not that Westinghouse will be charging the surrounding towns for treating their wastes, but that clean-up is being advanced. This may indeed be an important victory.

Thomas was off to a fast start. Within a period of 30 days, not only had Westinghouse capitulated and agreed to clean dump sites in Indiana, but on June 13th, under an agreement filed in Federal District Court in Cincinnati, 158 companies who had dumped PCBs, acids, resins, heavy metals, solvents, and cyanides at the 10-acre Chem-Dyne site in Hamilton, Ohio, had agreed to participate in its clean-up.

Chem-Dyne, 20 mi from Cincinnati, sits astride the lower Great Miami River Valley aquifer, one of the largest and most productive sources of groundwater in the Midwest. Clean-up of this dump is expected to take 10 years and cost $12 million. A list of companies that used this site to dispose of their waste reads like a "Who's Who" of American industry. If Thomas can sustain this momentum, significant sources of pollution will be curtailed.

III. HEALTH EFFECTS OF HAZARDOUS WASTES

Exposure and the fear of exposure to hazardous and toxic wastes have engendered concern for adverse health effects. Given the nature, number, and mixtures of esoteric chemicals, such behavior is reasonable and understandable. But is it warranted?

In an attempt to define the problem, Congress stated that hazardous waste (42 USC 6903, Section 1004 or the RCRA) "may cause, or significantly contribute to an increase in mortality or an increase in serious irreversible, or incapacitating reversible, illness; or pose a substantial threat or potential hazard to human health . . . when improperly treated, stored, transported or disposed of, or otherwise meanaged." The operational terms are "may" and "threat." Neither Congress nor their scientific consultants had a firm basis for judgment. What have we learned since the passage of the RCRA?

Unless present in unusually large quantities, most chemicals are degraded by biological and photolytic processes. They are also rapidly metabolized and excreted by the human body. It is the odd metal, polyhalogenated compound, that persists in either the environment or individuals. However, given the growing ability to detect substances at or below parts per trillion, the question arises as to the significance of such levels. More on this further on.

According to the Surgeon General, over 840 million tons of waste chemicals have been discarded by individual generators since 1958, approximately 30 million tons per year, in thousands of sites across the country. Exposed individuals include landfill operators and haulers, as well as those people living near the routes and those at various distances who drink contaminated groundwater. Families living on 97th and 99th Streets, abutting Love Canal in Niagara Falls, New York, are among the optimum examples of close proximity to hazardous wastes, and these families are thereby subject to the likelihood of chronic, long-term exposure.

Firefighters and others called to extinguish fires at dump sites are prime candidates for acute exposures. The typical exposure, however, is low level over the long term. Thus, distinct levels of exposure may be ascertained, and it is from these various levels (doses) that varying ill effects in populations may be elicited.

A. Epidemiologic Considerations

If we are to move beyond anecdotal expressions of untoward effects to either secondary associations or cause–effect relationships, convincing data or at least three factors must be forthcoming:

1. The type and degree of exposure must be known.
2. Those exposed must be distinguishable from the unexposed.
3. Illness or dysfunction in the exposed must be significantly* greater than in the unexposed. A dose–response relationship would be convincing evidence.

Love Canal may be an illuminating example of the difficulties and uncertainties inherent in relating exposure to ill effects. In 1978, officials of the New York State Department of Health urged and warned pregnant women living near the canal to move at once. In 1982, the Department of Health and Human Services affirmed its earlier decision that the area was in fact habitable. Then, in 1983, the Office of Technology Assessment concluded that "it was not possible to conclude either that unsafe levels or toxic contamination exist or that they do not exist." Love Canal is not only a dilemma, but a great source of confusion.

What evidence has since accumulated to assist in evaluating the relationship between hazardous (toxic?) waste and ill health? For Love Canal the source is unquestionably site specific. Beyond that, uncertainties begin to develop and the relationship becomes fuzzy.

The exposure pathways are believed to be via air, soil, and water.† Of the 400 chemicals identified as being present, 11% were classified as mutagens, 13% as carcinogens, 7% as fetal embryo toxins, 16% as neurotoxicants, 9.7% as hepatotoxicants, 10% as renal toxicants, and 9%

---

* Significance in the statistical sense refers to results of an experimental study that are not likely to have occurred by chance alone. This is quite different from the general meaning usually referring to importance.
† Following reports of skin and eye irritation, muscle weakness, shortness of breath, nausea, vomiting, diarrhea, and abdominal pain, the EPA in November 1978 advised residents living near a toxic waste dump (in Hardeman County, Tennessee, 60 mi north of Memphis) to stop drinking water from their wells. An analysis of waste water revealed the presence of hexachlorocyclopentadiene and hexachlorobiocycloheptadiene. Urine specimens of residents were positive for these two compounds. An investigation undertaken by the Department of Environmental Health, University of Cincinnati Medical Center, showed that exposure to the contaminated water produced "a subclinical transitory liver insult that appeared to be associated with consumption of water from wells contaminated by leachate from the chemical waste dump located in the area" (3).

as pulmonary toxicants. These classifications were made from either animal or other biological determinations such as in vitro tests. Furthermore, for over 50% of the identified chemicals, toxicological data were unavailable. Peter Voytek, former director of the EPA's Reproductive Effects Assessment Group, takes the uncertainties a step further. He maintains that experience has shown that for those chemicals for which some toxicological evidence is at hand, much of it is inadequate to permit any decisions as to human toxicity in either qualitative or quantitative terms.* The EPA's Carcinogen Assessment Group has sufficient cancer data on less than 4% (15 of 400) of the chemicals detected at Love Canal to make anything but the most questionable estimate of risk (5,6).

1. A Study of Birth Weights

Given the above, we are indebted to Nicholas Vianna and Adele Polan of the Division of Health Risk Control, New York State Department of Health, for their thorough and well-designed study of low birth weight infants in the Love Canal area (7).

To explore potential causal connections between an exposure and illness (which posits the presumed independent variable—illness or dysfunction), ingenuity on the part of the investigator is often called for, particularly because of the long latency periods that can be anticipated with chronic illness.

To avoid the extended end-points as well as being able to establish a quantitatively objective and identifiable end, they chose to investigate the incidence of low birth weight infants (less than 2500 g at normal term) born to parents living near Love Canal, compared to the incidence in other areas of New York State. Low birth wight, a readily identifiable end-point, requiring a relatively short waiting period, is an appropriate marker.

Data on birth weight were exabmined for all infants born in the Love Canal area between 1940 (prior to dumping) and 1978, well after the dumping which spanned the years 1942–1953.

The results of their analysis are shown in Figure 6. The three trend lines represent the swale area (natural shallow depressions that serve as drainage ways that can produce ponding during periods of heavy rain), other areas of the canal, and upstate New York, which is in fact a control population.

---

* There is a *potential* epidemiologic connection between chemicals ingested by mothers and their children after birth. Human milk has been known to be a repository for fat-soluble organic chemicals. DDT and PCBs were among the first to be identified. These and others are known to be tightly bound to fatty tissue. Once absorbed they appear to be stored in the body fat for a lifetime. The scientific and medical literature is replete with published studies documenting the worldwide contamination of breast milk. At least six commonly used pesticides have been identified. Exposure has been shown to come from food, water, air, and soil. However, and this is of transcendent importance, no illnesses occurring in breast-fed children have as yet been attributed to these contaminants (4). DDT and benzene hexachloride have had 30 years or more in which to have been implicated.

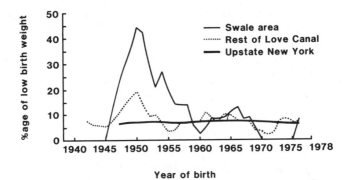

Figure 6. Five-year moving averages for percentages of low birth weight among infants born in the Love Canal swale area, the rest of Love Canal, and upstate New York.

The upstate curve indicates that low birth weight is endemic and occurs with an expected frequency. Clearly, Love Canal generally, but the swale area particularly, during the dumping of the 1940s and 1950s, produced low weight infants at a frequency well above the expected. The curves for both the swale and the rest of the canal are reminiscent of continuous point source epidemic curves. Exposure and low weight are clearly spatially related. "Residents of the swale area had significantly more low-birth-weight infants than the rest of the Canal." Vianna and Polan also accounted for education level, smoking, occupation, and past medical history to avoid confounding with hazardous waste.

Of immense importance was the finding that between 1960 and 1978 (when the alarm was sounded and pregnant women were urged to leave) infants born in the study area "were at no greater risk of low birth weight than were those born in upsate New York." True, this study was retrospective and worked from birth certificates rather than interviews, and spatial relationships can be misleading; nevertheless, the data are highly suggestive of a direct relationship. It therefore seems that with the cessation of dumping at the end of the 1950s the incidence of low birth weights returned to normal.

If instruction can be drawn from this study, perhaps it is that assumptions from animal data can be applied to human populations only with the utmost caution. In Italy, after the Meda/Seveso accident (cf. Chapter 5), pregnant women were advised to abort, based on expectations of birth defects from studies of animals treated with dioxin. Pathological changes were not found in any of the aborted fetuses.

The above notwithstanding, David Axelrod, Health Commissioner, State of New York, declared on September 27, 1988 that half the neighborhoods abandoned around the Love Canal waste dump 10 years ago could be rehabilitated. In an act of political courage, but scientific reasonableness, Axelrod noted that about 250 homes near the canal could be resettled. His decision was based on a 5-year environmental review.

Although the momentous decision leaves lingering doubts with many residents, in point of fact there has never been an official finding linking the many waste chemicals with illness complaints.

Given the passage of sufficient time, additional testing, and another act of political forthrightness, Times Beach, Missouri, could be resurrected from oblivion.

2. Dioxin

What with the widespread use of pesticides such as 2,4,5-T and the manufacture of hexachlorophene, dioxin (TCDD), an inadvertent contaminant of both, has become a widely disseminated environmental pollutant for well over 20 years. Nevertheless, human exposure to large doses over the short term and small doses over the long term has not resulted in the type of ill effects predicted by animal experiments. And as detailed earlier (cf. Chapter 5), the species of anmal employed may be critical. None has yet been designated as most reflective of human responses.

It is thus to be lamented that an entire town, Times Beach, Missouri, was expunged by federal fiat. Times Beach will be removed from all maps. No one lives there anymore because decisionmakers believed there was a one-to-one conjunction between animal data and human health effects.

If cancer of any body site from dioxin is involved, it has continuously eluded detection. The best that can be said is that the concept of hazard as a "potential" threat to health is just that—potential.

At this point an idea noted earlier bears repeating. If adverse health effects are unobservable—cannot be found—for chemicals undetectable at the parts per million level, can we expect the appearance of adverse effects at the parts per billion or parts per trillion level—concentrations 1000 and 1,000,000 times *less*, respectively? Of course not. Yet the growing ability to detect smaller and smaller amounts of contaminants, moving inexorably toward a single molecule, appears to some more related to health effects than (as it should) simple fascination with the creative ability of analytical chemists.

IV. WASTE MANAGEMENT TECHNOLOGIES

A. Alternatives to Land Disposal

Surprisingly enough, the technologies for dealing effectively with hazardous wastes are varied, numerous, and effective, more than one would have imagined, given the predilection for dumping and running. I say American, because other Western countries have hazardous wastes, and lots of it, but they have managed to deal effectively with the problem. That we are so slow to pick up on the available effective methods suggests that action can come only with a crisis.

For the most part, the technologies can be grouped into three general categories: storage, treatment, and disposal. Given the possibilities being advanced, as well as considerations of suitability and economic feasibility, a brief description of the field will precede a detailed discussion of several with a more certain future.

1. *Storage.* The holding of wastes prior to transport, treatment, or disposal is one of the least desirable and effective procedures. Storage in drums, cans, barrels, or portable tanks may be suitable for short periods prior to treatment, but leakage into both groundwater and surface water, and air emissions, as well as fires and explosions, make this an unsafe method. Its only redeeming feature is ease and low cost. Unfortunately, these have been overriding features.

2. *Treatment*

a. *Physical Treatment.* This category includes nonchemical methods designed to reduce the volume of waste and separate out the liquid portion. It does not result in the destruction of toxic components should they exist. Some processes are capable of recovering valuable, reusable constituents prior to discarding. Among these procedures are: filtration, sedimentation, distillation, and absorption.

b. *Chemical Treatment.* Among the principal chemical processes used to treat or reduce wastes are the following:

Neutralization converts acids and alkalis into substances no longer corrosive. In the process, acid wastes are mixed with bases such as lime or sodium hydroxide to form neutral salts. Alkaline wastes are mixed with hydrochloric or sulfuric acids, again forming neutral salts. Neutral salts may or may not be hazardous.

Precipitation is often used for wastes containing dissolved metal salts. Precipitation can remove these from solution, thereby forming materials that can be managed more effectively.

Solidification is a process that converts liquid and semiliquid wastes into a solid mass. By conversion to a solid, the waste is no longer available for migration away from a landfill. The WPC* VRS process (VRS = volume reduction and solidification), for example, removes water and combines the waste with asphalt, completely stabilizing the waste. Concrete and mortar, as in the Petrifix process, can also be used as a solidifying agent.

c. *Biological Treatment.* This technology is similar to that utilized in municipal waste water treatment systems. It relies on microbial metabolism of chemicals, rendering them innocuous or transforming them into carbon dioxide and water.

* Werner & Pfleiderer Corporation

3. *Disposal*

a. *Incineration.* The controlled burning of waste at very high temperatures (up to 3000°F) for a specified time to destroy combustible constituents is one of the most preferred technologies. Incineration is designed to convert gaseous, liquid, or solid wastes into gaseous emissions and solid ash. For many hazardous wastes, including liquid organics, incineration provides an environmentally preferable alternative to landfills.

B. Landfills

Excavations in which wastes are buried are ordinarily referred to as landfills. Land disposal, because we had so much land available, has been the preferred method for as much as 80% of hazardous waste. Landfilling does not assure permanent containment of wastes. Subsidence, the sinking of the land surface, can occur at landfills as a consequence of collapse of containers. This can easily disrupt the landfill cap. Additionally, maintenance and security of landfills can be measured in decades. Here too, ease, simplicity, and relative cheapness have made this the preferred method of disposal.

C. Preferred Technologies

1. Incineration

High-temperature reduction of hazardous waste is becoming more attractive to many communities. Increasingly stringent landfill regulations, rising public objections even to well-managed landfills, and in many communities the lack of available land are sharply reducing the landfill method of disposal.

By far the largest proportion of hazardous wastes are organic and thus amenable for destruction via high-temperature incineration. For a waste to sustain combustion it has to have a minimum calorific value of 8000–10,000 Btu* per pound and, of course, be pumpable.

The two types of incineration system cited as most versatile and useful for hazardous wastes are liquid injection systems and rotary kilns. The versatility of the rotary kiln system is seen in its ability to deal with solids, liquids, slurries, and gaseous combustible wastes. Combustion temperatures range from 1300 to 3000°F (764–1600°C) depending on the waste material and the burning characteristics.

One of the most efficient rotary kiln systems has proved its effectiveness in Denmark, and land is precious in Denmark, a country no bigger than the state of Maryland. They have better things to do with a little land they have than give it over and away for disposal of hazardous wastes.

---

* A Btu is a British thermal unit, which is the amount of heat required to raise the temperature of 1 lb of water 1°F.

Denmark's alternative to pollution was the establishment of Kommun-nekemi a/s in 1971. They have been at it for over 15 years.

Kommunnekemi (the word literally means "municipal chemical") is located in Nyborg on the island of Fyns. There, wastes are detoxified by incineration or chemical treatment. The energy value of the waste is utilzied in a boiler for the production of steam, 20% of which is used at the plant with the rest sold to Nyborg for municipal heating. In terms of oil conservation, the quantity of heat produced is equal to the burning of some 1.5 tons of oil per hour.

Figure 7 shows in flowchart form the movement of the waste stream entering Kommunnekemi. Waste oil moves to an oil treatment plant (Figure 8), liquid and solid organic waste move to the rotary kilns, and inorganic waste moves directly to an inorganic treatment plant (Figure 9).

The rotary kilns, with temperature of 2200°F, achieve a destruction of the wastes of approximately 99.99%. Waste water is piped to the municipal sewers, and metals are buried in their own secure landfill to await such time as new recovery processes make it economical to retrieve them. Harmless gases are vented to the atmosphere. Solid residues amounting to 25% of the original bulk, now no longer hazardous, are disposed of at their landifll. The facilities at Nyborg are shown in Figure 10.

Denmark is not alone in using this type of waste disposal. Sweden, West Germany, The Netherlands, and Finland have similar processes. From them we learn that it works, and works well. They avoid our worst problems: the regulatory battles that must decide what qualifies as regulated toxic waste and how they should be dealt with to ensure no harm to people or the environment.

The lesson to be learned from these European examples is straightforward. Detoxification facilities exist that virtually eliminate the need for land disposal of untreated chemical waste. The process does not require sophisticated professional expertise to operate, and over the years, with the sale of steam, it is as economical as our least expensive methods. What our communities lack is motivation and political will. Few mayors want to be in what they see as the unenviable position of being the one "who raised taxes."

2. Fluidized Bed Combustion Plants

A variation on incineration is fluidized bed technology, which provides another means of converting such feed stocks as solid waste, tar sands, petroleum cake hazardous waste, sludges, and industrial wastes generally into low-cost energy.

A fluid or "fluidized bed" is actually a "bed" or hot sand that exhibits liquid or fluid characteristics. The hot sand is suspended by air injection and the waste is incinerated within the moving sand. Due to the turbulence within the fluid bed, rabble arms to push the waste into the

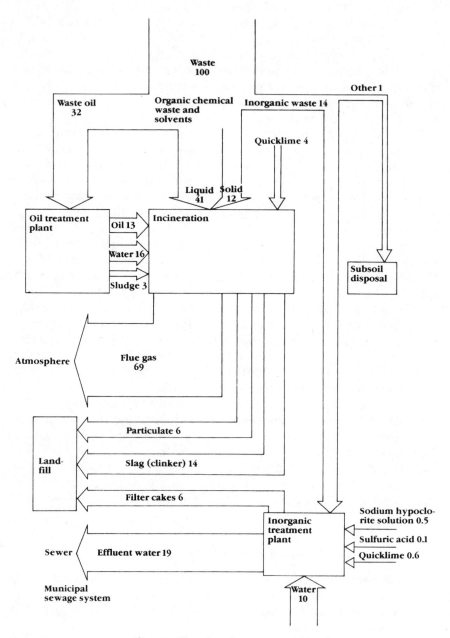

Figure 7. Flowchart for waste streams.

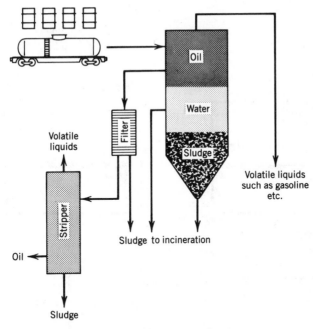

Figure 8. Oil treatment plant.

hottest layers are unnecessary. Depending on the design, all the waste obtains equally hot treatment.

As a consequence, mixing of all materials in the bed is nearly perfect. Each waste particle is subjected to the same residence time and oxygen contact within the bed, thus assuring complete destruction of organics in either solids or gases. The fluidized bed itself acts as a dry scrubber when lime is added. This reduces the opportunity for formation of such polychlorinated organics as TCDD and PCB, as well as hydrochloric acid. Additional information on fluid bed incineration is found in Section V on municipal waste.

3. Wetox*

Wet air oxidation (Wetox) is an established technology that can effectively treat industrial wastes too dilute to be incinerated economically yet too toxic to treat biologically. It is particularly effective in the destruction of oxidizable, organic contaminants in concentrations between 1 and 20% by weight in industrial waste waters.

Essentially, wet air oxidation is the aqueous phase oxidation of dissolved or suspended organic substances at elevated temperatures and

* Wetox is a process developed by the Ontario Research Foundation.

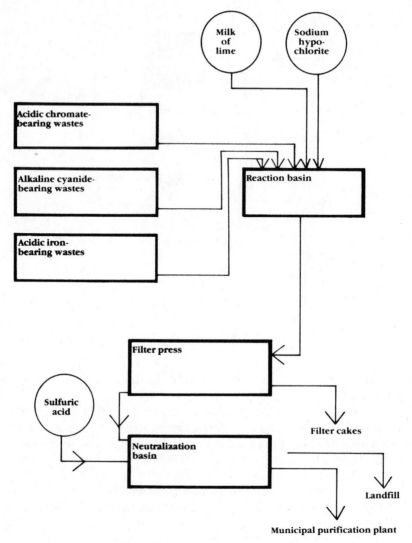

Figure 9. Inorganic treatment plant.

pressures. Water, which makes up the bulk of the aqueous phase, catalyzes the oxidation reactions so they proceed at relatively low temperatures (350–650°F) and at the same time serves to moderate the oxidation rates, removing excess heat by evaporation. The oxygen needed is provided by air bubbled through the liquid phase in a reactor used to contain the process.

Process pressure is maintained at a level high enough to prevent excessive evaporation of the liquid phase, generally maintained between 300 and 3000 psig.

(a)

(b)

Figure 10. (a) Tank farms at Nyborg. (b) Rotary kiln incinerator for treating oil wastes.

A significant advantage of wet air oxidation is the minimal air pollution produced by the process. Contaminants tend to remain in the aqueous phase. It also provides an efficient means for energy recovery from relatively dilute waste streams (8).

4. Technology in Motion

With options for disposal of hazardous wastes becoming progressively more limited and expensive, with environmental regulations increasingly restrictive, and costs of hauling and disposal to offsite facilities increasing, versatile onsite management becomes desirable. The idea is to bring state-of-the-art technology to the waste site.

Mobile activated carbon units, mobile certrifuges, belt filter presses, dredges, auxiliary generators, pumps, and neutralization and mobile clarifiers are becoming available where needed. Companies such as International Technology Corporation provide services via 40-ft tractor-trailer-mounted, fully equipped mobile laboratories that can be deployed at pits, ponds, lagoons, and/or spill sites. These mobile labs go directly to any site in the continental United States to furnish complete organic and inorganic analyses. A mobile preloaded activated carbon system can pull up to a water source, drop a hose in, and treat at the rate of 300 gal/h. Mobile systems can be a preferred approach for specific sites. Of particular importance is the fact that the EPA issued a Record of Decision (ROD), which specifies the remedy the government has chosen along with the reasons for doing so, indicating that a mobile thermal destruction unit will be used to destroy and remove the dioxin-contaminated sewer and creek sediments from Love Canal.

The mobile unit has a dioxin-destroying efficiency of 99.9999%. The cost of this treatment is double that of land burial, but the ROD selected thermal destruction on the basis of its ability to meet statutory requirements by eliminating the factors of toxicity and mobility. In addition, the high efficiency of thermal destruction by mobile units has been amply demonstrated.

5. Waste Exchanges

A unique form of recycling is industry to industry transfer. Once a company generates hazardous waste, it may not be able to utilize it or recycle it "in house." But some other company may have use for it. Unfortunately, secrecy being what it is in industry, it has been next to impossible for one company to know what another has available, or what it may need. This "need to know" is being addressed through establishment of Waste Exchanges, which are nonprofit organizations relaying information between waste generators.

These exchanges publish bulletins listing waste generated and waste wanted. The exchanges do not become engaged in negotiations; rather,

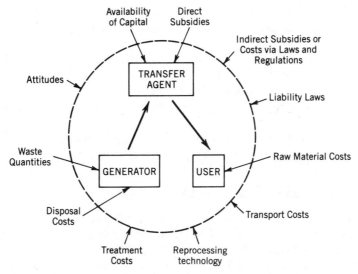

Figure 11. Basic transfer system and its external influences.

they act as middlemen or brokers, bringing two interested parties together. The exchanges are funded by state subsidies, private organizations, or both. Currently, some 15 operate in the United States and Canada.

During 5 years of operation, the United Kingdom Materials exchange has fostered the transfer of some 156,000 tons of waste per year, and 196,000 tons as a single transaction. Similarly, in Canada, the Waste Materials Exchange has been able to effect transfers totaling 80,000 tons/yr. In the United States, it has been estimated that some 8 million tons are available for exchange. Figure 11 describes the major "partners" involved and the forces that impinge on the process (4).

By a recent extension, the RCRA's onsite storage limit of 90 days was extended to 180. This allows small-scale generators to enter the exchange market. Previously, their waste generation was too small to accrue rapidly enough to come within the 90-day holding period. With the extension they will be able to take advantage of exchange trading.

6. Dumping in Underdeveloped Countries

With increasing disposal costs—up to $2500 per ton—the ever tightening regulations, and the lack of fluidized bed incinerators, waste contractors have begun exploiting the underdeveloped countries of Africa.

Recently, a reporter for The New York Times exposed this traffic in hazardous waste. European and American brokers have sought to dispose of hundreds of thousands of tons of chemical and radioactive wastes in countries from Morocco to the Congo. Many of these countries, such as

Guinea-Bissau, Nigeria, Equatorial Guinea, Liberia, and Sierra Leone, need additional revenue and are tempted. But they also lack sophistication about these chemicals, which are often purposely mislabeled.

Many of the disposal sites are geologically unsuitable, consisting of porous volcanic soil which permits offshore seepage that can threaten fishing grounds. In addition, much of the waste was being diposed of in open sites or simply on the ground, with no regard for the surrounding communities (9).

Until appropriate thermal destruction facilities become more readily available, these sharp practices can be expected to continue.

7. Regional Underground Monolith Disposal

Regional underground monolith disposal (RUMOD) is primarily for disposal of solid materials. In this process, heavy metal wastes such as copper, lead, chromate, mercury, nickel, and cadmium are mixed with cement. The alkalinity of the cement stabilizes these metals in the form of their hydroxides. Adding additional strength to the cement–metal matrix are the asbestos fibers and metal filings often present in wastes. If these solids are buried beneath the frostline, they are not subjected to regular thawing and freezing stresses.

At Oak Ridge National Laboratory in Tennessee, wastes will be mixed with cement and pumped into caverns 300–2000 m underground. Within the cavern the slurry will harden and be unable to migrate. Groundwater contamination should not occur. The economics of this process, which will play a major role in its development, will turn on the costs of excavation and drilling. This, of course, will depend on the type of strata being used. However, all the necessary technology is currently available.

Clearly, it is time to move beyond land disposal of hazardous waste. The technologies exist to make the move entirely possible and reasonable to do so. At this point, the people must speak.

Unfortunately, intelligent alternatives are sometimes stymied by emotional reactions. A New Jersey company recently proposed joining with Kommunekemi to build a hazardous waste treatment incinerator in the city of Perth Amboy. Never mind that this type of treatment plant has been successfully employed for almost 15 years in Denmark and that half a dozen other European countries also use similar processes (usually with stricter safety codes than our own), the residents of Perth Amboy said it loud and clear, "Not in our backyard."

At a recent town meeting, the message given by a coalition of residents was that a "toxic waste disposal plant was not welcome in the city. Everyone agrees there is a need for such a facility, but the greatest need is in an area that is completely safe." The fact that incinerators have played a successful role in waste disposal in many American cities seems to have been overlooked or dismissed. The adjectives hazardous and toxic have added new stresses to the disposal problem.* NIMBY operates every-where.†

What in fact is the worst thing that could occur at such a traetment plant? After years of operation and study, an uncontrolled fire is likely to be the greatest risk. But a fire at the heat of operation would only produce complete combustion and destruction of the waste, with no explosions or hazardous air pollutants. The risk to individuals in the community would be no greater than each of them smoking 1.4 cigarettes once in their lives. But how is this conveyed to a community, and what will it require for them to believe it? Furthermore, who will they believe? These questions are pursued in Chapter 15.

V. MUNICIPAL WASTE

Philadelphia, the birthplace of the nation, has the distinction of initiating the country's first system of waste disposal. In the 1790s, domestic servants and others hired for the purpose were sent wading into the Delaware River to toss garbage into its current. Some might argue that not much has changed.

Not all waste carries the epithet hazardous. No more than 20% by most estimates can be so classified. The remainder is innocuous. Nevertheless, state and local officials are bedeviled by the problem of how to manage the 200 million tons of solid waste produced in the United States each year. Some 5 lb/person per day.

What do we throw away in greatest profusion? The composition of our waste is presented in Figure 12.

Almost 80% consists of grass clippings, waste paper, food, glass and metal. The remaining 20% is made up of billions of plastic caps and containers, junked automobiles, tires, refrigerators, furniture, bed springs, bathtubs, radios, TV sets, lamps, dishes, screens, sneakers, and on and on, to the tune of 1650 lb/person per year—920 million pounds per day. Note that "fast-food" packaging waste is not at or near the top of the list. Most people believe it constitutes as much as 35% of the waste stream. In fact, it represents less than 0.5%—closer to 0.3%. The reality and the perception are inversely related. Nevertheless, with some 900 million pounds per day and climbing, to an expected 1.145 billion by the year 2000, the dimensions of the problem become imaginable. Where and how can such quantities be disposed of? Is it any wonder that "everyone wants us to pick up his garbage, but nobody wants us to put it down."

---

* Clearly, public concern for nuclear power generation and radioactive waste disposal affect decisionmaking about solid waste.
† Perhaps the solution to the public's refusal to accept much-needed disposal plants and areas is to require everyone to dispose of their own waste in their own backyards.

CONTENT OF OUR GARBAGE

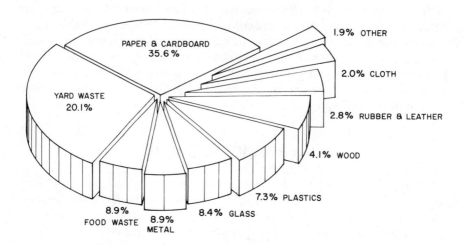

PRIMARY DISPOSAL METHODS

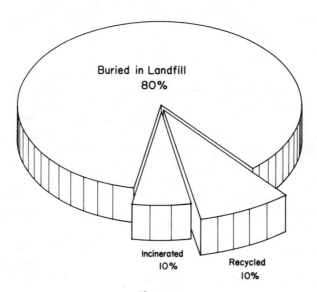

Figure 12.

Wallace Stegner said it for all time in his ballad, The Town Dump:

> I think I learned more from the town dump than I learned from school: more about people, more about how life is lived, not elsewhere but here, not in other times but now. If I were a sociologist anxious to study in detail the life of any community, I would go very early to its refuse piles. For a community may as well be judged by what it throws away, what it has to throw away and what it chooses to, as by any other evidence.

Littering the landscape along with garbage and refuse are trash, rubbish, litter, junk, and scrap, all interchangeable terms for solid waste. Garbage used to mean food waste, and rubbish was any waste without garbage, but these fine distinctions appear to have disappeared. In the discussion that follows these terms will be used interchangeably for everything but liquids.

A. Sanitary Landfills

For most of our history the attitude toward waste was burn it, bury it, or dump it. It is the same mentality that suggests shooting it up into space. Anything, as long as it is out of sight. Burying waste can be a satisfactory solution if land is available and is properly managed to prevent leaching of water-soluble materials into groundwater. An outgrowth of the open dump is the sanitary landfill. Unlike the open dump it is a planned and supervised operation requiring trained personnel. Properly run, offensive odors are barely noticeable, rodents are nonexistent, and it is appropriately economically.

Basically, it requires the deposition of refuse, compaction by bulldozer, and covering with earth at the conclusion of each day, with a final covering and compaction to prevent burrowing animals from penetrating. Figure 13 shows the anatomy of a modern landfill. A detailed section of the figure depicts the many layers with their functions.

1. Types of Landfill

Depending on the site available, four major landfilling methods have been developed: area, trench, slope/ramp, and canyon/pit or quarry.

For the area method, the site is usually flat and the refuse is deposited in horizontal layers on relatively flat ground, compacted, and covered over, sides and top, with soil or another inert material, layer-cake fashion. With the trench method, land is excavated in the shape of a trench to a depth, length, and width of 10–15 ft, and widths up to 20ft are widely employed. Refuse is deposited in the trench and compacted by bulldozer; then the trench is filled in with earth. In both the area and trench

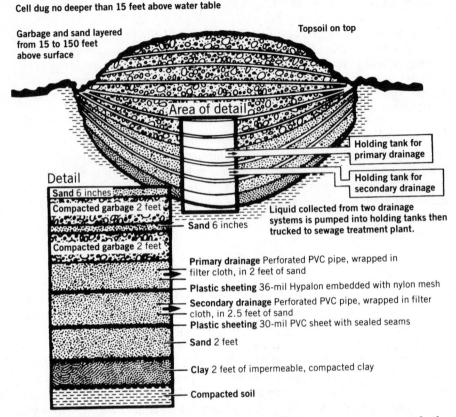

Cell dug no deeper than 15 feet above water table

Garbage and sand layered from 15 to 150 feet above surface

Topsoil on top

Area of detail

Holding tank for primary drainage

Holding tank for secondary drainage

Detail

Sand 6 inches

Compacted garbage 2 feet

Compacted garbage 2 feet

Sand 6 inches

Liquid collected from two drainage systems is pumped into holding tanks then trucked to sewage treatment plant.

Primary drainage Perforated PVC pipe, wrapped in filter cloth, in 2 feet of sand

Plastic sheeting 36-mil Hypalon embedded with nylon mesh

Secondary drainage Perforated PVC pipe, wrapped in filter cloth, in 2.5 feet of sand

Plastic sheeting 30-mil PVC sheet with sealed seams

Sand 2 feet

Clay 2 feet of impermeable, compacted clay

Compacted soil

Figure 13. Garbage cover-up: anatomy of a modern landfill. New Jersey environmental rules define how new garbage landfills must be layered to avoid contaminating nearby soil and water. This schematic drawing shows one cell of a typical landfill; exact depths of individual layers vary with local conditions.

procedures, the compaction of refuse and depth of the soil covering act as a barrier to prevent fly eggs from hatching and rats from boring. Another important feature is that the separation of refuse by walls of soil, as shown in Figure 14, keeps fires from spreading beyond a single cell. The soil cover, from 18 to 24 in. thick, virtually eliminates odors.

Both these procedures usually require about 1 acre per 10,000 people per year. Therefore, a community must set aside large tracts of land for them. In the face of the increasing demand for living space, the two needs come into conflict. A solution is the slope or ramp method of land filling, which usually employes existing ravines in which refuse is deposited on an angle, against the side of the ravine. As with the area and trench methods, inert cover is placed on the sides and top at regular intervals. In this procedure, cover material is excavated from the toe of the slope,

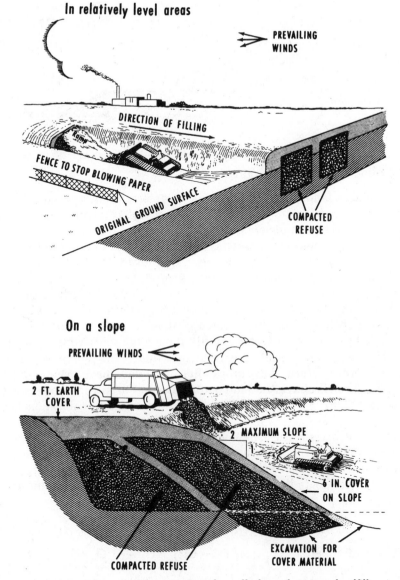

Figure 14. Schematic representation of a well-planned sanitary landfill.

opening a cavity into which wastes are inserted for compaction. After the first lift, the procedure is similar to that of the area method.

The canyon/pit or quarry method, shown in Figure 15, takes advantage of the availability of existing depressions and quarries into which waste can be dumped and dozed. It begins by compacting layers stepwise from the bottom and working to the surface.

Canyon/pit/quarry

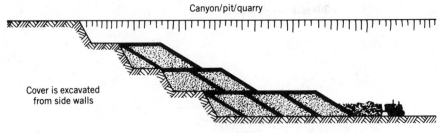

Cover is excavated
from side walls

Figure 15.

2. Biological Decomposition

The sanitary landfill is a biological waste treatment procedure. More specifically, it is microbiological in that a variety of microbes use the waste as sources of nutrients. Given the compaction and the absence of free oxygen, anaerobic organisms decompose the organic materials to more stable forms. That is, energy is removed from the waste as a consequence of microbial metabolism.

Refuse arriving at a fill site normally contains a lively flora of saprophytic organisms and, at times, pathogenic bacteria, viruses, yeasts, and fungi, which are a potential source of disease to the community. In a controlled, properly designed landfill, the energy released by the chemical changes taking place is so great that there is a precipitous rise in internal termperature. Recordings after 7–10 days have been as high as 150–160°F, well above the temperature needed to kill the heat-sensitive pathogens. This offers a fair degree of assurance that the fill is not a health hazard.

Microbial activity in a newly filled area is primarily that of aerobic bacteria, and it will continue until the residual oxygen contained in crevices of the refuse is exhausted. At this point, anaerobic organisms continue the degradation. The generalized expressions of these two reactions are

(1) organic matter + oxygen + organisms $\rightarrow$ CO_2, H_2O, NH_3 + energy

(2) organic matter + organisms $\rightarrow$ CH_4 + H_2S + CO_2 + H_2O + energy

A more specific example linked with the production of heat is seen in the degradation of the amino acid leucine, present in the protein of many foods:

$$(CH_3)_2\!\!-\!\!CH\!\!-\!\!CH_2\!\!-\ \overset{\displaystyle H}{\underset{\displaystyle NH_2}{\vert\ \ C\ \ \vert}}\ \!\!-\!\!COOH + 150 \rightarrow$$

$$6CO_2 + 5H_2O + NH_3 + 755 \text{ Cal}$$

The greater the amount of organic matter present, the higher the temperatures will be. When the anaerobic organisms enter the scheme, their metabolic reactions yield different end-products. Gases such as hydrogen sulfide (H_2S) and methane (CH_4) are produced.

The rate of gas production can be controlled by varying the size of refuse particles as well as by the amount of moisture contributed. Low gas production can be maintained by low moisture, large particle size, and high density as a consequence of increased compaction. Unwanted gas has been allowed to escape by provision of vent pipes distributed around the fill area. Methane can be a desirable by-product, fetching handsome profits by supplying gas to nearby industries, thereby further reducing operation costs. However, we now know that methane can contribute to the greenhouse effect. Thus, trapping and containing it may become a requirement.

3. Community Response to Landfills

The most severe by-product of landfill operation is the formation of leachates, solutions of solubilized substances, resulting from the passage of water through the landfill. These leachates can seep through the underlying soil, as shown in Figure 16, with their load of dissolved and suspended chemicals to reach groundwater aquifers.

Nevertheless, landfills in many communities have been, and continue to be, appropriate disposal sites. Precipitous closing of landfills, as has

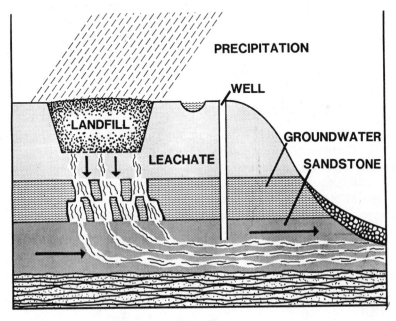

Figure 16.

been occurring throughout the Northeast, is senseless, especially where sites have 10–20 years of additional capacity and do not have or take chemical waste.

In Philadelhia between 1984 and 1985, solid waste cartage bills for commercial establishments, restaurants, retail shops, and the like jumped 300%, primarily because of increased hauling distances and higher permit costs for the few remaining landfills. Of 17 sites available to carters in 1984, only four remained open in the Philadelphia County area in 1985.

There will always be need for landfills because nonburnables such as metals, glass, bricks, cinder block, and dry-wall (plaster) require disposal. Even the most efficient incinerators will produce ashes that require burial. It will be another 25 years (at the current rate of progress) before the heat of fusion, as detailed further on, can be available to reduce all materials to their original and recoverable elemental form.

The national hysteria surrounding garbage, landfills, and groundwater was nowhere more evident than in the incident dubbed Islip's folly.

Because its landfill was full, or so it seemed, the Islip, New York, town supervisor, Frank Jones, hired a garbage scow to move 3200 tons of cardboard, plastics, paper, and tires to a landfill in another city. From the day it left Islip the scow was charted and reported on by helicopter. For 162 days it traveled the length of the East Coast, around Florida, up its west coast over the Louisiana, down the Gulf of Mexico to the Caribbean and back. During its months at sea, six states and three countries refused it permission to dispose of its cargo. Midway in its voyage to nowhere, New York City health inspectors boarded the *Mobro* for 2 days of inspection and found no evidence of toxic materials present. That made no difference. The scow was kept moving. "The garbage" and its location became a nightly TV spectacle. Politicians up for election in 1988 had a made-to-order issue. Was anyone served by this affair? Hardly. Was anything learned from it? Time will tell.

B. Incineration and Resource Recovery

In point of fact, incineration is more a means of waste reduction than a method of disposal. Incinerated materials are converted to gases by controlled burning at high temperature. A small percentage remains as residue that requires disposal, usually at landfills. Depending on the ratio of combustibles to noncombustibles, the percentage can be increased or decreased. Currently, volume reductions of 90% are obtained with little difficulty. New designs are expected to achieve even greater reductions.

Currently, mass burning and fluid bed combustion are the two major types of incinerator processes gaining wide use. Both systems burn trash and produce steam. The trash becomes ash, thus increasing the lifetime of existing landfills, and the steam can be used for heating or turned into electricity.

1. Mass Burning

The City and County of Baltimore's new mass burning incinerator is a model system. Figure 17 is generalized view of the process. It is quite simple. Incoming trucks dump refuse into an enclosed receiving pit. Without prior sorting, or shredding, refuse is transferred from the pit to the feed hopper of each furnace. On reciprocating grates, the refuse is moved through the furnace over a period of 45 min at combustion temperatures between 2500 and 2800°F. This temperature is high enough to vaporize a tire and its steel belting and destroy chemical pollutants such as dioxin.

Air from the refuse pit is blown in above and below the grates to fuel complete combustion in the furnace and maintain negative pressure over the pit, preventing escape of dust. A water well boiler (tubes of water) above the grate produces superheated steam, which drives a turbine generator which produces electricity for the Baltimore Gas and Electric Company.

An electrostatic precipitator for each furnace assures that stack emissions meet the city's standards.

When the plant is operating at full capacity, it processes 2300 tons/day and produces 500,000 lb of steam an hour.

The four electrostatic precipitators electrically charge the particulates moving up the stacks, causing them to migrate, under the influence of a strong electric field, toward the collecting electrodes, in this case the oppositely charged plates. Figure 18 shows the trapping of particulates as they move past the electrically charged plates. Every 20 min an automatic hammer bangs the plates to shake off the particles. This process removes particles as small as 0.05 μm. The cleansed air then passes up the stack.

When the garbage is reduced to ash, it falls off the edge of the grate and is loaded on conveyor belts, where magnets remove scrap iron and steel for delivery to scrap dealers. The remaining ash is taken to a landfill specifically designed for this waste.

Incineration of the garbage now being dumped offshore would remove the blight of wastes washing up on coastal beaches. The closing of beaches during the summers because of the threat of contamination by infectious wastes such as used hyperdermic syringes, needles, and containers with infected blood would be a thing of the past.

Preventing the siting of incinerators because of some perceived threat to health from chemicals in stack gases, at parts per billion or parts per trillion levels, is a disservice to the community.

2. Fluid Bed Combustion

The fluidized bed principle was first applied on an industrial scale about 30 years ago and has become a standard operation of chemical processing in Europe.

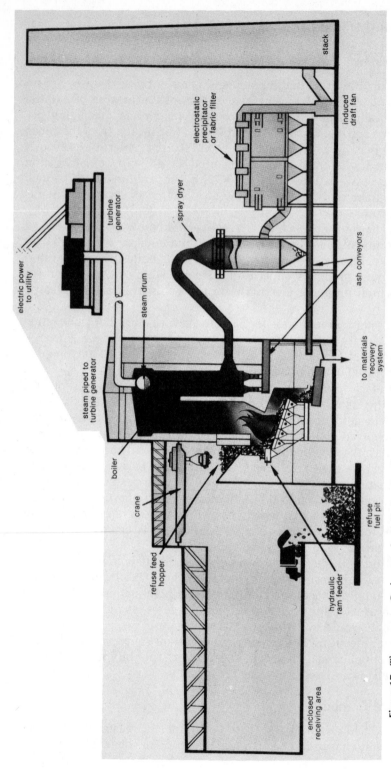

Figure 17. The process. Garbage is deposited at the plant. It is moved to an inclined grate where it is burned. Air is pumped through it to speed burning. Hot gases produced heat water in the walls and at the top of the combuster, generating steam for heating and for generating electricity. Dry scrubbers treat smoke with alkaline dust to neutralize acids, then this dust is trapped in a filtering unit. The hot gases rise through the cooler stack and dissipate into the atmosphere.

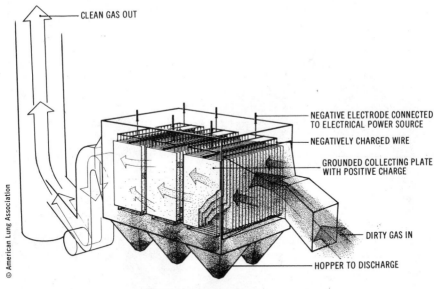

CLEAN GAS OUT

NEGATIVE ELECTRODE CONNECTED
TO ELECTRICAL POWER SOURCE

NEGATIVELY CHARGED WIRE

GROUNDED COLLECTING PLATE
WITH POSITIVE CHARGE

DIRTY GAS IN

HOPPER TO DISCHARGE

© American Lung Association

ELECTROSTATIC PRECIPITATOR

The dirty air flows between negatively charged wires and grounded metal collecting plates. The particles in the airstream become charged and are attracted to the
plates. The plates hold the accumulated dust until it is periodically knocked into
hoppers. The clean air is pumped out through a stack.

Figure 18. Electrostatic Precipitator. The dirty air flows between negatively charged wires
and grounded metal collecting plates. The particles in the airstream become charged and are
attracted to the plates. The plates hold the accumulated dust until it is periodically knocked
into hoppers. The clean air is pumped out through a stack.

A fluid bed is a bed or layer of sand or other granular material
(depending on what is being incinerated) in a combustion chamber fitted
with tiny holes through which compressed air is continuously forced. The
rising air lifts and separates the granules, creating a turbulent mixture of
air and granules, which behaves as though it were a liquid or fluid.

When this bed is heated to 2500–3000°F, it becomes fluidlike and any
type of material fuel—garbage, tires, chemicals, wet sewage—can be
added to it and burned at extremely high efficiency. With the addition of
powdered limestone to the burning bed, the exhaust gases emitted are free
of sulfur oxides. Figure 19 is a generalized view of the operation.

In addition to the fluid bed, combined multiple hearth/fluidized bed
incinerators are also coming on-line equipped with a predrying and
distributing zone which can hold up to six hearths, with the fluidized bed
beneath them. Forty to 60% of the water contained in the waste is
evaporated in the predrying zone. The vapors flow through the upper
section of the combustion zone where all the volatile organics are burned.
Because of the efficient drying operation, the cross-section of the fluidized

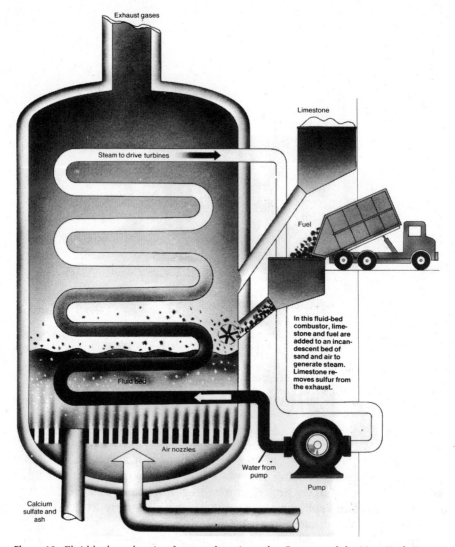

Figure 19. Fluid bed combustion furnace: how it works. Courtesy of the New York Times.

bed zone can become smaller, making the process unusually economical for many types of waste.

C. Microbial Fungal Metabolism

Although incineration is a certain approach to waste disposal, it can be expensive. The recent discovery that a ligase enzyme secreted by a common soil mold, *Phanerochaete chrysosporium*, can rend asunder the sturdiest organic molecules offers a new dimension and approach to decontaminating soil and water, as well as a measure of harvesting valuable chemicals from agricultural and industrial waste water cheaply.

Phanerochaete chrysosporium, a Basdiomycete and a distant relative of edible mushrooms, is a voracious consumer of dead wood and thereby earns its other name of white rot fungus. Recently, John A. Bumpus of Michigan State University reported that dead wood was only a part of this mold's diet (10). It also attacks DDT, dioxin, lindane, and a number of other organohalides. Once the complex carbon bonds are split, other soil organisms can begin their meals, which ultimately leads to the production of harmless carbon dioxide.*

Bumpus believes that potentially carcinogenic chemicals polluting soil and water can be rendered harmless by seeding effluents with the white rot fungus, a far simpler and cheaper procedure than current methods. He found that after a 30-day exposure to the fungus, 50% of the DDT has been destroyed.

The enzyme (or group of enzymes) that oxidizes carbon bonds is called ligase because of its ability to break down lignin, a heteropolymer of coniferyl alcohol, whose complexity yet defies elucidation. This may be the solution for the stubborn problems faced by paper mills and manufacturers of plastics, explosives, synthetic fibers, and other products derived from cellulose.

The breakdown of lignin is useful because it can remove an expensive nuisance from the processing of cellulose, but the resulting fragments of the lignin molecule are rich in aromatics (molecules containing rings of carbon atoms). These make excellent raw materials for the manufacture of needed chemicals.

In manufacturing even the cheapest grades of paper, the lignin in wood pulp is ordinarily bleached. Lignin gives paper bags their brown color and is the reason why paper exposed to sunlight yellows rapidly. Bleaching paper yields high chlorine and sulfur wastes, which can produce pollution problems of another sort.

Where Bumpus isolated an organism that would oxidize organic bonds, the U.S. Patent Office recently granted V. R. Srinivasan of Louisiana State University a patent for developing a new bacterium. This one was genetically engineered to enable it to break down wood and other crop fibers. It can metabolize agricultural and paper mill wastes.

And there is yet another approach to waste disposal. Make the product that will ultimately end up at a landfill nutritious—the source of food for microbes.

Laboratories and factories across the country are developing a host of products, including molded plastics from corn, the country's largest and most important crop. By making products from biodegradable farm

---

* Given the magnitude of the waste that could be treated in this manner, would be the carbon dioxide generated contribute to the greenhouse effect? Would we then simply be trading one risk for another? Are we locked into a tight system? These questions are worth considering.

products, corn, soybeans, and cotton, instead of from petroleum-based plastics, garbage piles would shrink and along with it pollution.

Because corn is so abundant, and because virtually anything that can be made from the carbon and hydrogen that form petroleum can also be manufactured from corn starch with its long chains of carbon, hydrogen, and oxygen atoms, corn is seen as replacing petroleum-based products.

Efforts are currently being focused on the 14 billion pound per year disposable packaging industry with a product that is part corn starch part polyethylene. Agric-Tech Industries of Illinois is developing a plastic resin for disposable bottles. Not only would these be less expensive, but microbes will eat the starch, converting it to carbon dioxide, and the polyethylene chain will rupture allowing insects to feed on it.

Clearly, our waste disposal will be solved by sheer creativity.

D. Composting

Composting is the controlled aerobic microbial degradation of organic materials to a sanitary, nuisance-free, humuslike material. The breakdown of refuse by microorganisms is probably the oldest biological waste treatment method. People have been burying unwanted waste for centuries, and our allies the microbes have split the complex organic compounds into simpler substances, which are then taken up by plants and converted back to complex organic matter. Refuse disposal by microbial action occurs both in the sanitary landfill and in composting. Although composting is not new, only within recent years have major improvements in the process been made that allow its use for large-scale disposal.

Sanitary landfills, as noted earlier, degrade waste via anaerobic metabolism. For rapid aerobic metabolism to occur, it is necessary only to ensure an optimum balance between available air and moisture. This can be enhanced by grinding the refuse for a landfill site to a size that will create optimum surface area for the organisms.

A primary requisite for proper functioning of a composting operation is the removal of such noncompostibles as glassware, metals, and ceramic items. Oxygen needs are provided by tumbling or aerating the compost. Figure 20 is a schematic representation of a typical composting plant. The moisture requirement of not less than 30% can be provided by sewage sludge, thus disposing of the sludge as well. Temperature within the compost range above 150°F, which is sufficient to destroy pathogenic bacteria such as the streptococci, *Mycobacterium tuberculosis*, fly larvae, and parasitic worms often found in human waste and garbage. Some dozen composting methods are available. Nevertheless, all attempts to compost municipal refuse on a large scale in the United States have failed.

One of the more important reasons for failure has been the inability of compost producers to dispose of their products given the readily available

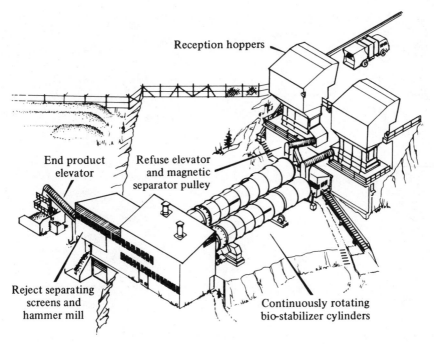

Reception hoppers

End product
elevator

Refuse elevator
and magnetic
separator pulley

Reject separating
screens and
hammer mill

Continuously rotating
bio-stabilizer cylinders

Figure 20. Schematic diagram of a compost plant.

and inexpensive inorganic fertilizers. In addition, farmers who would normally be the primary users are not generally in need of soil conditioners. Consequently, profitability is low and composting plants become noncompetitive.

VI. THE FUTURE: BRIGHT PROMISE

A. Solid Waste and Fusion Energy

1. Fusion Reactions

At Princeton University's Plasma Physics Laboratory, scientists are conducting an experiment as part of a national effort to develop nuclear fusion as a safe and economical means of generating electricity. The Princeton Large Torus (PLT) belongs to a class of fusion devices known as tokamaks, a name derived from the Russian acronym for "toroidal magnetic chamber."

In a tokamak, an extremely hot ionized gas called a plasma is confined within a stainless steel doughnut-shaped vacuum vessel. Powerful magnetic fields are used to keep the plasma from the steel walls, which would melt on contact with the incendiary temperatures of the plasma.

Figure 21.

The magnetic fields are generated by massive magnetic coils wrapped around the vessel and by use of an electric current made to flow inside the plasma itself. Figure 21 shows the size and shape of the vessel. The flow of charged particles with and without a magnetic field is shown in Figure 22. Schematic representations of the magnetic field and the plasma held within it are shown in Figure 23.

The most readily attainable fusion process is the combination of deuterium and tritium nuclei. The products of their reaction, shown in Figure 24, are helium-4 (^{4}He) the common type of helium (an alpha particle—cf. Chapter 9) and a more highly energetic free neutron (n).

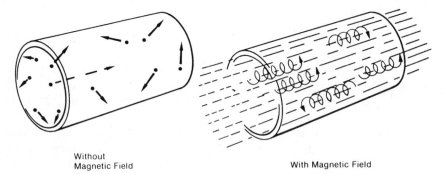

Without
Magnetic Field   **With Magnetic Field**

Figure 22. Motion of charged particles. Courtesy of Princeton University's Plasma Physics Laboratory.

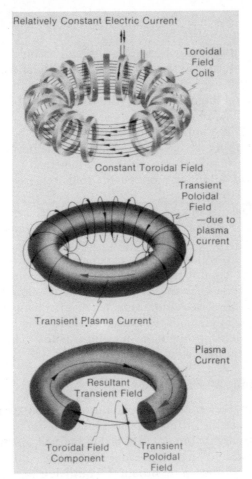

Figure 23. Tokamak plasma confinement. Courtesy of Princeton University's Plasma Physics Laboratory.

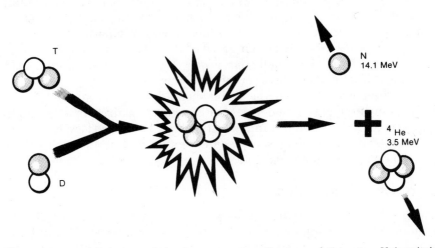

Figure 24. Generalized view of the fusion reaction. Courtesy of Princeton University's Plasma Physics Laboratory.

Since nuclei carry positive charges, they normally repel one another. The higher the temperature, the faster the atoms or nuclei move. When thy collide at their furious speeds, they overcome the force of repulsion of the positive charges, and the nuclei fuse, with the release of large amounts of energy.

The difficulties in producing fusion energy have been the slippery problems of developing devices that can heat the deuterium–tritium fuel to sufficiently high temperatures while simultaneously confining it for a long enough period, allowing more energy to be released through fusion than is used for heating.

In order to release sufficient energy for production of electricity and heat, the gaseous deuterium–tritium fuel must be heated to about 100 million degrees celsius. This is six times hotter than the interior of the sun, which has been estimated at approximately 15 million °C.

The problem of confinement must also be overcome. No metal or combination of metals known today can withstand such incredibly high temperatures. The hot gases containing negatively charged ions make up the plasma. In a magnetic field the ions, as noted in Figure 22, are forced to follow spiral paths through the magnetic fields. This prevents the plasma from striking the vessel walls.

2. Disposal of Solid Waste

Plasma-containing devices with temperatures between 20 million and 50 million °C are seen as the ultimate form of incineration for the disposal of waste. This incomparable form of disposal reduces waste to their elemental inorganic forms. Such a system would not release pollutants into the air, nor would its residues or by-products require burial. It would provide valuable raw materials that could be used again and again.

In the heat of a fusion (the plasma), an automobile, for example, would completely vaporize, leaving the elemental iron, tin, zinc, lead, chromium, and other substances from which it was originally made. Since natural materials can only be composed of some 93 elements, there would be no more than 93 (usually far fewer) substances that would need separation from any combination of solid wastes.

Unfortunately, although the theory behind fusion is admirable, the engineering problems for a devices that can handle garbage practically are staggering. There is comfort in the fact that our best minds are beginning to think about waste disposal. I would venture that by the year 2010 a workable model will be available.

3. Personal Responsibility

There is another approach to solid waste that is far less sophisticated. It requires a good deal more personal involvement and personal discipline, which may render it unworkable. This process requires that each house-

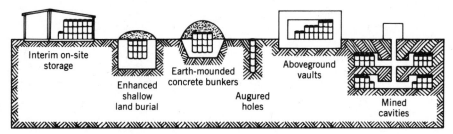

Figure 25. Types of burial for low-level waste.

hold shift from the traditional all-purpose (single) waste container to four: one for metal, one for glass, one for waste food and paper, and one for industrial chemicals and drugs (old medicines, household cleaners for wood, metal, glass, shoes, etc.).

These containers must either be brought to appropriate facilities or picked up by equally appropriate carters. If this can be made economically compelling, recycling and recovery could further reduce the need for burial. But this seemingly simple system requires motivation and dedication. For garbage, that may be asking too much.

B. Radioactive Low-Level Waste

Radioactive wastes present problems of disposal similar to other hazardous wastes, but due to their far greater half-lives and penetrating characteristics (cf. Chapter 9) additional precautions are necessary.

Figure 25 indicates current and future types of burial for low-level wastes (LLW).* The concrete bunker is currently being experimented with in France. Each of these, including the augured holes for storage of concrete containers, will add substantially to the cost of disposal but may not provide significantly greater protection beyond that currently afforded by shallow land burial.

Most of these engineered approaches are responses to sociopolitical pressures spawned by anxiety over proximity to burial sites. In addition, because of rising burial and transportation costs, options for final packaging of LLWs are being vigorously investigated. Figure 26 presents three major types of volume reduction (VR) procedures. Incineration with a 30-fold reduction ratio capability seems most appropriate to the task.

---

* As defined in the 1980 law, low-level waste is any material not classified as spent reactor fuel, high-level waste (from reprocessing reactor fuel), transuranic waste (containing significant amounts of such long-lived isotopes as those of americium or plutonium), or uranium or thorium by-product materials such as mill tailings. The Nuclear Regulatory Commission carries the definition one step further, specifying that LLW should be acceptable or underground disposal and defining subclassifications according to the degree of care required for disposal.

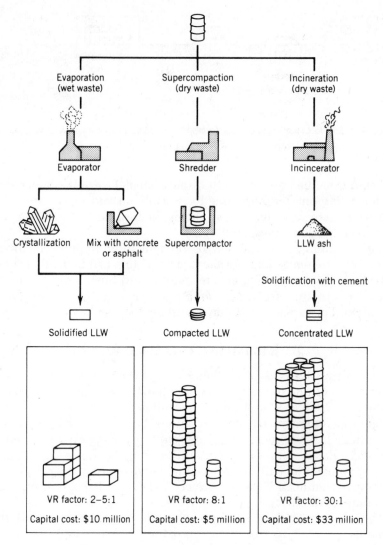

Figure 26. Reproduced with permission of *EPRI Journal.*

However, it also costs three to six times more in initial capital investment than either of the others. Initial cost, however, may not be the limiting factor—if reason prevails.

REFERENCES

1. A. G. Levine, *Love Canal: Science, Politics and People.* Lexington Books/D.C. Heath & Company, Lexington, MA, 1982.

2. U.S. EPA, Criteria, Identification, Methods and Listing of Hazardous Wastes. Draft of Proposed Rules Issued Under Section 301 RCRA Legislation, Washington, DC, June 1978.

3. C. R. Meyer, Liver Dysfunction in Residents Exposed to Leachate from a Toxic Waste Dump. *Environ. Health Perspect.* **48**(2):9–13, 1983.

4. W. Rogan and B. Gladen, Monitoring Breast Milk Contamination to Detect Hazards from Waste Disposal. *Environ. Health Perspect.* **48**:87–91, 1983.

5. *Waste Exchanges: Background Information.* U.S. EPA, Washington, DC, 1983, SW-887.1.

6. P. E. Voytek, Aspects of Risk Assessment Strategy. In *Assessment of Health Effects at Chemical Disposal Sites* (William W. Lowrance, Ed.). Proceedings of a Symposium June 1–2, 1981, New York City. Distributed by William Kaufman, Inc., Los Altos, CA 94022.

7. N. J. Vianna and A. K. Polan, Incidence of Low Birth Weight Among Love Canal Residents. *Science* **226**:1217–1219, 1984.

8. R. G. W. Laughlin, T. Gallo, and H. Robey, Wet Air Oxidation for Hazardous Waste Control. *J. Haz. Mater.* **8**:1–9, 1983.

9. J. Brooke, Waste Dumpers Turning to Africa. *The New York Times*, July 17, 1988.

10. J. A. Bumpus, M. Tien, D. Wright, and S. D. Aust, Oxidation of Persistent Environmental Pollutants by a White Rot Fungus. *Science* **228**:1434–1436, 1985.

SUGGESTED READINGS

Are We Cleaning Up? 10 Superfund Case Studies. A Special Report of OTA's Assessment on Superfund Implementation. Congress of the United States, Office of Technology Assessment, Washington, DC, June 1988.

Davis, C. E. and Lester, J. P. (Eds.). *Dimensions of Hazardous Waste Politics and Policy.* Greenwood Press, New York, 1988.

Haas, C. S. Incentives for the Treatment and Disposal of Hazardous Wastes by Alternative Methods. Report prepared by EPA/HAAS, 1984.

1986 National Screening Survey of Hazardous Waste Treatment, Storage, Disposal and Recycling Facilities. EPA/530-SW-88-035. September 1988. Office of Solid Waste, USEPA, Washington, DC.

Public Law 94-580. 94th Congress Resource Conservation and Recovery Act of 1976. Title II—Solid Waste Disposal. Office of the Federal Register, National Archives and Records Service, Washington, DC 20408.

Wynne, B. *Risk Management and Hazardous Waste: Implementation and the Dialectics of Credibility.* Springer-Verlag, New York, 1987.

14

Population and Patterns
of Disease

The world is deluged with panaceas, formulas, proposed laws,
machineries, ways out, and myriads of solutions. It is
significant that every one of these . . . deals with . . . the
structure of society, but none concerns the substance
itself—the people. This despite the eternal truth of the
democratic faith that the solution always lies with the people.
—Saul D. Alinsky

There are real and growing dangers to our simple and most precious possessions: the air we breathe; the water we drink; and the land which sustains us. The rapid depletion of irreplaceable minerals, the erosion of topsoil, the destruction of beauty, the blight of pollution, the demand of increasing billions of people, all combine to create problems which are easy to observe and predict but difficult to resolve. Acknowledging the physical realities of our planet does not mean a dismal future of endless sacrifice. In fact, acknowledging these realities is the first step in dealing with them.

This was President Carter's message to the American people on the occasion of his farewell address in January 1981. "The demand of increasing billions of people:" that may indeed be the crux of the environmental pollution dilemma.

On Sunday, July 1, 1983, the U.S. population stood at 234,799,480. Ten years before, it had been 10% less—211,909,032. On that same July Sunday in 1983, the population of the world was just short of its fifth billion. World population had itself risen by a relentless 7%. By the summer of 1987, that fifth billion became a reality. And by the year 2000, we will number 6 billion. The inexorable growth of World population is seen in Figure 1.

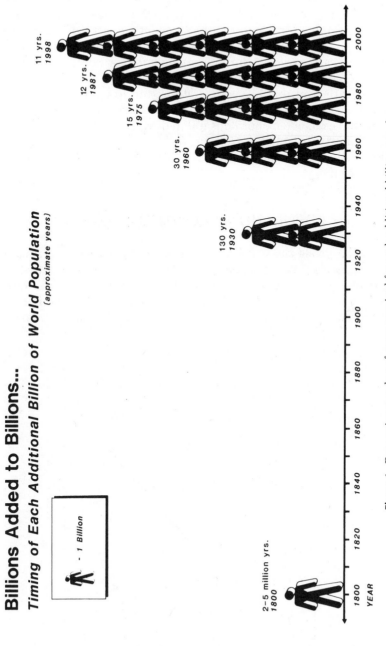

Figure 1. Decreasing number of years required for each additional billion people.

Famine, pestilence, war, and death, the four apocalyptic horsemen, are credited with great human destructive potential. Yet they do not seem so great a force as our ability to procreate. During the desperate famine in Ireland in the 19th century, the population trebled. Even after the decimation of Europe's population by bubonic plague in the 14th century (estimates vary up to 25 million, one-quarter of Europe's total population), the population fully rebounded in 20 years.

With little warning, a cyclone struck the low-lying coastal plain of western Pakistan in November 1970. The fearsome wall of water moved inland and exacted an enormous toll in lives. Estimates of the dead ranged from 300,000 to 500,000. Nothing in human history could match that deadly toll in so short a time. Yet by January 1971, the population had rebuilt itself.

I. POPULATION GROWTH AND STRUCTURE

Population levels can usually be ascribed to three factors: births, deaths, and migration. Figure 2, a generalized representation, shows that increasing population is primarily a consequence of birth and death rates, and although birth rates and death rates are declining, the far higher rate of birth forces population upward. Figure 3 shows this phenomenon for

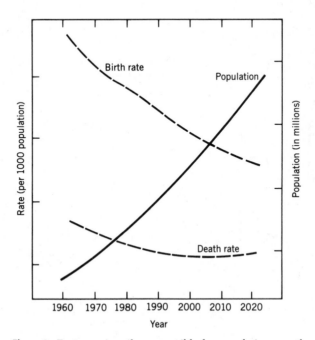

Figure 2. Factors primarily responsible for population growth.

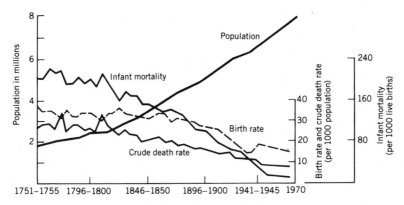

Figure 3. Growth of Sweden's population 1750–1970.

Sweden over the 220 years from 1750 to 1970. In fact, these curves tell the population story for all countries since World War II. Figure 4 shows this for the period 1905–1973.

A. U.S. Statistics

For the United States, infant mortality rates, age-adjusted death rates, and birth rates are unusually illuminating. In 1900, the age-adjusted* death rate stood at 15.1/1000. That is, for every 1000 people, approximately 15 could be expected to die that year. By 1950, it had dropped to 8.4—a decrease of 44%, and between 1950 and 1986, another 36% to 5.4. Thus, between 1900 and 1986, the death rate had declined by a striking 64%. Less than 6 people of every 1000 can now be expected to die in a year. This dramatic decrease in mortality rate is shown in Figure 5. Bear in mind that much of this decline occurred well before the 1980s.

B. Infant Mortality Rates

Infant mortality rates (IMRs), always a sensitive barometer of environmental health, were no less remarkble in their downward trend. In 1900 the IMR was 162.4 per 1000 live births. By 1950 it had dropped to 29.2, and by 1986 it was at 10.4. Astounding is not an excessive superlative to describe this singular public health achievement. Figure 6 requires more than a hurried glance. It has much to tell us. True, 10.4 is not as low as

---

* Age adjustment is a means of accounting for the increasing aging of a population. As greater numbers of deaths can be expected in a population with a larger number of elderly, no matter how salubrious the environment, it is necessary to adjust for this, otherwise the rates among the elderly would swing the rate for the entire population upward toward its rate. This would distort the true picture of a community's death rate. To avoid this, crude death rates are age adjusted.

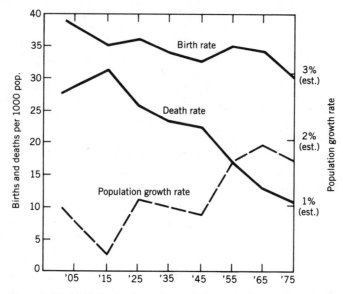

Figure 4. World birth, death, and population growth rates, 1905–1973.

Sweden's 6.2, but it may be all the more remarkable given the size and polyglot nature of the U.S. population. Sweden, with a shade over 8 million people, half that of New York State, and with an almost totally homogeneous population of white Lutherans living in three major population centers, is in fact a poor comparison for the United States. The Soviet Union, with an IMR of 30, may be better.

Table 1 shows the wide variation in IMRs to be expected around our country. Recall that the IMR for the United States is an average of the IMRs of the 50 states. The substantially higher rates in the south Atlantic

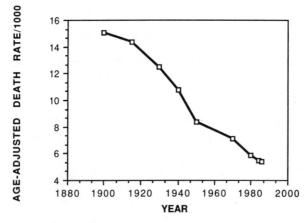

Figure 5. Age-adjusted death rates (per 1000 population) for 1900–1986.

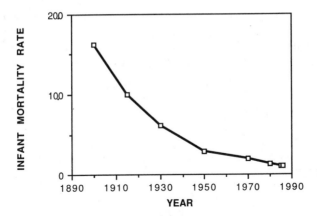

Figure 6. Infant mortality rate in the United States (per 1000 live births) for 1900–1986.

states and the District of Columbia suggest that health care policies may find their greatest impact here. It is surprising that Maryland and the District, so close to the halls of political power, have the country's highest rates. That the District, our nations capital, has the highest IMR in the

Table 1
Infant Mortality Rates Vary by Region

A. Ten States with IMRs Less than 10

Rhode Island	8.2
Vermont	8.5
North Dakota	8.5
Minnesota	8.8
Hawaii	8.8
Maine	9.1
Wisconsin	9.1
Kansas	9.3
Iowa	9.5
Nebraska	9.6

B. Ten States with IMRs Greater than 10

North Carolina	11.8
Louisiana	11.9
Delaware	12.1
Wyoming	12.2
Georgia	12.7
Mississippi	13.7
Alabama	13.7
South Carolina	14.2
Maryland	14.8
District of Columbia	20.8

nation, as well as one of the highest in the Western world, is both startling and depressing. On reflection, however, it is evident that the District and all the states below the Mason–Dixon line are heavily populated by blacks with exceedingly low incomes. Income levels below the poverty line ($12,500 per year) may be a primary reason for the far higher U.S. rate compared with Sweden or Japan, where low-income high-risk groups do not occur.

Furthermore, among white infants, the average IMR in 1986 was 9.3. For blacks it was 18.2. Even more revealing, the rate for white females was 8.0. Clearly, there is room for a substantial reduction of the current national IMR.

C. Population

For the most part, birth rates have been almost double that of death rates. In 1986, the birth rate was 15.5 per 1000 women. In 1950 and 1900 the rates were 76 and 104, respectively. Even with the purposeful decline in birth rates that began in the 1960s, births far exceeded deaths. Accordingly, total population had to increase. This says nothing about the effect of immigration. For the United States, it has always been positive. Many enter, few depart. Thus, another factor pushing population up is the steady and heavy legal and illegal entry of new people.

By 1988, the estimated total U.S. population was 245,302,000. And by the year 2000, it is expected to be just under 270,000,000. Figure 7 shows the trend in growth of our population between 1790, the year of the first census, and the anticipated level in the year 2050.

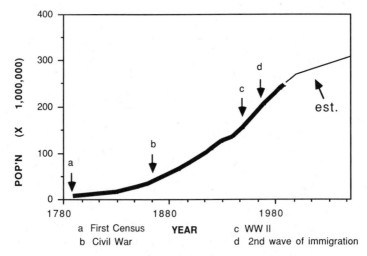

Figure 7. U.S. population for 1790–2050.

D. Aging

Another distinguishing characteristic of the U.S. population is its continuous aging. The median age, which was 26.9 years in 1970, increased to 31.7 in 1986. It is projected to become 33 by 1990 and 36.3 in 2000. Apparently, the environment is conducive to long and productive life.

Table 2 displays the median age of the population for 20 countries. The United States with a median age of 31.7 years (1986) is far from being one of the youngest but is fast approaching the older. To be an "old" country, the population under age 15 is less than 20%, while the population over 65 is above 10%. Note too that the "young" countries are the undeveloped or underdeveloped. "Young" really means that few people live on to the

Table 2
The Ten Youngest and Ten Oldest Countries

Rank	Country	Population Median Age	Percent of Population Under Age 15	Percent of Population Aged 65 or More
		Youngest Countries		
1.	Botswana	14.2	49.9	2.6
2.	Kenya	14.3	51.8	2.6
3.	Liberia	15.8	48.4	2.6
4.	Malawi	15.9	48.2	2.6
5.	Syria	16.0	48.0	2.9
6.	Nigeria	16.1	47.8	2.5
7.	Zimbabwe	16.1	47.7	2.7
8.	Algeria	16.3	47.1	3.3
9.	Ghana	16.4	47.1	2.3
10.	Libya	16.7	46.4	2.8
		Oldest Countries		
1.	Sweden	37.6	17.8	16.9
2.	West Germany	37.6	15.8	13.6
3.	Luxembourg	37.4	16.5	13.6
4.	Switzerland	36.5	17.6	13.7
5.	Denmark	35.7	19.3	14.7
6.	Italy	35.7	19.6	13.5
7.	United Kingdom	35.5	18.6	15.0
8.	Austria	35.5	18.2	14.1
9.	Belgium	35.3	15.8	13.4
10.	East Germany	35.2	18.7	14.2

Source: United Nations, Demographic Indicators of Countries: Estimates and Projections as Assessed in 1980 (New York: 1982), data for 1985.

older ages. Infant mortality is extremely high, and infectious diseases are a major public health problem. It is not until a country joins the "old" category that chronic diseases such as cancer become a major concern. The United States was a "young" country around 1900. The pattern of diseases it was concerned with then are the diseases the "young" countries are concerned with now.

In the next 20 years, there will be a much larger population of elderly both in absolute and relative terms. The number of people over 65, which amounted to 20 million in 1970 (9.8% of the total population) became 27 million in 1980 (11.9%) and is expected to reach 32 million in 1990. It will exceed 13% of the population by the year 2000. Seven states now have a million people 65 and over. In 1960, there were only three. Two more will shortly have that million. The nine states with their numbers of people over 65 are listed in Table 3. Displayed in Table 4 are the median ages of selected states, as well as the range of median ages. Florida with 36 is clearly the "oldest," and Utah at 25.5 is the "youngest." In terms of a maturing population, the number over 50 years may be illuminating. The large increases anticipated in this category are shown in Figure 8.

The population over 50 and 65 is one thing, the population over 85 is quite another. In March 1984, the National Center for Health Statistics (NCHS) made a startling announcement. Using a "declining mortality" assumption with 1978 as the base year, they predicted that the nation's 85 and older population would increase from 2.1 million to 6.7 million, or from 1% to 2.4% of the total population by the year 2003. In the 25 years from 1953 to 1978, the number of people 85 and older tripled from 700,000 to 2.1 million. But most important of all, the prediction of a tripling of this age group is "the result of improved health status." This statement is fraught with meaning and implication for us and must be given appropriate attention.

In 1900 there were only 3,080,000 people over 65. By 1930 that number had risen to 6,634,000 and represented only 5.4% of the population. It was about this time, as a consequence of the Great Depression, that discussion

Table 3
States with Largest Population 65 and Over
(July 1, 1983)

California	2,615,000
New York	2,223,000
Florida	1,867,000
Pennsylvania	1,639,000
Texas	1,470,000
Illinois	1,331,000
Ohio	1,250,000
Michigan	983,000
New Jersey	922,000

Table 4
Median Ages of Population for Selected States, 1986

Florida	36
New Jersey	34.1
Connecticut	33.9
Pennsylvania	33.8
New York	33.3
Utah	25.5
United States	31.7

of programs to deal with old age, survivors, and disability insurance began to be heard in the land. On August 14, 1935, with President Franklin Delano Roosevelt's signature, the Social Security Act became law. The growing national problem of old-age dependency was to be dealt with.

At the time, however, the biblical injunction of "three score and ten" as the years of life everyone could be expected to attain was little more than a prophetic vision. And because it was, planning and provision for a financially sound system was based on projections of the average longevity of a white male. Relatively few were expected to survive beyond 65 to collect their accumulated benefits!

The long, steady decline of the infectious diseases is shown in Figure 9. Although it describes the pattern of decrease in New York City, it is similar to the pattern of the entire country.

E. Life Expectancy

As noted earlier, by 1980 12% of the population was 65 and older. And the Social Security System was feeling the pressure as more and more

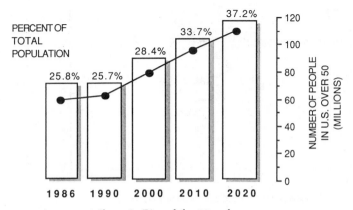

Figure 8. Rise of the 50 and over.

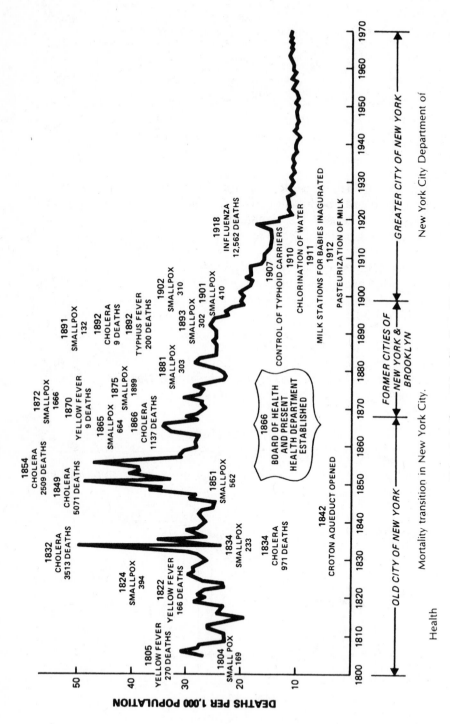

Figure 9. Is there "Mortality Transition"

Mortality transition in New York City.

Health
New York City Department of

people were not only arriving at retirement age but were doing so in a healthy, vigorous condition. By 1980, Congress was discussing and pressing for an increase in the age of retirement to 70. "Three score and ten" had been attained and passed in 1969. By 1986, it had climbed to just under 75.* That meant that a person born in 1980, on average, could be expected to live approximately 75 years. It is the "on average" that holds even greater meaning.

Breaking the 1986 average into its constituent parts we find the following:

White women	78.9
Nonwhite women	73.6
White men	72.0
Nonwhite men	65.5

Clearly, women have a greater life expectancy than men, but the difference between white women and black men is remarkble. White females in the United States are fast approaching the biblical four score years of life expectancy, and more than likely will attain it by the year 2000. Recall that this too is an average. Many white females already exceed that number. If as a group they continue to increase their life expectancy at the current rate of 0.186 yr/yr, they will have gone well over 80 by the year 2000.

If nothing changes their present rate of increase of 0.152 yr/yr, black men will not attain such length of days until the year 2081. Nevertheless, we are clearly witnessing an increase in life expectancy of all segments of the population. With aging and life expectancy rising to such unprecedented levels, is it reasonable to believe that our environment is hazardous to children and other growing things?

Increased life expectancy can be viewed from another, larger perspective. In 1775, at this country's inception, life expectancy was 45 years. In 1986, 211 years later, it had increased by 30 years. Little of that increase had occurred by 1900. Significant increases began to be seen in the 20th century—and toward the second half at that: the period most people would tag as the era of environmental pollution.

1900	47.3 years
1940	62.9 years
1980	73.7 years
1985	74.9 years

* In the Soviet Union, the average life expectancy in 1987 was 69 years.

But life expectancy is only one measure of the vigor of a people. The fact that so many are attaining 65 and over—anxious to continue working and quite able to do so—with little loss in creativity and mental acuity, is perhaps far more meaningful.

At an international meeting in Washington, DC, in May 1987, Alvor Svanborg of the University of Goteborg, Sweden, showed that 70-year-olds today are healthier, more vital, and "intellectually significantly more capable" than 70-year-olds of a decade ago (1).

II. PATTERNS OF DISEASE

Perhaps even more to the point would be insight into the current patterns of disease in the United States. Just what are the conditions with which people must contend, and again for our purposes, are they related to a polluted environment? An examination of the current leading causes of death seems an appropriate and realistic point of departure. Table 5 displays the major causes of death in the United States for 1985, and Table 6 compares 1985 with 1900. Figure 10 again shows the leading causes of death in chart form. It is evident that infectious, communicable diseases have given way to a set of stubborn, noncommunicable disorders which can be grouped as failures of organ competence or performance, consequences of trauma, and excessive exposure to alcohol.

A. Cancer

Although deaths from heart disease are not quite double the number of cancer deaths, it would be difficult to find a group of 100 people wherein the majority did not believe cancer to be the major health problem in the

Table 5
Leading Causes of Death in the United States, 1985 (per 100,000 Population)

Rank	All Causes 873.9	100.0%
1. Diseases of the heart	323.0	37.0
2. Malignant neoplasms	193.3	22.1
3. Cerebrovascular diseases (stroke)	64.7	7.4
4. Accidents	39.1	4.4
5. Chronic obstructive lung disease	31.3	3.6
6. Pneumonia/influenza	28.3	3.2
7. Diabetes	15.5	1.8
8. Suicide	12.3	1.4
9. Liver disease/cirrhosis	11.2	1.3
10. Atherosclerosis	10.0	1.1
All others	145.8	16.7

Table 6
Leading Causes of Death in the United States, 1900 and 1985 (Percentage of Total)

1900		1985	
Pneumonia/influenza	11.2%	Heart disease	37. %
Tuberculosis	11.2	Cancer	22.1
Heart disease	9.4	Stroke	7.4
Stroke	7.6	Accidents (all types)	4.4
Diarrhea/enteritis	6.3	Chronic obstructive pulmonary disease	3.6
Nephritis	5.9	Pneumonia/influenza	3.2
Cancer	4.5	Diabetes	1.8
Accidents	4.2	Suicide	1.4
Diphtheria	1.9	Liver disease	1.3
Other	37.2	Atherosclerosis	1.1
		All other	16.7
	100.0%		100.0%

United States. Certainly, it is the most feared of diseases. Given that perception, let us look at it first.

Is cancer one disease or many? If it is one, the search for a causative factor should be relatively easy. If, as suspected, cancers of various anatomic sites are the consequences of different insults, then multiple causes must be sought. Thus, the search must be for multiple needles in multiple haystacks. Surely cancer of the lung, for example, is not solely the result of excessive smoking.

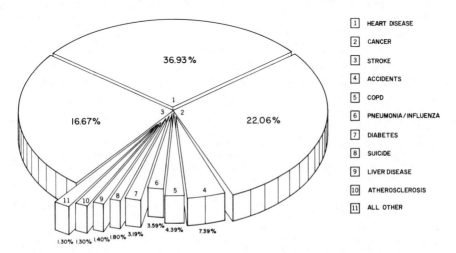

LEADING CAUSES OF DEATH, 1985

Figure 10. Leading causes of death in the United States, 1985.

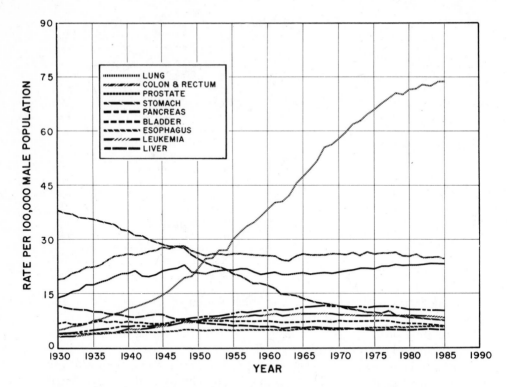

Figure 11. Male cancer death rates by site for the United States, 1930–1985. Rate for male population standardized for age on the 1970 U.S. population. Data from the National Vital Statistics Division and Bureau of the Census, United States.

Figures 11 and 12 portray the secular trends and the age-adjusted* death rate per 100,000 people for cancer of eight major sites for men and women.

By examining these long-term trends, insights into their causes may be forthcoming. For example, a first impression of the trend for women could well suggest, contrary to popular opinion, that cancer does not appear to be increasing. In fact, it has remained fairly constant for 50 years. That has to be a remarkable and contrary view. The largest majority of any 1000 people would never believe so heretical an idea.

* As cancer is a disease of old age, an aging population can be expected to yield more cancers. To account for these age-related cancers, and not be mislead into believing other risks are at work, cancer rates are age adjusted. That is, the rate for any year is calculated by applying the age-specific mortality rates to the age distribution of a standard reference population, such as the U.S. population in 1940 or 1960. The rate obtained represents a hypothetical value that would have been observed if the 1980 or 1988 population was similar in age distribution to the reference population. This means that had the age distributions been similar, rates also would be expected to be similar or risk factors other than age had not been present or active. If, after age-adjustment, an excess remains, there would be justification for expecting something other than age as being involved or responsible for the increase. The current excess can thus readily be explained as lung cancers due to smoking, not some unknown environmental risk(s) abroad in the land.

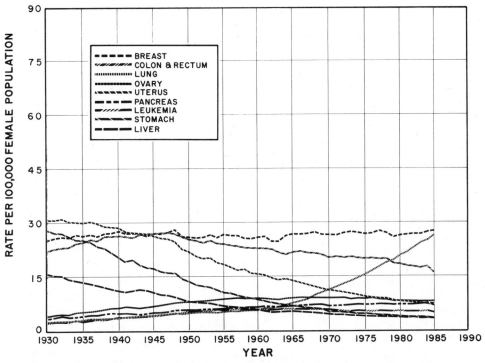

Figure 12. Female cancer death rates by site for the United States, 1930–1985. Rate for female population standardized for age on the 1970 U.S. population. Data from the National Vital Statistics Division and Bureau of the Census, United States.

Closer inspection reveals that a number of the major cancers have been decreasing over time. Cancer of the stomach and uterus have been declining for longer than most of us reading this page are old: stomach for 50 years and uterus for about 40, long before the "Pap test" was introduced. The most striking upward trend is for cancer of the lung. Obviously, lung cancer must be the result of a different risk factor than either stomach or uterus. The same pollutant could not be expected to increase one while decreasing another: fairly good evidence for multiple causative factors—the multiple haystacks again. The question to bear in mind is whether lung cancer is a consequence of polluted air or other factors.

Breast cancer rates have remained constant for 50 years. Has the environment? Is the United States in 1980 or 1988 the same or similar to the United States circa 1930 or 1920? Hardly. Remember, the cancers observed in 1930 were most likely the result of risk factors acquired 20–30 years earlier—when the United States was truly an agricultural country. For breast cancer, or any other for that matter, is it of any import whether the United States is an industrial or agricultural society?

In a recent review of ionizing radiation's contribution to cancer mortality, Jablon and Bailor of the National Academy of Sciences and the

National Cancer Institute, respectively, remarked that "although ideas about the effects of small doses are largely speculative, a great deal is known about the effects of large doses . . . from studies of Hiroshima and Nagasaki, repeated diagnostic X-rays and from laboratory experimentation." They go on to say that "it is perhaps ironic that despite this knowledge, and in the face of widespread public concern and apprehension about radiation-induced cancer, only a very small part of the cancer burden which affects humanity—less than 3%—may actually be attributed to radiation . . . more than half of this hypothetical burden is caused by radiation that we can not do everything about—that resulting from natural background. And finally, if our object is to lessen appreciably the impact of environmentally induced cancer, ionizing radiation is not a promising target (2). (For more detailed discussion of ionizing radiation see chapter 9.)

Clearly, in our overzealous concern for environmental pollution, as the insult seen as most directly related to community morbidity and mortality, we have lost sight of the target, which is underfoot. If our concern really is for the health of the community, we have been running in the wrong race. It is perhaps time to take note of this. As time permits, we might ask how this occurred, but that is of far less urgency. The incisive questions should be: How do we proceed? Where do we begin?

There is yet another message in these many figures. It would appear and eminently so, that if effort and resources are to be appropriately utilized to effect reductions in death and disability, the environment with which we ought to be concerned is the microenvironment—our lifestyles. That is where the problem lies. To place our strongest efforts elsewhere would be mistaken and nonproductive, if it is quality of life and life span that really concern us. But this means individual involvement and individual willingness to make changes, and that may be another matter entirely.

Similarly with cancers diagnosed in 1980, the "20-year rule" applies. It would be a mistake to look to medical care for a suitable explanation of this trend. Medical care has yet to make a dent in the death rates.

Given the levels of air and water pollution, food additives, radiation, pesticides, and hazardous and toxic wastes, and given the long-term decline or steady-state of cancer deaths, there is more than a little difficulty in reasonably ascribing one to the other.

Looking at the rates and trends for men, the overall impression is much the same, except for the astounding increase in lung cancer. But there is little problem here. Cigarette smoking notably accounts for 90% of all lung cancers, and 25% of all cancers. Imagine, a quarter of all cancer, 25%, could be effectively done away with simply by cessation of smoking. In February 1985, the American Cancer Society announced that for the first time more women would die of lung cancer than breast cancer. Blame was placed squarely on increased and excessive smoking by women since World War II.

A particularly startling revelation is the finding that if lung cancer deaths were to be subtracted from the total number of cancers (all sites), the trend for cancer in the United States would be downward. Cancer has not been rocketing uncontrollably upward as the media and most people would have us believe.

But what of the 10% of lung cancers not attributable to cigarette smoking? The evidence not only incriminates asbestos for a portion of this, but especially workers who smoke. Compared to asbestos workers who do not smoke, those who do have a tenfold greater risk. The relationship is multiplicative, not additive. (For more on asbestos, see Chapter 8).

Increased risk of lung cancer has also been strongly associated with such heavy metals as nickel, chromium, and arsenic and with ionizing radiation. There is little question that people differ in their susceptibility to these insults. Not everyone who smokes develops cancer. It is also evident that dose plays a major role. It is not an all or none phenomenon. The probability of lung cancer increases sharply above 20 cigarettes per day.

High fat consumption has been implicated in the etiology of breast cancer. The mechanism by which this occurs is not currently known, but hypotheses envisage alteration of hormone metabolism promoting tumor growth and development (3).

There is yet another salient feature about cancer. I refer to future expectations about its magnitude. Cancer is a disease of the older ages. This does not mean that children do not get cancer. Unfortunately, they do. But for the majority of people, cancer appears during the fifth, sixth, and seventh decades of life. Considering that life expectancy is increasing, that the population is aging, and that our population will continue to increase, one can anticipate that even with a pristine environment, with every risk factor neutralized or removed, cancer will increase. It must. And it will continue until such time as oncologists, physiologists, and biochemists understand the mechanism, the cellular and molecular basis for the malignant transformation that allows cells to proliferate uncontrollably. Only after that is thoroughly understood will there be opportunity for reversing the process. Until that time, probably not before 1995, we can expect the trend lines for cancer to continue in their current directions. The fact that lung cancer persists in its exponential rate of increase testifies to the intransigence and perverseness of people. Most fear and detest cancer, yet knowing the cause of the most pernicious of all cancers, they refuse to employ the absolute preventative—cessation of smoking.

To further compound this refractory behavior, people persist in looking for cancer in all the wrong places. Food additives are an instructive example (see Chapter 3). For years, food additives have been attacked as though their elimination would banish cancer. Thus, we have a situation in which people willfully refuse to prevent a cancer responsible for the yearly death of over 100,000 people, preferring to speculate on the possible or imagined carcinogenicity of substances whose use in foods has not been known to have produced a single human cancer.

Look again at Figures 11 and 12. Stomach cancer in both men and women has had the most substantial declines over the past 50 years, a period of increased use of food additives. Colon and rectal cancers have been declining in women since the late 1940s and have remained fairly constant in men for the same period. If food additives are cancer provoking, they have proved to be the most elusive of culprits. Even more elusive is comprehending the behavior of people.

1. The Environment as a Factor

This may be the appropriate point at which to ask what we mean by environment. The *Oxford American Dictionary** is brief in its definition, stating that environment is "surroundings, especially those affecting peoples lives." The *Random House Dictionary of the English Language* (unabridged)† goes a shade further and adds "conditions or influences especially affecting the existence or development of someone." Those are general, all encompassing definitions and may be just what are needed. It may be that visions of the environment have been too restricted: especially so if we are concerned with which haystack to search in for those slippery "needles" we call risk factors.

While environment may denote the human-devised physical and chemical environment of the industrial revolution and its evolution into a "high-tech" society, it is becoming more inclusive, referring to any influence other than that of the genetic material inherited from one's parents. Given the facts of smoking, and its effects on the body, motor vehicle accidents and their effects, alcohol, diet, drug abuse, and a sedentary life-style, any definition of "environment" should include our close personal and individual microenvironment(s) as well as the larger global (physical) environment.

If we include the microenvironment of life-style, answers to environmental problems that have heretofore defied solution become more reconcilable. Failure to appreciate the different senses in which environment is used has led to misinterpretations of statements concerning the proportion of cases environmentally induced.

John Higgenson's remark that 60–90% of all cancers were environmentally determined was misinterpreted for some 25 years. During the 1960s, Higgenson, the founder and director of the World Health Organization's International Agency for Research on Cancer, compared the incidence of various cancers among African and American blacks and concluded that about two-thirds of all cancers had an environmental component and were therefore amenable to preventive measures. Higgenson had no idea

---

* Oxford University Press, New York, 1980.
† Random House, New York, 1967.

how that statement would be used and abused over the ensuing quarter of a century.

Recently, Higgenson stepped forward and clarified misinterpretations of his words. "When I used the term environment," he explained, "I was considering the total environment, cultural as well as chemical" (4). His meaning and message were clear. If we meant to find preventive solutions to the chronic illnesses that ail us, we must look to ourselves. Chemicals in air, water, and food may only be a minor part of the total problem.

One further piece of information may be useful in placing the cancer–environment dilemma in clearer perspective. We may, for example, glean some interesting clues from an international comparison of cancer rates.

From Figures 13 and 14, we learn immediately that, for men, Luxembourg (of all places) has the highest combined cancer rate. Could it be that their exceedingly high lung cancer rate has carried all other rates upward along with it. Statisticians refer to this as skewing.

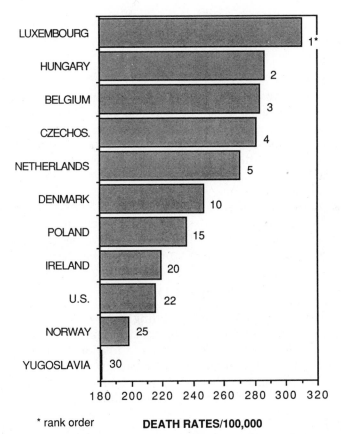

Figure 13. International comparisons of age-adjusted death rates for cancer—all sites (male), 1986.

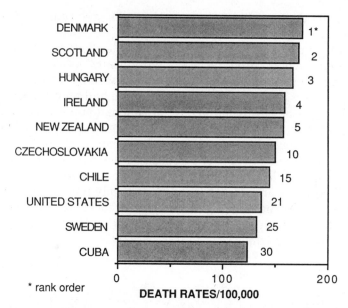

Figure 14. International comparisons of age-adjusted death rates per 100,000 for cancer—all sites combined (female).

Another glance and we see that the United States is far from being a front-runner with respect to cancer. For men and women alike, we rank among the countries with the lowest rates. Perhaps that comes as another surprise. I would have some reservations about data gathering in Yugoslavia. The low rate is a little suspect considering the rates in Hungary and Czechoslovakia. Other than that, all the countries included have well-established reporting systems.

Dwelling on possible reasons for the differences among these countries, as well as breaking out the rates for specific sites, may be unusually rewarding. The point, however, is to try to determine the relationship between cancer of any site and environmental risk factors.

B. Heart Disease

Coronary heart disease (CHD) is the leading cause of death in the United States. The average, apparently healthy, American man stands a 1 in 5 chance of suffering a myocardial infarction* and about a 1 in 10 chance of dying of CHD before the age of 60. Men are 20–25 times as likely of sustaining a "heart attack" as women of comparable age. These same grim statistics apply to most countries in western Europe, Canada, Australia,

---

* An infarct is an area of necrosis produced by the sudden reduction in coronary blood flow to a segment of myocardium.

and New Zealand. On the other hand, Polish men are more likely to incur CHD than American men, except for American black men, whose death rates from CHD are among the highest in the world.

Do these comparative figures suggest that clues to prevention and control should be sought in the macro- or microenvironment? To "muddy the waters" further, a decline in deaths from CHD has been evident to epidemiologists since the 1960s. In fact, the decline attained substantial proportions, approximately 25% between 1968 and 1980.

A number of possible or potential factors have been advanced to account for the decline. Among these are:

- Reduction in smoking (cancer is not the only illness smoking has an impact on)
- Decrease in high cholesterol foods.
- Improved medical care—specialized rescue teams and CCUs (coronary care units).
- Increased exercise.
- Health education campaigns.
- Increased detection and treatment of hypertension.

Each of these should be scrutinized to ascertain whether it can in truth account for the decline, either alone or in combination. For example, given the increases in lung cancer among both men and women, is it likely that decreased smoking can stand up as a factor? What about increased exercise? When did the country take to the streets with the jogging mania? An estimated 30 million runners pound their way from dawn to dusk through city streets and along country roads seeking fitness and health. Are these the high-risk groups? The period 1968 to 1980 seems a bit premature for "Nautilus" and "Nikes" to have had an effect. Each of the other factors has similar problems. No single one, or combination, appears to fit. Is there less stress in our lives? Hardly. In fact, stress at work is a growing national concern.

For the past two decades much has been written about a possible link between CHD and hardness of the drinking water. Those who support the association propose that the relationship may be due to specific trace elements such as calcium and magnesium which could affect homeostasis. Critics of the association have argued that the proportion of total mineral intake from water is far too small to produce the types of biological changes that lead to myocardial damage (5,6), and animal feeding studies offer no support for such a relationship.

At this point, scientists are unable to explain the decline. If we are doing something right, it would be nice to know what it is. Whatever it turns out to be, the environment, as we usually conceive of it, seems not to be implicated.

C. Cerebrovascular Illness

Cerebrovascular accidents (CVAs), organic disease of the brain, account for approximately 7% of all deaths in the United States. In 1985, 65 of every 100,000 deaths were due to CVAs. In 1975, the figure was 91.8, a decline of some 29%, quite a substantial decrease.

Various terms are used synonymously for CVA, including apoplexy, shock, and stroke. It is the most common cause of neurologic disability in the Western world.

The brain, of course, is an uncommonly sensitive organ and cannot function without copious amounts of oxygen. As such, the brain consumes some 20% of the total cardiac output to satisfy its thirst for oxygen-rich blood. Most CVAs are the result of atherosclerotic disease. Again, if the environment contributes, it must be related to hardening of the arteries or increases in blood pressure. At this time, evidence of that, although scanty, suggests nutritional excesses and a sedentary life-style. More likely, genetic effects may be contributing factors.

Although not classified as a CVAs, a number of degenerative diseases beset the central nervous system. Alzheimer's disease, characterized by premature mental deterioration, is one of these. Recall that acid rain, with its release of aluminum salts from the soil into waterways, may be directly responsible for the unusual pile-up of aluminum in brain tissue. Current evidence indicates that the high concentrations of aluminum in brain tissue of patients with Alzheimer's disease may be nothing more than a fortuitous association (7).

Pellegra, a nutritional deficiency in which individuals do not obtain sufficient niacin (one of the B vitamins), can produce a form of dementia. The dementia or psychosis results from degeneration of neurons within the cerebral cortex. Fortunately, supplying the necessary niacin in the diet can reverse the process.

D. Chronic Obstructive Pulmonary (Lung) Disease

Chronic obstructive pulmonary or lung disease (COPD/COLD) is a generalized airway obstruction associated with varying degrees of chronic bronchitis and emphysema. Less than 4% of all deaths are due to COPD.

Emphysema is seen as a progressive loss of elasticity and eventual rupture of the alveoli and is usually accompanied by labored breathing, a husky cough, and labored heart action. It is the most common form of COPD and results in more disability than any other pulmonary disorder. Typically, the condition occurs in cigarette smokers past the age of 40. As women continue their smoking, emphysema will become part of their lives as well. Indeed, this is a consequence of an environmental pollutant—one we unnecessarily inflict upon ourselves.

E. Accidents

As detailed in Chapter 6, accidents, especially by motor vehicle, are the fourth leading cause of death in the United States, contributing 4.5% of all deaths. Unquestionably, environment plays a major role, but it is primarily related to the microenvironment.

The unholy diad of alcohol and tobacco appears to be responsible for more death and disability than any other toxins combined. Think of it. Tobacco and alcohol are responsible for a major proportion of cancer, a substantial amount of heart disease, accidents, suicide, homicide, chronic liver disease, and congenital anomalies. It is not necessary to go far afield to reduce this country's burden of death by 50%. It boggles the mind that so vast a reduction can be achieved by attention to these two risk factors.

F. A Variety of Other Causes

Together, diabetes, suicide, cirrhosis, homicide, and atheroslcerosis are responsible for approximately 7% of all deaths in this country.

Although diabetes contributes less than 2% of the deaths, between 2 and 3% of the entire population has one form or another of "sugar" diabetes, diabetes mellitus. Diabetes develops when the hormone insulin fails to control blood sugar adequately. How or why this occurs has remained a medical mystery for more than 2000 years. It is not new to civilization and certainly not a consequence of industrialization.

Suicide, homicide, cirrhosis, and atherosclerosis have also been associated with civilization for long years. Homicide began with Cain and Abel and the end is not yet in sight. Cirrhosis is directly related to alcohol consumption. Of course, the wine-producing countries of the world have the highest rates, but the nongrape countries strive mightily to catch up.

Recently, Kenneth G. Manton of Duke University found that white men are at highest risk of suicide, and that among them suicides are highest at ages 45–50 and again at 80. Among white women the suicide rate peaks between ages 45 and 55—about the age of menopause—and the time children leave home. For black men, the suicide rate is highest at age 30 and declines slowly through late middle age. Curiously enough, suicide rates for black women do not appear to have any regular pattern (8).

If there is a message in the leading causes of death, it is that smoking, alcohol, drug abuse, fatty foods, and a sedentary life-style are the real enemies of the people. Nevertheless, if we take the long view, it does appear that we are becoming a healthy people in spite of our efforts to the contrary.

It may be of more than passing interest to learn that at no time did AIDS—auto immune deficiency syndrome—join the list of major causes of mortality. At no time did AIDS displace any of the top 20 causes of death—although the response of the communications media made it seem as though it had. From the start, however, AIDS had a political constituency that was able to mobilize action for it, at a time when many other illnesses, with higher death rates, could not do so. Recall that over the period 1981–1988, AIDS was responsible for less than 5000 deaths per year.

Perhaps the real health problems in the United States today, and for the foreseeable future, are the dismal findings that 40% of boys ages 6–12 cannot do more than one pull-up and 25% cannot do any. Imagine, fully one-quarter of all boys do not have the muscle power to pull their weight once!

Girls are not much better. Seventy percent of all girls cannot do more than one pull-up and 50% cannot do any. Half of the girls age 6–17 cannot run a mile in less than 10 min, nor can 30% of the boys.

Forty percent of the boys cannot touch their toes without bending their knees.

Much of this poor physical condition is attributed to inactivity—a sedentary life-style. Apparently, much of their time is spent watching TV; up to 40 h/wk in some cases, while munching all manner of tasty goodies, gaining weight while losing muscle.

III. PREVENTION: A PAINLESS APPROACH TO HEALTH

There is yet another way of looking at the condition of health of our contemporary communities. A reasonable and rational individual ought to be able to ask how much of the illness and death in any city or town is in fact unnecessary and can, with appropriate attention, be removed or substantially reduced?

In November 1984, a national conference dealing with this issue was convened at the Carter Center of Emory University (9). Involved were representatives of public, private, voluntary, and academic institutions. Their focus was the health concerns that would be patently unnecessary if currently available knowledge were put to use.

They found that the 14 conditions listed in Table 7 accounted for some 70% of all hospitalization days, 85% of direct expenditures for personal health, 80% of all deaths, and 90% of potential years of life lost prior to age 65. Approximately 66% of the total mortality in the United States today is due to potentially preventable causes (9). This is an astonishing statistic immensely worthy of pause and consideration. Such a statistic must be pondered in light of the hue and cry over untoward effects of environmental pollutants on human health. Note carefully the conditions contained in the table. Which of them are of a type or category generally equated with or considered to be affected by an environmental pollutant?

Consider too that fully 1.2 million deaths result and 8.4 million years of

Table 7
Health Problems for Which Preventive Measures Are Available

Infant mortality	Respiratory diseases
Infectious diseases	Dental diseases
Arthritis	Homicide/suicide
Drug abuse	Cancer
Alcoholism	Depression
A variety of injuries	Diabetes
Coronary heart disease	Digestive diseases

life are lost before age 65. A reduction of the magnitude noted in both mortality and morbidity would dissipate the current burden of ill health: a burden defined in medical, economic, and human terms. The staggering amounts of resources saved would easily find application for other pressing community needs.

The immediacy of such a course, as well as the motivation for its pursuit, is nothing less than the availability of knowledge. Additional information need not be gathered. We have what is needed. Implementation is what is required.

REFERENCES

1. C. Holden, Adjusting to an Aging Population. *Science* **236**:772, May 15, 1987.

2. S. Jablon and J. C. Bailor, The Contribution of Ionizing Radiation to Cancer Mortality in the United States. *Prev. Med.* **9**:219–226, 1980.

3. C. W. Welsch, Can Dietary Fat Influence Hormone-Induced Growth Responsiveness of the Mammary Gland? *Preventive Med.* **16**:475–480, 1987.

4. J. Higgenson, Present Trends in Cancer Epidemiology. *Proc. Can. Cancer Conf.* **8**:40–75, 1969.

5. S. Heyden, The Hard Facts Behind the Hard-Water. Theory and Ischemic Heart Disease. *J. Chronic Dis.* **29**:149–157, 1976.

6. National Research Council: Water Hardness and Health. In *Drinking Water and Health.* National Academy of Sciences, Washington, DC, 1977, pp. 431–447.

7. M. A. Benarde, Health Effects of Acid Rain: Are There Any? *J. R. Soc. Health* **104**(4):139–145, 1987.

8. K. G. Manton, D. G. Blazer, and M. A. Weber, Suicide in Middle Age and Later Life: Sex and Race Specific Life Table and Cohort Analysis. *J. Gerontol.* **42**(2):219–227, 1987.

9. W. H. Foege, R. W. Amler, and C. D. White, Closing the Gap. Report of the Carter Center Health Policy Consultation. *J. Am. Med Assoc.* **254**(10):1355–1358, 1985.

SUGGESTED READINGS

1986 Annual Cancer Statistics Review. National Cancer Institute, Division of Cancer Prevention and Control, December 1986. U.S. Department of Health and Human Services, NIH, Bethesda, MD 20892.

Gwatkin, D. R., and Brandel, S. H. Life Expectancy and Population Growth in the Third World. *Sci. Am.* **246**(5):57–66, 1982.

Health United States, 1987, DHHS Pub. (PHS) 88-1232 U.S. Department of Health and Human Services Hyattsville, MD, March 1988.

U.S. News and World Report. What the Next 50 Years Will Bring. Special Supplement, May 9, 1983.

U.S. Bureau of the Census. Projections of the Population of the United States: 1982 to 2050 (Advance Report). Current Population Report, Series P-25, No. 922, 1982. World Health (WHO) Population and Health June 1984.

15

Risk Assessment, Communication, and the Politics of Pollution

The great danger today is of slogans, collective opinions, ready-made trends of thought. We have to be able to resist individually, to criticize, to distinguish between what is proven and what is not.

—Jean Piaget

Hotspur was right. It is out of the nettle, danger, that we pluck the flower, safety.

—Peter Huber

Chapter 14 contains some of the truly remarkable demographic advances that have been made over the past 50 years. Life expectancy has never been higher, nor longevity longer. Illness and death have declined significantly. Yet there is despair over perceived dangers to life and health from environmental degradation. Radiation, pesticides, industrial chemicals and food additives, and contamination of air and water are held to be life-threatening risks. Quantitatively these should be of least concern. Smoking, homicide, suicide, motor vehicle accidents, and alcohol and drug abuse are the proven toxins of our time. These are the risks people take unto themselves and to others. But for reasons not yet clear, the public and the scientific "community" are clearly "out of synch" with respect to their estimations of real risks to life. Each has a different list.

Is it possible that the thoroughly documented advances in human health are a well-kept secret? Is it possible, given our vast communications network, that these advances have remained unknown to a vast majority of the public? Are these advances and perceptions of ill effects of environmental pollution regarded as mutually exclusive parallel phenomena? Or is something else at work? Do people want to know, to understand the numbers? In this case the "body count" is entirely favorable. Would knowing destroy a grand illusion? Yet the people need to know.

Toward the end of Chapter 13, I raised several questions about hazardous waste and its potential for adverse health effects. How is this information to be conveyed to a community? What will it take for them to believe it? Will they believe it, and, furthermore, who will be the messenger?

Informing the public is at the heart of all our environmental issues. When William Ruckelshaus returned to the EPA in 1983, he believed the public needed to know and needed to participate in policy decisionmaking. In a widely heralded speech he resurrected an idea Thomas Jefferson believed to be at the core of a democracy. "If we think the people are not enlightened enough," Jefferson and Ruckelshaus remarked, "to exercise their control with a wholesome discretion, the remedy is not to take it from them, but to inform their discretion" (1).

Twelve stormy months later he was not so sure. Harkening back to Jefferson at a conference one evening at Princeton University, a chastened Ruckelshaus opined: "Easy for him to say. As we have seen, informing discretion about risk has itself a high risk of failure" (2).

What is the problem? Do people really want to share in community decisionmaking? Given the complexity of the problems currently at issue, are they able to participate? Is the problem intractable? Are people asking for more than can realistically be delivered?

The precipitous change from an agricultural to a highly technological country following World War II seems not yet to have been fully absorbed. Too many people are educationally unprepared for the new high-tech world. Interest in science, mathematics, and engineering, never high, are barely on anyone's wish list. Our educational system (some call it a wasteland) has allowed students to graduate without even confronting the biological or chemical world. Somewhere along the way the idea of a risk-free life, absolutely safe from harm, has laid hold of people's minds. And during the 1960s and 1970s there arose a relentless hostility to both government and the corporate world. To say that a substantial erosion of trust has occurred would not be an overstatement. Once trust is lost, it is extremely difficult to retrieve.

It is into this uncertain world that regulatory decisionmakers must thread an uneasy passage in their attempt to quantify risks of a broad spectrum of environmental and occupational exposures.

Figure 1.

Attempts to quantify risks have engendered the new discipline of risk assessment. Figure 1 describes in detail the range of Risk Assessment activities.* Test animals are fed the substances under investigation. Results are obtained, and a decision about the safety of the source is rendered.

In a few areas the public may define the risks and benefits fairly easily. For example, epidemiological and clinical studies have identified a

---

* The figure should pose difficulties for those trying not to smile. If ever a sense of humor was needed, now is the time.

number of activities and conditions related to cardiovascular disease and have estimated their severity. Heavy cigarette smoking, elevated blood pressure, and elevated cholesterol are the three major risk factors originally identified in the Framingham Study* and since confirmed by others.

Among people who do not smoke and have neither elevated blood pressure nor elevated cholesterol, the average death rate from cardiovascular disease per 1000 individuals is 13. Among those with one elevated risk factor, the death rate is 23. With two elevated, it is 44, and with three, it is 82: more than 64-fold higher than with none. The "risk" side of the equation is fairly well documented. The risk of death from cancer would, of course, increase the risk side even more.

Individuals must decide for themselves whether perceived *benefits* of smoking outweigh the risks. Similarly, people must also decide whether the perceived benefits of their other numerous actions that may contribute to high blood pressure and cholesterol levels (eating an excessively rich diet, not exercising, etc.) outweigh the risks.

Not all risk–benefit decisions are as easily assessed. The risk of death from cancer as a consequence of DDT residues in food cannot be calculated because there has not been a documented case of such cancer. There is experimental evidence that DDT induces cancer in mice, but none for people, even for industrial workers exposed for long periods to levels much higher than generally encountered. Nonetheless, the risk of death from many activities can be calculated. Tables 1–4 each present daily risks. Table 1 indicates the increase in risk of death by one in a million if a certain activity is performed.† Table 2 lists similar types of activity in terms of days of life lost by performing these actions, while Table 3 shows risks in perspective. Table 4 describes the risks in terms of number of deaths per 100,000 per year. Note that once past auto accidents, the numbers decline rapidly.

Interestingly enough, when people were asked to rate every day risks, the more common were grossly underestimated, while the more rare were overestimated. Figure 2 shows this curious dichotomy.

I. PRELUDE TO RISK ASSESSMENT

"The subject of risk in a complex society involves a set of interrelated questions about how much risk we should be prepared to accept as individuals and how much risk a society should be allowed to impose on,

---

* The U.S. Public Health Services Framingham Study investigated the incidence of coronary heart disease in a group of men and women aged 30–60, in Framingham, Massachusetts. The study began in 1949 and is still in progress.

† A cancer risk of 1 in 1,000,000 per year, if exposure is nationwide and if all people are equally exposed, and are equally susceptible, would result in 240 excess deaths in a population of 240,000,000.

Table 1
You Will Increase Your Chance of Death by One in a Million If You . . .

Risk	Cause of Death
Spend 1 h in a coal mine	Black lung disease
Spend 3 h in a coal mine	Accident
Travel 6 min by canoe	Accident
Travel 10 min by bicycle	Accident
Travel 300 mi by car	Accident
Travel 1000 mi by jet	Accident
Travel 6000 mi by jet	Cancer caused by natural radiation
Live 2 months in Denver on vacation from New York	Cancer caused by natural radiation
Live 2 months in an average stone or brick building	Cancer caused by natural radiation
Have one chest x-ray taken in a good hospital	Cancer caused by radiation
Eat 40 tablespoons of peanut butter	Cancer caused by aflatoxin
Drink Miami drinking water for 1 year	Cancer caused by chloroform
Live 5 years at site boundary of typical nuclear power plant	Cancer caused by radiation
Drink 1000 24-oz. soft drinks from recently banned plastic bottles	Cancer from acrylonitrile monomer
Live 20 years near a PVC plant	Cancer caused by vinyl chloride (1976 standard)
Live 150 years within 20 mi of a nuclear power plant	Cancer caused by radiation
Eat 100 charcoal broiled steaks	Cancer from benzopyrene
Live within 5 mi of a nuclear reactor for 50 years	Cancer caused by radiation from large accident
Work for 1.5 weeks in typical factory	Accident
Rock climb for 1.5 min	Accident
Spend 20 min being a man aged 60	Mortality from all causes

Source: Based on a table in Analyzing the Daily Risks of Life by Richard Wilson in Technology Review, Feb. 1979.

or remove from, individual choice."* If this is true, then it follows that the decisionmaking process is in large measure a political one, involving many disparate opinions, often contending interests.

It is into this often contentious arena that scientific data, often incomplete, often inadequate, meet economic and sociopolitical realities. And let us be clear about it initially: scientific data by itself has little community or societal application. It takes on meaning when it becomes part of public policy. Consequently, scientists bear a heavy responsibility,† if the public is to be served.

---

* In 1979, the Marsh and McLennon Companies, Inc., a leading insurance broker in the United States, asked Louis Harris and Associates to conduct a public opinion survey to ascertain attitudes about risk in a complex society. The quote was contained in the introduction to the survey document.
† This is as it should be, but whether it is always the case will be discussed further along.

Table 2

Average Life Expectancy Reductions Associated with Various Risks

Risk	Average Days of Life Lost
Cigarette smoking—male	2250
Being 30% overweight	1300
Being a coal miner	1100
Cigar smoking	330
Pipe smoking	220
Dangerous job—accidents	300
Motor vehicle travel	207
Alcohol (U.S. average)	130
Legal drug misuse	90
Average job—accidents	74
Job with radiation exposure	40
Being a pedestrian—accidents	37
Safest jobs—accidents	30
Illicit drugs (U.S. average)	18
Exposure to natural radiation	8
Medical x-rays	6
Oral contraceptives	5
Nuclear reactor accidents	0.02–2
Exposure to radiation from nuclear industry	0.02

Source: Based on a table by Bernard L. Cohen and I-Sing Lee, A Catalog of Risks, *Health Phys.* **36:** 707, 1979.

Table 3

Annual Risk of Death from Selected Common Human Activities[a]

	Number of Deaths in Representative Year	Individual Risk/Year
Coal mining		
Accident	180	1.3×10^{-3} or 1/770
Black lung disease	1,135	8×10^{-3} or 1/125
Fire fighting	—	8×10^{-4} or 1/1,250
Motor vehicle	46,000	2.2×10^{-4} or 1/4,500
Truck driving	400	10^{-4} or 1/10,000
Falls	16,339	7.7×10^{-5} or 1/13,000
Football (averaged over participants)		4.10^{-5} or 1/25,000
Home accidents	25,000	1.2×10^{-5} or 1/83,000
Bicycling (assuming one person per bicycle)	1,000	10^{-5} or 1/100,000
Air travel: one transcontinental trip/year		2×10^{-6} or 1/500,000

[a]Selected from X. Hutt *Food, Drug, Cosmetic Law J.* **33,** 558–589, 1978.

Table 4
Some Common Risks to Life in the United States[a]

Causes of Death	Number of Deaths per Year	Number of Deaths per Year per 100,000 People
Heart Disease	716,200	336
Cancer	365,700	173
Stroke	194,000	91.9
Auto accidents	45,900	21.8
Suicide	27,100	12.8
Homicide	21,300	10.1
Falls	14,900	7.1
Drowning	8,000	3.79
Poisoning	6,300	2.99
Fires	6,100	2.89
Surgical and medical "complications and misadventures"	3,200	1.52
Inhalation and ingestion of food	2,200	1.04
Airplane travel	1,600	0.76
Struck by falling object	1,100	0.52
Railway accidents	600	0.28
Excessive cold	360	0.17
Electrocution from home wiring and appliances	260	0.12
Suffocation in bed or cradle	210	0.10
Lightning	120	0.06
Cataclysm (tornado, flood, earthquake, etc.)	100	0.05
Bites and stings, venomous animals and insects	50	0.02

[a]Based on 1975 data from the National Center for Health Statistics.

Protection of the health of people is a traditional function of government. Laws prohibiting air pollution and adulteration of food were promulgated as early as the 12th and 13th centuries. Although protection of the public health and well-being remains a fundamental and essential principle of government, it becomes increasingly difficult to achieve in an adversarial climate.

The idea of estimating or assessing possible or potential hazards posed by environmental contaminants includes a constant comparison between risks and benefits. Everyone, even though they may not think in these terms, makes many day-to-day decisions on the basis of a judgment of the benefits to be derived in comparison to the risks inherent in the action being contemplated.

Currently, risk assessment (RA) and risk management (RM) are being widely considered and used as a strategy for enhancing regulatory actions, and thereby more reliably protecting the public from the multi-

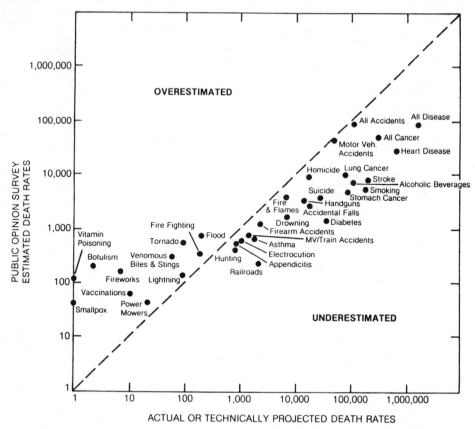

Figure 2. Public perception of U.S. annual death rates from selected causes. Adapted from Slovic et al., 1980.

plicity of risks almost endemic in our communities (3). Bear in mind that risk is a potential for harm and that part of the process requires the calculation of the probability or odds that harm will occur, as well as how often and at what magnitude.

II. RISK ASSESSMENT AND RISK MANAGEMENT

Currently, RA and RM are concerned with evaluating the risks of cancer, impotence, and birth defects which may be associated with exposure to any number of toxic substances. However, decisions of federal agencies— EPA, FDA, OSHA, HHS—which are by law mandated with responsibility for protecting the health of the public via regulatory actions, have been bitterly controversial.

The roots of the controversy lie in several areas: increased ability to detect minuscule amounts of potentially hazardous chemicals; the public's expectations about health protection, as well as their perception of *acceptable* risk; and in the costs and benefits of policies that fall unequally on different interest groups in the country.

A. Risk Assessment

RA refers to the qualitative and quantitative characteristics of the potential adverse health effects of exposure to environmental hazards. In short, it refers to evidence or data which may indicate that exposure and disease are related.

RA generally distinguishes four areas of concern:

- Hazard identification
- Dose–response relationships
- Exposure assessment
- Risk characterization

1. Hazard Identification

This is the process of determining whether exposure to some risk factors can produce an increase in the rate of cancer, birth defects, or whatever condition is being investigated. Although this should be a "yes" or "no" decision, there are unfortunately few chemicals for which human data are so definitive. Consequently, the question is often restated in terms of effects in laboratory animals or other test systems. Positive responses are typically taken as evidence for human oncogenesis.

To obtain the most comprehensive documentation for hazard identification, data from four types of studies are assembled:

- Epidemiologic
- Animal bioassay
- In vitro effects
- Comparison of molecular structure

a. Epidemiology. Epidemiology is the study of the factors that cause disease, illness, or other conditions. The epidemiologist can study cholera, collisions, or crime, but the subjects are people—always people. For me, the most pithy definition of epidemiology is the study of why some people get sick and some do not.

Epidemiologic data are accepted as the most convincing evidence of *human* risk, but they can be difficult to acquire. Five, 10, 20 years is not an unusual time to obtain data, and careful and sophisticated interpretation is also required. As a consequence, there is *pressure* for less demanding and direct methods. Several essential ideas to bear in mind are human, direct, and pressure.

b. Animal Bioassay. The inference that animal studies are applicable to people has long been a fundamental tenet of toxicologic research. Even with all their uncertainties they are seen by some as reliable. Currently, they play a major role in efforts to identify human carcinogens in air, water, food, and pesticides, as well as from radiation. They are used to identify specific abnormalities that a chemical may produce, the specific site or organ that may be vulnerable, and the mechanism by which the damage is produced.

Which animal species is appropriate and what concentration or dose of test chemical is translatable to people are sources of controversy among both scientists and policymakers.

c. In Vitro Effects. Considerable experimental evidence supports the proposition that chemical carcinogens are mutagens and that many mutagens are carcinogens. Therefore, positive response in a mutagenicity assay, such as the Ames test,*is supportive evidence that a chemical is likely to be carcinogenic. In the absence of positive animal bioassays, in vitro tests (test tubes and other types of glassware) are rarely sufficient to support a conclusion of carcinogencity. But they are rapid, inexpensive, and do lend support to both animal and epidemiologic studies.

d. Comparison of Molecular Structure. Here, of course, a suspect chemical's physical and chemical properties are compared with those known to induce cancerous or other tissue changes. This type of comparison can often provide evidence of potential harmfulness. By itself, little decisionmaking value would be attached to it.

2. Dose–Response Relationships

In some instances, epidemiologic studies do yield suitable data from which a dose–response relationship can be developed. For example, in the cigarette smoke–lung cancer association, not smoking, smoking less than half-a-pack per day, one pack per day, and more than one pack per day produce a gradient of concomitant numbers of lung cancer.

Similarly, in animal studies, usually with the smaller species such as rats, mice, guinea pigs, and hamsters (where it is possible to assemble hundreds), three, four, or five dose levels, rather than simply high or low, can be utilized with a similar end-point (other than death) if the particular animal species is susceptible. The purpose of the study is to identify the

---

* The Ames test, also referred to as the *Salmonella*/microsome assay, is a short-term in vitro test that utilizes specifically constructed *Salmonella typhimurium* strains to detect mutations and genetic damage (B. N. Ames et al., *Mutat. Res.* **31**:347–364, 1975.)

nature of the injury and the dose that produces the injury. When the data are collected, a number of pertinent questions must be considered: What mathematical model or models fit the data appropriately? Is extrapolation from the high doses used to the low doses typically ingested warranted? Is there a problem with interspecies extrapolation? What if one species is positive and one negative? Which way would or will people respond? As a matter of policy, some government agencies go with the positive test as overriding. Does this conservative approach serve the best interests of the public?

Carcinogens vary in potency. Some, like cigarette smoke and saccharin, are low-potency carcinogens, while others such as aflatoxin, beta-naphthalene, and dioxin are high-potency oncogens. A potent oncogen is one that produces many malignant tumors at low dose within a short period. This type of information can be obtained from animal studies. However, to make a statement about human risk requires not only the acceptance of the interspecies translation* but extrapolation from high dose to low dose, to the types of level people are most often exposed to, and for which animal data do not exist. Recall too that because of their relatively short lives, rodents are fed inordinately high doses, which can produce metabolic overload. Swamping animals with exorbitantly high doses can itself produce overt toxicity. Furthermore, translation to human populations often assumes a linear relationship down to zero dose. This means that there is no threshold, no point or level higher than zero at which adverse effects do not exist. All levels are held to be toxic. Can this be tested for? No. Any such experiment would require tens of millions, even hundreds of millions of test animals, if the extremely rare adverse effects are to be detected. Consequently, they are impossible to undertake. At this point mathematical models are invoked.

Without the ability to test the low-dose effect experimentally, projections are made which indicate *probabilities* of harm. It is these calculated estimates that are the bases for such lists as contained in Tables 1–3. Perhaps at this point the difficulties and complexities of determining harm begin to emerge.

Recently, David B. Clayson of the Health Protection Board, Health and Welfare Canada, stated that it is necessary "to identify situations in which animal carcinogens may not be relevant for human safety." He noted too that "even a minimal consideration of the biology or mechanism of carcinogenesis may indicate that individual chemical carcinogens identified in animals may not be effective in humans"(4).

---

* The debate over use of animal data proceeds from the concern that prudence dictates that *regulation* must not permit 20–40 years of human exposure to potential carcinogens while regulators await the outcome of epidemiologic studies. Policy dictates, so the argument goes, that substances found carcinogenic in animals be regarded as carcinogenic in people. This has the ring of high ethical purpose. However, as in the Meda/Seveso accident, pregnant women were unnecessarily aborted on the basis of animal data.

Similarly, Edward J. Calabrese, University of Massachusetts, concluded that "the highly varied susceptibility of humans to environmental agents can be related to the fact that humans are a highly outbred species that follow a broad variety of dietary patterns and divergent life styles. In contrast, the responses of humans to toxic agents are currently predicted from highly imbred rodent strains raised on the standardized diets and housed under similar environmental conditions." He went on to note that "current practice assumes that the homogeneity of response to toxic agents, including carcinogens, in the rodent model will be representative of humans. The issue than becomes, to which of the broad spectrum of human responses are specific animal models likely to be related" (5). Indeed, is the rat, mouse, rabbit, hamster, dog, monkey, or goldfish most reminiscent of the human physiological system and its biochemical pathways? As noted in the case of dioxin, do we pick the animal most susceptible to the most minute amounts of the chemical and assume that people will respond similarly? Clearly, this is a less than satisfactory response. We can do better.

One of the real problems to be discussed at length is the public's seeming need for certainty. "Is it or isn't it?" "Is it black or white?" Clearly, biological systems do not lend themselves to such a view. Shades of gray is nature's way. Yet, the public mistakenly sees the world in absolute terms.

III. MODELS FOR ESTIMATING RISK

In an attempt to predict both outcome of exposure (harm) as well as the magnitude of the outcome (how many people can be expected to develop X), complex mathematical models have been developed. Six have attained prominence:

- One-hit*
- Probit
- Multistage
- Weibull
- Multihit
- Logit

Each is based on an aspect of the general theory of carcinogenesis rather than on data for a specific chemical. None of them has a strong biological

---

* The one-hit model is also referred to as the "no threshold" or linear model. In this concept, there is no level at which a chemical found to be carcinogenic at one level poses no threat at any lower level. For these chemicals, there is no safe level, no threshold.

basis, and none can be verified in the range of verifiable exposure, which can be six logs (each log being a factor of 10) below the dose used in animal assays. Currently, the EPA and FDA lean toward the one-hit, multistage, and probit models.

Table 5 shows the wide variation in risk estimates from exposure to a food additive at a level of 1.0 mg/kg·day. The difference between the risk estimated by the one-hit and the probit models is 100,000 times.

Another example of the wide variations obtainable comes from the National Academy of Sciences. They estimated the number of cases of human bladder cancer that could result between the years 1983 and 2053 from continued use of the nonnutritive sweetener saccharin. Using the models noted above, they concluded that the incidence of cancer could be between 0.22 and 1,144,000. That is, over the 70 years there could be an excess of from less than one to over a million. Again, which end of the spectrum is closest to human reality? If that remains unknown, what degree of conservatism does a decisionmaker employ in protecting the public? These are both real and legitimate questions and are the causes of premature baldness and lack of sleep for policymakers. This substantive difference in outcome employing the same original data is shown in a humorous dimension in Figure 3.

The EPA has tended to be highly conservative in favoring the one-hit model, which appears to yield the highest estimates of low-dose risk. The probit tends to yield lower risk estimates. Is it in the public's best interest to project and predict the worst estimates?

Given the preference for the one-hit model, a closer look at its appropriateness is warranted.

Low-dose carcinogenic risk estimates are based on mathematical equations that attempt to characterize the unknown underlying relationship between exposure and risk. To estimate the risk of low exposures, linear extrapolation procedures are employed.

The one-hit model is a member of a class of models referred to as mechanistic. Its name derives from an assumption that the mechanism or

Table 5
Spectrum of Risks (Predicted Deaths Forecast by
Five Current Models)

Model Applied	Lifetime Risk[a] at 1.0 mg/kg · day
One-hit	6.0×10^{-5} (1 in 17,000)
Multistage	6.0×10^{-6} (1 in 167,000)
Multihit	4.4×10^{-7} (1 in 230,000)
Weibull	1.7×10^{-8} (1 in 59 million)
Probit	1.9×10^{-10} (1 in 5.2 billion)

[a] All risks are for a full lifetime of daily exposure. The lifetime is the unit of risk measurement because the experimental data reflect the risk experienced by animals over their full lifetime (6).

"HEY, I THOUGHT WE WERE WORKING WITH THE SAME DATA..."

Figure 3. Reproduced with permission from *National Wildlife Magazine*, August/September 1984.

process of carcinogenesis—or tumorigenesis—is a consequence of damage to a single cell or structure within the cell. The one-hit model is the simplest of the mechanistic models and assumes that damage to the target cell sufficient to yield a tumor results from a single "chemical hit" to a DNA molecule. This model is linear in the low-dose region.

In addition, the EPA and other regulatory agencies (often taking their lead from the EPA) have made yet another assumption. They have assumed that the mechanism of chemical damage is the same as that induced by radioactive particles, especially gamma (cf. Chapter 9). To establish by regulatory fiat that chemicals and ionized particles move through tissues and cells the same way, because both can induce cancer, unnecessarily strains the available data.*

As if two were not sufficient, a third major assumption is invoked. This one so departs from established toxicologic and pharmacologic doctrine that a grand controversy has ensued. As noted earlier (cf. Chapter 5), an underlying and unifying principle of these disciplines is the concept of a dose–response relationship. Paracelsus realized this over 400 years ago. If there is a dose and a consequent response, there must be a point between exposure and the observation of an effect, and no effect. Toxicologists and pharmacologists refer to this point or area as the threshold. It is generally defined as the maximum dose of a substance which produces no effect, or

---

* Even with radiation, which is energy rather than matter, a threshold seems an appropriate assumption. The fact of natural background levels of radiation of 75 millirem per person per year in New Jersey, 190 in Colorado Springs, and 1300 in Kerala, India, without increased cancer incidences do indeed suggest no effect levels (cf. Chapter 9).

the minimum dose at which there is an observable effect—in effect, a safe level.

Although a threshold is unique for each chemical, it varies with type of animal and by sex within a species. Nevertheless, it is a well-established parameter. Not only is it well established, but it is of critical importance in judging risk to people from experiments in animals. If this were not the case, such translations would be meaningless.

Enter the EPA. The third assumption incorporated is that potentially carcinogenic chemicals are unsafe at any level. Thus, they do not, it is maintained, have a threshold. It is therefore possible to postulate a mechanism whereby a single molecule of a chemical is sufficient to initiate the malignant transformation. One hit by one molecule is all that is needed.

Let us extend this reasoning a bit further, and let us use BAP (benzene-α-pyrene), a known carcinogen in tobacco and other smoke, as our example.

It is a fact that only a small proportion of all smokers develop lung or other cancer. That, unfortunately, is a major impediment to most anti-smoking campaigns. The smokers the campaign is aimed at invariably point to those who have been "smoking for 50 years."

Although smokers are at far greater risk of lung cancer than non-smokers, not all smokers get cancer. Most do not. In fact, as the dose (number of cigarettes smoked) drops below two packs per day, the number of lung cancers declines proportionately. That is of overriding importance to this discussion. BAP is not only present in cigarette, cigar, and other tobacco smoke, it is present in smoked foods, charcoal-broiled foods, as well as in the air around wood-burning fires. BAP has been present in our environment for from 500,000 to 1 million years, ever since *Homo erectus*, Peking man, learned to control fire and cook food.

Consider too that for every mole of BAP, as with every mole of any chemical, there are, according to Avogadro,* 6×10^{23} molecules. If one molecule were all it required to produce a cancer, *Homo erectus* would never have made it to *Homo sapiens*. In more contemporary times, with the deep inhalation of smoke into the depths of our lungs, down to the alveoli and into the pulmonary blood supply, the opportunity for molecules to encounter and strike sensitive targets became even greater.

Take another look at Figure 11, Chapter 14. Stomach cancer has been declining in the United States for the past 50 years, at a time when smoking and consumption of smoked and charcoal-broiled foods have increased considerably. Trillions upon trillions of BAP molecules are

---

* Amadeo Avogadro, an Italian physician (1776–1856), was the first to enunciate the concept that equal volumes of gases under the same temperature and pressure have the same number of molecules. Actually, his number is 6.02×10^{23}.

available for every man, woman, and child on this planet. If the one-hit model even remotely reflected reality, we should all have died of cancer long ago.

We must be for protection, and we must be for safety. But excessively conservative estimations suggest a modicum of overkill.

Norman Gravitz of the Epidemiological Study Section, California Department of Health, recently commented on the one-hit model. Referring to the idea that the one-hit, no-threshold model dealt with the ability of a *single molecule* of a potentially oncogenic chemical to evade the body's defenses and set off a process that eventually produces cancer, Gravitz noted that "benzene is a moderately potent, proven human and animal carcinogen. Low-dose extrapolation from occupational studies using the one-hit, no threshold model suggests that persons drinking water containing 1 part per billion of benzene throughout their lives might have an additional risk of cancer (excess cancer risk) as high as approximately 2×10^{-6} (two additional cases of cancer for every million people so exposed)." He then went on to calculate the probability of cancer from a single molecule of benzene:

> If one assumes that the average person weighs 70 kilograms (154 pounds), and drinks 2 liters of water per day for a lifetime, the excess carcinogenic risk of drinking water contaminated with one molecule of benzene per liter, a lifetime consumption of about 51,000 molecules of benzene, is approximately 10^{-22}. This risk is more than 16 orders of magnitude (logs) smaller than the most stringent state or federal regulatory standard for an allowable risk level of 1×10^{-6} (one in a million excess lifetime risk of cancer). Assuming that the present total world population is 5 billion people, and that it consumes this "contaminated" water, one would not expect even one additional case of cancer from this contaminated water since the probability of one excess case of cancers occurring in the world's population is 5×10^{-15}. . . . According to the no-threshold, one-hit model, there is a finite probability that one molecule of a carcinogen will cause cancer: However the opponents of this theory are correct in expressing their incredulity at this possibility. For all practical purposes, the probability of this occurring is so slight as to make this skepticism reasonable. (7)

A highly conservative approach may be seen as a prudent way to protect people, but it also creates a distressing psychological problem. People are quick to assume the worst. The world is seen as a far more noxious and hazardous place than it obviously is. If, as is shown in Chapter 14, people in the United States have never been healthier, is it necessary to create a false impression in order to protect them?

A multihit approximation is far more realistic. But that too fails to deal adequately with the magnitude of cancer in the world today—especially if cigarette smoking is diminished.

IV. EXPOSURE ASSESSMENT

The first task of exposure assessment is the determination of the concentration to which people are exposed—concentrations of chemicals in foods, water, and air. This may be known from direct measurement but more often it must be estimated. In food, for example, assessments are complicated by variations in diet and personal habits among different subgroups in the population, as well as the frequency of consumption, to say nothing of storage conditions.

Exposure measurements made on a small group of workers in a specific factory can be misleading if applied to other segments of the population. In addition, exposure assessments must determine which groups may be at greatest risk of exposure. Some, such as pregnant women, the very old, and the very young, as well as those with impaired health, may be especially susceptible. Who is really at risk?

There is also the problem of mixed exposures. Smoking and asbestos are a formidable example. That unwholesome combination yields a cancer rate far in excess of that produced by either smoking or asbestos alone. This creates additional assessment conundrums.

V. RISK CHARACTERIZATION

This is the estimate of the anticipated magnitude of the public health problem. At this point, one must be aware of all the uncertainties. How can they be factored in? Which exposures should be used? Which population should be used? Is there an average population? Which population should be the primary target for protection? Furthermore, estimating the number of people exposed to a substance and identifying the type, route, and intensity of exposure are extremely difficult. Given all the foregoing, the "data" now pass to policymakers for a decision. The amount of data needed is enormous and the sources of uncertainty and error can be substantial.

VI. RISK MANAGEMENT

The process of sorting through the mounds of data and choosing between possible alternatives is referred to as risk management.

Policymakers have their own set of impediments. They are faced with five major problems:

- Uncertainty
- Limited analytic resources

- Complexity
- External pressures
- Visible economic interests

A. Uncertainty

As previously noted, this condition permeates the entire analytic process. Although scientists would argue otherwise, uncertainty exists every step of the way in experimental studies from obtaining samples to recording the data, including feeding and dosing animals to standardizing equipment. Data may well be incomplete and, as noted, exposure uncertain. Conclusive, direct evidence of a threat to health is extremely rare. Fewer than 30 chemicals of the millions cataloged are definitely linked to human cancer. Some 1500 chemicals are reportedly carcinogenic in animal bioassays and they include studies of questionable experimental design. Only some 7000 over the over 5 million known have ever been tested for carcinogenicity. We know even less about health effects other than cancer.

B. Limited Analytic Resources

Simply stated, there are more chemicals to be tested than animals, cages, laboratories, and scientists to test them. With the advent of lifetime inutero carcinogenicity testing in two rodent species, at $1,000,000 per chemical, and up to 3 years of testing per chemical, the idea may be totally impractical.

The FDA must itself contend with over 2000 chemicals directly related to food and another 12,000 indirect additives. Should the newer invitro bioassays being tested prove suitable, they would herald a new day for screening potentially toxic substances.

C. Complexity

For the most part, decisionmakers are nonscientists. They must deal with masses of scientific data from a host of disciplines if they are to comprehend the problem and use it efficaciously. Obviously, they must rely on and seek guidance and opinions, often conflicting, from a variety of scientists.

They must also be concerned about the effectiveness of existing technologies to reduce exposure. Do they exist? How good are they? Are they better than what is currently used? And what of the economic impact of a decision: Will it be overly burdensome? Should that be a factor in the decision? Perhaps the question raised earlier about why decisions take as long as they do begins to have meaning.

D. External Pressure

With such possible end-points as birth defects, impotence, and cancer, emotion can be expected to run high. Various groups who see themselves at risk—Vietnam veterans, for example, with their concern about the effects of Agent Orange on their health and that of their offspring—insist that regulatory agencies need not wait for conclusive, direct cause–effect proof of harm. Regulatory action, they insist, need not be based exclusively on the most advanced testing procedures or controlled epidemiological studies that can take years to complete.

They are voters and command a substantial constituency. Public sympathy is often on their side. Why shouldn't it be? But how should these concerns be factored into a decision? Should they be?

E. Visible Economic Interests

Groups, industries, and companies having to bear the cost of control of some exposure may be unmoved by data derived from assumptions based on sparse and indirect evidence. They demand to be shown direct proof that a chemical they may have manufactured did indeed produce the harmful effects claimed. They resist decisionmaking on the grounds that the economic burden of clean-up falls disproportionately on them. Nor do city, county, or state governments want bills sent to them. Decisions may be held up simply to avoid this dilemma.*

The needs are great. If agencies are to discharge their mandated responsibility and do it economically, while avoiding adversarial relationships that eventually are decided by batteries of lawyers and judges, new and Solomonic insights may be needed.

Given the assumptions, uncertainties, pressures, and policy choices, estimates of risk generally overstate the consequences and magnitude of the problems posed by environmental hazards such as radiation and chemicals. But a nagging question remains. Are they safe?

VII. THE CONCEPT OF SAFETY: WHAT IS SAFE?

Up to this point, the concept of safety has only been treated in passing. Risk, as shown in Tables 1–4, is a statement of probability. Although an event may be expected or estimated to occur only once per 100,000 people or once in 5 years, whether it can be said to be safe is something else again.

Safety consists of two components: measuring risk, a quantitative scientific activity, with all its shortcomings, and a *judgment* on the part of

* Ultimately, it will be the people who foot the bill, either in higher taxes or higher prices for goods and services.

individuals or groups or communities as to the acceptability of the stated risk.

That is the key: safety is a personal evaluation about the acceptability of risk. It is a value judgment of how much risk we as individuals, or collectively as a society, will accept. Consequently, a chemical, a process, a job, a drive to the beach, or a plane trip from London to Athens is safe if it is judged to be acceptable, or unsafe if it is judged to be unacceptable.

This definition is not the same as that in most dictionaries which define "safe" as "free from risk," a definition both unacceptable and misleading. Nothing, as shown in the four tables, can be totally without risk: not even taking a bath. There are only degrees of risk and thereby degrees of safety. There are no absolutes. A trip, a chemical, or any activity can never be "all white or all black." "Is it or isn't it?" is not nature's way. In nature, all is shades of gray. Life is a continuum of risks. But that is not new. Risk is an integral part of life and has been ever since Eve offered an apple to Adam—who, I understand, took it. As long as there is human activity there must be risk.

Failure to appreciate this concept gives rise to the false expectation that scientists can measure whether something is safe. They can not. Deciding whether people might or should be willing to bear an estimated risk is a "judgment call" that scientists are no more qualified to make than any other citizen. Safety is a relative attribute that changes from time to time and place to place. And society—people—clearly show acceptance and rejection of risk. It has been estimated that 2000 additional deaths over the next 20 years, 100 additional (excess) deaths per year, may be related to radiation effects from all our nuclear power plants. If the public outcry against nuclear energy is any bellwether, a larger segment of the public is unalterably opposed to such a risk. The estimated 100 additional deaths per year is nothing more than a guesstimate—a calculation, a mathematical construct based on equations that have been plugged with hypothetical numbers. I am not at all sure that the public understands or appreciates this. On the other hand, 50,000 known and documented deaths per year from motor vehicle accidents is quite acceptable; 100 estimated deaths per year from radiation apparently is not.

Take a second look at Figure 2 (page 592). I noted that in the perception of risk, people generally tend to overestimate the probability of unfamiliar, catastrophic, and well-publicized events while minimizing or underestimating the probability of unspectacular familiar events that claim one victim at a time—motor vehicle accidents, for example. Fear of flying keeps many people away from the commercial airlines, but most of these same people have no fear of driving, a "victory of subjectivity over actuarial statistics."

In addition, response to risks seems most contradictory when the degree of risk is unknown and the consequences are particularly dreaded—nuclear radiation, for example. People, communities, will band together and vehemently protest the placement of an incinerator or landfill in their neighborhood—not in my backyard (NIMBY)—despite all assurances of freedom from harm. At the same time people living under immense dams located on earthquake faults pay little heed to warnings of real danger.

"Safe" then is anything people decide is safe. It is the idea or concept of acceptability which creates problems for policymakers and communities alike. Acceptable for whom? In what terms? In whose view? Safe may not be only "in the eye of the beholder," but it may have little to do with "the facts in the case."

There is yet another dimension: fear—an abiding sense of trepidation.

Recently, while traveling by train between Washington, DC, and Princeton, a women took the seat next to me. She was on her way to Stamford, Connecticut, and had not traveled this route before. She was afraid of missing her station. We got to talking about fear. I asked her if she felt safe in her community. Of course my mind was on environmental issues. What else was there? She said she no longer left her house doors unlocked. She made sure she locked her car and continued on with a litany of concerns about crime in the streets, rape, and the threat of nuclear war. Not once did she mention food additives, nuclear power, waste disposal, pesticides, or water pollution. I mentioned this to her. Were these of no concern? Her response took me by surprise. "Well, it's all part of it." If that is a widely held attitude, the process of education will be far more difficult than Jefferson ever imagined.

Before I got up to leave at my station, I said that we are part of the healthiest generation that ever existed. To that her response was an incredulous stare followed by "Do you really think so?" "Are you sure?" "That's comforting, but I'll still lock my doors."

I decided to conduct a small survey. The 12 people I accosted on the train between Philadelphia and Princeton would hardly qualify as a survey by the Gallup Organization, Opinion Research, or Louis Harris. But I did get a feel for what some folks are thinking.

I learned that my lady of the train was not atypical. Environmental issues barely ever mentioned are indeed scrambled together in the omelette of disturbing contemporary issues: rampant crime, rape, loss of freedom (unable to walk out at night or even during the day in out-of-the-way places), and the threat of an annihilating nuclear disaster have shaped attitudes. Environmental issues simply get a "bad press" as a consequence of "guilt by association."

Remember that my 12 respondents were not environmental activists. They would undoubtedly separate the issues. On the other hand, my group was more likely to reflect views of the general population. Although I do not want to push this survey too far, the attitude is buttressed by responses offered by my students in both undergraduate and graduate courses.

It must be inordinately difficult to believe that we are as healthy as the U.S. Surgeon General says we are, given the overlay of wretched community problems. For those concerned with the public's participation in environmental decisionmaking, this poses a daunting challenge. Unfortunately, the news media, which should be a source of support, are not. Often they exacerbate the problem. We shall get to that shortly.

Early in his second tenure as administrator of the U.S. EPA, William D. Ruckelshaus added another dimension to risk management. In an address to the National Academy of Sciences he said: "We must search for ways to describe risk in terms that the average citizen can comprehend. Telling a family living close to a manufacturing facility . . . that according to our linear model, their risk is only 10^{-6}, is not very reassuring . . . we need to help them compare that risk to those with which they are more familiar." He then went on to add that "to effectively manage the risk, we must seek new ways to involve the public in the decision-making process . . . they need to become involved early, and they need to be informed if their participation is to be meaningful." At this point Jefferson was invoked.

Four years before Ruckelshaus proposed that all federal agencies involved in environmental decisionmaking move into the RA/RM arena, Judge David L. Bazelon (Senior Circuit Judge, U.S. Court of Appeals for the District of Columbia) in a talk to the American Bar Association reminded the audience that "considering all relevant data and viewpoints is essential to good decisions." He noted too that "calm reflection, full debate and mature decision more than compensates for the additional costs it imposes." His closing remarks were as follows: "By strengthening that administrative process we provide a constructive and creative response to the inherent uncertainties of risk regulation. Approaching the decision to take or to step back from risks such as nuclear power," he reminded them, "is like coming to a busy intersection with our view partially obscured. Our instincts tell us to proceed with caution, because intersections are dangerous. Ultimately, the importance of our journey and the desirability of our goal may lead us to brave the traffic and pull into the highway. But even when we decide to proceed, we should not omit the moment of reflection to observe passing cars, and look both ways" (9).

RA and RM are inherently uncertain processes. Involving the public in decisionmaking may be seen as participatory democracy but it can also be seen as adding an additional degree of uncertainty given the public's lack of substantive knowledge of the issues.

The real problem and challenge are communication. If we are to have an informed public, what mechanisms are available for reaching the greatest proportion of our 240 million people in 50 states—spanning a continent and beyond?

Communication is another way of saying education. That is actually what is sought—an educated public—informed on issues they can tackle when confronted with important choices so that decisions can be reached based on knowledge rather than perceptions, or worse, untutored opinion—bias.

But this cannot be a one-way encounter. It is incumbent upon those seeking to impart information to a wary, untrusting, often hostile public, to determine what it is the public wants, needs, and can manage. This may have to be the first step.

Judge Bazelon's equating approval of nuclear energy with driving through a highway intersection, with its need to proceed with caution, cannot be faulted. However, if the popular din can be interpreted, large segments of our population will not or will refuse to enter the "intersection" of environmental issues.

Confused by fear and misled by ignorance, the popular stance is "no," rather than deal with the issue. I see three problems currently afflicting the American people. They are lack of accurate, trustworthy information, conflicting information, and continuous exposure to misinformation—a babel of tongues and, as a consequence, mental pollution.

The "public," the American people, suggests a unity. It is not. We appear to have become a fragmented nation, of individual groups concerned with promoting special causes or vested interests and talking at cross-purposes. This may just be the shoal upon which communication founders.

A. The Media

Ours is an electronic age. Communications wizardry is capable of instantly transmitting information from border to border and coast to coast. Nevertheless, the public has been ill-served by its television, radio, newspapers, and magazines.

1. Radio

On June 23, 1985, a radio news broadcast reported that coffee consumption may be related to lung cancer. Dr. Leonard Schuman of the University of Minnesota in a voice-over statement said that smoking was the primary cause of lung cancer and that coffee's role was not all that clear. The announcer had raised the specter of coffee being unsafe, and Dr. Schuman then cast doubt on a report that emanated from his laboratory. Why this report reached the wire services is difficult to understand. It not only offered nothing new, it certainly clarified nothing. It did confuse, and it raised an unnecessary warning "flag" about the safety of coffee.

I had listened to this report along with six other people while at the beach that Saturday. Everyone turned to me for an explanation. I could only shrug. There was no way I could clarify that confusing report. They were left wondering.

How could that have happened? A researcher at a university completes a study. He can submit his work to a professional journal in the form of a manuscript for possible publication. At the same time, he may inform the public relations office at his university about the study. The PR person anxious to obtain coverage and show that she is doing her job, rewrites or thinks she understands the thrust of the investigator's report. A news release is prepared and sent—or distributed to newspaper editors, radio, and TV news bureaus. There, the story may catch someone's attention and is once again rewritten to accommodate the 15–20 s it may get on the air. It then "goes to press," and what comes out can be an important item or a garbled message. Obviously with respect to coffee and lung cancer, we got the latter.

Since coffee has never been implicated as a risk factor in lung cancer, that could be an important observation—if it were true. Such a finding would require careful scrutiny. The story was released prematurely. The media people who look upon themselves as "gatekeepers" can surely do better.

Although "talk shows" abound and could be vehicles for transmitting important ideas, there is, however, no mechanism for separating fact from fancy: primarily because the moderators or hosts either do not have the background to pose the hard or pointed questions, or entertainment is the name of the game. But whoever said it was necessary to be dull in order to inform. Unfortunately, too many of the"entertainers" are simply promoting themselves or their products.

Unfortunately too, given the objectively verifiable notion that disaster sells, we have for years been treated to a steady diet of programs promoting the illusion that our environment is deadly—when all evidence says otherwise. Do we prefer bad news? I doubt anyone would admit to that. But the steady "drum beat" of the communications media fosters that impression.

2. Newspapers

Michael Ryan of the School of Communications at Houston reported on a study of attitudes of scientists and journalists about their perceptions of how science information was covered in the press. (10). He found that scientists strongly agreed, but that science writers disagreed, that scientists should have an opportunity to read—prior to publication—the article in which they are being quoted. They also disagreed on the point that a science writer should not reinterpret a scientist's conclusions. Others have also found that whereas scientists, writers, and readers considered a news story of value regardless of whether it was thought to be "exciting," news editors' judgments of what constitutes a valuable story is highly colored and correlated with their idea of its excitement value.

Cancer sells papers and air time. In fact, print and broadcast news media are the public's major sources of information about cancer. But what has the public gotten from the media?

In 1980, Vicki S. Freimuth and her colleagues of the University of Maryland mounted a study of the coverage cancer received in newspapers (11). Five questions were of concern to them:

1. To what extent is news coverage of cancer, by body site, commensurate with the recorded incidence of cancer?
2. To what extent does newspaper coverage of cancer provide information that individuals need to know in order to understand or seek help about the disease?
3. To what extent does news about cancer coincide with identified public conceptions about the disease?
4. How do the headlines of cancer news stories reflect their content?
5. Do health columns provide different kinds of cancer information than news stories?

Their sources included the 50 newspapers with the highest circulation in the country. Articles were obtained from a clipping service which was instructed to select all stories that contained the key words: cancer, carcinogens, tumors, lump, malignant, or terminal illness.

According to the investigators, "news stories did not provide statistics on the incidence of cancer in general, and they particularly underreported the recorded incidence of colon-rectum cancer relative to other types of cancer." They went on to report that "neither did news stories provide information on the topics of prevention, risks, detection and treatment of cancer. Information on these topics is crucial because the means of prevention, detection and treatment are different for the four major sites—lung, colon-rectum, breast and uterus."

As for question 3, Freimuth found that, generally, news stories did not focus on those cancer issues about which the public has been shown to be uninformed. For Americans, no disease raises such harrowing fear. A diagnosis of cancer is widely perceived as a death sentence.

Measures that could be taken to minimize risks received little coverage. Cigarette smoking, the single most important risk factor for lung cancer, and the cancer whose increasing magnitude distorts upward the total cancer rate, was discussed in 7% of the stories on the causes of cancer.

As far as headlines are concerned, they were "generally accurate and neutral in tone." Articles, they felt, "about drugs, chemicals, and additives in cosmetics, household cleaners, clothing, pesticides, and air and water pollutants had fear arousing headlines over 60 percent of the time."

And finally, what of health care columns? Did they provide different kinds of cancer information than "hard news stories"?

Six of ten columns were written by physicians, and seven of ten were nationally syndicated. Most focused on either causes or detection. Less than 20% carried information on how to detect cancer.

Minimal coverage was assigned ways of coping with cancer and the social resources available.

Freimuth and her collegues believe that improvement in news coverage "must begin with those who disseminate the scientific information to the press." I was surprised that they believed it was the scientists, the researchers, who had the best grasp of the type of information the public needs, as well as the emphasis the story should contain. I do not believe that research scientists are currently in touch with public perceptions, nor are they aware of community needs. They could be. But that will take time and a different set of motivations.

Because of the lack of well-trained science and medical writers, Freimuth opts for a "middleman," a science information specialist who would act as intermediary between scientists and writers both to clarify issues and to improve accuracy of news coverage. Certainly newspapers are not going to add such additional employees to their staffs. Newspaper personnel and writers are among the lowest paid professionals in the country. If it were not for the perception of "glamour" surrounding the fourth estate the available pool of writers would rapidly evaporate. Besides, why develop another level when science writers need only upgrade their science backgrounds.

I suspect that it will fall to the scientists to be more forceful and demanding in determining how the information they supply is used. That, however, is not going to occur quickly. To reach the public with the type of information that scientists consider important, other channels of mass media must be used. Newspapers simply have not been, and do not seem to be, doing the job. They have failed the public.

Dr. Edward Lawless of the Midwest Research Institute (12) also looked at the media's treatment of scientific information. He found that, on the whole, "the news media tends to overdo the bizarre or the scare aspect at the beginning of a case and seldom follow through to summarize adequately the resolution of an issue." He also noted that "media attention is not necessarily proportional to the numbers of people threatened, the severity of the threat, or even the certainty that it exists. The dangers of alcohol, cigarettes, and the auto receive relatively little attention compared to the speculation over the meaning of . . . Love Canal."

For me, the October 14, 1985 issue of Time magazine was a disgrace. The issue featured toxic wastes, "the poisoning of America." The cover showed a person up to his nose in toxic waste, and below the surface the body was a skeleton. The accompanying article contained a litany of sites around the United States and tonnages of waste. It left the reader to infer illness and death. That may sell, but is there no requirement to educate? This is a blatant example of toxic journalism.

They may not like the color of the waste nor its consistency. They may be unhappy with its being there. I can understand that, and I would agree with their dislike. But is it toxic? Do they know anything of its toxicity? Probably not. They assumed it must be. To get *their* message out, they used colorful language: "We wanted to show that above ground things may look o.k., but underneath it's death."

Their thinking of how the article should be pitched is also revealing. Senior correspondent Peter Stoler tells us how he experienced a strong sense of déjà vu as they updated their 1980 reporting. Stoler was reminded of John Brunner's 1972 science-fiction novel *The Sheep Look Up*, which described a world that was poisoning its air. "I thought it farfetched," recalls Stoler. "Now I wonder if Brunner shouldn't have been a bit more hysterical." Why not? Why not get the folks worked up another notch or two. Since the health effects are so little known, a dash of hysteria could go a long way.

We are told they visited Times Beach, Missouri, for this story. "Times Beach looks as if a neutron bomb hit it." Houses were standing, windows were unbroken, toys still scattered around, "but nobody walks the streets." Indeed, they do not. But who is to blame for that? The news media. Hysteria over the possible contamination by dioxin—a chemical shown to be toxic to laboratory animals (cf. Chapter 5) and to this day shown to be the cause of nothing more than chloracne in human beings—forced the EPA to declare Times Beach unfit for human habitation. Times Beach was literally abolished—and abandoned. Love Canal was luckier. Hysteria never overtook it. Toxic journalism did not envelope it as it did Times Beach.

Time editor Kurt Andersen has reason to be depressed by the sight. Unfortunately, he has not yet understood that his reporting helped dispatch the town unnecessarily. That is what is really depressing.

During the 1960s and 1970s, many students graduated from high school without a math or science course. Even college and university students managed to arrange their programs to avoid science courses. As adults, now living in a high-tech world, they are without "the tools" for even the most rudimentary interpretation of environmental events. How are these people to interpret news reports such as the *Time* article, participate in community decisionmaking, or evaluate risks?

If Prof. Jon D. Miller, director of Public Opinion Laboratory, Northern Illinois University, is correct, calling for public participation in environmental policy decisionmaking may be inherently self-defeating. His surveys find a decided lack of scientific/technologic literacy in our country. Only 1 in 14 Americans meet the minimum definition of scientific literacy—the ability to read, comprehend, and communicate about science at a level permitting a person to function effectively in a scientific culture.

"How", he wonders, "can a citizen who does not understand the nature of radiation, for example, participate usefully in a discussion of the disposal of nuclear powerplant wastes?" Nevertheless, he goes on to say that "given our commitment to democratic government, the data argue for the urgency of the task of increasing and improving technologic literacy in the U.S." (13). Prof. Miller and Thomas Jefferson appear to be standing on similar ground. However, neither offers a hint as to how this is to occur. That is the real challenge.

To these handicapped can be added those students in both elementary and high school who for years have been taught science by history, English, or "shop" teachers. Although I have been a long-suffering Giant fan in need of a winning season, Erich Schubert, a one-time Giants place kicker and field goal artist, unfortunately provides a pernicious example of the teacher who teaches anything and everything. After kicking five field goals to beat Tampa Bay, our man was the darling of the sports writers. A few lines from The New York Times tells the sad story: "Small for a football player at 5 feet 7 inches and 190 pounds, he had also been distant for a football player, working until a few days ago as a $36-a-day substitute high school teacher at Lakeland Regional [N.J.] taking over classes in French or Latin or Auto Shop or mechanical drawing or anything else, whether familiar with the subject or not, usually not." This is unfair to students, and unfair to teachers. Does the Lakeland Regional really believe it is getting a bargain?

The countrywide scene with respect to sophistication in science is dismal. We have been ill-served by our institutions and by the news media.

3. Advertising

Advertising, a gray area with respect to communication, has served its own purposes more than it has served the public. For example, given its toll of human illness, smoking must be placed at the head of all lists of adverse environmental risks. Given that hundreds of thousands of new cases of lung cancer will occur each year at least until 1996, considering the current level of cigarette smoking, cigarette smoke must be seen as the most notorious environmental problem. None of the problems we generally think of, air and water pollution, pesticides, radiation, chemicals in foods, or hazardous wastes, is even remotely close. Motor vehicle accidents are the only risk capable of challenging cigarette smoke for "top billing" in the arena of detrimental health risks. And both are fostered by advertising.

Take another look at the cigarette ads on billboards, in magazines, and in newspapers. What do we see? Camel has a handsome, robust young man with a cigarette firmly planted between his lips, dangling effortlessly from a rope on the side of a mountain. Cigarettes and fitness, that's the message.

Vantage shows us a skilled cyclist negotiating a difficult course and notes that "performance counts" and that Vantage provides the "thrill of real cigarette taste." A Benson & Hedges ad has a debonair young man in a hammock, and a lovely woman with tennis racquets. "She wants to play. He'd rather sway. But there's one thing they agree upon." The taste . . . you know what.

Of course, the Surgeon General's warning that "Quitting smoking now greatly reduces serious risks to your health" is strategically isolated from the serious business at hand.

Advertisers are paid handsomely to promote smoking (and alcohol consumption) and thereby illness (and accidents). They are experts in appealing to our desires and fantasies. Cigarette ads now appear on the sports page of newspapers. This is a new source of additional revenue. Are the newspapers going to jeopardize this income by discussing the perils of smoking? Hardly. The public is thus doubly undercut. Some would say "between a rock and a hard place." We are undermined by clever ads on the one side and provided little or no reliable in-depth information on the other. A no win situation.

4. Television

Television, TV, has literally taken over communications. It is exceedingly difficult not to watch—no matter the fare. An audience is always there. TV has the power and potential to reach the greatest number of people with tremendous impact. Unfortunately, it has yet to "flex its muscles" and demonstrate its inherent capabilities for public education.

TV news is a disaster. The 15–20-s snippets served up as information can barely convey a message. What currently passes for news is no more than glorified headlines. As with newspapers, editors prefer "excitement." Keep viewers glued to their sets, to get the ads across. The environment rarely falls in that category unless an oil-carrying tanker breaks up in high seas near the shore, or a truck overturns while transporting hazardous chemicals. That's news. The "big eye" can focus on a graphic event and bring it into our living rooms. That's where it shines.

A recent survey found that viewers preferred stories dealing with the economy, and how to be a wiser consumer. Unexpected events such as plane high-jackings, train wrecks, and overturned trucks get high marks for audience approval—as did the "soaps."

Producers of TV are convinced that the great majority of viewers will not "sit still" for discussions of environmental issues. They are not "sexy" topics. There are no sponsors for expensive "prime time" discussions of the health effects of acid rain, nutritional status of "junk" food, or the uses and benefits of food additives. If such topics do make it past the producers, they are "on" after midnight so as not to interfere with important programs or ball games. TV is a powerful medium. It can do a lot better. Whether we can get more from it is up to the viewers. They usually get what they prefer—or what the companies believe they prefer.

B. Perception of Risk

Any number of writers have commented on the apparent contradictions in the public's assessment of risk. "How," it is often asked, "can a risk such as that posed by motor vehicles with 45,000 to 50,000 deaths annually, or smoking with 350,000 deaths per year be acceptable, while nuclear power plants, food additives and air or water pollution with barely, if any, deaths at all, be unacceptable?" The answer may not be as elusive as some think.

Psychologists and psychiatrists who have studied fear, feelings, risk, and attitudes have sifted their research data and have come up with a set of fundamental perceptions:

1. Is it a single negative event, or is it spread over time? A crash of an airliner with 52 or 212 dead makes flying appear a far more serious risk than automobile accidents with 130 people killed *around the country* everyday.

In this situation the media plays a central role. A plane crash is the type of event they do best. Cameras and reporters are on the scene quickly, covering it from every angle "in living color." Fifty thousand automobile deaths a year is a statistic. It has little coverage value or appeal. It does not sell. Imagine, however, the coverage, if on one day of every year all the deaths from motor vehicle accidents, or all the smoking-related deaths, occurred. That would be a media event! That would sell. That would be news. And it would send an unambiguous message about which risks are in fact at the top of the list.

2. Who controls the risk? If the individual can control the risk, then any risk becomes acceptable. On the other hand, if an outside source controls the risk, no risk is acceptable.

Clearly, this explains the motor vehicle–nuclear power plant dichotomy and the cigarette smoking–air pollution asymmetry.

There is of course a contradiction here. With 50,000 deaths per year and millions of injured, many maimed for life, control is an illusion. The same can be said for smoking and substance abuse. Thus, we have another mythical perception. In fact, the public has far more control of nuclear power plants and siting of incinerators than they realize. They clearly do not have control of the road. That needs clarification.

In Chapter 6, Accidents, I noted that AIDS had a powerful political constituency and thereby could move the political process in their preferred direction. I noted too that between 1981 and 1988, for each death from AIDS, there were ten on the road. And I indicated that even with those horrific numbers of deaths, especially among young people, there is no political constituency to move for regulation or additional funds for accident research. There is not, and there will not be, because of the control exerted by drivers. Would any driver or combination of drivers importune their elected representatives to enact legislation restricting the manner of their driving? Never. Would our elected representatives take it upon themselves to move for legislation making for safer driving? Not if they want to be reelected they won't. So, in this sense, drivers have control. But on the road, no.

3. Is it a familiar or unfamiliar event? Being totally familiar with automobiles, motorcycles, firearms, and the like, death from accident is held to be quite acceptable. Nuclear power plants and nuclear energy being totally unfamiliar, they become unacceptable risks. Is this subject to change?

4. Is the risk perceived to be needed or not needed? This is similar to item 3. Automobiles, smoking, firearms, drugs, and noise are seen as necessary attributes—for the type of life (some) people prefer to live. Nuclear energy, on the other hand, is perceived as having no place in society. Is this subject to change? Can the media help?

Given the fact that there are over 400 operating nuclear power plants in 30 countries and four continents without a fatal accident in over 30 years, isn't it extraordinary that so many people hold so negative a view of nuclear energy? With the four criteria above, seemingly irrational fear of such things as nuclear power becomes comprehendible.

The idea of a no-risk world was well illustrated in a recent court decision. The Delaney Clause of the Food and Drug Act specifies that any chemical shown to cause cancer in animals or people be banned from all use. Because the FDA determined their risks to be minimal, four coloring agents (dyes) were approved for use in cosmetics and drugs. Suits brought by consumer advocacy groups challenged the FDA decisions, maintaining that no risk is permissible. The Federal Appeals Court of the District of Columbia "with reluctance" agreed.

On the basis of scientific review, risk assessments estimated that Red dyes #8, #9, and #19 and Orange dye #17 would have the following risks of cancer based on animal tests:

Red #8   1 in 60,000,000
Red #9   1 in 60,000,000
Red #19   1 in 9,000,000
Orange #17   1 in 19,000,000,000

For Red #8 and #9, that could mean four additional cases of cancer in a population of 240,000,000 and for Red #19, 27 additional cases. For

Orange #17, it would be necessary to more than triple the current world population before a single additional case could occur—if people responded the way animals do. Of course, the RA models have error factors that cut both ways. The risks could be still lower or higher. Either way they are minimal. The court understood this but had no latitude: the Delaney Clause is the law. And for those seeking absolute safety, there is no such thing as risk.

Clearly, the public decides the risks it will accept. It is also clear that there is a commonality of thought around the country, at least on some issues. Guns, no; pesticides, yes. Motor vehicle speed, no; nuclear power plants, yes. Retin A, yes. Obviously there is widespread interest in looking young.

Recently, newspapers, TV, and radio carried stories about Retin A, a new antiacne drug with vague claims for erasing skin wrinkles. Warnings that this was a potent drug to be used sparingly, and with the advice of a physician, were also carried.

The message the public selected out was that a new drug for smoothing out wrinkled skin was available, and all that was needed was to rub it on—all over. The fact that this drug had little more than 4 months of testing, along with warnings from the FDA about adverse effects of long-term use, mattered little. Sales soared in the face of continued warnings and equivocal results.

When do people believe a government agency? Is this another example of personal control, or is this simply the acceptance of risk when the benefits are deemed highly desirable? Are warnings not seen or heard, merely blocked out, given the magnitude of desire? Is there a message here? Can images be reversed and perceptions modified? Would incinerators, nuclear power plants, handguns, and driving while inebriated profit from repackaging—made to be seen in a more, or less, favorable light?

C. Pugwash for Students

Recently, I attended the Student Pugwash International Conference at Princeton University. The topic being considered was communicating risk to the public. At a session devoted to the responsibility of the media, students applauded the suggestion that the public not be given information directly. They preferred the notion that it was incumbent upon the public to "dig out" the available information. It was, they felt, a less paternal approach. They really believed that to be in the public's best interest.

Irritation followed bemusement. How, I inquired, was the public to do this? "Education" and "the library," were some of the glib responses. I realized this conference would produce little to further perception and comprehension of environmental issues.

The current unhappy state of general ignorance of environmental issues at the community level stems from the fact that people get little in-depth presentation of issues from newspapers, radio, or TV, the "big-three" sources of public information.

Going beyond media snippets requires time—fair amounts of time to read and digest specialized magazine and journal articles which most newsstands and libraries normally do not carry. Also required in the reading of semiprofessional and professional journals is a modicum of background in several areas of biology and chemistry. Most undergraduate students do not have this type of knowledge or information. Graduate students take years to acquire it. But all too often they become so narrowly specialized as to neither appreciate nor know the wider implications of their chosen specialty. What, then, can be expected of the general public?

Most people simply cannot, or will not, take the time to dig out "the facts." Evenings and weekends are devoted to household chores and recreation. Many are glued to TV sets hour upon endless hour. Diversion is craved and needed. Knowledge of environmental issues comes only with each "crisis" reported on TV or in newspapers. No, I cannot agree with the idea that people must find the appropriate information for themselves. It cannot and will not be done. If the public is to have its environmental consciousness raised, it must be done for them—in painless ways. I propose two approaches.

VIII.   APPROACHES TO CHANGE

A. Institutions of Higher Learning

Because it is high time that our institutions of higher learning begin to repay their long-standing debt to society—their surrounding communities (taxpayers)—one day twice-a-year should be set aside for discussions of environmental issues. Each university and college in the United States would, on these days, offer day-long programs in which faculty and students would present issues of concern in an open forum during which there would be free exchange of ideas. These "environmental days" would be held off-campus in locations and times accessible to most people, and they would of course be dejargonized, comprehendible, people to people discussions. Given the thousands of schools in the country, the amount of ongoing information transfer would be astounding. In time, relationships may even become friendly and all involved come to enjoy it—even look forward to it.

Of course, universities and colleges would have to encourage their faculty to become involved, to participate. This means that the Vice President for Academic Affairs will have to add community service (the external community) to research, teaching, and publications as an appropriate activity suitable for upward mobility. Without such recognition, nothing will change. Many faculty members are ready to take this on. Many have—and have paid a price. The public has been left to the doomsayers and purveyors of purple prose far too long.

The variations on this theme are limited only by creative imagination. Industry, government, and foundations would join in supporting such an effort.

B. Popularization

Given the observable fact that we are living in the age of celebrities,* an offspring of TV, I propose that celebrities and TV and radio be wedded to send clear messages to the public. These would be of course be environmentally related.

Is there any reason why Bruce Springsteen could not write a song about automobile accidents (especially under the influence of alcohol) or people getting involved in community issues? There really isn't. Rarely do we ask our celebrities to throw their weight behind issues which affect their lives, as they do ours.

Bill Cosby, Larry Hagman, Angel Cordero, "Dr. J," Miss America, Frank Guilford, Johnny Carson, and on and on, could offer 30–60-s "spots" on topics of general and specific environmental concern. I believe the funding for this could be worked out should these people become involved.

An offshoot of the celebrity concept is the comic book with its host of celebrity heroes: Superman, Conan, Little Alex, and Batman as well as the daily and Sunday newspaper cartoons: Peanuts, Shoe, Bloom County, and Doonesbury. I believe that the writers of these popular series can easily have their heroes offer advice and suggestions on problems such as nuclear energy, drinking and driving, chemicals in foods, noise, air pollution, or any other of consequence if they themselves appreciated the evidence.

If comic books are what millions read, and if TV and radio are the national pastime, let us put the message and the media together to do the most good. The important thing is to get the message out. Currently, it is not being done—to everyone's misfortune. Environmental issues must become institutionalized. They must become part of our daily lives. They must lose their crisis quality and become mundane subjects requiring the ongoing attention of the community. With this type of community participation, I believe that "gloom and doom" can be overcome and that people can appreciate their world rather than fear it.

---

* The idea of using celebrities to change attitudes did not originate with me. My son, Scott, a feature-writer for the *Palm Beach Post*, is "into" the world of celebrities. The idea is his. However, father and son agree completely.

IX. RECAPITULATION

For me the greatest problem in the country is fear. It is fear, I believe, that motivates the all-encompassing need for security—protectionism.

The protectionist economic policies put forward during the recent presidential campaign appear symptomatic of a more widespread protectionist attitude that has enveloped the country. My specific concern is for health protectionism that has all but hobbled many of our citizens. One aspect of this protectionism shows itself as an unreasonable fear of living side by side, as it were, with chemicals: chemicals in foods, in agriculture, and in industry. Chemicals are perceived as noxious substances inimical to life and health. The fear is pervasive.

Appraising the effects of chemicals on the environment, Prof. C. F. Wilkinson of the Institute of Comparative and Environmental Toxicology, Cornell University, remarked that "our world is heavily dependent upon chemicals. If we look around our homes, our offices, and our transportation vehicles; if we consider the quantity, the variety, and quality of food available to us; if we consider the quality of our health and the drugs available to cure our ills; if we can see in total the thousands of chemicals that touch our lives; then we will begin to appreciate the enormous impact chemicals have on each of us. Chemicals are an integral part of our lives. They have been irreversibly built into our technology and, by and large, have immeasurably improved the quality of our lives" (14). Most scientists would agree with that appraisal. The majority of the public would not.

X. PERSONAL PROTECTIONISM

For 20 years scientists in government, academia, and industry have sought diligently for indications of the harmful effects these chemicals were supposed to have inflicted. But search as they may, their efforts have gone unrewarded. Ordinarily, this would be cause for rejoicing. In the current protectionist climate this lack of adverse effect is translated as loss of trust in scientists, physicians, government, and academia. Industry had fallen from public grace years before.

If only we had spend more money or made a greater effort, (those) ill effects would surely come to light. Apparently, few people look around to see who among them are ill. This failure of observation has its counterpart in the AIDS epidemic that appeared to engulf the country from 1981 to 1988.

As the number of cases and deaths were tallied, and the numbers of deaths rose beyond 35,000—over the 8-year period—people were gripped by the chilling fear that AIDS was spreading out of control. No one was safe. The fact that not a single proven case of heterosexual transfer was documented meant nothing. Few were listening. The fact that neither friends, neighbors, nor heterosexual family members were ill with AIDS made a difference. There was an epidemic, and it was spreading every-where. Never mind that it was not. It was.

With respect to the perceived ill effects of environmental chemicals, where in the population are the illnesses? Are there groups at unusually high risk, as in the case of AIDS? Who among us harbors these environ-mentally related diseases?

Having opened the available records, we see that the problem is certainly not among infants. Infant illness and death rates have been dropping steadily. Birth defects, present since the time of the pharaohs, continue to hover between 4 and 6%. Childhood illness and death appear to be the lowest in our history. Our adolescents and teenagers are inordinately healthy except for excessive use of cigarettes, alcohol, controlled drugs, and their predilection for high speed, which exacts its fearsome toll in death and injury on our highways. Apathy, anomie, and ennui appear more their problems than hazardous chemicals, food additives, DDT, or air pollution.

And our population of adults ages 21–65 are healthy physically. If these people—especially between 25 and 45*—have a health problem, it may just be stress and anxiety incurred in their relentless pursuit of status, a sybaritic life-style, and ever higher income.

The 60-, 65-, and 70-year-olds who have come to represent an ever-increasing portion of our population are more vigorous and desirous of working. Retirement, the Golden Years, holds no lure for them. In addition, we are witnessing the rise of a new class of elderly: the very old—above 70, to 80, 90, and beyond. Since 1969, this group has literally tripled its numbers. Can such increased life span, vigor, and reduced death rates and illness exist side by side with a noxious environment?

Heart disease, cancer, and stroke, which together account for 66% of current deaths, are each declining. Indeed, heart disease, the nation's number one cause of death has been declining since 1968. Well before jogging, dieting and smoke-enders became national pastimes. And as noted earlier, when lung cancer is omitted from the tabulation of cancer death rates, cancer is seen to have been declining for several years.

Stroke, cerebrovascular accidents, although the most common cause of neurologic disability in the Western world, and the third leading cause of death in the United States, has also been declining. All this is at a time of increased use of agricultural and industrial chemicals.

The protectionist view of health not only does not permit a glance at vital statistics, which indicate the levels of health and illness, it must also

* These are the MOSS—middle-aged, overstressed, semiaffluent suburbanites.

view illness and death as an abstraction—without regard to real people—the public. They look ahead to the years 2010, 2020, and beyond hoping—perversely—that their belief in the ill effects of chemicals will be vindicated, that adverse effects will express themselves in increased birth defects, cancers, and other human blemishes to justify their protectionism.

There is nothing mysterious about this view of life. Health protectionism and economic protectionism are threads of the same cloth: loss of national will. We need to protect what we have. We do not take risks. Security is all important. In fact, the pioneering spirit, the creativity, and the inventiveness that made this country great have been lost: perhaps only for a time. But loss of national will certainly does describe our current national climate. This is the climate that has not built a major airport in over 20 years. Not built a bridge of any consequence, nor planned a high-speed train. It is in this climate of loss of will that our space program has stalled.

Risk and vitality have given way to fear—to playing it safe at a time when all indices clearly show the American people to be the healthiest in history. When the protectionists grasp the idea that they are a healthy bunch, the outpouring of pent-up energy will move this country forward as nothing since the opening of the West!

REFERENCES

1. W. D. Ruckelshaus, Science, Risk and Public Policy. *Science* **221**: 1026–1028, 1983.
2. W. D. Ruckelshaus, Risk in a Free Society. *Risk Anal.* **4** (3):157–162, 1984.
3. *Risk Assessment in the Federal Government: Managing the Process.* Commission on Life Sciences, National Research Council, National Academy Press, Washington, DC, 1983.
4. D. B. Clayson, Needs for Biological Risk Assessment in Interspecies Extrapolation. *Environ. Health Perspect.* **77**: 93–97, 1988.
5. E. J. Calabrese, Comparative Biology of Test Species. *Environ. Health Perspect.* **77**: 55–62, 1988.
6. J. Rodricks, and M. R. Taylor, Application of Risk Assessment to Food Safety Decision Making. *Regul. Toxicol. Pharmacol.* **3**: 275–307, 1983.
7. N. Gravitz, Models of Carcinogenesis. *Science* **226**: 1022, 1985.
8. W. W. Lowrence, *Of Acceptable Risk: Science and the Determination of Safety.* William Kaufman, Los Altos, CA, 1986
9. D. L. Bazelon, Risk and Responsibility. *Science* **205**: 277–280, 1979.
10. M. Ryan, Attitudes of Scientists and Journalists Toward Media Coverage of Science News. *Journalism Q.* **56**(Spring): 18–26, 53–54, 1979.
11. V. S. Friemuth, R. H. Greenberg, J. De Witt, and R. M. Romano, Covering Cancer: Newspaper and the Public Interest. *J. Commun.* **34**(1): 62–73, 1984.
12. E. W. Lawless, M. V. Jones, and R. M. Jones, Comparative Risk Assessment: Toward an Analytical Frame work. MRI Report, Midwest Research Institute, Kansas City, MO, 1984.

13. J. D. Miller, A National Survey of Public Attitudes Toward Science and Technology. Northern Illinois University, Dekalb, IL, 1982.

14. C. F. Wilkinson, The Science and Politics of Pesticides. In *Silent Spring Revisited* (G. J. Marco, R. M. Hollingsworth, and W. Durham, Eds.), American Chemical Society, Washington, DC, 1987.

SUGGESTED READINGS

Black, B. Evolving Legal Standards for the Admissibility of Scientific Evidence. *Science* **239:** 1508–1512, 1988; and the responses to this article in Letters to the Editor, *Science* **241:** 1413–1414, 1988.

Colin N. P., and Snee, R. D. Quantitative Risk Assessment: State-of-the-Art for Cancinogenesis. *Fund. Appl. Toxicol.* **3:** 320–333, 1983.

Hallenbeck, W. H., and Cunningham, K. M. *Quantitative Risk Assessment for Environmental and Occupational Health.* Lewis Publishers, Inc. Chelsea, MI, 1986.

Miller, J. D. Scientific Literacy: A Conceptual and Empirical Review. *Daedalus* **112**(2): 29–48, 1983.

Miller, J. D. *Technological Literacy: Some Concepts and Measures.* Public Opinion Laboratory, Northern Illinois University, DeKalb, IL, February 14, 1986.

Peto, R. Distorting the Epidemiology of Cancer: The Need for a More Balanced Overview. *Nature (London)* **284:** 297–300, 1980.

Risk Assessment, Management, Communication: A Guide to Selected Sources. U.S. EPA, Washington, DC, May 1987, EPA/IMSD/87-002-a.

Weinberg, A. M. Science and Its Limits: The Regulator's Dilemma. *Issues Sci. Technol.,* 59–71, Fall 1985.

Wildavsky, A. *Searching for Safety. Studies in Social Philosophy and Policy No. 10.* Transaction Books, New Burnswick, NJ, 1988.

Index